Jean Pierre Fouassier and
Jacques Lalevée

Photoinitiators for Polymer Synthesis

Related Titles

Ramamurthy, V.

Supramolecular Photochemistry

Controlling Photochemical Processes

2011
ISBN: 978-0-470-23053-4

Allen, N. S. (ed.)

Handbook of Photochemistry and Photophysics of Polymeric Materials

2010
ISBN: 978-0-470-13796-3

Albini, A., Fagnoni, M. (eds.)

Handbook of Synthetic Photochemistry

2010
ISBN: 978-3-527-32391-3

Wardle, B.

Principles and Applications of Photochemistry

2010
ISBN: 978-0-470-01493-6

Stochel, G., Stasicka, Z., Brindell, M., Macyk, W., Szacilowski, K.

Bioinorganic Photochemistry

2009
ISBN: 978-1-4051-6172-5

Jean Pierre Fouassier and Jacques Lalevée

Photoinitiators for Polymer Synthesis

Scope, Reactivity and Efficiency

WILEY-VCH Verlag GmbH & Co. KGaA

The Authors

Prof. Jean Pierre Fouassier
formerly University of Haute Alsace
Ecole Nationale Supérieure de Chimie
3 rue Alfred Werner
68093 Mulhouse Cedex
France

Prof. Jacques Lalevée
University of Haute Alsace
Institut Science des Matériaux
IS2M-LRC 7228, CNRS
15 rue Jean Starcky
68057 Mulhouse Cedex
France

All books published by **Wiley-VCH** are carefully produced. Nevertheless, authors, editors, and publisher do not warrant the information contained in these books, including this book, to be free of errors. Readers are advised to keep in mind that statements, data, illustrations, procedural details or other items may inadvertently be inaccurate.

Library of Congress Card No.: applied for

British Library Cataloguing-in-Publication Data
A catalogue record for this book is available from the British Library.

Bibliographic information published by the Deutsche Nationalbibliothek
The Deutsche Nationalbibliothek lists this publication in the Deutsche Nationalbibliografie; detailed bibliographic data are available on the Internet at <http://dnb.d-nb.de>.

© 2012 Wiley-VCH Verlag & Co. KGaA, Boschstr. 12, 69469 Weinheim, Germany

All rights reserved (including those of translation into other languages). No part of this book may be reproduced in any form – by photoprinting, microfilm, or any other means – nor transmitted or translated into a machine language without written permission from the publishers. Registered names, trademarks, etc. used in this book, even when not specifically marked as such, are not to be considered unprotected by law.

Cover Design   Adam-Design, Weinheim
Typesetting   Laserwords Private Limited, Chennai, India
Printing and Binding   Markono Print Media Pte Ltd, Singapore

Print ISBN: 978-3-527-33210-6
ePDF ISBN: 978-3-527-64827-6
ePub ISBN: 978-3-527-64826-9
mobi ISBN: 978-3-527-64825-2
oBook ISBN: 978-3-527-64824-5

To Geneviève F., Gaëlle, Hugo and Emilie L.

for their patience and understanding

To our colleagues over the world for the marvellous time we spent together during all these years.

Contents

Abbreviations *XIX*

Introduction *XXV*

Part I   Basic Principles and Applications of Photopolymerization Reactions *1*

1   **Photopolymerization and Photo-Cross-Linking** *3*
         References *6*

2   **Light Sources** *11*
2.1   Electromagnetic Radiation *11*
2.2   Characteristics of a Light Source *12*
2.3   Conventional and Unconventional Light Sources *13*
2.3.1   Xenon Lamp *13*
2.3.2   Mercury Arc Lamp *14*
2.3.3   Doped Lamps *14*
2.3.4   Microwave Lamps *14*
2.3.5   Excimer Lamps *14*
2.3.6   Light-Emitting Diodes (LEDs) *16*
2.3.7   Pulsed Light Sources *17*
2.3.8   Laser Sources *17*
2.3.9   Sun *18*
2.3.10   Household Lamps *18*
2.3.11   UV Plasma Source *19*
         References *20*

3   **Experimental Devices and Examples of Applications** *21*
3.1   UV Curing Area: Coatings, Inks, Varnishes, Paints, and Adhesives *21*
3.1.1   Equipment *21*
3.1.2   End Uses *21*

3.1.3	Coating Properties 24	
3.2	Conventional Printing Plates 25	
3.3	Manufacture of Objects and Composites 25	
3.4	Stereolithography 25	
3.5	Applications in Microelectronics 26	
3.6	Laser Direct Imaging 26	
3.7	Computer-to-Plate Technology 27	
3.8	Holography 27	
3.9	Optics 28	
3.10	Medical Applications 28	
3.11	Fabrication of Nano-Objects through a Two-Photon Absorption Polymerization 29	
3.12	Photopolymerization Using Near-Field Optical Techniques 29	
3.13	Search for New Properties and New End Uses 30	
3.14	Photopolymerization and Nanotechnology 32	
3.15	Search for a Green Chemistry 33	
	References 34	

4 Photopolymerization Reactions 41

4.1	Encountered Reactions, Media, and Experimental Conditions 41	
4.2	Typical Characteristics of Selected Photopolymerization Reactions 45	
4.2.1	Film Radical Photopolymerization of Acrylates 45	
4.2.2	Film Cationic Photopolymerization 48	
4.2.3	Thiol-ene Photopolymerization 49	
4.2.4	Photopolymerization of Water-Borne Light Curable Systems 52	
4.2.5	Photopolymerization of Powder Formulations 52	
4.2.6	Charge-Transfer Photopolymerization 53	
4.2.7	Dual Cure Photopolymerization 54	
4.2.8	Hybrid Cure Photopolymerization 55	
4.2.9	Anionic Photopolymerization 55	
4.2.10	Metathesis Photopolymerization 56	
4.2.11	Controlled Photopolymerization Reactions 56	
4.2.11.1	Radical Photopolymerization Reactions 56	
4.2.11.2	Cationic Photopolymerization Reactions 58	
4.2.11.3	Anionic Photopolymerization Reactions 59	
4.2.12	Hybrid Sol–Gel Photopolymerization 59	
4.2.13	Photo-Cross-Linking Reactions in the Presence of Photobases or Photoacids 59	
4.3	Two-Photon Absorption-Induced Polymerization 60	
4.4	Remote Curing: Photopolymerization without Light 60	
4.5	Photoactivated Hydrosilylation Reactions 61	
	References 61	

5	**Photosensitive Systems** 73
5.1	General Properties 73
5.2	Absorption of Light by a Molecule 74
5.2.1	Absorption 74
5.2.2	Molecular Orbitals and Energy Levels 74
5.2.3	Absorption of Light and Optical Transitions 74
5.2.4	Reciprocity Law 76
5.2.5	Multiphotonic Absorption 77
5.3	Jablonski's Diagram 78
5.4	Kinetics of the Excited State Processes 78
5.5	Photoinitiator and Photosensitizer 80
5.6	Absorption of a Photosensitive System 81
5.7	Initiation Step of a Photoinduced Polymerization 82
5.7.1	Production of Initiating Species 82
5.7.2	Competitive Reactions in the Excited States 83
5.7.3	Reactivity in Bulk versus Solution: Role of the Diffusion 84
5.7.4	Cage Effects 85
5.8	Reactivity of a Photosensitive System 86
	References 87

6	**Approach of the Photochemical and Chemical Reactivity** 89
6.1	Analysis of the Excited-State Processes 89
6.1.1	Nanosecond Laser Flash Photolysis 89
6.1.2	Picosecond Pump-Probe Spectroscopy 90
6.1.3	Photothermal Techniques 90
6.1.4	Time-Resolved FTIR Spectroscopy 91
6.1.5	Direct Detection of Radicals 92
6.1.6	CIDNP, CIDEP, and ESR Spectroscopy 92
6.2	Quantum Mechanical Calculations 93
6.3	Cleavage Process 94
6.4	Hydrogen Transfer Processes 95
6.5	Energy Transfer 96
6.6	Reactivity of Radicals 98
	References 99

7	**Efficiency of a Photopolymerization Reaction** 103
7.1	Kinetic Laws 103
7.1.1	Radical Photopolymerization 103
7.1.2	Cationic Photopolymerization 105
7.1.3	Dependence of the Photopolymerization Rate 106
7.1.4	Laser-Induced Photopolymerization 107
7.1.5	Kinetics of the Photopolymerization in Bulk 108
7.2	Monitoring the Photopolymerization Reaction 109
7.2.1	FTIR Analysis 109
7.2.2	Photocalorimetry 110

x | Contents

7.2.3	Optical Pyrometry 110
7.2.4	Other Methods 110
7.3	Efficiency versus Reactivity 111
7.4	Absorption of Light by a Pigment 112
7.5	Oxygen Inhibition 114
7.6	Absorption of Light Stabilizers 115
7.7	Role of the Environment 117
	References 118

Part II   Radical Photoinitiating Systems 123

8	**One-Component Photoinitiating Systems** 127
8.1	Benzoyl-Chromophore-Based Photoinitiators 127
8.1.1	Benzoin Derivatives 129
8.1.2	Benzoin Ether Derivatives 132
8.1.3	Halogenated Ketones 136
8.1.4	Dialkoxyacetophenones and Diphenylacetophenones 136
8.1.5	Morpholino and Amino Ketones 137
8.1.6	Hydroxy Alkyl Acetophenones 137
8.1.7	Ketone Sulfonic Esters 140
8.1.8	Thiobenzoate Derivatives 140
8.1.9	Sulfonyl Ketones 144
8.1.10	Oxysulfonyl Ketones 147
8.1.11	Oxime Esters 148
8.2	Substituted Benzoyl-Chromophore-Based Photoinitiators 148
8.2.1	Benzoin Ether Series 149
8.2.2	Morpholino Ketone and Amino Ketone Series 149
8.2.3	Hydroxy Alkyl Acetophenone Series 151
8.2.3.1	Oil- or Water-Soluble Compounds 151
8.2.3.2	Difunctional Compounds 152
8.2.4	Modified Sulfonyl Ketones 154
8.2.5	Limit of the Substituent Effect 154
8.2.6	Macrophotoinitiators 156
8.2.7	Supported Cleavable Photoinitiators 157
8.3	Hydroxy Alkyl Heterocyclic Ketones 157
8.4	Hydroxy Alkyl Conjugated Ketones 158
8.5	Benzophenone- and Thioxanthone-Moiety-Based Cleavable Systems 158
8.5.1	Benzophenone Phenyl Sulfides 158
8.5.2	Ketosulfoxides 159
8.5.3	Benzophenone Thiobenzoates 159
8.5.4	Benzophenone-Sulfonyl Ketones 159
8.5.5	Halogenated Derivatives 160

8.5.6	Cleavable Benzophenone, Xanthone, and Thioxanthone Derivatives *160*	
8.6	Benzoyl Phosphine Oxide Derivatives *161*	
8.6.1	Compounds *161*	
8.6.2	Excited State Processes *162*	
8.6.3	Absorption Properties and Photolysis *164*	
8.6.4	Bis-acyl phosphine oxide/Phenolic Compound Interaction *165*	
8.7	Phosphine Oxide Derivatives *165*	
8.8	Trichloromethyl Triazines *165*	
8.9	Biradical-Generating Ketones *166*	
8.10	Peroxides *166*	
8.11	Diketones *167*	
8.12	Azides and Aromatic Bis-Azides *168*	
8.13	Azo Derivatives *168*	
8.14	Disulfide Derivatives *168*	
8.15	Disilane Derivatives *169*	
8.16	Diselenide and Diphenylditelluride Derivatives *170*	
8.17	Digermane and Distannane Derivatives *170*	
8.18	Carbon–Germanium Cleavable-Bond-Based Derivatives *170*	
8.19	Carbon–Silicon and Germanium–Silicon Cleavable–Bond-Based Derivatives *172*	
8.20	Silicon Chemistry and Conventional Cleavable Photoinitiators *172*	
8.21	Sulfur–Carbon Cleavable-Bond-Based Derivatives *173*	
8.22	Sulfur–Silicon Cleavable-Bond-Based Derivatives *173*	
8.23	Peresters *173*	
8.24	Barton's Ester Derivatives *174*	
8.25	Hydroxamic and Thiohydroxamic Acids and Esters *174*	
8.26	Organoborates *176*	
8.27	Organometallic Compounds *176*	
8.27.1	Titanocenes *177*	
8.27.2	Chromium Complexes *177*	
8.27.3	Aluminate Complexes *178*	
8.28	Metal Salts and Metallic Salt Complexes *178*	
8.29	Metal-Releasing Compound *178*	
8.30	Cleavable Photoinitiators in Living Polymerization *179*	
8.30.1	Cleavable C–S- or S–S-Bond-Based Photoiniferters *179*	
8.30.2	TEMPO-Based Alkoxyamines *182*	
8.31	Oxyamines *183*	
8.31.1	Alkoxyamines *183*	
8.31.2	Silyloxyamines *184*	
8.32	Cleavable Photoinitiators for Two-Photon Absorption *184*	
8.33	Nanoparticle-Formation-Mediated Cleavable Photoinitiators *185*	
8.34	Miscellaneous Systems *185*	
8.35	Tentatively Explored UV-Light-Cleavable Bonds *185*	
	References *187*	

9	**Two-Component Photoinitiating Systems** *199*
9.1	Ketone-/Hydrogen-Donor-Based Systems *199*
9.1.1	Basic Mechanisms *199*
9.1.2	Hydrogen Donors *201*
9.1.2.1	Amines *201*
9.1.2.2	Thio Derivatives *203*
9.1.2.3	Benzoxazines *204*
9.1.2.4	Aldehydes *204*
9.1.2.5	Acetals *204*
9.1.2.6	Hydroperoxides *205*
9.1.2.7	Silanes *205*
9.1.2.8	Silylamines *207*
9.1.2.9	Metal-(IV) and Amine-Containing Structures *208*
9.1.2.10	Silyloxyamines *210*
9.1.2.11	Germanes and Stannanes *210*
9.1.2.12	Phosphorus-Containing Compounds *211*
9.1.2.13	Borane Complexes *211*
9.1.2.14	Alkoxyamines *212*
9.1.2.15	Monomers *213*
9.1.2.16	Photoinitiator Itself *214*
9.1.2.17	Alcohols and THF *214*
9.1.2.18	Polymer Substrate *214*
9.1.2.19	Silicon-Hydride-Terminated Surface *214*
9.1.3	Benzophenone Derivatives *215*
9.1.3.1	Benzophenone *215*
9.1.3.2	Aminobenzophenones *218*
9.1.3.3	Other Benzophenones *218*
9.1.3.4	Photopolymerization Activity *219*
9.1.4	Thioxanthone Derivatives *219*
9.1.4.1	Thioxanthone *219*
9.1.4.2	Substituted Thioxanthones *220*
9.1.5	Diketones *223*
9.1.5.1	Aromatic Diketones *223*
9.1.5.2	Camphorquinone *224*
9.1.6	Ketocoumarins *225*
9.1.7	Coumarins *225*
9.1.8	Alkylphenylglyoxylates *225*
9.1.9	Other Type II Ketone Skeletons *228*
9.1.9.1	Anthraquinones *228*
9.1.9.2	Fluorenones *229*
9.1.10	Aldehydes *229*
9.1.11	Aliphatic Ketones *229*
9.1.12	Cleavable Ketones as Type II Photoinitiators *229*
9.1.13	Tailor-Made Type II Ketones *229*

9.1.13.1	Low- and High-Molecular-Weight Compounds and Macrophotoinitiators *229*	
9.1.13.2	Water-Soluble Compounds *233*	
9.1.13.3	Two-Photon Absorption Photoinitiators *235*	
9.1.13.4	Photomasked Photoinitiator *235*	
9.1.13.5	Oxygen Self-Consuming Thioxanthone Derivatives *236*	
9.1.13.6	Low-Molecular-Weight One-Component Systems *236*	
9.2	Dye-Based Systems *238*	
9.2.1	Dye/Amine Systems *238*	
9.2.2	Dye/Coinitiator Systems *239*	
9.2.3	Improvement of Dye/Amine Systems *240*	
9.2.4	Dye/Amine Water-Soluble Systems *241*	
9.2.5	Kinetic Data *241*	
9.3	Other Type II Photoinitiating Systems *241*	
9.3.1	Maleimide/Amine and Photoinitiator/Maleimide *241*	
9.3.2	Donor/Acceptor Systems *243*	
9.3.3	Bisarylimidazole Derivative/Additive *244*	
9.3.4	Pyrylium and Thiopyrylium Salts/Additive *245*	
9.3.5	Ketone/Ketone-Based Systems *246*	
9.3.6	Organometallic Compound/Ketone-Based Systems *247*	
9.3.7	Organometallic Complex/Additive *248*	
9.3.7.1	Organometallic Derivatives *248*	
9.3.7.2	Ferrocenium Salts *248*	
9.3.8	Metal Carbonyl/Silane *250*	
9.3.9	Photosensitizer-Linked Photoinitiator or Coinitiator-Based Systems *250*	
9.3.10	Photoinitiator/Peroxide- or Hydroperoxide-Based Systems *251*	
9.3.11	Photoinitiator/Disulfide or Group 4B Dimetal Derivatives *252*	
9.3.12	Photoinitiator/Phosphorus-Containing Compounds *253*	
9.3.13	Photoinitiator/Onium Salts *254*	
9.3.14	Photosensitizer/Titanocenes *254*	
9.3.15	Type I Photoinitiator/Additive: Search for New Properties *255*	
9.3.16	Nanoparticle-Formation-Mediated Type II Photoinitiators *256*	
9.3.17	Miscellaneous Two-Component Systems *257*	
9.3.17.1	Light Absorbing Amine/Monomer *257*	
9.3.17.2	Nitro Compound/Amine *257*	
9.3.17.3	Hydrocarbon/Amine *257*	
9.3.17.4	Photosensitizer/Triazine Derivative *257*	
9.3.17.5	Other Systems *258*	
	References *258*	
10	**Multicomponent Photoinitiating Systems** *269*	
10.1	Generally Encountered Mechanism *269*	
10.1.1	Ketone/Amine/Onium Salt or Bromo Compound *269*	
10.1.2	Dye/Amine/Onium Salt or Bromo Compound *270*	

10.1.3	Ketone/Amine/Imide Derivatives 270
10.1.4	Ketone/Ketone/Amine 271
10.1.5	Dye/Amine/Triazine Derivative 271
10.2	Other Mechanisms 271
10.2.1	Triazine-Derivative-Containing Three-Component Systems 272
10.2.2	Dye/Ketone/Amine 272
10.2.3	Dye/Amine/Metal Salt 272
10.2.4	Dye/Borate/Additive 273
10.2.5	Photosensitizer/Cl-HABI/Additive 273
10.2.6	Metal Carbonyl Compound/Silane/Hydroperoxide 275
10.2.7	Photosensitizer/Amine/HABI Derivatives/Onium Salt 276
10.2.8	Dye/Ferrocenium Salt/Hydroperoxide 277
10.2.9	Coumarin/Amine/Ferrocenium Salt 277
10.2.10	Dye/Ferrocenium Salt/Amine/Hydroperoxide 278
10.2.11	Organometallic Compound/Silane/Iodonium Salt 279
10.3	Type II Photoinitiator/Silane: Search for New Properties 279
10.4	Miscellaneous Multicomponent Systems 281
	References 281

11 **Other Photoinitiating Systems** 283
11.1 Photoinitiator-Free Systems or Self-Initiating Monomers 283
11.2 Semiconductor Nanoparticles 284
11.3 Self-Assembled Photoinitiator Monolayers 284
       References 285

Part III Nonradical Photoinitiating Systems 287

12 **Cationic Photoinitiating Systems** 289
12.1 Diazonium Salts 289
12.2 Onium Salts 289
12.2.1 Iodonium and Sulfonium Salts 289
12.2.1.1 Compounds 289
12.2.1.2 Photopolymerization Reaction 290
12.2.1.3 Role of the Anion 292
12.2.1.4 Absorption Properties 293
12.2.1.5 Decomposition Processes of Iodonium Salts 294
12.2.1.6 Decomposition Processes of Sulfonium Salts 299
12.2.1.7 Acylsulfonium Salts 301
12.2.1.8 Substituted Iodonium and Sulfonium Salt Derivatives 301
12.2.2 Other Onium Salts 303
12.2.3 Search for New Properties 304
12.2.3.1 Absorption 304
12.2.3.2 Solubility 305
12.2.3.3 Stability 305
12.2.3.4 Benzene Release 307

12.2.3.5	Odor	*307*
12.2.3.6	Toxicity	*307*
12.2.3.7	Amphifunctionality	*308*
12.3	Organometallic Derivatives	*308*
12.3.1	Transition Organometallic Complexes	*308*
12.3.2	Inorganic Transition Metal Complexes	*310*
12.3.3	Non-Transition-Metal Complexes	*310*
12.4	Onium Salt/Photosensitizer Systems	*311*
12.4.1	Photosensitization through Energy Transfer	*312*
12.4.2	Photosensitization through Electron Transfer	*312*
12.4.2.1	Photosensitizers	*312*
12.4.2.2	Photoinitiation Step	*317*
12.5	Free-Radical-Promoted Cationic Photopolymerization	*318*
12.5.1	Radical/Onium Salt Interaction	*318*
12.5.2	Radical Source/Onium Salt Based Systems	*318*
12.5.2.1	Radical Source/Iodonium Salt Two-Component Systems	*318*
12.5.2.2	Radical Source/Iodonium Salt Three-Component Systems	*320*
12.5.2.3	Role of the Onium Salt	*329*
12.5.2.4	One-Component Radical Source-Onium Salt System	*330*
12.5.2.5	Photoinitiation Step	*330*
12.5.2.6	Recent Applications	*331*
12.5.3	Radical/Metal-Salt-Based Systems	*331*
12.5.4	Addition/Fragmentation Reaction	*331*
12.6	Miscellaneous Systems	*332*
12.7	Photosensitive Systems for Living Cationic Polymerization	*332*
12.8	Photosensitive Systems for Hybrid Cure	*333*
	References	*333*
13	**Anionic Photoinitiators**	*343*
13.1	Inorganic Complexes	*343*
13.2	Organometallic Complexes	*343*
13.3	Cyano Derivative/Amine System	*344*
13.4	Photosensitive Systems for Living Anionic Polymerization	*344*
	References	*345*
14	**Photoacid Generators (PAG) Systems**	*347*
14.1	Iminosulfonates and Oximesulfonates	*347*
14.2	Naphthalimides	*348*
14.3	Photoacids and Chemical Amplification	*349*
	References	*349*
15	**Photobase Generators (PBG) Systems**	*351*
15.1	Oxime Esters	*351*
15.2	Carbamates	*351*
15.3	Ammonium Tetraorganyl Borate Salts	*351*

15.4	N-Benzylated-Structure-Based Photobases 352
15.5	Other Miscellaneous Systems 353
15.6	Photobases and Base Proliferation Processes 354
	References 354

Part IV Reactivity of the Photoinitiating System 357

16 Role of the Experimental Conditions in the Performance of a Radical Photoinitiator 359
- 16.1 Role of Viscosity 360
- 16.2 Role of the Surrounding Atmosphere 361
- 16.3 Role of the Light Intensity 361
- References 364

17 Reactivity and Efficiency of Radical Photoinitiators 367
- 17.1 Relative Efficiency of Photoinitiators 367
- 17.1.1 Photopolymerization of MMA in Organic Solvents 367
- 17.1.2 Photopolymerization of Various Monomers in Organic Solvents and Bulk Media 367
- 17.1.3 Photopolymerization of TMPTA in Film 371
- 17.1.4 Photopolymerization of Monomer/Oligomer Film 373
- 17.1.5 Photopolymerization of Acrylamide in Water 376
- 17.1.6 Photopolymerization of Oil-Soluble Monomers in Direct Micelles 376
- 17.1.7 Photopolymerization of Water-Soluble Monomers in Reverse Micelles 377
- 17.2 Role of the Excited-State Reactivity 377
- 17.2.1 Generation of Radicals 377
- 17.2.2 Role of Monomer Quenching 378
- 17.3 Role of the Medium on the Photoinitiator Reactivity 381
- 17.3.1 Reactivity in Solution 381
- 17.3.2 Reactivity in Microheterogeneous Solution 383
- 17.3.3 Reactivity in Bulk 386
- 17.4 Structure/Property Relationships in Photoinitiating Systems 388
- 17.4.1 Role of the Bond Dissociation Energy in Cleavable Systems 388
- 17.4.2 Role of the Bond Dissociation Energy in Noncleavable Systems 389
- 17.4.3 Role of the Initiating Radical in the Initiation Step 391
- 17.4.3.1 Generation of the First Monomer Radical 391
- 17.4.3.2 Addition Rate Constants to Monomer 392
- 17.4.3.3 Description of the Radical/Monomer Reactivity 392
- References 394

18	**Reactivity of Radicals toward Oxygen, Hydrogen Donors, Monomers, and Additives: Understanding and Discussion** *399*	
18.1	Alkyl and Related Carbon-Centered Radicals *399*	
18.2	Aryl Radicals *401*	
18.3	Benzoyl Radicals *402*	
18.4	Acrylate and Methacrylate Radicals *403*	
18.5	Aminoalkyl Radicals *404*	
18.5.1	Reactivity *404*	
18.5.2	Role of the Class of the Amine *408*	
18.5.3	N-Phenyl Glycine Derivatives *408*	
18.5.4	Chain Length Effect *409*	
18.5.5	Regioselectivity of the Hydrogen Abstraction Reaction *409*	
18.5.6	Aminoalkyl Radicals and the Halogen Abstraction Reaction *410*	
18.5.7	Reactivity under Air *411*	
18.6	Phosphorus-Centered Radicals *412*	
18.7	Thiyl Radicals *413*	
18.8	Sulfonyl and Sulfonyloxy Radicals *417*	
18.9	Silyl Radicals *418*	
18.9.1	The Particular Behavior of the Tris(Trimethylsilyl)Silyl Radical *421*	
18.9.2	Reactivity and Photoinitiation under Air *422*	
18.9.3	Other Sources of Silyl Radicals *424*	
18.10	Oxyl Radicals *425*	
18.11	Peroxyl Radicals *426*	
18.11.1	Interaction with H-Donors *428*	
18.11.2	Interaction with Monomers *430*	
18.11.3	Interaction with Triphenylphosphine *430*	
18.11.4	S_H2 Substitution *430*	
18.11.5	Other Oxyls and Peroxyls *431*	
18.12	Aminyl Radicals *431*	
18.13	Germyl and Stannyl Radicals *432*	
18.13.1	Reactivity *433*	
18.13.2	Reactivity under Air *434*	
18.13.3	Reactivity and Structure of $(TMS)_3Ge^\bullet$ versus $(TMS)_3Si^\bullet$ *434*	
18.14	Boryl Radicals *435*	
18.14.1	Reactivity *436*	
18.14.2	Reactivity under Air *438*	
18.14.3	Photoinitiation under Air *438*	
18.15	Lophyl Radicals *439*	
18.16	Iminyl Radicals *439*	
18.17	Metal-Centered Radicals *440*	
18.18	Propagating Radicals *442*	
18.19	Radicals in Controlled Photopolymerization Reactions *443*	
18.19.1	Photoiniferters and Dithiocarbamyl Radicals *443*	
18.19.2	Light-Sensitive Alkoxyamines and Generation of Nitroxides *445*	

	18.20	Radicals in Hydrosilylation Reactions *446*
		References *447*
	19	**Reactivity of Radicals: Towards the Oxidation Process** *455*
	19.1	Reactivity of Radicals toward Metal Salts *455*
	19.2	Radical/Onium Salt Reactivity in Free-Radical-Promoted Cationic Photopolymerization *456*
		References *459*

Conclusion *461*

Index *465*

Abbreviations

AA	acrylamide
ABE	allylbutylether
ABP	aminobenzophenone
ADD	acridinediones
AH	electron/proton donor
AIBN	azo bis-isobutyro nitrile
ALD	Aldehydes
ALK	Alkoxyamines
AN	acrylonitrile
AOT	bis-2 ethyl hexyl sodium sulfosuccinate
APG	alkylphenylglyoxylates
AQ	anthraquinone
ATR	Attenuated reflectance
ATRP	atom transfer radical polymerization
BA	butylacrylate
BAc	[2-oxo-1,2-di(pheny)ethyl]acetate
BAPO	Bis-acyl phosphine oxide
BBD	Benzoyl benzodioxolane
BBDOM	bisbenzo-[1,3]dioxol-5-yl methanone
BC	borane complexes
BD	Benzodioxinone
BDE	bond dissociation energy
BE	benzoin esters
BIP-T	bis-(4-tert-butylphenyl) iodonium triflate
BMA	butylmethacrylate
BME	benzoin methyl ether
BMS	benzophenone phenyl sulfide
BP	benzophenone
BPO	benzoyl peroxide
BPSK	1-Propanone,1-[4-[(4-benzoylphenyl)thio]phenyl]-2-methyl-2-[(4-methylphenyl)sulfonyl]
BTTB	4-Benzoyl(4$'$-tert-butylperoxycarboxyl) tert-butylperbenzoate
BVE	butylvinylether

Bz	benzil
BZ	benzoin
C1	7-diethylamino-4-methyl coumarin
C6	3-(2′-benzothiazoryl)-7-diethylaminocoumarin
CA	cyanoacrylates
CD	cyclodextrin
CIDEP	chemically induced electron polarization
CIDNP	chemically induced nuclear polarization
CL	caprolactone
CNT	photopolymerized lipidic assemblies
co-I	co-initiator
CPG	cyano N-phenylglycine
CQ	camphorquinone
CT	charge transfer
CTC	charge transfer complex
CTP	computer-to-plate
CTX	chlorothioxanthone
CumOOH	cumene hydroperoxide
CW	continuous-wave
DB	deoxybenzoin
DCPA	dicylopentenyl acrylates
DDT	diphenyldithienothiophene
DEAP	2,2-dietoxyacetophenone
DEDMSA	N,N-diethyl-1,1-dimethylsilylamine
DEEA	2-(2-ethoxy-ethoxy) ethyl acrylate
DFT	density functional theory
DH	hydrogen donor
DMAEB	dimethylamino ethyl benzoate
DMPA	2,2-dimethoxy-2 phenyl-acetophenone
DMPO	5,5′-dimethyl-1-pyrroline N-oxide
DPA	diphenyl acetylene
dPI	difunctional photoinitiators
DSC	differential scanning calorimetry
DTAC	dodecyl trimethylammonium chloride
DUV	deep UV
DVE	divinylether
EA	electronic affinity
EAB	diethyl amino benzophenone
EDB	ethyl dimethylaminobenzoate
EHA	2-ethyl hexyl ester
EL	ethyl linoleate
EMP	N-ethoxy-2-methylpyridinium
EMS	epoxy-modified silicone
Eo	Eosin Y
EP	epoxy acrylate

EpAc	epoxy acrylate; see Section 16 p359
EPDM	ethylene-propylene-diene monomers
EPHT	electron/proton hydrogen transfer
EPOX	3,4-epoxycyclohexane)methyl 3,4-epoxycyclohexylcarboxylate
EPT	ethoxylated pentaerythritol tetraacrylate
ERL	exposure reciprocity law
ESO	epoxidized soybean oil
ESR	electron spin resonance
ESR-ST	Electron spin resonance spin trapping
ET	energy transfer
eT	electron transfer
EtBz	ethylbenzene
EUV	extreme UV
EVE	ethylvinylether
FBs	fluorescent bulbs
Fc(+)	ferrocenium salt derivative
FRP	free radical photopolymerization
FRPCP	free-radical-promoted cationic polymerization
FTIR	Fourier transform infrared
FU	fumarate
GRIN	gradient index
HABI	2,2',4,4',5,5'-hexaarylbiimidazole
HALS	Hindered amine light stabilizer
HAP	2-hydroxy-2- methyl-1- phenyl-1- propanone
HCAP	1-hydroxy- cyclohexyl-1- phenyl ketone
HCs	hydrocarbons
HDDA	hexane diol diacrylate
hfc	hyprefine splitting
HFS	hyperfine splitting
HOMO	highest occupied molecular orbital
HQME	hydroquinone methyl ether
HRAM	Highly reactive acrylate monomers
HSG	hybrid sol–gel
HT	hydrogen transfer
IP	ionization potential
IPNs	Interpenetrating polymer networks
IR	infrared
ISC	intersystem crossing
ITX	isopropylthioxanthone
JAW	julolidine derivative
K-ESR	kinetic electron spin resonance
KC	ketocoumarin
2K-PUR	two-component polyurethane
LAT	light absorbing transients
LCAO	linear combination of atomic orbitals

LCD	liquid crystal display
LDI	Laser direct imaging
LDO	limonene dioxide
LED	light-emitting diode
LFP	laser flash photolysis
LIPAC	laser-induced photoacoustic calorimetry
LS	light stabilizers
LUMO	lowest unoccupied molecular orbital
MA	methylacrylate
MA	monomer acceptor
MAL	maleate
MB	methylene blue
MBI	mercaptobenzimidazole
MBO	mercaptobenzoxazole
MBT	mercaptobenzothiazole
MD	monomer donor
MDEA	methyldiethanolamine
MDF	medium-density fiber
MEK	methyl ethyl ketone
MIR	multiple internal reflectance
MK	Mischler's ketone
MMA	methylmethacrylate
MO	molecular orbitals
mPI	Multifunctional photoinitiators
MPPK	2-benzyl-2-dimethylamino-1-(4-morpholinophenyl)-1-butanone
MWD	molecular weight distribution
NAS	2-[p-(diethyl-amino)styryl]naphtho[1,2-d]thiazole
NHC	N-heterocyclic carbene
NIOTf	N-(trifluoromethanesulfonyloxy)-1,8-naphthalimide
NIR	near-IR reflectance
NMP2	nitroxide-mediated photopolymerization
NMP	nitroxide-mediated photopolymerization
NMR	Nuclear magnetic resonance
NOR	norbornenes
NP	nanoparticles
NPG	N-phenyl glycine
NQ	naphthoquinone
NVET	nonvertical energy transfer
NVP	N-vinylpyrolidone
OD	optical density
OLED	organic light-emitting diode
OMC	organometallic compounds
On$^+$	onium salt derivative
OrM	organic matrixes
P$^+$	pyrilium salt derivative

PAG	Photoacid generators	
PBG	photobase generator	
PBN	phenyl-*N*-tertbutyl nitrone	
PC	photocatalyst	
PCBs	printed circuit boards	
PCL	polycaprolactone	
PDO	1-phenyl 2-propanedione-2 (ethoxycarbonyl) oxime	
PEG	polyethyleneglycol	
PES	potential energy surface	
PETA	pentaerythritol tetraacrylate	
PHS	Poly(hydrosilane)s	
PHT	pure hydrogen transfer	
PI	photoinitiator	
PIS	photoinitiating system	
PLA	Polylactic acid	
PLP	pulsed laser polymerization	
PLP	Pulsed laser-induced polymerization	
PMK	2-methyl-1-(benzoyl)-2-morpholino-propan-1-one	
PMMA	polymethylmethacrylate	
POH	phenolic compounds	
PPD	1-phenyl-1,2-propanedione	
PPK	2-benzyl-2-dimethylamino-1-(phenyl)-1-butanone	
PS	photosensitizer	
PS/PI	photosensitizer/photoinitiator	
PSAs	Pressure-sensitive adhesives	
PVC	polyvinylchloride	
PWBs	printed wiring boards	
PYR	pyrromethene	
RAFT	reversible addition-fragmentation transfer	
RB	Rose Bengal	
RFID	radiofrequency identification	
ROMP	Ring-opening metathesis photopolymerization	
ROOH	peroxide derivative	
ROOH	hydroperoxide derivative	
ROP	ring-opening polymerization	
RP	radical pair	
RPM	radical pair mechanism	
RSH	mercaptan	
RT-FTIR	real-time Fourier transform infrared	
SCM	solvatochromic comparison method	
SCRP	spin-correlated radical pair	
SDS	sodium dodecyl sulfate	
SG1	N-(2-methylpropyl)-N-(1-diethylphosphono-2, 2-dimethylpropyl)-N-oxyl	
SHOMO	singly highest occupied molecular orbital	

SOMO	singly occupied molecular orbital
STY	styrene
SU	suberone
SWNT	single-wall carbon nanotube
TEA	triethyl amine
TEMPO	2,2,6,6, tetramethylpiperidine N-oxyl radical
THF	tetrahydrofuran
ThP	thiophene
TI	titanocene derivative
TIPNO	2,2,5-tri-methyl-4-phenyl-3-azahexane-3-nitroxide
TLS	thermal lens spectroscopy
TM	triplet mechanism
TMP	2,2,6,6-tetramethylpiperidine
TMPTA	trimethylolpropane triacrylate
TP$^+$	thiopyrilium salt derivative
TPA	two-photon absorption
TPGDA	tetrapropyleneglycol diacrylate
TPK	1-[4-(methylthio) phenyl]-ethanone
TPMK	2-methyl-1-(4-methylthiobenzoyl)-2-morpholino-propan-1-one
TPO	2,4,6-trimethyl benzoyl-diphenylphosphine oxide
TPP	triphenylphosphine
TR-ESR	time resolved electron spin resonance
TR-FTIR	time-resolved Fourier transform infra red
TR-S2FTIR	Laser-induced step-scan FTIR spectroscopy
TS	transition state
TST	transition state theory
TTMSS	tris(trimethyl)silylsilane
TX	thioxanthone
TX-SH	2-mercaptothioxanthone
Tz	triazine derivative
ULSI	ultra large scale integration
UV	ultraviolet
UVA	UV absorbers
VA	vinyl acetate
VC	vinylcarbazole
VE	vinyl ethers
VE	vinylacetate
VET	vertical energy transfer
Vi	violanthrone
VIE	vitamin E
VLSI	very large scale integration
VOC	volatile organic compounds
VP	vinylpyrrolidone
VUV	vacuum ultraviolet
XT	xanthones

Introduction

Light-induced polymerization reactions are largely encountered in many industrial daily life applications or in promising laboratory developments. The basic idea is to readily transform a liquid monomer (or a soft film) into a solid material (or a solid film) on light exposure. The huge sectors of applications, both in traditional and high-tech areas, are found in UV curing (this area corresponds to the largest part of radiation curing that includes UV and electron beam curing), laser imaging, microlithography, stereolithography, microelectronics, optics, holography, medicine, and nanotechnology.

UV curing represents a green technology (environmentally friendly, nearly no release of volatile organic compounds (VOCs), room temperature operation, possible use of renewable materials, use of convenient light sources (light-emitting diodes (LEDs), household lamps, LED bulbs, and the sun) that continues its rapid development. The applications concern, for example, the use of varnishes and paints (for a lot of applications on a large variety of substrates, e.g., wood, plastics, metal, and papers), the design of coatings having specific properties (for flooring, packaging, release papers, wood and medium-density fiber (MDF) panels, automotive, pipe lining, and optical fibers), the development of adhesives (laminating, pressure sensitive, and hot melt), and the graphic arts area (drying of inks, inkjets, overprint varnishes, protective and decorative coatings, and the manufacture of conventional printing plates).

Other applications of photopolymerization reactions concern medicine (restorative and preventative denture relining, wound dressing, ophthalmic lenses, glasses, artificial eye lens, and drug microencapsulation), microelectronics (soldering resists, mask repairs, encapsulants, conductive screen inks, metal conductor layers, and photoresists), microlithography (writing of complex relief structures for the manufacture of microcircuits or the patterning of selective areas in microelectronic packaging using the laser direct imaging (LDI) technology; direct writing on a printing plate in the computer-to-plate technology), 3D machining (or three-dimensional photopolymerization or stereolithography) that gives the possibility of making objects for prototyping applications, optics (holographic recording and information storage, computer-generated and embossed holograms, manufacture of optical elements, e.g., diffraction grating, mirrors, lenses, waveguides, array illuminators, and display devices), and structured materials on the nanoscale size.

Photopolymerization reactions are currently encountered in various experimental conditions, for example, in film, gas phase, aerosols, multilayers, (micro)heterogeneous media or solid state, on surface, in ionic liquids, *in situ* for the manufacture of microfluidic devices, *in vivo*, and under magnetic field. Very different aspects can be concerned with gradient, template, frontal, controlled, sol–gel, two-photon, laser-induced or spatially controlled, and pulsed laser photopolymerization.

As a photopolymerization reaction involves a photoinitiating system, a polymerizable medium, and a light source, a strong interplay should exist between them. The photoinitiator has a crucial role as it absorbs the light, converts the energy into reactive species (excited states, free radicals, cations, acids, and bases) and starts the reaction. Its reactivity governs the efficiency of the polymerization. A look at the literature shows that a considerable number of works are devoted to the design of photosensitive systems being able to operate in many various (and sometimes exotic) experimental conditions. This research field is particularly rich. Fantastic developments have appeared all along the past three decades. Significant achievements have been made since the early works on photopolymerization in the 1960s and the traditional developments of the UV-curing area. At present, high-tech applications are continuously emerging. Tailor-made photochemistry and chemistry have appeared in this area. The search for a safe and green technology has been launched. Interesting items relate not only to the polymer science and technology field but also to the photochemistry, physical chemistry, and organic chemistry areas.

We believe that the proposed book focused on this exciting topic related to the photosensitive systems encountered in photopolymerization reactions will be helpful for many readers. Why a new book? Indeed, in the past 20 years, many aspects of light-induced polymerization reactions have been obviously already discussed in books and review papers. Each of these books, however, usually covers more deeply selected aspects depending first on the origin (university, industry) and the activity sector of the author (photochemistry, polymer chemistry, and applications) and second on the goals of the book (general presentation of the technology, guide for end users, and academic scope). Our previous general book published more than 15 years ago (1995) and devoted to the three photoinitiation-photopolymerization-photocuring complementary aspects already provided a first account on the photosensitive systems.

For obvious reasons, all these three fascinating aspects that continuously appear in the literature cannot be unfortunately developed now (in 2011) in detail in a single monograph because of the rapid growth of the research. A book that mostly concentrates on the photosensitive systems that are used to initiate the photopolymerization reaction, their adaptation to the light sources, their excited state processes, the reactivity of the generated initiating species (free radicals, acids, and bases), their interaction with the different available monomers, their working out mechanisms, and the approach for a complete understanding of the (photo)chemical reactivity was missing. This prompted us to write the present book. It aims at providing an original and up-to-date presentation of these points together with a discussion of the structure/reactivity/efficiency relationships observed in photoinitiating systems usable in radical, cationic, and anionic photopolymerization as well as in acid and base catalyzed photocrosslinking reactions. We wish to focus on the necessary role of the basic research toward the progress of the applied research through the large part we have devoted to the involved mechanisms. In fact, everybody is aware that there is no real technical future development without a present high-quality scientific research. In our opinion, such an extensive and complete book within this philosophy has never been written before.

Science is changing very fast. During the preparation of a book, any author has the feeling of walking behind the developments that unceasingly appear. It is rather difficult to have the latest photography of the situation by the end of the manuscript; this is also reinforced by the necessary delay to print and deliver the book. Therefore, we decided here to give not only the best up-to-date situation of the subject but also to take time to define a lot of basic principles and concepts, mechanistic reaction schemes, and examples of reactivity/efficiency studies that remain true and are not submitted to a significant aging on a 10-year timescale.

The book is divided into four parts. In Part I, we deliver a general presentation of the basic principles and applications of the involved photopolymerization reactions with a description of the available light sources, the different monomers and the properties of the cured materials, the various aspects and characteristics of the reactions, and the role of the photosensitive systems and the typical examples of applications in different areas. The part especially concerned with the polymer science point of view (as other books have already dealt in detail with this aspect) focuses on general considerations and latest developments and to what is necessary to clearly understand the following parts. Then, we enter into the heart of the book.

In Parts II and III, we give (i) the most exhaustive presentation of the commercially and academically used or potentially interesting photoinitiating systems developed in the literature (photoinitiators, co-initiators, photosensitizers, macrophotoinitiators, multicomponent combinations, and tailor-made compounds for specific properties), (ii) the characteristics of the excited states, and (iii) the involved reaction mechanisms. We provide an overview of all the available systems but we focus our attention on newly developed photoinitiators, recently reported studies, and novel data on previous well-known systems. All this information is provided for radical photopolymerization (Part II) and cationic and anionic photopolymerization and photoacid and photobase catalyzed photocrosslinking (Part III).

In Part IV, we gather and discuss (i) a large set of data, mostly derived from time-resolved laser spectroscopy and electron spin resonance (ESR) experiments, related to both the photoinitiating system excited states and the initiating radicals (e.g., a complete presentation of the experimental and theoretical reactivity of more than 15 kinds of radicals is provided); (ii) the most recent results of quantum mechanical calculations that allow probing of the photophysical/photochemical properties as well as the chemical reactivity of a given photoinitiating system; and (iii) the reactivity in solution, in micelle, in bulk, in film, under air, in low viscosity media, or under low light intensities.

The book also outlines the latest developments and trends for the design of novel molecules. This concerns first the elaboration of smart systems exhibiting well-designed functional properties or/and suitable for processes in the nanotechnology area. A second direction refers to the development of an evergreen (photo)chemistry elaborating, for example, safe, renewable, reworkable, or biocompatible materials. A third trend is related to the use of soft irradiation conditions for particular applications, which requires the design of low oxygen sensitivity compounds under exposure to low-intensity visible light sources, sun, LEDs, laser diodes, or household lamps (e.g., fluorescence or LED bulbs).

When questioning the Chemical Abstract database, many references appear. We have not intended to give here an exhaustive list of references or a survey of the patent literature. We used, however, more than 2000 references. Pioneer works are cited but our present list of references mainly refers to papers dispatched during the past 15 years. The selection of the articles is most of the time a rather hard and sensitive task. We have done our best and beg forgiveness for possible omissions.

This research field has known a fantastic evolution. We would like now to share the real pleasure we had (and still have) in participating and contributing to this area. Writing this book was really a great pleasure. We hope that our readers, R and D researchers, engineers, technicians, University people, and students involved in various scientific or/and technical areas such as photochemistry, polymer chemistry, organic chemistry, radical chemistry, physical chemistry, radiation curing, imaging, physics, optics, medicine, nanotechnology will appreciate this book and enjoy its content.

And now, it is time to dive into the magic of the photoinitiator/photosensitizer world!

Part I
Basic Principles and Applications of Photopolymerization Reactions

In this first part of the book, we give a general presentation of the photopolymerization reactions, the light sources, the experimental devices and the applications, the role of the photosensitive systems, the evaluation of the practical efficiency of a photopolymerizable medium, and the approach of the photochemical and chemical reactivity. As stated in the introduction, the different aspects related to photopolymerization reactions are currently presented and discussed in books (see, e.g., [1–31]) and review papers (see, e.g., [32–96]). As a consequence, some of the above-mentioned topics that have formerly received a deeper analysis are not exhaustively treated here. In that case, we only provide a basic and rather brief description that should allow an easy understanding of the three subsequent parts.

1
Photopolymerization and Photo-Cross-Linking

Everybody knows that a polymerization reaction [97] consists in adding many monomer units M to each other, thereby creating a macromolecule (Eq. (1.1)).

$$M \rightarrow M-M \rightarrow \rightarrow (M)_n \qquad (1.1)$$

The initiation step of this reaction corresponds to the decomposition of a molecule (an initiator I) usually obtained through a thermal process (Eq. (1.2)). This produces an initiating species (e.g., a free radical $R^\bullet$) able to attack the first monomer unit. Other units add further to form the macromolecule.

$$I \rightarrow R^\bullet(\Delta) \quad \text{and} \quad R^\bullet + M \rightarrow RM^\bullet \qquad (1.2)$$

Instead of a thermal activation of the polymerization, other stimuli such as light, electron beam, X-rays, γ-rays, plasma, microwaves, or even pressure can be used [85]. Among them, the exposure of a resin (monomer/oligomer matrix) to a suitable light appeared as a very convenient way for the initiation step: in that case, the reaction is called a *photopolymerization reaction* (Eq. (1.3)).

$$\text{Monomer/oligomer formulation} \rightarrow \text{Polymer } (h\nu) \qquad (1.3)$$

Owing to their absorption properties, monomers or oligomers are usually not sensitive to the available lights (except a few cases involving specifically designed light-absorbing structures). The addition of a photoinitiator (PI) is at least necessary (Eq. (1.4)). Excited states are generated under the light exposure of PI (Chapters 2 and 3). Then, an initiating species is produced. Its nature – radical ($R^\bullet$), cationic (C^+), and anionic (A^-) – is dependent on the starting molecule.

$$\text{Photoinitiator (PI)} \rightarrow \text{Excited states } (h\nu) \text{ and excited states}$$
$$\rightarrow \text{Initiating species}$$
$$\text{Initiating species} + \text{monomer/oligomer} \rightarrow \text{Polymer} \qquad (1.4)$$

Accordingly, the usual types of photopolymerization reactions (radical, cationic, and anionic photopolymerization or acid and base catalyzed photo-cross-linking reactions) can be encountered (Eq. (1.5)) in suitable resins.

$$\text{Photoinitiator (PI)} \rightarrow \text{Radicals } (h\nu)$$
$$\text{Photoinitiator (PI)} \rightarrow \text{Acids, cations, cation radicals } (h\nu)$$
$$\text{Photoinitiator (PI)} \rightarrow \text{Bases, anions } (h\nu) \qquad (1.5)$$

Photoinitiators for Polymer Synthesis: Scope, Reactivity and Efficiency, First Edition.
Jean-Pierre Fouassier and Jacques Lalevée.
© 2012 Wiley-VCH Verlag GmbH & Co. KGaA. Published 2012 by Wiley-VCH Verlag GmbH & Co. KGaA.

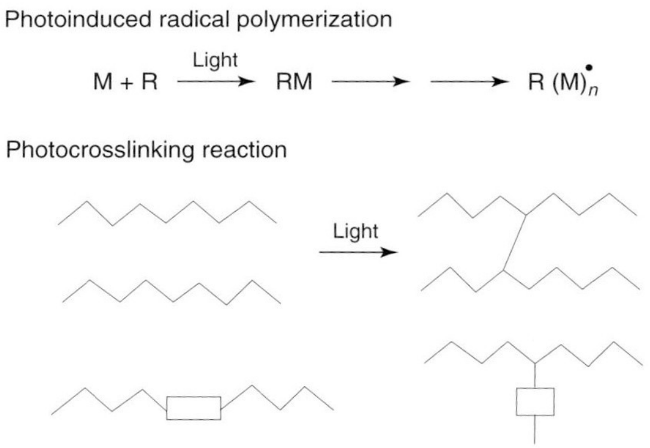

Photoinduced radical polymerization

$$M + R \xrightarrow{Light} RM \longrightarrow \longrightarrow R(M)_n^\bullet$$

Photocrosslinking reaction

Scheme 1.1

The term *photopolymerization* is very general and relates to two different concepts (Scheme 1.1). A photoinduced polymerization reaction is a chain reaction where one photon yields one initiating species and induces the incorporation of a large number of monomer units. A photo-cross-linking reaction refers to a process involving a prepolymer or a polymer backbone in which a cross-link is formed between two macromolecular chains. This kind of polymer can be designed in such a way that it contains pendent (e.g., in polyvinylcinnamates) or in-chain photo-cross-linkable moieties (e.g., in chalcone-type chromophore-based polymers).

A monomer (**1**) is a rather small molecule having usually one or several chemical reactive functions (e.g., acrylates), whereas an oligomer (**2**) is a large molecular structure consisting of repetitive units of a given chemical structure that constitutes the backbone (e.g., a polyurethane) and containing one or more reactive chemical functions. The oligomer skeleton governs the final physical and chemical properties of the cured coating.

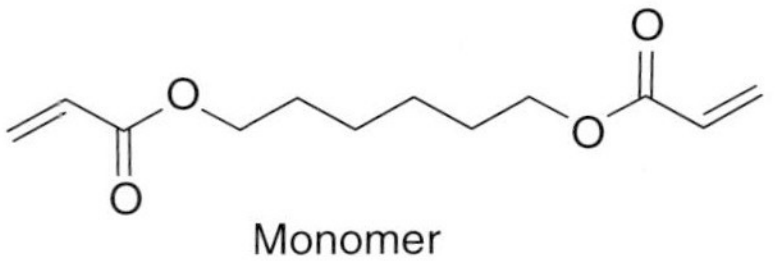

Monomer **1**

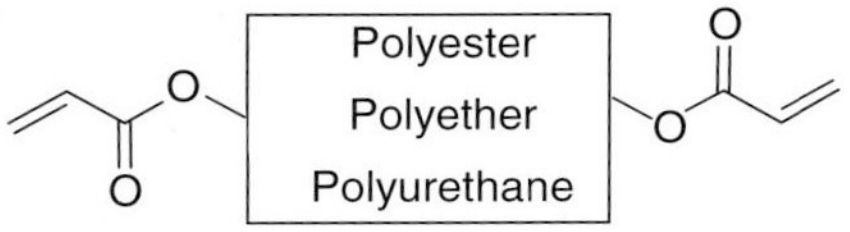

Oligomer **2** (Polyester / Polyether / Polyurethane)

When using multifunctional monomers or oligomers, the photoinduced polymerization reaction does not obviously proceed to form a linear polymer. As it develops in the three directions of space, it also leads to a cross-linking reaction, thereby creating a polymer network (see, e.g., Scheme 1.2 for a free radical reaction). Sometimes, the reaction is depicted as a cross-linking photopolymerization.

A photopolymerizable formulation [25] consists of (i) a monomer/oligomer matrix (the monomer plays the role of a reactive diluent to adjust the viscosity of the formulation; it readily copolymerizes), (ii) a PI or a photoinitiating system (PIS) (containing a PI and other compounds), and (iii) various additives, for example,

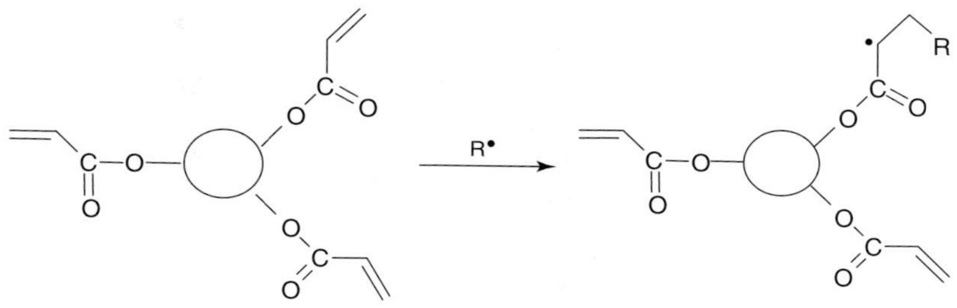

Scheme 1.2

flow, slip, mist, wetting, dispersion agents, inhibitors for handling and fillers, plasticizers, matting or gloss agents, pigments, and light stabilizers according to the applications.

UV curing is a word that defines an ever-expanding industrial field [6, 10, 16, 25] where the light, often delivered by a mercury lamp, is used to transform a liquid photosensitive formulation into an insoluble solid film for coating applications through a photopolymerization reaction. Photocuring is a practical word that refers to the use of light to induce this rapid conversion of the resin to a cured and dried solid film. Film thicknesses typically range from a few micrometers to a few hundred micrometers depending on the applications. In photostereolithography, the idea consists in building up a solid object through a layer-by-layer photopolymerization procedure.

In the imaging area, an image is obtained according to a process largely described in the literature [8, 9, 17, 90]. The resin layer is irradiated through a mask. A reaction takes place in the irradiated areas. Two basically different reactions can occur: (i) a photopolymerization or a photo-cross-linking reaction that renders the film insoluble (using a suitable solvent allows to dissolve the monomer present in the shadow areas; after etching of the unprotected surface and a bake out of the polymerized film, a negative image is thus formed) and (ii) a depolymerization or a hydrophobicity/hydrophilicity change (that leads to a solubilization of the illuminated areas, thereby forming a positive image). Free radical photopolymerization and photo-cross-linking (Scheme 1.3) lead to a negative image through (i). The acid or base catalyzed reaction (Scheme 1.3) leads either to a negative image (i) or a positive image (ii).

In microelectronics, such a monomer/oligomer or polymer matrix sensitive to a light source is named a photoresist. In imaging technology, the organic matrix is

Monomer/low-molecular-weight polymer $\xrightarrow{R^\bullet \quad C^+ \quad A^-}$ High-molecular-weight polymer (1)

High-molecular-weight polymer $\xrightarrow{C^+ \quad A^-}$ Low-molecular-weight polymer (2)

Scheme 1.3

called a *photopolymer*. Strictly speaking, photopolymer refers to a polymer sensitive to light but this word is often used to design a monomer/oligomer matrix that polymerizes under light exposure. The new term *photomaterial* refers to an organic photosensitive matrix that leads, on irradiation, to a polymer material exhibiting specific properties useful in the nanotechnology field; it could also design the final material formed through this photochemical route.

References

1. Roffey, C.G. (1982) *Photopolymerization of Surface Coatings*, John Wiley & Sons, Inc., New York.
2. Rabek, J.F. (1987) *Mechanisms of Photophysical and Photochemical Reactions in Polymer: Theory and Practical Applications*, John Wiley & Sons, Ltd, New York.
3. Hoyle, C.E. and Kinstle, J.F. (eds) (1990) *Radiation Curing of Polymeric Materials*, ACS Symposium Series, Vol. 417, American Chemical Society.
4. Fouassier, J.P. and Rabek, J.F. (eds) (1990) *Lasers in Polymer Science and Technology: Applications*, CRC Press, Boca Raton, FL.
5. Bottcher, H. (1991) *Technical Applications of Photochemistry*, Deutscher Verlag fur Grundstoffindustrie, Leipzig.
6. Pappas, S.P. (1986) *UV-Curing: Science and Technology*, Technology Marketing Corporation, Stamford, CT; Plenum Press, New York (1992).
7. Fouassier, J.P. and Rabek, J.F. (eds) (1993) *Radiation Curing in Polymer Science and Technology*, Chapman & Hall, London.
8. Krongauz, V. and Trifunac, A. (eds) (1994) *Photoresponsive Polymers*, Chapman & Hall, New York.
9. Reiser, A. (1989) *Photoreactive Polymers: The Science and Technology of Resists*, John Wiley & Sons, Inc., New York.
10. Fouassier, J.P. (1995) *Photoinitiation, Photopolymerization, Photocuring*, Hanser, Münich.
11. Allen, N.S., Edge, M., Bellobono, I.R., and Selli, E. (eds) (1995) *Current Trends in Polymer Photochemistry*, Ellis Horwood, New York.
12. Scranton, A.B., Bowman, A., and Peiffer, R.W. (eds) (1997) *Photopolymerization: Fundamentals and Applications*, ACS Symposium Series, Vol. 673, American Chemical Society, Washington, DC.
13. Olldring, K. and Holman R. (eds) (1997) *Chemistry and Technology of UV and EB Formulation for Coatings Inks and Paints*, vol. I - VIII, John Wiley & Sons, Ltd and Sita Technology Ltd, London.
14. Fouassier, J.P. (1998) *Photoinitiated Polymerization: Theory and Applications*, Rapra Review Reports, Vol. 9(4), Rapra Technology Ltd, Shawbury.
15. Holman, R. (ed.) (1999) *UV and EB Chemistry*, Sita Technology Ltd, London.
16. Davidson, S. (1999) *Exploring the Science, Technology and Application of UV and EB Curing*, Sita Technology Ltd, London.
17. Neckers, D.C. (1999) *UV and EB at the Millennium*, Sita Technology Ltd, London.
18. Fouassier, J.P. (ed.) (1999) *Photosensitive Systems for Photopolymerization Reactions*, Trends in Photochemistry and Photobiology, Vol. 5, Research Trends, Trivandrum.
19. Crivello, J.V. and Dietliker, K. (1999) *Photoinitiators for Free Radical, Cationic and Anionic Photopolymerization*, Surface Coatings Technology Series, vol. III (ed. G. Bradley), John Wiley & Sons, Inc.
20. Fouassier, J.P. (2001) *Light Induced Polymerization Reactions*, Trends in Photochemistry and Photobiology, Vol. 7, Research Trends, Trivandrum.
21. Dietliker, K. (2002) *A Compilation of Photoinitiators Commercially Available for UV Today*, Sita Technology Ltd, London.
22. Drobny, J.G. (2003 and 2010) *Radiation Technology for Polymers*, CRC Press, Boca Raton, FL.
23. Belfied, K.D. and Crivello, J.V. (eds) (2003) *Photoinitiated Polymerization*, ACS

Symposium Series, Vol. 847, American Chemical Society, Washington, DC.
24. Fouassier, J.P. (ed.) (2006) *Photochemistry and UV Curing*, Research Signpost, Trivandrum.
25. Schwalm, R. (2007) *UV Coatings: Basics, Recent Developments and New Applications*, Elsevier, Oxford.
26. Schnabel, W. (2007) *Polymer and Light*, Wiley-VCH Verlag GmbH, Weinheim.
27. Lackner, M. (ed.) (2008) *Lasers in Chemistry*, Wiley-VCH Verlag GmbH, Weinheim.
28. Mishra, M.K. and Yagci, Y. (eds) (2009) *Handbook of Vinyl Polymers*, CRC Press, Boca Raton, FL.
29. Allen, N.S. (ed.) (2010) *Photochemistry and Photophysics of Polymer Materials*, John Wiley & Sons, Inc., Hoboken, NJ.
30. Fouassier, J.P. and Allonas, X. (eds) (2010) *Basics of Photopolymerization Reactions*, Research Signpost, Trivandrum.
31. Green, W.A. (2010) *Industrial Photoinitiators*, CRC Press, Boca Raton, FL.
32. Fouassier, J.P. (2000) *Recent Res. Devel. Photochem. Photobiol.*, **4**, 51–74.
33. Fouassier, J.P. (2000) *Recent Res. Devel. Polym. Sci.*, **4**, 131–145.
34. Fouassier, J.P. (1999) *Curr. Trends Polym. Sci.*, **4**, 163–184.
35. Fouassier, J.P. (1990) *Prog. Org. Coat.*, **18**, 227–250.
36. Paczkowski, J. and Neckers, D.C. (2001) *Electron Trans. Chem.*, **5**, 516–585.
37. Schnabel, W. (1990) in *Lasers in Polymer Science and Technology*, vol. II (eds J.P. Fouassier and J.F. Rabek), CRC Press, Boca Raton, FL, pp. 95–143.
38. Urano, T. (2003) *J. Photopolym. Sci. Technol.*, **16**, 129–156.
39. Fouassier, J.P., Allonas, X., and Burget, D. (2003) *Prog. Org. Coat.*, **47**, 16–36.
40. Fouassier, J.P., Ruhlmann, D., Graff, B., Morlet-Savary, F., and Wieder, F. (1995) *Prog. Org. Coat.*, **25**, 235–271.
41. Fouassier, J.P., Ruhlmann, D., Graff, B., and Wieder, F. (1995) *Prog. Org. Coat.*, **25**, 169–202.
42. Fouassier, J.P. (1993) in *Radiation Curing in Polymer Science and Technology*, vol. 1 (eds J.P. Fouassier and J.F. Rabek), Elsevier, Barking, pp. 49–118.
43. Fouassier, J.P. (1993) in *Radiation Curing in Polymer Science and Technology*, vol. 2 (eds J.P. Fouassier and J.F. Rabek), Elsevier, Barking, pp. 1–62.
44. Lissi, E.A. and Encinas, M.V. (1991) in *Photochemistry and Photophysics*, vol. IV (ed. J.F. Rabek), CRC Press, Boca Raton, FL, pp. 221–293.
45. Fouassier, J.P. (1990) in *Photochemistry and Photophysics*, vol. II (ed. J.F. Rabek), CRC Press, Boca Raton, FL, pp. 1–26.
46. Timpe, H.J. (1993) in *Radiation Curing in Polymer Science and Technology*, vol. 2 (ed. J.P. Fouassier), Elsevier, Barking, pp. 529–554.
47. Cunningham, A.F. and Desobry, V. (1993) in *Radiation Curing in Polymer Science and Technology*, vol. 2 (eds J.P. Fouassier and J.F. Rabek), Elsevier, Barking, pp. 323–374.
48. Green, W.A. and Timms, A.W. (1993) in *Radiation Curing in Polymer Science and Technology*, vol. 2 (eds J.P. Fouassier and J.F. Rabek), Elsevier, Barking, pp. 375–434.
49. Carlini, C. and Angiolini, L. (1993) in *Radiation Curing in Polymer Science and Technology*, vol. 2 (eds J.P. Fouassier and J.F. Rabek), Elsevier, Barking, pp. 283–322.
50. Li Bassi, G. (1993) in *Radiation Curing in Polymer Science and Technology*, vol. 2 (eds J.P. Fouassier and J.F. Rabek), Elsevier, Barking, pp. 239–282.
51. Murai, H. and Hayashi, H. (1993) in *Radiation Curing in Polymer Science and Technology*, vol. 2 (eds J.P. Fouassier and J.F. Rabek), Elsevier, Barking, pp. 63–154.
52. Steiner, U.E. and Wolff, H.J. (1991) in *Photochemistry and Photophysics*, vol. IV (ed. J.F. Rabek), CRC Press, Boca Raton, FL, pp. 1–130.
53. Hayashi, H. (1990) in *Photochemistry and Photophysics*, vol. I (ed. J.F. Rabek), CRC Press, Boca Raton, FL, pp. 59–136.
54. Fouassier, J.P. and Lougnot, D.J. (1990) in *Lasers in Polymer Science and Technology*, vol. II (eds J.P. Fouassier and J.F. Rabek), CRC Press, Boca Raton, FL, pp. 145–168.
55. Jacobine, A. (1993) in *Radiation Curing in Polymer Science and Technology*, vol. 3

56. Peeters, S. (1993) in *Radiation Curing in Polymer Science and Technology*, vol. 3 (eds J.P. Fouassier and J.F. Rabek), Elsevier, Barking, pp. 177–218.
57. Broer, D.J. (1993) in *Radiation Curing in Polymer Science and Technology*, vol. 3 (eds J.P. Fouassier and J.F. Rabek), Elsevier, Barking, pp. 383–444.
58. Higashi, M. and Niwa, M. (1993) in *Radiation Curing in Polymer Science and Technology*, vol. 3 (eds J.P. Fouassier and J.F. Rabek), Elsevier, Barking, pp. 367–382.
59. Hasegawa, M., Hashimoto, Y., and Chung, C.M. (1993) in *Radiation Curing in Polymer Science and Technology*, vol. 3 (eds J.P. Fouassier and J.F. Rabek), Elsevier, Barking, pp. 341–366.
60. Dworjanyn, P.A. and Garnett, J.L. (1993) in *Radiation Curing in Polymer Science and Technology*, vol. 1 (eds J.P. Fouassier and J.F. Rabek), Elsevier, Barking, pp. 263–328.
61. Dreeskamp, H. and Palm, W.U. (1990) in *Lasers in Polymer Science and Technology*, vol. II (eds J.P. Fouassier and J.F. Rabek), CRC Press, Boca Raton, FL, pp. 47–56.
62. Mc Lauchlan, K.A. (1990) in *Lasers in Polymer Science and Technology*, vol. I (eds J.P. Fouassier and J.F. Rabek), CRC Press, Boca Raton, FL, pp. 259–304.
63. Phillips, D. and Rumbles, G. (1990) in *Lasers in Polymer Science and Technology*, vol. I (eds J.P. Fouassier and J.F. Rabek), CRC Press, Boca Raton, FL, pp. 91–146.
64. Mc Gimpsey, W.G. (1990) in *Lasers in Polymer Science and Technology*, vol. II (eds J.P. Fouassier and J.F. Rabek), CRC Press, Boca Raton, FL, pp. 77–94.
65. Bortolus, P. (1993) in *Radiation Curing in Polymer Science and Technology*, vol. 2 (eds J.P. Fouassier and J.F. Rabek), Elsevier, Barking, pp. 603–636.
66. Timpe, H.J., Jockush, S., and Korner, K. (1993) in *Radiation Curing in Polymer Science and Technology*, vol. 2 (ed. J.P. Fouassier), Elsevier, Barking, pp. 575–602.
67. Crivello, J.V. (1993) in *Radiation Curing in Polymer Science and Technology*, vol. 2 (eds J.P. Fouassier and J.F. Rabek), Elsevier, Barking, pp. 435–472.
68. Hacker, N.P. (1993) in *Radiation Curing in Polymer Science and Technology*, vol. 2 (eds J.P. Fouassier and J.F. Rabek), Elsevier, Barking, pp. 473–504.
69. Sahyun, M.R., De Voe, R.J., and Olofson, P.M. (1993) in *Radiation Curing in Polymer Science and Technology*, vol. 2 (ed. J.P. Fouassier), Elsevier, Barking, pp. 505–529.
70. Davidson, R.S. (1993) in *Radiation Curing in Polymer Science and Technology*, vol. 3 (eds J.P. Fouassier and J.F. Rabek), Elsevier, Barking, pp. 153–176.
71. Decker, C. and Fouassier, J.P. (1993) in *Radiation Curing in Polymer Science and Technology*, vol. 3 (eds J.P. Fouassier and J.F. Rabek), Elsevier, Barking, pp. 1–36.
72. Castle, P.M. and Sadhir, R.K. (1993) in *Radiation Curing in Polymer Science and Technology*, vol. 3 (eds J.P. Fouassier and J.F. Rabek), Elsevier, Barking, pp. 37–72.
73. Cabrera, M., Jezequel, J.Y., and André, J.C. (1993) in *Radiation Curing in Polymer Science and Technology*, vol. 3 (eds J.P. Fouassier and J.F. Rabek), Elsevier, Barking, pp. 73–97.
74. Allen, N.S. (2005) *Photochemistry*, vol. 35, The Royal Society of Chemistry, London, pp. 206–271.
75. Decker, C. (1999) in *Macromolecules*, vol. 143 (ed. K.P. Ghiggino), Wiley-VCH Verlag GmbH, Weiheim, pp. 45–63.
76. Decker, C. and Hoang Ngoc, T. (1998) in *Functional Polymers, Modern Synthetic Methods and Novel Structures*, ACS Symposium Series (eds A.O. Patil, D.N. Schultz, and B.M. Novak), American Chemical Society, pp. 286–302.
77. Decker, C. (1998) *Des. Monom. Polym.*, 1, 47–64.
78. Decker, C. and Elzaouk, B. (1997) in *Current Trends in Polymer Photochemistry* (eds N.S. Allen, M. Edge, I. Bellobono, and E. Selli), Ellis Horwood, New York, pp. 131–148.
79. Decker, C. (1997) in *Materials Science and Technology*, vol. 18 (ed. H.E.H. Meijer), Wiley-VCH Verlag GmbH, Weinheim, pp. 615–657.
80. Decker, C. (1996) in *Polymeric Materials Encyclopedia*, vol. 7 (ed. J.C. Salamone),

CRC Press, Boca Raton, FL, pp. 5181–5190.
81. Decker, C. (2001) in *Specialty Polymer Additives* (eds S. Al Malaika, A. Golovoy, and C.A. Wilkic), Backwell Science, Oxford, pp. 139–154.
82. Decker, C. (2003) in *Photoinitiated Polymerization*, ACS Symposium Series, Vol. 847, Chapter 23 (eds K.D. Belfield and J. Crivello), American Chemical Society, p. 266.
83. Sipani, V. and Scranton, A.B. (2004) in *Encyclopedia of Polymer Science and Technology* (ed. H.F. Mark), Wiley-Interscience, New York. doi: 10.1002/0471440264.pst491
84. Cai, Y. and Jessop, J.L.P. (2004) in *Encyclopedia of Polymer Science and Technology* (ed. H.F. Mark), Wiley-Interscience, New York. doi: 10.1002/0471440264.pst490
85. Fouassier, J.P., Allonas, X., and Lalevée, J. (2007) in *Macromolecular Engineering: from Precise Macromolecular Synthesis to Macroscopic Materials Properties and Applications*, vol. 1 (eds K. Matyjaszewski, Y. Gnanou, and L. Leibler), Wiley-VCH Verlag GmbH, Weinheim, pp. 643–672.
86. Allonas, X., Crouxte-Barghorn, C., Fouassier, J.P., Lalevée, J., Malval, J.P., and Morlet-Savary, F. (2008) in *Lasers in Chemistry*, vol. 2 (ed. M. Lackner), Wiley-VCH Verlag GmbH, pp. 1001–1027.
87. Arsu, N., Reetz, I., Yagci, Y., and Mishra, M.K. (2008) in *Handbook of Vinyl Polymers* (eds M.K. Mishra and Y. Yagci), CRC Press, pp. 50–75.
88. Fouassier, J.P., Allonas, X., Lalevée, J., and Dietlin, C. (2010) in *Photochemistry and Photophysics of Polymer Materials* (ed. N.S. Allen), John Wiley & Sons, Inc., Hoboken, NJ, pp. 351–420.
89. Kahveci, M.U., Gilmaz, A.G., and Yagci, Y. (2010) in *Photochemistry and Photophysics of Polymer Materials* (ed. N.S. Allen), John Wiley & Sons, Inc., Hoboken, NJ, pp. 421–478.
90. Ivan, M.G. and Scaiano, J.C. (2010) in *Photochemistry and Photophysics of Polymer Materials* (ed. N.S. Allen), John Wiley & Sons, Inc., Hoboken, NJ, pp. 479–508.
91. Muftuogli, A.E., Tasdelen, M.A., and Yagci, Y. (2010) in *Photochemistry and Photophysics of Polymer Materials* (ed. N.S. Allen), John Wiley & Sons, Inc., Hoboken, NJ, pp. 509–540.
92. Gisjman, P. (2010) in *Photochemistry and Photophysics of Polymer Materials* (ed. N.S. Allen), John Wiley & Sons Inc., Hoboken, NJ, pp. 627–680.
93. Lalevée, J., El Roz, M., Allonas, X., and Fouassier, J.P. (2009) in *Organosilanes: Properties, Performance and Applications*, Chapter 6 (eds E. Wyman and M.C. Skief), Nova Science Publishers, Hauppauge, NY, pp. 164–193.
94. Lalevée, J., Tehfe, M.A., Allonas, X., and Fouassier, J.P. (2010) in *Polymer Initiators*, Chapter 8 (ed. W.J. Ackrine), Nova Science Publishers, Hauppauge, NY, pp. 203–247.
95. Yagci, Y. and Retz, I. (1998) *Prog. Polym. Sci.*, **23**, 1485–1538.
96. Lalevée, J. and Fouassier, J.P. (2011) *Polym. Chem.*, **2**, 1107–1113.
97. Matyjaszewski, K., Gnanou, Y., and Leibler, L. (eds) (2007) *Macromolecular Engineering: From Precise Macromolecular Synthesis to Macroscopic Materials Properties and Applications*, vol. 1, Wiley-VCH Verlag GmbH, Weinheim, pp. 643–672.

2
Light Sources

2.1
Electromagnetic Radiation

An electromagnetic radiation, usually referred to as a *light*, is characterized by a frequency ν (in Hz) or a wavelength λ (in m), which are inversely proportional to one another (Eq. (2.1) where $c = \sim 3 \times 10^8$ m s^{-1} is the speed of light in vacuum); ν is proportional to the wave number ϖ. In photochemistry, λ is very often used in nanometers (nm).

$$\nu = \frac{c}{\lambda} = c\,\varpi \tag{2.1}$$

The electromagnetic spectrum spans a large frequency range that varies by many orders of magnitude. Owing to the absorption properties of the photosensitive systems, the light used in photopolymerization reactions typically consists of (i) ultraviolet (UV) light (200–400 nm range classified as UVA: 320–400 nm, UVB: 290–320 nm, and UVC: 190–290 nm); (ii) visible light (400–700 nm); and (iii) sometimes near-infrared (IR) light (700–1000 nm). Light has a dual wave–particle nature: it behaves like a wave, for example, in interference phenomena and like a particle (the photon), for example, in the photoelectric effect.

A monochromatic radiation is considered as carrying an energy E_m that corresponds to a photon (Eq. (2.2) where $h = 6.62 \times 10^{-34}$ J s is the Planck's constant). The energy of 1 mol of photons E_{mole} (called an *Einstein*) is defined by Eq. (2.3), where N_a is the Avogadro's number (6.02×10^{23}).

$$E_m = h\nu_m \tag{2.2}$$
$$E_{mole} = N_a E_m \tag{2.3}$$

The intensity of the radiation received by a sample at a given wavelength I_0 (λ_0) is usually defined in photochemistry by the number of photons delivered by the source at this wavelength, in a defined wavelength interval, per surface unit and time unit. The amount of energy transported by the light beam $E_{beam}(\lambda_0)$ in J cm^{-2} s^{-1} is expressed in Eq. (2.4): it corresponds to a power density (W cm^{-2}). In the case of a laser source, I_0 (λ_0) usually refers to the number of photons per time unit, the irradiated area having the dimension of the laser spot; E_{beam} (λ_0), which

Photoinitiators for Polymer Synthesis: Scope, Reactivity and Efficiency, First Edition.
Jean-Pierre Fouassier and Jacques Lalevée.
© 2012 Wiley-VCH Verlag GmbH & Co. KGaA. Published 2012 by Wiley-VCH Verlag GmbH & Co. KGaA.

is thus expressed in $J\ s^{-1}$, corresponds to a power P (in W).

$$E_{beam}(\lambda_0) = I_0(\lambda_0)\,h\nu_0 \tag{2.4}$$

The overall intensity I_0 (or the total number of photons) delivered over a part of the electromagnetic spectrum is thus defined by an integral (Eq. (2.5)). The total energy E_{beam} is also an integral.

$$I_0 = \int I_0(\lambda)\,d\lambda \quad \text{and} \quad E_{beam} = \int E_{beam}(\lambda)\,d\lambda \tag{2.5}$$

2.2
Characteristics of a Light Source

A light source is characterized by an emission spectrum (Figure 2.1), which represents the photon distribution as a function of the wavelength, that is, $I_0(\lambda) = f(\lambda)$. The ordinate of such a spectrum is often presented in arbitrary units or as a relative intensity. Some absolute irradiance measurements can also be available when using a calibrated spectrometer. According to the type of light source, it will be shown later that (i) either a polychromatic light (such as what is called a *white light*) or a monochromatic light (e.g., a green light at 532 nm using a Nd:YAG laser) is delivered, (ii) the intensity can vary over several orders of magnitude (e.g., $1-10\,000$ mJ cm^{-2}), and (iii) the light can also be emitted either continuously in time or as a pulse or a series of pulses, thereby leading to a wide range of luminous power densities (e.g., $10^{-3}-10^8$ W cm^{-2}). This results in a large panel of situations for the energy absorbed by PI, the concentration of the initiating species, and the presence of biphotonic effects.

When using a polychromatic light, a careful selection of wavelengths (Figure 2.1) can be achieved through the use of various filters [1]:

1) **Interference and passband filters**: they transmit wavelengths in a given interval; they are narrow (a few angstrom for the former) or broader (a few nanometers or a few tens of nanometers for the latter). This allows getting a quasi-monochromatic light source.
2) **Cutoff filters**: they eliminate a part of the spectrum, for example, the UV light (when only a visible light exposure is required) or the IR light (if heating of the exposed sample has to be avoided).

The measurement of the light intensity is done in practice by radiometers, which provide a calibrated flat output response in a given wavelength range in terms of the luminous energy per second and surface unit ($J\ s^{-1}cm^{-2}$ or W cm^{-2}). As usually encountered in photochemistry [1–5], chemical actinometers can be also used to calculate the number of photons.

2.3
Conventional and Unconventional Light Sources

Typical light sources are conventional artificial light sources, laser beams, and the sun [6–10].

2.3.1
Xenon Lamp

The emission spectrum is continuous in wavelengths (Figure 2.1). Xenon (Xe) as well as mercury (Hg) (Section 2.3.2) lamps have high electrical power requirements and give off a lot of heat but emit a high light intensity over a wide spectral range. For laboratory equipment, the light intensities are relatively low compared to the Xe–Hg lamp (e.g., typically 200 and ~60 mW cm^{-2} for Hg–Xe and Xe lamps, respectively).

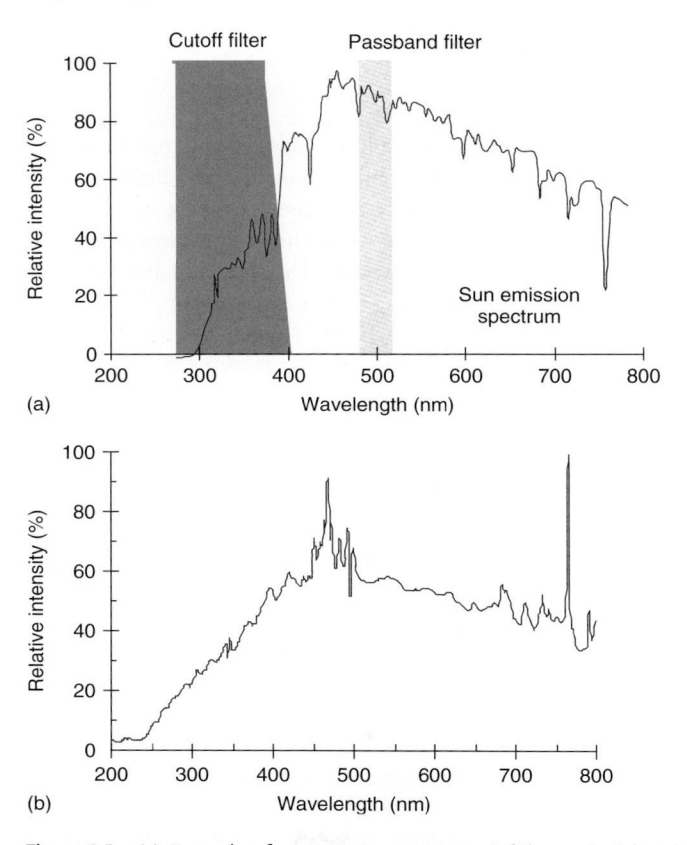

Figure 2.1 (a) Example of an emission spectrum (of the sun). Schematic transmission curves of typical filters (cutoff filter: $T = 0\%$ for $\lambda < 390$ nm and passband filter: $T = 100\%$ for 500 nm). (b) Emission spectrum of a xenon lamp.

2.3.2
Mercury Arc Lamp

An electrical discharge in a Hg vapor produces excited atomic energy levels. Characteristic narrow transitions occur between some of these levels and the ground state. A set of particular wavelengths is thus emitted: 254, 313, 366, 405, 435, 546, and 579 nm. The reflector used also affects the output spectrum and can be selected for an appropriate wavelength (254 and 366 nm). The relative intensity of the different lines also depends on the Hg vapor pressure. Commercial medium pressure Hg lamps for the UV-curing technologies (electrical power $\sim$60–240 W cm^{-1}; length of the lamp: 100–2300 mm) exhibit the spectrum shown in Figure 2.2c. High fluxes of photons (typically > 1.2 W cm^{-2} in the 280–445 nm range) are available. Hg lamps emit heat (>50%).

Low-pressure Hg lamps or germicidal lamps mostly deliver photons around 254 nm. They can be used to produce a wrinkled surface cure.

2.3.3
Doped Lamps

Mercury lamps can be doped (e.g., by xenon, gallium, indium, and iron halides) in order to change the emission spectrum and generally to increase the light emission in the near-UV-visible region around 410–420 nm, as seen in Figure 2.2d.

Low-pressure fluorescent lamps are Hg lamps that contain a fluorescent agent: the spectrum is changed and the intensity is very low. They are used as pregelification lamps to increase the core polymerization.

2.3.4
Microwave Lamps

Contrary to usual Hg lamps, where the Hg atoms are excited by an electric field, the excitation in the radio-frequency-excited electrodeless lamps is provided by microwaves. Doped lamps are also available. This allows to mostly adjust the emission around, for example, 420 or 380 nm as shown in Figure 2.2a,b. These lamps have a high input power (300–600 W in^{-1}; lamp design with 6 or 10 in width) and emit a lower IR amount.

2.3.5
Excimer Lamps

Excimer lamps are discharge lamps. The emitted wavelength is a function of the gases that form the excimer. In particular, they can work at 308 nm or even at 222 or 172 nm in vacuum ultraviolet (VUV) devices (see, e.g., in [11] and references therein). They have a limited commercial use. Applications in photopolymerization are under development.

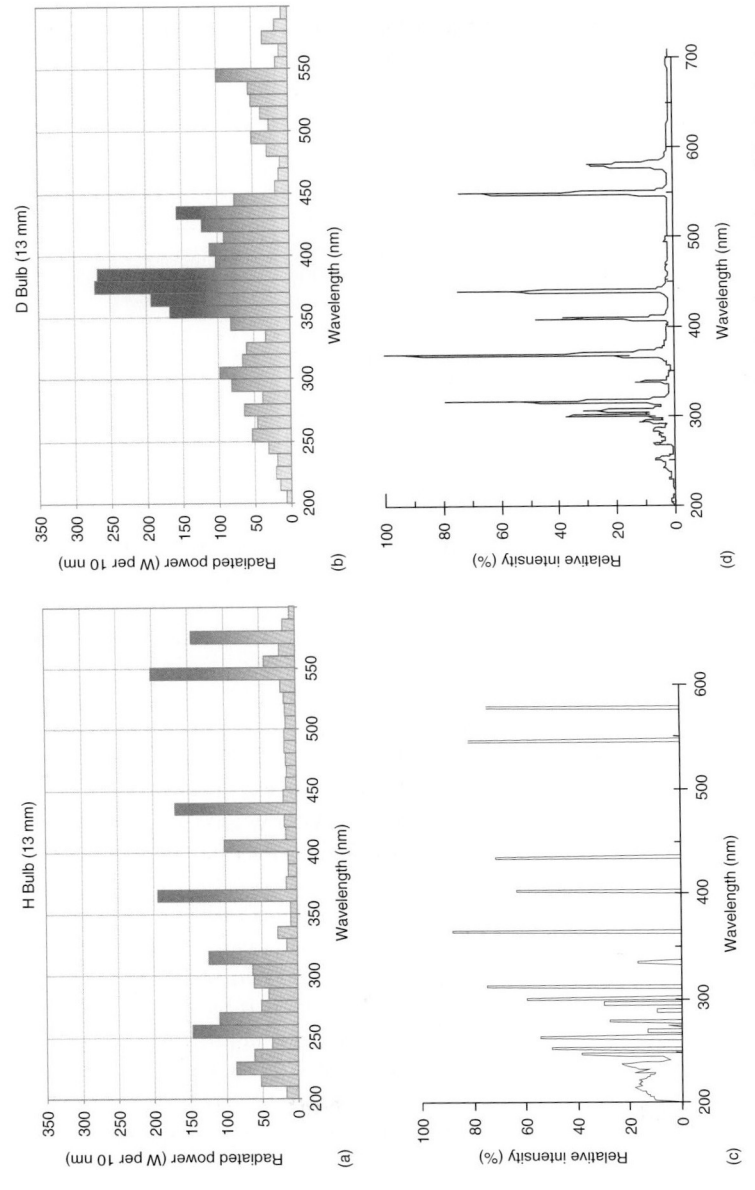

Figure 2.2 (a,b) Typical emission spectra of different microwave lamps. (Source: Reproduced with permission of Fusion Inc.) (c) Typical emission spectrum of a mercury lamp. (d) Emission spectrum of a doped mercury lamp (Xe–Hg lamp) equipped with a 366 nm reflector.

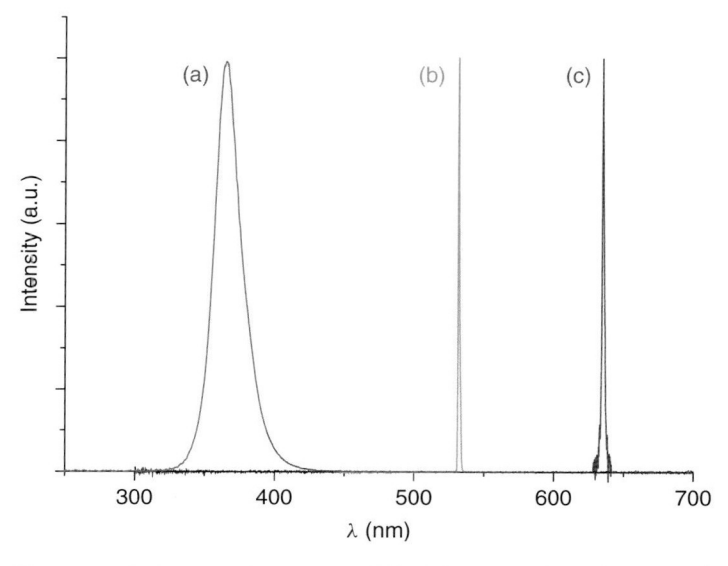

Figure 2.3 Typical emission spectra of (a) 365 nm LED, (b) 532 nm diode laser, and (c) 635 nm diode laser (J. Lalevée unpublished data).

2.3.5
Light-Emitting Diodes (LEDs)

A light-emitting diode (LED) is based on a semiconductor material device [12, 13]. At present, according to the technical developments, the proposed LEDs emit a light with an almost Gaussian distribution in a narrow wavelength range centered at 365 nm (345–385 nm) or 395 nm (380–420 nm) with an intensity about a few 10–100 mW cm^{-2} (Figure 2.3). Highly packed arrays of LED are possible (up to 400 LED per cm^2). This allows light intensities up to 2 and 8 W cm^{-2} at 365 and 395 nm, respectively. As the light is delivered within a small wavelength interval, the output luminous power density per nanometer is typically 0.1 and 0.5 W cm^{-2} nm^{-1} (at 365 and 395 nm) compared to 0.2 W cm^{-2} nm^{-1} for a 100 W short arc lamp in the 365–440 nm range. However, the total power density (in W cm^{-2}) over the whole emission spectrum is obviously noticeably lower for an LED arrangement than for a mercury lamp.

The main benefits of LEDs are (i) low heat generation (no IR light); (ii) low energy consumption; (iii) low operating costs, less maintenance, a ~50000 h life and portability; and (iv) a possible incorporation in programmed robots that can move the lamps to improve the curing of shadow areas. LEDs have a high potential for digital inkjet printing, adhesive curing, curing of thin heat-sensitive plastic foils, spot curing (automotive topcoat repair), and dentistry (UV handhold for curing teeth inlay). Development of the LED technology is fully under way and holds great promise. A spray gun equipped with LEDs allows the resin application and the photocuring step in a single operation [14]. The performance of LEDs versus conventional light sources in UV-curing experiments has been evaluated [15, 16] and is discussed in the following chapters. The emission spectra exhibit a higher full width half maximum (FWHM) (~20–30 nm) for LED compared to a diode laser irradiation (<1.5 nm), as seen in Figure 2.3.

2.3.7
Pulsed Light Sources

While all the lamps described above emit a light whose intensity is constant as a function of time, flash lamps and pulsed UV light sources deliver the photons in a single flash or pulse of light. Examples can be found in [17]. Pulsed Xe lamps deliver lower heat. They are suited to specialty applications. The pulsed laser polymerization (PLP) techniques can be useful compared to continuous irradiations and lead to specific properties, that is, the initiation takes place only during the irradiation, and the propagation and termination processes can be observed in the dark period so that the molecular weights and the molecular weight distributions (MWDs) can be modulated by the frequency as well as the pulse width.

2.3.8
Laser Sources

Details on lasers [18] and laser applications in photopolymerization can be found in [19–21]. Laser beams present specific characteristics: (i) a monochromatic light (this allows a control of the light absorption and then the photochemical reactions; side reactions and local heating are reduced), (ii) large possibilities in wavelength selection, (iii) a high energy concentration onto a small surface (this ensures a quasi-instantaneous curing and a reduced oxygen inhibition), (iv) a high spatial resolution, spectral selectivity, narrow bandwidth of the emission (these properties are useful in holography), (v) an easy focalization (the photochemical event can be produced in a sample placed at a long distance from the light source), and (vi) a very short exposure time allowing a possible scanning of a surface by the laser spot, which behaves as a pencil of light (the direct imaging is affordable; high-resolution images can be obtained).

As in conventional sources, many lasers deliver the light continuously as a function of time. Some lasers can also emit the light as a (very) short pulse. In that case, very high power P or power density DP can be attained as the amount of energy is divided by the pulse duration: for example, a continuous-wave (CW) laser (100 mJ s^{-1}) gives P values, for example, ~0.1 W, whereas a low-intensity pulsed laser (e.g., 100 ns, 1 mJ) leads to P~10 kW. Compared to a conventional light source (typically DP ~ 0.2 W cm^{-2}), this CW laser yields higher DP (DP ~ 10 W cm^{-2}) because of its small beam diameter. In the same way, the pulsed laser considered above would lead to DP ~ 1000 kW cm^{-2}.

Many different lasers are available in the market; among them are the following:

1) **CW lasers.** Argon-ion lasers emitting at 488 or 514 nm appeared in the 1980s. The high power (up to a few watts) and the high quality of this light beam

were beneficial for the development of many fields such as holography, laser direct imaging (LDI), and computer-to-plate (CTP) technology. The krypton-ion lasers (406.7, 413.1, 415.4, 468.0, 476.2, 482.5, 520.8, 530.9, 568.2, 647.1, and 676.4 nm) are sometimes used. He–Ne lasers (at 633 nm) were used for some applications such as holography despite their rather low light powers. He–Cd lasers emit light in the UV region (325 and 442 nm) with a power high enough to ensure the development of applications such as stereolithography. Solid-state continuous lasers are available, for example, the Nd:YAG lasers at 1064, 532, and 355 nm.

2) **Pulsed lasers.** The Q-switched Nd:YAG laser could also be used, but no strong development has been reported so far in the industry. Pulsed excimer lasers (157, 193, and 248 nm) appeared on the microelectronics market for the manufacturing of integrated circuits.

3) **Femtosecond lasers (e.g., Ti:sapphire lasers).** They recently opened up new opportunities such as in two-photon stereolithography (see below).

4) **Semiconductor-based lasers or diode lasers.** They originally emit in the red region (e.g., at 630, 680, and 720 nm) and deliver moderate light powers (hundreds of megawatts to watts). By the end of the 1990s, several wavelengths were accessible in the near-UV-visible region such as at 405 nm. Diode lasers (with output DPs $\sim$ 10–100 mW cm^{-2}) operating at selected wavelengths (405, 457, 473, 532, and 635 nm) are now available (Figure 2.3). Driving factors are (i) simple design (no complicated laser cavity), (ii) high power output, and (iii) easy stacking and arrangement (allowing the curing of complex forms). The typical power density can average 10 kW cm^{-2} [22]. Diode lasers become an alternative to usual lasers. The possibility of electronically driving the diode lasers will certainly lead to important future developments in the LDI or CTP fields, for example, (see below).

2.3.9
Sun

The sun is a very convenient and inexpensive source of light but delivers a low intensity (typically <5 mW cm^{-2} in the near-UV-visible wavelength range; see the emission spectrum in Figure 2.1). This intensity is strongly affected by the weather, the location, and the year period. It might be of interest for particular outdoor or daylight indoor applications. Visible-light-induced polymerization is well documented, as seen in Parts II and III, but works devoted to sunlight curing itself are rather scant.

2.3.10
Household Lamps

With the actual need for green technologies, the development of new (photo)chemical systems and new soft irradiation conditions for photopolymerization reactions is ever required. For example, the basic interest in household

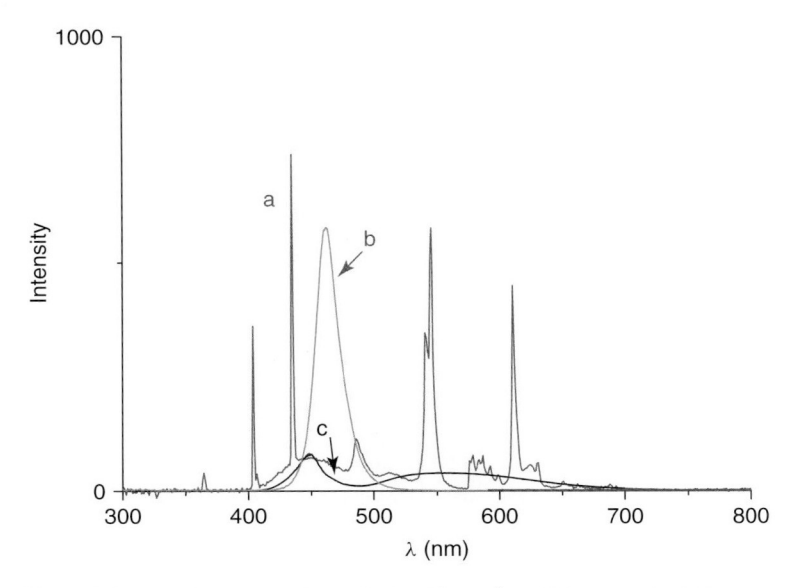

Figure 2.4 Emission spectra (intensity in µW cm^{-2}) of (a) a household fluorescent bulb, (b) a household blue LED bulb, and (c) a household white LED bulb.

fluorescent bulbs (FBs) (Figure 2.4), which already appeared in organic synthesis [23], might be (i) the available higher luminous power in the 380–800 nm range (~15 mW cm^{-2} compared to <5 mW cm^{-2} for solar irradiation); (ii) the stability of this irradiation source, which is not affected by the weather and location (iii); the disposal of very safe light sources; and (iv) the use of commercial and cheap standard devices.

In the same way, compared to FBs, household LED bulbs (Figure 2.4) emit 7–20 mW cm^{-2} in a selected spectral range (e.g., $\lambda_{max} \sim 462$ nm for a blue LED bulb) with no IR or UV radiation and are more rugged and damage resistant. Other advantages are (i) lower electrical energy consumption, (ii) longer lifetimes, (iii) safe and environmentally friendly lamps (no mercury; cool to the touch), (iv) no frequency interference, (v) large range of color (no filter; white lights can be delivered in a variety of color temperatures), and (vi) relatively low divergence of the light beam (their directional output can be exploited by clever designs such as light strips or concentrated arrays).

2.3.11
UV Plasma Source

It is known that UV lights can be generated from plasma produced by a microwave excitation of suitable gases. This was recently applied [14] to the complete and regular curing of complicated objects (even a car body) in a plasma chamber with an integrated roller bed.

References

1. Braun, A., Maurette, M.T., and Oliveros, E. (1986) *Technologie Photochimique*, Presses Polytechniques Romandes, Lausanne, CH.
2. Wayne, R.P. (1988) *Principles and Applications of Photochemistry*, Oxford Science and Publications, Oxford.
3. Turro, N.J. (1978 and 1990) *Modern Molecular Photochemistry*, Benjamin, New York.
4. Gilbert, A. and Baggot, J. (1991) *Essentials of Molecular Photochemistry*, Blackwell Scientific Publications, Oxford.
5. Suppan, P. (1994) *Chemistry and Light*, The Royal Society of Chemistry.
6. Fouassier, J.P. (1995) *Photoinitiation, Photopolymerization, Photocuring*, Hanser, Münich.
7. Davidson, S. (1999) *Exploring the Science, Technology and Application of UV and EB Curing*, Sita Technology Ltd, London.
8. Drobny, J.G. (2010) *Radiation Technology for Polymers*, CRC Press, Boca Raton, FL.
9. Schwalm, R. (2007) *UV Coatings: Basics, Recent Developments and New Applications*, Elsevier, Oxford.
10. Rabek, J.F. (1993) in *Radiation Curing in Polymer Science and Technology*, vol. 1 (eds J.P. Fouassier and J.F. Rabek), Chapman & Hall, London, pp. 453–502.
11. Vicente, J.S., Gejo, J.L., Rothenbacher, S., Sarojiniamma, S., Gogritchiani, E., Miranda, M.A., Wörner, M., Oliveros, E., Kasper, G., and Braun, A.M. (2010) in *Basics of Photopolymerization Reactions*, vol. 3 (eds J.P. Fouassier and X. Allonas), Research Signpost, Trivandrum, pp. 47–66.
12. Mills, P. (2005) Proceeding of the RadTech Europe Conference, http://www.radtech-europe.com/.
13. Karsten, R. (2009) Proceeding of the RadTech Europe Conference, http://www.radtech-europe.com/.
14. Dietliker, K., Braig, A., and Ricci, A. (2010) *Photochemistry*, **38**, 344–368.
15. Anyaogu, K.C., Ermoshkin, A.A., Neckers, D.C., Mejiritski, A., Grinevich, O., and Fedorov, A.V. (2007) *J. Appl. Polym. Sci.*, **105**, 803–808.
16. Brandl, B., IST Metz GmbH (2007) Proceeding of the RadTech Europe Conference., Germany, http://www.radtech-europe.com/.
17. Kinpling, K. (2008) Proceeding of the RadTech USA, Chicago, http://www.radtech.org/.
18. Duarte, F.J. (ed.) (2009) *Tunable Laser Applications*, CRC Press, Bocca Raton, FL.
19. Fouassier, J.P. and Rabek, J.F. (eds) (1990) *Lasers in Polymer Science and Technology: Applications*, CRC Press, Boca Raton, FL.
20. Lackner, M. (ed.) (2008) *Lasers in Chemistry*, Wiley-VCH Verlag GmbH, Weinheim.
21. Allonas, X., Crouxte-Barghorn, C., Fouassier, J.P., Lalevée, J., Malval, J.P., and Morlet-Savary, F. (2008) in *Lasers in Chemistry*, vol. 2 (ed. M. Lackner), Wiley-VCH Verlag GmbH, pp. 1001–1027.
22. Kress, K. (2009) Proceeding of the RadTech Europe Conference, http://www.radtech-europe.com/.
23. Angioni, S., Ravelli, D., Emma, D., Dondi, D., Fagnoni, M., and Albini, A. (2008) *Adv. Synth. Catal.*, **350**, 2209–2214.

3
Experimental Devices and Examples of Applications

3.1
UV Curing Area: Coatings, Inks, Varnishes, Paints, and Adhesives

3.1.1
Equipment

The heart of the equipment is represented by the usual UV-curing conveyor [1] where the sample is exposed to the lamp during a time defined by the belt speed (the sample is placed on the belt). On the industrial scale, several arrangements depending on the applications are found. For example, a UV line for flat substrates is a big machine with a device allowing the application of the coating onto the substrate (manual or automatic machine spray, roller coating, and curtain coating), a flash time zone (to remove bubbles), a pregelification area, and a UV unit (consisting of a single lamp, an array of multiple lamps, or multiple arrays of lamps). This static arrangement allows curing of the sample in a fixed position in a given exposure time. Inerting (i.e., eliminating the air atmosphere usually by replacing with nitrogen; now, the use of carbon dioxide has been introduced) and extracting the ozone produced by the lamps are possible. Very fast curing speed can be achieved, for example, 300 m mn^{-1} for the curing of silicone release coatings or the drying of printing inks. The curing of nonflat substrates, cylindrical pieces, or complex objects (e.g., a chair, a car body) exhibiting shadow areas, can be achieved with (i) a static arrangement using an array of fixed lamps mounted at different angles to get an optimal uniform irradiation, the curable systems being able to move around the lamps or (ii) a dynamic arrangement in robotic UV curing where the lamps (Hg or LED lightweight lamps) are moved as close as possible around the object.

3.1.2
End Uses

Well-known applications are found in the industrial UV-curing area (see, e.g., in the many papers of the RadTech Conference proceedings [2–5]). The benefits of this technology are [6] (i) rapid through-cure; (ii) low energy requirements; (iii) low temperature treatments suitable for heat-sensitive

Photoinitiators for Polymer Synthesis: Scope, Reactivity and Efficiency, First Edition.
Jean-Pierre Fouassier and Jacques Lalevée.
© 2012 Wiley-VCH Verlag GmbH & Co. KGaA. Published 2012 by Wiley-VCH Verlag GmbH & Co. KGaA.

substrates (pinewood and plastics); (iv) use of nonpolluting and solvent-free formulations (almost no volatile organic compounds (VOCs)); (v) fast cure speeds; (vi) enhanced product durability; (vii) high surface quality for coatings; (viii) reduced throughput times in production, application versatility, small space requirements, relatively low investment, easy machine integration, and easy maintenance; (ix) low costs of material, production process, and waste management; and (x) suitable adaptation for small series, high productivity, and very good price/performance ratio. LED curing is the subject of many works and presents a high potential of development, requiring a careful adaptation of the photoinitiating system. Sunlight curing is a green technology (Section 3.15) that presents the great advantage to combine both radiation curing and a sun-assisted processing, the main benefits being no VOC, no energy consumption, no irradiation device, and possibility of curing large dimension pieces or surfaces. Curing under soft irradiation conditions might be developed (Section 3.15).

Many innovations are currently making headlines. Short reviews are continuously focusing on the latest developments and potential interests (e.g., of new photoinitiating systems, monomers, oligomers, devices, as exemplified in [1, 7–16]), or describing the current industrial challenging orientations (e.g., [17]). The penetration in the market is well established and high. Owing to the environmental pressures, UV curing really appears now as an ever-expanding technology. Driving forces are performance, economy, and ecology. UV curing represents a green technology for the twenty-first century. Applications of coatings, inks, varnishes, paints, and adhesives are encountered in many fields of daily life and in many industrial sectors. Examples include the following:

1) Varnishes and coatings on a large variety of substrates:
    a. Paper: coatings for folding box boards and paper label overprints.
    b. Plastic: vinyl flooring; polycarbonate molding; coatings on electric sheaths, polyethylene, polypropylene, polymethylmethacrylate (PMMA), and polytetrafluoroethylene; protective coatings for compact discs and video discs; decorative and protecting coatings for fishing rods, skis, and crash helmets.
    c. Metal: beer and beverage cans, caps, and closures; protective coatings on metal substrates such as temporary anticorrosion coatings on steel pipes or thin pigmented coatings on aluminum or steel coils (referred to as *coil coating*); metal packaging; and primers for metal.
    d. Wood: decorative and protecting coatings for wooden boards in rubber boats, wood finishing industry (e.g., fillers, UV-cured topcoats, laminates, veneers and panel boards, and wooden floor tiles).
    e. Rubber.
    f. Glass: primers glass coatings (see below).
    g. Optical fibers.
    h. Polylactic acid (PLA) films for packaging.
    i. Miscellaneous substrates: barrier coatings; coatings requiring a specific property (coatings for inorganic oxides, magnetic media or fragile substrates, and flexible coatings); light reflective coatings for road signs or

marks; erasably markable luminescent articles through the fabrication of multilayer coatings; sound reduction coatings; coatings with stereoscopic pattern; abrasion-resistant coatings; nanocomposite coatings; antistatic coatings; photoluminescent coatings; photografted coatings on macro- or microporous substrates such as membranes and thermosetting coatings.
2) Coatings for durable products in construction such as house wraps and roofing; sun cross-linkable systems for façade paints and waterproof paints for exterior applications (crack-bridging paints with good low-temperature properties, elastic wall paint exhibiting a low soiling tendency); UV-curable mastics for resealing and smoothing of supports; antifogging films (e.g., for window glasses, bathroom mirrors, greenhouses, goggles, and helmet shields); coatings for glass, nails, and screws in exterior building panels; glossy artificial stone plates; clear varnishes for concretes (to prevent cement hydration and formation of unaesthetic spots on the surface); weather-resistant coatings (e.g., water-based compositions, sol-coatings, or multilayer coatings); adhesives, laminating adhesives (e.g., for the bonding of glass sheets to obtain laminated security glasses or for plastic film lamination on printed cartons).
3) Pressure-sensitive adhesives (PSAs) and structural adhesives where applications are shown in automotive door shields, laminated glass plates, laminated safety glasses, nameplates, special adhesives for burns and skins, polycarbonate to (non)polycarbonate bonding for medical applications, glass sheet bonding, polyacrylate component bonding of automotive headlamp housings, and lens bonding.
4) UV drying of offset, silk screen, flexographic or gravure inks; inkjet printing for coding, marking, addressing, printed plastic electronics (e.g., radiofrequency identification (RFID) tags, electrodes, and chip connections); UV inkjet drop-on-demand printing; dry color UV curable toners for digital printing; printing on metal (cans, boxes, bottle tops, and screw caps), paper (book covers, record album covers, posters, identity papers, currency papers, bank notes, photographic prints, pricing and promotional labels, holographic labels, shopping bags, and newspapers), and plastic (credit cards and telephone cards); plastic decoration (e.g., toothpaste tubes), packaging, security printing; paper upgrading (e.g., gift wrap, specialty papers, packaging papers, release papers, e.g., caul sheets (release against melamine in impregnated decorative paper foils), casting papers (release against polyurethane and PVC), and labels (release against PSA)) such as in silicone release papers.
5) Coatings for automotive trim, car light reflectors, car light covers, automotive windshields, rear window defrosters, headlamp bezels, lenses and brightworks, repair automotive coatings, exterior automotive coatings (they are under development), and automotive clear coats.
6) Water- and oil-repellent photocurable resins for coatings in electronics; manufacture of UV shielding circuit substrates; electrically conductive UV curable coatings containing Sb-doped and Sn-oxide-coated mica particles; conformal coatings for protective covering (from humidity, dirt, dust, and chemical

contaminants in the end-use environment) over military and aerospace electronic printed wire assemblies; insulation layers for printed circuits; conductive and insulating inks; screen printable solder masks; cover coats for flexible printed wiring; plate and etch resists, notation inks, dry-film photoresists; electrical coatings and encapsulants for field effect transistors; conformal coating spot repair and rework; conductive screen inks; soldering resists; transparent, crack resistant, and low stress formulations for sealing photosemiconductors; wire and cable coatings; thin-wall insulated wires; and UV resins in liquid crystal display(LCD) devices (side-sealing or gasketing, charge port sealing, metal pin terminal bonding, and terminal sealing for active matrix LCD).

7) Coatings for bioapplications: UV blocking materials obtained by coating a breathable fabric comprising a web of loosely intermeshed fibers or loosely woven threads with UV absorbers; UV-polymerized microporous coatings as barrier materials, for example, toward bacteria (both in air-borne and solution-borne media) or water, usable in sterile, breathable disposable medical nonwovens (for hospital end uses, medical packaging, and wound dressings) or in breathable, water- or liquid-impermeable materials (disposable consumer or industrial products, curable coatings for textile substrates such as sportswear and leather and suedelike coatings).

3.1.3
Coating Properties

Depending on the applications of the polymerized coating in the UV-curing area, a lot of specific and very different properties can be searched, for example:

1) **Physical, mechanical, and surface properties**: glass transition temperature, hardness (relative to glass), impact resistance, flexibility (tensile strength, flexural strength, modulus, elongation at break, 180° fold-bend test), abrasion, scratch or scuff resistance, tackiness (or cure grade), surface energy (pen reticulation), surface slip, adhesion (cross-hatch and scratch tests), subsequent adhesion (for release coatings), stiffening, heat resistance, wetting, hydrophilicity, hydrophobicity, bond strength (for adhesives), adequate dielectric constant, thermal conductivity, resistivity, specific heat or thermal diffusivity, low shrinkage, and moisture permeability.
2) **Chemical properties**: resistance to solvents (methyl ethyl ketone (MEK) double rub test), water, acids, and bases.
3) **Optical properties**: matting, gloss, transparency, color change, yellowing, blistering, fluorescence, and photochromism.
4) **Usage properties**: low odor, low skin irritation, low volatility, weatherability, wear properties, antisoiling properties, antibacterial properties, superabsorbancy, and gas barrier properties.

3.2
Conventional Printing Plates

A printing plate used in the graphic art area is based on the formation of positive or negative images in a photosensitive resin. The processes used to carry inks and transfer it to the substrate are lithography, flexography, letterpress, and gravure. For example, in an offset (or lithographic) printing plate, the resin is deposited on an anodized aluminum base. The plate is then exposed to a UV light through a mask and developed. The remaining cured polymer forms the image area. This area is oleophilic and will take up the printing ink, which is transferred to a printing blanket and finally to the paper (the paper is "offset" from the plate) [18].

3.3
Manufacture of Objects and Composites

As thick coatings can be photopolymerized, the creation of an object becomes suitable. Such objects are manufactured in static devices, for example, bottles, filament-wound tanks, and molded articles. This is encountered now in the composite industry for the construction of doors, flat panels, boat components, roofing, surfboards, and underground pipes. The composites are here UV-curable formulations where fillers and pigments have been introduced. Examples are found [19] in glass-reinforced unsaturated polyester prepregs (thin sheets already impregnated into fibers), lightweight prepregs, glass fiber mats, carbon-reinforced composites, and silica-filled composites [20]. They are also concerned with UV-cured gelcoats.

3.4
Stereolithography

Rapid prototyping technologies have appeared for the manufacture of molds, patterns, and prototypes directly formed from computer-aided design models without tooling and machining [21–25]. Stereolithography involves a liquid UV-sensitive formulation that is polymerized layer by layer on exposure to a laser beam, for example, a He–Cd laser (emitting at 325 nm), an argon-ion laser (351 nm), or a frequency-tripled Nd:YAG laser (355 nm) [26]. A layer (x–y stage) is polymerized according to the spatially controlled irradiation; then, the layer is moved down (z stage) and a new layer is polymerized. One important question is the shrinkage that often occurs in the polymer during the process (particularly when using acrylates).

Stereolithography provides a good way for producing microfluidic devices that show a fast development in various applications such as microanalysis systems, drug delivery devices, microfuel cells, and microchemical reactors. Rapid prototyping also includes the dynamic mask-based stereolithography (the fabrication speed is faster as it allows exposing of the whole layer in one step). Applications are also encountered in medicine and surgery (for tissue engineering, blood vessel substitutes, and medical modeling) [27], inkjet-based system technology, or two-photon polymerization (see below).

3.5
Applications in Microelectronics

The microlithography process used in the semiconductor industry for more than 30 years [28, 29] is similar to that depicted in Scheme 1.3. A photoresist is applied onto a substrate and exposed to UV radiation through a mask. The exposed area is either cross-linked (negative image resist) or solubilized (positive image resist). After removal of the unexposed resist and etching, a conductive circuit is formed on the substrate: a board (for printed circuits) or a silicon wafer (for integrated circuits).

Nonchemically amplified photoresists such as the Novolak systems leading to positive images are still used for the manufacture of integrated circuits [30]. New materials are under development [31]. Examples of negative photoresists are known. Common exposure wavelengths are 365 and 436 nm using the I and G lines of the Hg lamps. The increase in the resist photosensitivity was possible through the development of the chemically amplified photoresists [26, 30, 32] (evoked in Part III). The future of these systems follows the increasing demand of the deep UV (DUV), vacuum UV (VUV), and extreme UV (EUV) microlithography for the production of ultra large scale integration (ULSI) and very large scale integration (VLSI) circuits. DUV and VUV use excimer lasers: KrF laser (enabling sub-100 nm feature scale at $\lambda = 248$ nm), ArF laser ($\lambda = 193$ nm, 65 nm feature scale), and F_2 laser ($\lambda = 157$ nm). The development of sources for the EUV technology that will deliver wavelengths (typically 13.4 nm) 10–20 times shorter than that of the excimer lasers should allow production of features sizes about 32 nm. The use of two-photon processes (Section 3.11) coupled with suitable lithographic devices should allow a 16 nm resolution [33].

3.6
Laser Direct Imaging

Laser direct imaging (LDI) can be seen as an essential process [26, 34, 35] in the manufacture of printed circuit boards (PCBs) or printed wiring boards (PWBs). It avoids a photomask and an exposure to conventional lamps. The irradiation of the photosensitive resist is achieved by scanning the panel surface with a laser beam. This technology used high power argon-ion lasers at 488 nm in the 1980s and then UV lasers (argon lasers operating at 351 and 364 nm or frequency-tripled

Nd:YAG laser at 355 nm) in the 1990s. At present, it is mostly based on diode lasers (gallium nitride) emitting at 405 nm. Sensitivities are currently better than 10 mJ cm^{-2} compared to 30 mJ cm^{-2} with conventional contact printing using photomasks and flood exposure. The major benefits are no use of silver halide films, minimum lead times, flexible manufacturing, automatic alignment, high resolution, cost savings and improvements over the traditional production, high quality, and high productivity.

3.7
Computer-to-Plate Technology

The concept in computer-to-plate (CTP) device is basically similar to that shown for LDI [26]. A typical CTP photopolymer plate [36] consists in a photosensitive layer (a few microns thick) deposited onto a surface-grained aluminum substrate and protected by a polyvinyl alcohol film (to overcome the oxygen inhibition effect). The light sources were successively based on infrared semiconductor lasers (700–1400 nm; output power $\sim$ 100 mW; required sensitivity: 10–300 mJ cm^{-2}); argon-ion lasers (green printing plates; 488 nm line; scanning speeds $\sim$ 700 m mn^{-1}; sensitivities around or less than 1 mJ cm^{-2}) and at the moment, the 405 nm blue-violet diode laser (blue printing plates).

3.8
Holography

A holographic recording allows storing of both the phase and the amplitude of the electromagnetic wave [26]. In the interference pattern constructed from the superposition of two laser beams (formed from a unique laser beam), bright and dark fringes appear. This leads to a spatially modulated polymerization in the bright areas of the photosensitive plate.

Holography finds applications in different fields such as integrated optics, information recording and storage, interferometry, nondestructive testing, computer-generated holograms, and holographic films for multicolor recording using several laser lines [37–40]. Among the media that can be used to record holograms, photopolymers present the main advantage to be self-developing materials. Many studied systems (based on acrylates [41] or sol–gel nanocomposites [42]) are able to work under laser lines at 488, 514, and 532 nm. The actual trend of development is devoted to three-dimensional bit-oriented data storage (with capacities of 200 GB to 1 TB) [43]. Using cationic resins (that exhibit a low shrinkage), it was possible to store data with a density of 100 bits µm^{-1} in volume holograms with a thickness >500 µm; up to 90 holograms were recorded [44].

3.9
Optics

Many examples of applications concern the manufacture of, for example, microlenses, diffraction gratings, volume phase gratings, waveguides, gradient index (GRIN) materials, and nanorods. The search for specific optical properties for a material and the design of optical devices are the subject of intense research works such as the formation of anisotropic photopolymers (that shows a selective reflection), high refractive index transparent coatings [45, 46], large refractive index change materials (allowing a photo-optical control) [47], Bragg reflection gratings [48], surface relief gratings [49, 50], and waveguides [51].

Specific properties in smart photopolymers are currently obtained; for example, photoalignment in homopolymers or block copolymers [52] and photo-cross-linkable side-chain liquid crystalline copolymers (using a linearly polarized light [53]), optically induced birefringence (in photoresponsive polymers [54–56]), photoinduced chirality (in a starting achiral liquid crystalline polymer irradiated with a circularly polarized light [57, 58]), optical activity (in photochromic polymers [59]), quadratic optical properties (in doped photopolymers [60]), electro-optical properties (owing to the uniform orientations obtained in a polymerization-induced phase-separation process of monomer molecules in liquid crystal droplets under a magnetic field [61], polarized light-induced enhancement [62]), assembly of photopolymerized structures by optical trapping and manipulation [63], high aspect ratio surface relief structures by photoembossing (this technique involves a thin film formation, a patterned UV exposure through a mask that produces a latent image in the exposed areas, development at elevated temperature and complete photopolymerization by applying a flood exposure) [64], and nonlinear optical properties [65].

3.10
Medical Applications

Applications in biosciences [66] are found in:

1) **Medicine** [67]: tissue engineering, rebuilding of a bone assisted by the replacement of the missing part by a scaffold created by stereolithography, biocompatible polymer materials for artificial blood vessels [27, 68, 69] formed by microstereolithography, photocurable materials for mouthguards [70], hydrogels for drug delivery, UV adhesives for anesthesia masks, surgical needles, wound protective coatings, bonding piping, construction of surgical equipment, photografted polymers for catheters, surgical dressings, prosthetics and medical implants, and devices for lab-on-chip assay methodologies.
2) **Dentistry** [71, 72]: tooth care, restorative and preventative dental work, dental filling, denture relining, and full and partial denture occlusal splints and nightguards.

3) **Ophthalmology** [73]: fashioning of plastic lenses, lenses for protecting the eyes from UV light, contact lenses, photochromic lenses, laminates for lenses, and artificial eyelens materials for cataract operations.

Small lamps are used in specific areas, for example, for dental applications. The shrinkage of the resin is sometimes an important question (e.g., in dental filling) [74].

3.11
Fabrication of Nano-Objects through a Two-Photon Absorption Polymerization

The two-photon absorption (TPA)-induced polymerization is now recognized as a very promising laser technology (Section 5.3). The light absorption is confined within a nanometric volume localized at the focus point of the laser beam. This unique feature is worthwhile in nanostereolithography; the development of ultrashort lasers (femtosecond Ti:Sa lasers) enables intensities $\sim 10^{12}$ W cm^{-2} and confocal microscopes offer the possibility of observing and irradiating micro-objects with a 3D resolution [75].

Many different objects [76–79] can be produced by this technique, for example, sculptures, optical elements (such as photonic crystals, phase lenses, mirrors [80]), index-modulated grating couplers in polymeric waveguides [81], micromachines (such as rotors), and microneedles in medicine. One key point is the development of new photoinitiators with high TPA cross sections (Section 8.32), the adaptation of the photopolymerizable resins, and the improvement of the experimental techniques to build complicated 3D objects.

3.12
Photopolymerization Using Near-Field Optical Techniques

Near-field optical techniques have been rapidly developed these past years, offering new opportunities to overcome the optical diffraction limit [82]. For example, evanescent waves that result from the submicrometric penetration of the light into a reflective medium allows the production of polymerized submicrometric thick films [83] or nano-objects (e.g., with a 1 μm in-plane resolution and 30 nm out-of-plane resolution) usable in imaging, holographic recording, or pixel-by-pixel addressing setup.

Recently, the field enhancement that occurs when irradiating a nanometric tip with a laser has attracted strong interest [84–86] as this phenomenon can induce a local photopolymerization. A potential interest appears in the high-density data storage applications. Very fine 20 nm photopatterned structures were obtained. Micro-optical devices (e.g., micropillars and microlenses) integrated at the end of an optical fiber can also be fabricated [87].

3.13
Search for New Properties and New End Uses

The design of new photopolymerizable systems exhibiting new properties is continuously under way. Many publications appear in the literature. They show the numerous perspectives that are now offered by the light-induced technologies. Examples of available polymer materials that are listed here outline the large variety of possibilities and the dynamism of this research area.

1) Hydrogels/gels/microgels/nanogels. Hydrogels are three-dimensional cross-linked polymeric structures that swell in suitable media. They retain a high fraction of water. For example, this makes them highly biocompatible and their mechanical properties mimic those of soft tissues so that applications can be found, for example, in tissue engineering and drug delivery. Other applications exist in the nanotechnology area. Examples of studies include hydrogels for DNA electrophoresis [88], starch hydrogels [89], biodegradable hydrogels [90–92], responsive hydrogels [93], superabsorbent hydrogels [94, 95], photoswitchable supramolecular hydrogels [96], kaolinite-poly(acrylic acid acrylamide) gels [97], slide-ring gels [98], polyacrylamide cryogels [99], UV-curable gel polymer electrolytes for Li-ion battery [100], core hair-type microgels with various hair lengths (their viscosity and dispersion ability are particularly suitable for applications in water-developable screen printing plates [101]), photoresponsive gels (with a control of the ionic conductivity [102]) and nanogels [103], nanostructured nanogels [104], nanogels with specific properties (e.g., a temperature-responsive core and pH-responsive arms [105, 106]), magnetic nanogels [107], reversibly photo-cross-linked nanogels [108], and face-to-face packing of macrocyclic organo gels containing diacetylene dicarboxamides [109].

2) Structured and gradient polymer brushes [110] and pH-responsive acrylamide/acrylate brushes on polymer thin films (through a surface-confined photopolymerization) [111].

3) Metallopolymers. They are encountered in photopolymerization reactions of metal-containing liquid crystalline monomers [112], photopolymerized films containing silver [113] or gold [114], metal polymer nanoparticles incorporated in multiwall carbon nanotubes [115], ion-conductive films (eventually anisotropic) for various electrochemical devices [116], silver nanoparticles containing photocurable conductive adhesives [117], conductive nanocomposites (involving double-walled carbon nanotubes [118]), conductive polymeric metal nanoparticles [119], paramagnetic nanoparticle containing coatings [120], silver- [121] or gold-epoxy [122] and gold-acrylate [123] nanocomposites, and nanocomposites for holographic recording and sensor applications [124].

4) Polymeric single-wall carbon nanotube (SWNT) liquid crystal composites [125].

3.13 Search for New Properties and New End Uses

5) Polymeric nanoparticles (formed by photopolymerization without microemulsion [126]), silver nanoparticle embedded polymers [127], functionalized silica nanoparticle containing polymers [128], core-shell-structured microcapsules [129], core-shell and multilayers nanoparticles [130], core-shell nanospheres [131], biopolymeric hollow nanospheres [132], photoreactive polymer nanosheets [133], polybasic nanomatrices [134], nanoporous polymer gratings [135], organic molecular crystal nanorods [136], carbon-containing nanotube epoxy coatings [137], nanomembranes [138], bicontinuous cubic structures generated from a coil–rod–coil molecule self-assembly [139], monomer/modified nanosilica systems [140], hydrodynamically shaped fibers having noncircular cross sections [141], and biphasic particles and fibers [142].

6) Monolithic polymers with controlled porous properties (such as monoliths containing hydrophilic, hydrophobic, or ionizable functionalities prepared in spatially defined positions using a photolithographic setup on microfluidic ships [143]).

7) Low shrinkage materials (formed from expanding monomers [144] or other photopolymers [145]), antishrinkage additives [146], polymer blends as hierarchically ordered defect-free materials [147], flame-retardant resins [148–150], coatings with enhanced wet adhesion [151], microporous membranes [152], novel membrane-based materials [153] optically active materials [154], 3D curved microstructures [155], novel interpenetrated networks [156], ε-caprolactone/L-lactide copolymers [157], or multiblock copolymers [158] (from polycaprolactone diol, polyethylene glycol, and cinnamoyloxy isophthalic acid) with shape memory properties usable in biomedical devices (surgical sutures, catheters, and smart stents), butyl-acrylate-based polymers with a narrow molecular weight distribution (synthesized in a narrow channel reactor) [159] or transparent inorganic polymer-derived microchannels [160], photopolymerized C_{60} molecules inside SWNTs [161], and control agents for living photopolymerization [162] (Section 8.30).

8) Matrices exhibiting a change of viscoelastic properties [163], nanosilica- and microcorundum-particle-containing acrylates for the matting of coatings [164], photopolymer nanowires having a giant elasticity [165], cross-linked polymers exhibiting photoinduced plasticity [166], and high modulus and low surface energy fluorinated polymer materials.

9) Photoelastomers as tailored biomaterials [167], biodegradable elastomers [168], compounds with an antibacterial activity [169], encapsulated biologically active compounds usable as biophotoresists that can be imaged on a surface at the level of pixel resolution (such as photopolymerizable glyphosate derivatives that exhibit a herbicidal activity [170]), and biofilm-resistant photopolymerizable coatings [171].

10) Photo-cross-linkable and de-cross-linkable materials, for example, modified polystyrenes [172], photodegradable polymers [173].

3.14
Photopolymerization and Nanotechnology

Nanotechnology is currently a magic word that everybody knows. The nanotechnology area corresponds both to (i) a technology that is able to create nano-objects and (ii) the use of nanoparticles in a lot of applications (see above); for example, nanoscale metal particles have interesting optical, electronic, magnetic, or catalytic properties and their encapsulation into a polymer increases their stability (against aggregation) or biocompatibility.

Photopolymerization in the nanotechnology area is a rather recent development field where academic and industrial interests are very important. A large growth is expected in the coming years. As seen, many aspects have been already explored for applications in high tech and more traditional sectors, for example,

1) Manufacture of nanocomposites with incorporated nanoparticles (such as carbon nanotube, titania, zirconia, zinc oxide, silver, gold, alumina, clay, or mica containing formulations), for example, for a modification of the physical properties, the electrical conductivity, or the polymerization profiles. The insertion of Ag(0) into a matrix is also highly worthwhile for the design of biocide- or bactericide-incorporating polymers. Luminophor nanoparticles allow the designing of pressure, temperature, or CO_2 sensors. The *in situ* incorporation of iridium in a photopolymerizable matrix can find applications in the organic light-emitting diode (OLED) technology. Organic–inorganic hybrid materials combine hardness, flexibility, resistance to chemicals, scratch and heat, good transparency and gloss; major applications include abrasion-resistant coatings, barrier-protective films, and novel optical devices. Such nanocomposite hybrid materials are created through three synthetic pathways: (i) sol–gel step and photopolymerization of an organo-alkoxysiloxane bearing a photopolymerizable group (in the case of an epoxide, a concomitant photoacid-catalyzed reaction can take place), (ii) dispersion of inorganic nanoparticles into a photocurable monomer, and (iii) *in situ* synthesis of metal nanoparticles into a polymer matrix through a redox process.
2) Use of nanoparticles as a photoinitiating system (e.g., semiconductor titanium dioxide; see Section 11.2).
3) Design of micro- or nanostructured materials and the creation of nanoscale features in microelectronics, optics, microlithography, biomedical engineering, and material science. The most commonly manufactured polymer nanostructures are one-dimensional fibers, wires, or rods.
4) Development of a wide range of reactions or processes such as
   a. the *in situ* topochemical photopolymerization of polydiacetylene/TiO_2 nanocomposites;
   b. the fabrication of nano- or microfluidic devices and multilayer microfluidic chips (<100 nm to several hundred micrometer channels for inkjet printheads, DNA chips, lab-on-a-chip technology, micropropulsion, and microthermal technologies);

c. the photopolymerization of organic molecular crystal nanorods, nanohybrids, and magnetic nanogels (they consist of a magnetic core and a polymer gel structure; applications are in biomedicine and bioengineering such as protein detection, magnetic-targeted drug carriers, bioseparation, and NMR imaging contrast agents);
   d. the fabrication of core-shell nanoparticles (they are composed of a hydrophobic polymer core coated with a hydrophilic polymer shell and have found applications in diagnostic testing, control release of drugs and biological agents, genetherapy, bioseparation, catalysis, and water-borne coatings);
   e. the elaboration of polymeric conductive nanomaterials (for applications as chemical- and biosensors, transistors, switches, photovoltaic cells, electrochromic devices, and optically conducting components);
   f. the manufacture of spring-shaped nanowires.
5) Local and confined photopolymerization using near-field optical techniques, for example, the creation of phase lenses, mirrors, index-modulated grating couplers, or nanoneedles by laser micro- or nano-fabrication.

3.15
Search for a Green Chemistry

The development of a green technology in the photopolymerization area occurs currently in several directions:

1) Use of new light sources or novel applications of existing light sources. Studies on sunlight curing are limited (see a recent review in [174] and references therein). Works have been carried out in the past to develop paints for crack-bridging applications and antisoiling properties in outdoor conditions of illumination (protection, renovation, or decoration of facades, roofs, walls, and terraces in order to improve the water proofing and the water resistance of the substrate together with less soiling). Interpenetrating polymer networks (IPNs) (Section 4.2.8) used as protective coatings, glues, or intercalates for safety glass laminates, glass-reinforced composites, various clearcoats, polymer-clay composites were readily obtained on sunlight exposure even under air.

Using sunlight is a dream that could come true for applications such as in fiber-reinforced composite boats, water storage, and delivery systems, in-field repair of equipment, building panels, antisoiling coatings, paints, protective coatings, strongly adhesive glass laminates, and quick-setting glues for the bonding of glass to a substrate.

The development of other artificial light sources, more economic and emitting less UV and less heat than the Hg lamps, such as LEDs (Section 2.3.6) as well as the recent use of household lamps (fluorescent bulbs, LED bulbs; Sections 2.3.10, 10.2.9, and the section titled Photoinitiator/Silane/Iodinium Salt in Chapter 12) is in progress. The problems related to the photoinitiation step encountered in LED bulbs and sunlight curing are discussed in Section 16.3.

2) Elaboration of renewable starting monomers/oligomers, biodegradable or photodegradable polymers, and reworkable materials. As in other sectors of polymer applications [173], this point now receives growing attention as (i) synthetic raw materials will become limited and (ii) the entire life cycle of the material has to be taken into account (production, durability, recycling, and waste). Extensive works are in progress for the design of new photopolymerizable compounds (Section 4.2).
3) Design of water-based formulations for inkjet printing, wood coating, screen printing of plastic articles, and adhesives [175]. They are of two types: (i) water-reducible (homogeneous solution of the resin in water) and (ii) water-based dispersions (conventional resins dispersed in water with the aid of emulsifiers/surfactants).

The advantages are (i) better toxicological properties, thanks to the absence of reactive diluents and the removal of any volatile organic solvents (that are still necessary, e.g., in specific applications of wood coatings via spraying or in inkjet printing with solvent-borne formulations), (ii) easy cleaning up of the equipment, (iii) reduced oxygen inhibition during the curing process, and (iv) interesting coating properties (generation of matt and low-gloss finishes, excellent adhesion of the cured coating to many types of substrates, less shrinkage owing to a lower cross-linked density in the coating). Possible disadvantages might be (i) special design of the resins, (ii) higher energy consumption (water must be removed before curing), and (iii) less coating durability (although this point is largely questioned).

References

1. Schwalm, R. (2007) *UV Coatings: Basics, Recent Developments and New Applications*, Elsevier, Oxford.
2. Technical Conference Proceedings, RadTech North America, Chevy Chase (1998), ibid (2000, 2002, 2004, 2006, 2008, 2010), http://www.radtech.org/.
3. Conference Proceedings, RadTech Europe Association, Hannover (1999), ibid (2001, 2003, 2005, 2007, 2009, 2011), http://www.radtech-europe.com/.
4. Technical Conference Proceeding, RadTech Asia, Japan (2003, 2007, 2011), http://www.radtechjapan.org/.
5. Technical Conference Proceeding, RadTech China (2005), http://www.radtechchina.com/.
6. Golden, R. (2009) Rad News, 1, pp. 22–26.
7. Dietliker, K. (2002) *A Compilation of Photoinitiators Commercially Available for UV Today*, Sita Technology Ltd, London.
8. Drobny, J.G. (2003 and 2010) *Radiation Technology for Polymers*, CRC Press, Boca Raton, FL.
9. Belfield, K.D. and Crivello, J.V. (eds) (2003) *Photoinitiated Polymerization*, ACS Symposium Series, Vol. 847, American Chemical Society, Washington, DC.
10. Fouassier, J.P. and Allonas, X. (eds) (2010) *Basics of Photopolymerization Reactions*, Research Signpost, Trivandrum.
11. Decker, C. (2003) in *Photoinitiated Polymerization*, ACS Symposium Series, Vol. 847, Chapter 23 (eds K.D. Belfield and J. Crivello), American Chemical Society, p. 266.
12. Sipani, V. and Scranton, A.B. (2004) in *Encyclopedia of Polymer Science*

and Technology (ed. H.F. Mark), Wiley-Interscience, New York. doi: 10.1002/0471440264.pst491

13. Cai, Y. and Jessop, J.L.P. (2004) in *Encyclopedia of Polymer Science and Technology* (ed. H.F. Mark), Wiley-Interscience, New York. doi: 10.1002/0471440264.pst490

14. Fouassier, J.P., Allonas, X., and Lalevée, J. (2007) in *Macromolecular Engineering: from Precise Macromolecular Synthesis to Macroscopic Materials Properties and Applications*, vol. **1** (eds K. Matyjaszewski, Y. Gnanou, and L. Leibler), Wiley-VCH Verlag GmbH, Weinheim, pp. 643–672.

15. Fouassier, J.P., Allonas, X., Lalevée, J., and Dietlin, C. (2010) in *Photochemistry and Photophysics of Polymer Materials* (ed. N.S. Allen), John Wiley & Sons, Inc., Hoboken, NJ, pp. 351–420.

16. Kahveci, M.U., Gilmaz, A.G., and Yagci, Y. (2010) in *Photochemistry and Photophysics of Polymer Materials* (ed. N.S. Allen), John Wiley & Sons, Inc., Hoboken, NJ, pp. 421–478.

17. Davidson, R.S. (2008) RadTech News, 1, pp. 9–12.

18. Valis, J. and Jasurek, B. (2010) in *Basics of Photopolymerization Reactions*, vol. **3** (eds J.P. Fouassier and X. Allonas), Research Signpost, Trivandrum, pp. 133–146.

19. Davidson, R.S. (2009) RadTech News, Citifluor, Ltd, London, UK, 1, pp. 13–16.

20. Sipani, V., Coons, L.S., Rangarajan, B., and Scranton, A.B. (2003) *Radtech Rep.*, **17** (3), 22–26.

21. Cabrera, M., Jezequel, J.Y., and André, J.C. (1993) in *Radiation Curing in Polymer Science and Technology*, vol. **3** (eds J.P. Fouassier and J.F. Rabek), Elsevier, Barking, pp. 73–97.

22. Jacobs, P.F. (1996) *Stereolithography and other RPM Technologies*, Society of Manufacturing Engineering, New York.

23. Scheffer, P., Bertsch, A., Corbel, S., Jejequel, A.J.Y., and Andre, J.C. (1997) *J. Photochem. Photobiol. A Chem.*, **107**, 283–290.

24. Chiu, S.H. and Wu, D.C. (2008) *J. Appl. Polym. Sci.*, **107**, 3529–3534.

25. Stampfl, J., Baudis, S., Heller, C., Liska, R., Neumeister, A., Kling, R., Ostendorf, A., and Spitzbart, M. (2008) *J. Micromech. Microeng.*, **18**, 125–134.

26. Lackner, M. (ed.) (2008) *Lasers in Chemistry*, Wiley-VCH Verlag GmbH, Weinheim.

27. Liska, R. and Schuster, M. (2010) in *Basics of Photopolymerization Reactions*, vol. **3** (eds J.P. Fouassier and X. Allonas), Research Signpost, Trivandrum, pp. 77–92.

28. Allen, R.D. (2007) *J. Photopolym. Sci. Technol.*, **20**, 453–455.

29. Shimizu, D., Maruyama, K., Saitou, A., Kai, T., Shimokawa, T., Fujiwara, K., Kikuchi, Y., and Nishiyama, I. (2007) *J. Photopolym. Sci. Technol.*, **20**, 423–428.

30. Ivan, M.G. and Scaiano, J.C. (2010) in *Photochemistry and Photophysics of Polymer Materials* (ed. N.S. Allen), John Wiley & Sons, Inc., Hoboken, NJ, pp. 479–508.

31. Chocos, C.L., Ismailova, E., Brochon, C., Leclerc, N., Tiron, R., Sourd, C., Bandelier, P., Foucher, J., Ridaoui, H., Dirani, A., Soppera, O., Perret, D., Brault, C., Serra, C.A., and Hadzioannou, G. (2008) *Adv. Mater.*, **10**, 1–5.

32. Takahara, S., Endo, T., Park, C.H., Tarumoto, N., Suzuki, S., Song, S., Miyagawa, N., and Yamaoka, T. (2010) in *Basics of Photopolymerization Reactions*, vol. **3** (eds J.P. Fouassier and X. Allonas), Research Signpost, Trivandrum, pp. 23–32.

33. Billone, P.S., Park, J.M., Blackwell, J.M., Bristol, R., and Scaiano, J.C. (2010) *Chem. Mater.*, **22**, 15–17.

34. Ezra, B.B. (1998) *Adv. Technol.*, **11**, 20–27.

35. Ichihashi, Y. and Kaji, M. (2004) *J. Photopolym. Sci. Technol.*, **17**, 135–138.

36. Suzuki, S., Urano, T., Ito, K., Murayama, T., Hotta, I., Takahara, S., and Yamaoka, T. (2004) *J. Photopolym. Sci. Technol.*, **17**, 125–129.

37. Barachevskii, V.A. (2006) *High Energy Chem.*, **40**, 131–141.

38. Garcia, C., Fimia, A., and Pascual, I. (2001) *Appl. Phys. B*, **72**, 311–316.

39. Carré, C. and Lougnot, D.J. (1990) *J. Opt.*, **21**, 147–152.

40. Waldman, D.A., Butler, C.J., and Raguin, D.H. (2003) in *Organic Holographic Materials and Applications*, vol. 5216 (ed. K. Meerholz), *SPIE*, pp. 178–184.
41. Noiret, N., Meyer, C., and Lougnot, D.J. (1994) *Pure Appl. Opt.*, **3**, 55–71.
42. Calvo, M.L. and Cheben, P. (2009) *J. Opt. A: Pure Appl. Opt.*, **11**, 024009/1–024009/11.
43. Jeon, M., Yoon, S.C., Lee, J., Han, M., and Lee, C. (2009) *J. Nanosci. Nanotechnol.*, **9**, 6912–6917.
44. Fernandez, E., Ortuno, M., Gallego, S., Garcia, C., Belendez, A., and Pascual, I. (2007) *Appl. Opt.*, **46**, 5368–5373.
45. Sangermano, M., Voit, B., Sordo, F., Eichhorn, K.J., and Rizza, G. (2008) *Polymer*, **49**, 2018–2022.
46. Nebioglu, A., Leon, J.A., and Khudyakov, I.V. (2008) *Ind. Eng. Chem. Res.*, **47**, 2155–2159.
47. Murase, S., Kinoshita, K., Horie, K., and Morino, S. (1997) *Macromolecules*, **30** (25), 8088–8090.
48. Natarajan, L.V., Brown, D.P., Tondiglia, V.P., Sutherland, R.L., Lloyd, P., Jakubiak, R., Vaia, R., and Bunning, T.J. (2005) *PMSE Prepr.*, **93**, 617–618.
49. Wang, G., Zhu, X., Wu, J., Zhu, J., Chen, X., and Cheng, Z. (2007) *J. Appl. Polym. Sci.*, **106**, 1234–1242.
50. Zhang, Y., Zhang, W., Chen, X., Cheng, Z., Wu, J., Zhu, J., and Zhu, X. (2008) *J. Polym. Sci., A: Polym. Chem.*, **46**, 777–789.
51. Jradi, S., Soppera, O., and Lougnot, D.J. (2008) *Appl. Opt.*, **47**, 3987–3993.
52. Jing, M., Lu, R., Yang, Q.X., Bao, C.Y., Sheng, R., Xu, T.H., and Zhao, Y. (2007) *J. Polym. Sci., A: Polym. Chem.*, **45**, 3460–3472.
53. Kawatsuki, N., Suehiro, C., and Yamamoto, T. (1998) *Macromolecules*, **31** (18), 5984–5990.
54. Choi, D.H., Hong, H.T., Cho, K.J., and Kim, J.H. (1999) *Korea Polym. J.*, **7**, 189–195.
55. Angiolini, L., Benelli, T., Giorgini, L., Salatelli, E., Bozio, R., Dauru, A., and Pedron, D. (2005) *Eur. Polym. J.*, **41**, 2045–2054.
56. Yu, L., Zhang, Z., Chen, X., Zhang, W., Wu, J., Cheng, Z., Zhu, J., and Zhu, X. (2008) *J. Polym. Sci., A: Polym. Chem.*, **46**, 682–691.
57. Hore, D.K., Natansohn, A.L., and Rochon, P.L. (2003) *J. Phys. Chem. B*, **107** (11), 2506–2518.
58. Zheng, Z., Su, Z., Wang, L., Xu, J., Zhang, Q., and Yang, J. (2007) *Eur. Polym. J.*, **43**, 2738–2744.
59. Angiolini, L., Benelli, T., Giorgini, L., and Salatelli, E. (2006) *Macromolecules*, **39**, 3731–3737.
60. Feuillade, M., Croutxe-Barghorn, C., Mager, L., Carré, C., and Fort, A. (2004) *Chem. Phys. Lett.*, **398**, 151–156.
61. Nicoletta, F.P., De Filpo, G., Cupelli, D., Macchione, M., and Chidichimo, G. (2001) *Appl. Phys. Lett.*, **79**, 4325–4329.
62. Lee, S.K., Cho, M.J., Jin, J.I., and Choi, D.H. (2007) *J. Korean Phys. Soc.*, **51**, 1668–1672.
63. Dam, J.S., Perch-Nielsen, I.R., Rodrigo, P.J., Kelemen, L., Alonzo, C.A., Ormos, P., and Glueckstad, J. (2007) *Proc. SPIE Int. Soc. Opt. Eng.*, 6644–6649.
64. Hermans, K., Wolf, F.K., Perelaer, J., Janssen, R.A.J., Schubert, U.S., Bastiaansen, C.W.M., and Broer, D.J. (2007) *Appl. Phys. Lett.*, **91**, 174103/1–174103/3.
65. Zhang, X., Sun, Y., Yu, X., Zhang, B., Huang, B., and Jiang, M. (2009) *Synth. Met.*, **159** (23–24), 2491–2496.
66. Davidson, R.S. (2007) *Radnews*, 3, pp. 6–14.
67. Nuttelman, C., Mortisen, D.J., Henry, S.M., and Anseth, K.S. (2001) *J. Biomed. Res.*, **57**, 217–223.
68. Castle, P.M. and Sadhir, R.K. (1993) in *Radiation Curing in Polymer Science and Technology*, vol. **3** (eds J.P. Fouassier and J.F. Rabek), Elsevier, Barking, pp. 37–72.
69. Baudis, S., Schuster, M., Heller, C., Turecek, C., Varga, F., Bergmeister, H., Weigel, G., Stampfl, J., and Liska, R. (2007) RadTech Europe 2007 Conference Proceedings.
70. Hoyle, C.E., Gould, T., Piland, S., Wei, H., Phillips, B., Nazarenko, S., Senyurt, A.F., and Cole, M. (2006) *Radtech Rep.*, **20**, 12–17.
71. Mozner, N. (2004) *Macromol. Symp.*, **217**, 63–75.

72. Bowman, C.N. and Cramer, N.B. (2009) US Patent Appl. Publ. US 2009 270,528.
73. Athanassiou, A., Kalyva, M., Lakiotaki, K., Georgiou, S., and Fotakis, C. (2003) *Adv. Mater. Sci.*, **5**, 245–251.
74. Mucci, V., Arenas, G., Duchowicz, R., Cook, W.D., and Vallo, C. (2008) *Dent. Mater.*, **25**, 103–114.
75. Malinauskas, M., Gilbergs, H., Purlys, V., Zukauskas, A., Rutkauskas, M., and Gadonas, R. (2009) *Proc. SPIE*, **7366**, 22/1–22/12.
76. Kawata, S., Sun, H.B., Tanaka, T., and Takada, K. (2001) *Nature*, **412**, 697–698.
77. Belfield, K.D., Ren, X.B., Van Stryland, E.W., Hagan, D.J., Dubikovsky, V., and Miesak, E.J. (2000) *J. Am. Chem. Soc.*, **122**, 1217–1218.
78. Zhou, W.H., Kuebler, S.M., Braun, K.L., Yu, T.Y., Cammack, J.K., Ober, C.K., Perry, J.W., and Marder, S.R. (2002) *Science*, **296**, 1106–1109.
79. Cumpston, B.H., Ananthavel, S.P., Barlow, S., Dyer, D.L., Ehrlich, J.E., Erskine, L.L., Heikal, A.A., Kuebler, S.M., Lee, I.Y.S., McCord-Maughon, D., Qin, J.Q., Rockel, H., Rumi, M., Wu, X.L., Marder, S.R., and Perry, J.W. (1999) *Nature*, **398**, 51–54.
80. Chen, Q.D., Wu, D., Niu, L.G., Wang, J., Lin, X.F., Xia, H., and Sun, H.B. (2007) *Appl. Phys. Lett.*, **17**, 171105/1–171105/3.
81. Dong, Y., Sun, Y.M., Li, Y.F., Yu, X.Q., Hou, X.Y., and Zhang, X. (2007) *Thin Solid Films*, **516**, 1214–1217.
82. Kim, W.C., Choi, H., Song, T., Park, N.C., Park, Y.P., and Lee, J.E. (2007) *Microsyst. Technol.*, **13**, 1–10.
83. Ecoffet, C., Espanet, A., and Lougnot, D.J. (1998) *Adv. Mater.*, **10**, 411–414.
84. Wurtz, G., Bachelot, R., D'Hili, F., Royer, P., Triger, C., Ecoffet, C., and Lougnot, D.J. (2000) *Jpn. J. Appl. Phys.*, **39**, 98–100.
85. Ploux, L., Anselme, K., Dirani, A., Ponche, A., Soppera, O., and Roucoules, V. (2009) *Langmuir*, **25**, 8161–8169.
86. Jradi, S., Soppera, O., and Lougnot, D.J. (2009) *Opt. Mater.*, **31**, 640–646.
87. Zeng, X., Plain, J., Jradi, S., Darraud, C., Louradour, F., Bachelot, R., and Royer, P. (2011) *Opt. Express*, **19**, 4805–4814.
88. Wang, J., Ugaz, V.M., and McFerrin, A. (2007) Proceedings of the 65th Annual Technical Conference for the Society of Plastics Engineers, pp. 1804–1817.
89. Li, J.M. and Zhang, L.M. (2007) *Starch/Staerke*, **59**, 481–422.
90. Lin-Gibson, S., Jones, R.L., Washburn, N.R., and Horkay, F. (2005) *Macromolecules*, **38**, 2897–2902.
91. Hao, J.Q., Li, H., and Woo, H.G. (2009) *J. Appl. Polym. Sci.*, **112**, 2976–2980.
92. Fairbanks, B.D., Singh, S.P., and Bowman, C. (2011) *Macromolecules*, **44**, 2444–2450.
93. Beines, P.W., Klosterkamp, I., Menges, B., Jonas, U., and Knoll, W. (2007) *Langmuir*, **23**, 2231–2238.
94. Ruan, W., Wang, X., Lian, Y., Yanqin, H., Huang, Y., and Niu, A. (2006) *J. Appl. Polym. Sci.*, **101**, 1181–1187.
95. Jockusch, S., Turro, N.J., Mitsukami, Y., Matsumoto, M., Iwamura, T., Lindner, T., Flohr, A., and di Massimo, G. (2009) *J. Appl. Polym. Sci.*, **111** (5), 2163–2170.
96. Tamesue, S., Takashima, Y., Yamaguchi, H., Shinkai, S., and Harada, A. (2010) *Angew. Chem. Int. Ed.*, **49**, 7461–7464.
97. Wan, T., Wang, X., Yuan, Y., and He, W. (2006) *J. Appl. Polym. Sci.*, **102**, 2875–2881.
98. Sakai, T., Murayama, H., Nagano, S., Takeoka, Y., Kidowaki, M., Ito, K., and Seki, T. (2007) *Adv. Mater.*, **19**, 2023–2025.
99. Kahveci, M.U., Beyazkilic, Z., and Yagci, Y. (2010) *J. Polym. Sci., A: Polym. Chem.*, **48**, 4989–4994.
100. Gerbaldi, C., Nair, J.R., Meligrana, C., Bongiovani, R., Bodoardo, S., and Penazzi, N. (2009) *Electrochem. Comm.*, **11**, 1796–1798.
101. Takahashi, T., Watanabe, H., Miyagawa, N., Takahara, S., and Yamaoka, T. (2002) *Polym. Adv. Technol.*, **13**, 33–39.

102. Tamada, M., Watanabe, T., Horie, K., and Ohno, H. (2007) *Chem. Commun.*, **39**, 4050–4052.
103. He, J. and Zhao, Y. (2010) *Dyes Pigm.*, **89**, 278–283.
104. Clapper, J.D. and Guymon, C.A. (2009) *Mol. Cryst. Liq. Cryst.*, **509**, 772–780.
105. Kuckling, D., Vo, C.D., and Wohlrab, S.E. (2002) *Langmuir*, **18** (11), 4263–4269.
106. He, J., Tong, X., and Zhao, Y. (2009) *Macromolecules*, **42**, 4845–4852.
107. Gong, P.J., Sun, H.W., Hong, J., Xu, D.M., and Yao, S.D. (2007) *Sci. China, Ser. B: Chem.*, **50**, 217–223.
108. Jin, Q., Liu, G., and Jian, J. (2010) *Eur. Polym. J.*, **46**, 2120–2128.
109. Nagasawa, J., Yoshida, M., and Tamaoki, N. (2011) *Eur. J. Org. Chem.*, **12**, 2247–2255.
110. Steenackers, M., Kueller, A., Stoycheva, S., Grunze, M., and Jordan, R. (2009) *Langmuir*, **25**, 2225, 2231.
111. Duner, G., Anderson, H., Myrskog, A., Hedlund, M., Aastrup, T., and Ramstroem, O. (2008) *Langmuir*, **24**, 7559–7564.
112. Marcot, L., Maldivi, P., Marchon, J.-C., Guillon, D., Ibn-Elhaj, M., Broer, D.J., and Mol, T. (1997) *Chem. Mater.*, **9** (10), 2051–2058.
113. Hodko, D., Gamboa-Aldeco, M., and Murphy, O.J. (2009) *J. Solid State Electrochem.*, **13**, 1077–1089.
114. Janovak, L. and Dekany, I. (2010) *Appl. Surf. Sci.*, **256**, 2809–2817.
115. Cai, X., Anyaogu, K.C., and Neckers, D.C. (2008) *Chem. Commun.*, **40**, 5007–5009.
116. Hoshino, K., Yoshio, M., Mukai, T., and Ohno, H. (2003) *J. Polym. Sci., A: Polym. Chem.*, **41**, 3486–3492.
117. Cheng, W.T., Chih, Y.W., and Yeh, W.T. (2007) *Int. J. Adhes. Adhes.*, **27**, 236–243.
118. Koizhaiganova, R., Kim, H.J., Vasudevan, T., Kudaibergenov, S., and Lee, M.S. (2010) *J. Appl. Polym. Sci.*, **115**, 2448–2454.
119. Cai, X., Anyaogu, K.C., and Neckers, D.C. (2009) *Photochem. Photobiol. Sci.*, **8**, 1568–1573.
120. Sangermano, M., Vescovo, L., Pepino, N., Chiolerio, A., Allia, P., Tiberto, P., Coisson, M., Suber, L., and Marchegiani, G. (2010) *Macromol. Chem. Phys.*, **211**, 2530–2535.
121. Balan, L., Malval, J.P., Schneider, R., Le Nouen, D., and Lougnot, D.J. (2010) *Polymer*, **51**, 1363–1369.
122. Yagci, Y., Sangermano, M., and Rizza, G. (2008) *Polymer*, **49**, 5195–5198.
123. Anyaogu, K.C., Cai, X., and Neckers, D.C. (2008) *Macromolecules*, **41**, 9000–9003.
124. Leite, E., Naydenova, I., Mintova, S., Leclercq, L., and Toal, V. (2010) *Appl. Opt.*, **49**, 3652–3660.
125. Kwon, Y.S., Jung, B.M., Lee, H., and Chang, J.Y. (2010) *Macromolecules*, **43**, 5376–5381.
126. Craparo, E.F., Cavallaro, G., Bondi, M.L., and Giammona, G. (2004) *Macromol. Chem. Phys.*, **205** (14), 1955–1964.
127. Balan, L., Jin, M., MalvalL, J.P., Chaumeil, H., Defoin, A., and Vidal, L. (2008) *Macromolecules*, **41**, 9359–9365.
128. Chemtob, A., Croute-Barghorn, C., Soppera, O., and Rigolet, S. (2009) *Makromol. Chem. Phys.*, **210**, 1127.
129. Lai, W., Li, X., Feng, H., and Fu, G. (2008) *J. Photopolym. Sci. Technol.*, **21**, 761–765.
130. Imroz, A.A.M. and Mayes, A.G. (2010) *Macromolecules*, **43**, 837–844.
131. Li, S., Deng, J., and Yang, W. (2010) *J. Polym. Sci., B: Polym. Chem.*, **48**, 936–942.
132. Yin, Y., Sha, S., Chang, D., Zheng, H., Li, J., Liu, X., Xu, P., and Xiong, F. (2010) *Chem. Commun.*, **46**, 8222–8224.
133. Sultana, S., Matsui, J., Mitsuishi, M., and Miyashita, T. (2008) *Polym. J.*, **40**, 953–957.
134. Fisher, O.Z. and Peppas, N.A. (2009) *Macromolecules*, **42**, 3391–3398.
135. Hsiao, V.K.S., White, T.J., Cartwright, A.N., Prasad, P.N., and Allan-Guymon, C. (2010) *Eur. Polym. J.*, **46**, 937–943.
136. Calvo, M.L. and Cheben, P. (2009) *J. Opt. A: Pure Appl. Opt.*, **11**, 1–11.
137. Sangermano, M., Pegel, S., Potschke, P., and Voit, B. (2008) *Macromol. Rapid Commun.*, **29**, 396–400.
138. Watanabe, H., Ohzono, T., and Kunitake, T. (2008) *Polym. J.*, **40**, 379–382.

139. Zhong, K., Chen, T., Yin, B., and Jin, L.Y. (2009) *Macromol. Res.*, **17**, 280–283.
140. Sadej-Bajerlain, M., Gojzewski, H., and Andrzejewska, E. (2011) *Polymer*, **52**, 1495–1503.
141. Thangawng, A.L., Howell, B.P., Spillman, C.M., Naciri, J., and Ligler, F.S. (2011) *Lab Chip*, **11**, 1157–1160.
142. Lee, K.J., Hwang, S., Yoon, J., Bhaskar, S., Park, T.H., and Lahann, J. (20011) *Macromol. Rapid Commun.*, **32**, 431–437.
143. Yu, C., Xu, M., Svec, F., and Frechet, J.M.J. (2002) *J. Polym. Sci., B: Polym. Chem.*, **40** (6), 755–769.
144. Chappelow, C.C., Pinzino, C.S., Jeang, L., Harris, C.D., Holder, A.J., and Eick, J.D. (2000) *J. Appl. Polym. Sci.*, **76** (11), 1715–1724.
145. Lu, B., Xiao, P., Sun, M., and Nie, J. (2007) *J. Appl. Polym. Sci.*, **104**, 1126–1130.
146. Duarte, M.L.B., Ortiz, R.A., Gomez, A.G.S., and Valdez, A.E.G. (2009) *React. Funct. Polym.*, **70**, 98–102.
147. Travasso, R.D.M., Kuksenok, O., and Balazs, A.C. (2006) *Langmuir*, **22**, 2620–2628.
148. Wang, Q. and Shi, W. (2006) *Eur. Polym. J.*, **42**, 2261–2269.
149. Liang, H., Asif, A., and Shi, W. (2005) *J. Appl. Polym. Sci.*, **97**, 185–194.
150. Yu, L., Zhang, S., Liu, W., Zhu, X., Chen, X., and Chen, X. (2010) *Polym. Degrad. Stab.*, **95**, 1934–1942.
151. Glass, P., Chung, H., Washburn, N.R., and Sitti, M. (2010) *Langmuir*, **26**, 17357–17362.
152. Yu, H.-Y., Li, W., Zhou, J., Gu, J.S., Huang, L., Tang, Z.Q., and Wei, X.W. (2009) *J. Memb. Sci.*, **343**, 82–89.
153. He, D., Susanto, H., and Ulbricht, M. (2009) *Prog. Polym. Sci.*, **34**, 62–98.
154. Angiolini, L., Giorgini, L., Mauriello, F., and Rochon, P. (2009) *Macromol. Chem. Phys.*, **210** (1), 77–89.
155. Chan-Park, M.B., Yang, C., Guo, X., Chen, L., Yoon, S.F., and Chun, J.H. (2008) *Langmuir*, **24** (10), 5492–5499.
156. Zhang, S., Feng, Y., Zhang, L., Sun, J., Xu, X., and Xu, Y. (2007) *J. Polym. Sci., A: Polym. Chem.*, **45**, 768–775.
157. Nagata, M. and Yamamoto, Y. (2010) *Macromol. Chem. Phys.*, **211**, 1826–1835.
158. Nagata, M. and Inaki, K. (2011) *J. Appl. Polym. Sci.*, **120**, 3556–3564.
159. Jachuck, R.J.J. and Nekkanti, V. (2008) *Macromolecules*, **41**, 3053–3062.
160. Min, K.-I., Park, J.H., Hong, L.Y., and Kim, D.P. (2009) *J. Nanosci. Nanotechnol.*, **9**, 7215–7219.
161. Saito, Y., Honda, M., Moriguchi, Y., and Verma, P. (2010) *Phys. Rev. B: Condens. Matter Mater. Phys.*, **81** (245416), 1–6.
162. Matyjaszewski, K. and Braunecker, A. (2007) in *Macromolecular Engineering: from Precise Macromolecular Synthesis to Macroscopic Materials Properties and Applications*, vol. 1 (eds K. Matyjaszewski, Y. Gnanou, and L. Leibler), Wiley-VCH Verlag GmbH, Weinheim, pp. 161–248.
163. Hosono, N., Furukawa, H., Masubuchi, Y., Watanabe, T., and Horie, K. (2007) *Colloids Surf., B: Biointerface*, **56**, 285–289.
164. Bauer, F., Flyunt, R., Czihal, K., Langguth, H., Mehnert, R., Schubert, R., and Buchmeiser, M.R. (2007) *Prog. Org. Coat.*, **60**, 121–126.
165. Nakanishi, S., Shoji, S., Kawata, S., and Sun, H.B. (2007) *Appl. Phys. Lett.*, **91** (63112), 1–3.
166. Scott, T.F., Schneider, A.D., Cook, W.D., and Bowman, C.N. (2005) *Science*, **308**, 1615–1617.
167. Baudis, S., Heller, C., Liska, R., Stampfl, J., Bergmeister, H., and Weigel, G. (2009) *J. Polym. Sci., A: Polym. Chem.*, **47**, 2664–2676.
168. Liu, J. and Sprague, J.J. (2008) in *PMSE Preprints*, ACS Polym. Div., vol. **98** (eds T. Samulski and V.V. Sheares), pp. 71–73.
169. Gozzelino, G., Dell'Aquila, G.A., and Tobar, D.R. (2009) *J. Appl. Polym. Sci.*, **112**, 2334–2342.
170. Bogdanova, A., Piunova, V., Berger, D., Fedorov, A.V., and Neckers, D.C. (2007) *Biomacromolecules*, **8**, 439–447.
171. Faulkner, R.A., Carstens, R.M., Omiwade, O., and Cavitt, T.B. (2010)

in *Basics of Photopolymerization Reactions*, vol. **3** (eds J.P. Fouassier and X. Allonas), Research Signpost, Trivandrum, pp. 93–100.
172. Okamura, H., Yamauchi, E., and Shirai, M. (2011) *React. Funct. Polym.*, **71**, 480–488.
173. Al-Malaika, S., Hewitt, C., and Sheena, H.H. (2010) in *Photochemistry and Photophysics of Polymer Materials* (ed. N.S. Allen), John Wiley & Sons, Inc., Hoboken, pp. 603–626.
174. Lalevée, J. and Fouassier, J.P. (2011) *Polym. Chem.*, **2**, 1107–1113.
175. Davidson, S. (1999) *Exploring the Science, Technology and Application of UV and EB Curing*, Sita Technology Ltd, London.

4
Photopolymerization Reactions

4.1
Encountered Reactions, Media, and Experimental Conditions

As in thermally initiated polymerization, light-induced photopolymerization reactions can be encountered in a lot of experimental conditions and/or with many different monomers, oligomers, or polymer media allowing large possibilities of applications or potential developments. The following list that outlines this large diversity of reactions might not be exhaustive.

1) Usual film radical or cationic photopolymerization and radical, acid- or base-catalyzed photo-cross-linking under conventional light sources [1–6], for example, (i) photopolymerization of acrylates, epoxides, vinyl ethers (VEs), thiol-enes, solvent- or water-borne systems, and powder formulations, (ii) dual cure, hybrid cure, and charge-transfer (CT) photopolymerization, (iii) sol–gel photopolymerization of organic–inorganic hybrid resins for the production of hybrid sol–gel (HSG) materials, (iv) photopolymerization of interpenetrated networks (IPNs) [7, 8], (v) photopolymerization of polydimethylsiloxanes [9], maleate esters of epoxidized plant oil triglycerides [10] or divinyl fumarates [11], (vi) copolymerization [12] and alternating copolymerization (using the CT absorption properties of suitable monomers under light), (vii) dendritic photopolymerization (the dynamic viscosity of dendritic acrylates is lower than that of the corresponding linear oligomers which facilitates the formulation handling; the cure speed is faster and the tensile strengths increase [13, 14]), (viii) photopolymerization of nanocomposites (incorporation of zeolites or clay derivatives [15–20] or titanium dioxide [21] in a photopolymerizable formulation allows getting better properties for the final material), (ix) photopolymerization of hydrogels (when polymerized using photoinitiator (PI)-free donor/acceptor monomer couples, these systems can find an interest in controlled release applications [22–24]) and nanocomposite gels in aqueous media [25], (x) photopolymerization on nanoparticle surfaces [26], (xi) redox-photoinitiated cross-linking polymerization [27], and (xii) photopolymerization of acid-containing monomers [28, 29] and liquid crystalline systems [30].

Photoinitiators for Polymer Synthesis: Scope, Reactivity and Efficiency, First Edition.
Jean-Pierre Fouassier and Jacques Lalevée.
© 2012 Wiley-VCH Verlag GmbH & Co. KGaA. Published 2012 by Wiley-VCH Verlag GmbH & Co. KGaA.

2) Formation of linear [31–33] and star [34] polymer block copolymers. They are obtained (see a review in [35]) from (i) a prepolymer Pn bearing a terminal A photolabile group (Pn-A) and a monomer M that creates a diblock or a triblock polymer according to the termination reaction, for example, Pn-A-(M)$_n$, (ii) a prepolymer having a mid-chain photoactive moiety Pn-A-A-Pn (it will behave as a macrophotoinitiator) and a monomer M still leading to Pn-A-(M)$_n$, and (iii) two homopolymers (having two photoreactive functions A and B) Pn-A and P'n-B where a further photoinduced coupling occurs to form Pn-A-B-P'n.

3) Photopolymerization in (micro)heterogeneous media [36–46] such as in microemulsions, inverse emulsions, emulsions, miniemulsions (they are a special class of emulsions stabilized by a surfactant and an osmotic pressure agent leading to 50–500 nm droplets), and dispersions. This allows forming of monodisperse microspheres (usable in electrophotographic toners, instrument calibration standards, column packing materials for chromatography, and biomedical analysis) [47, 48] or to photopolymerize hybrid colloids [49], amphiphilic monomers in micelles for encapsulation [50], and gold-nanoparticle-containing latexes [51].

4) Liquid-phase photopolymerization of polymers [52] and IPN hydrogels [53].

5) Photopolymerization of various systems under laser irradiation [54]. Pulsed laser-induced polymerization (PLP) is also largely used for the determination of the propagation and termination rate constants. As shown above, lasers allow creation of objects or writing of microcircuits. This method is different from the laser ablation process that can also generate microstructures, for example, microholes [55, 56]; in that case, the polymer is destroyed on exposure to an intense laser beam.

6) Surface photopolymerization of adsorbed or oriented molecules [57, 58], surface-confined photopolymerization [59], surface-initiated photopolymerization [60], for example, formation of polymer vesicles with well-defined PMMA brushes [61], surface modification through photopolymerization [62].

7) Gas-phase photopolymerization of monomers absorbed on a surface [63].

8) Photopolymerization in aerosols (for forming microcapsules [64]).

9) Photopolymerization of methacrylates and acrylates in ionic liquids. In that case, this leads to a large increase in the propagation rate constant (attributed to the polarity increase) and a decrease in the termination rate constant (ascribed to the increased viscosity, typically 10–500 mPa that is comparable to the viscosity of oils and two to three orders of magnitude higher than that of organic solvents) [65–69]. Advantages of ionic liquids are (i) an ability to dissolve a wide range of organic and ionic compounds, (ii) a negligible vapor pressure, and (iii) a high thermal and chemical stability. Several points have been reported: photosensitization by CdTe nanocrystals in ionic liquids, effect of a catalytic addition of an ionic liquid in a radical photopolymerization [70], formation of biomimetic ionic liquid crystalline hydrogels by self-assembly and photopolymerization [71], effect of the ionic liquid on the formulation and properties of the formed polymer materials [72], and photopolymerization of self-organizing ionic liquids [73].

10) Surface graft photopolymerization. It allows a modification of the surface of a polymer material [74, 75]. Grafting methods (Eq. (4.1)) are usually divided into (i) grafting from (1), (ii) grafting onto (2) and (iii) grafting through (3). Processes (1) and (3) are the most popular in surface photografting (process (2) is unknown). This has been applied to low-density polyethylenes (e.g., the wettability and adhesivity are considerably improved by grafting hydrophilic monomers [76]), thin polymer hydrogels [77], self-assembly monolayers on gold [78], cellulosic compounds [79], carbon-doped TiO_2 nanoparticles (under sunlight [80]), natural rubbers [81, 82], hydroxyapatite particle surfaces [83], microfiltration membranes (tailored grafting from functionalization [84]), surface modification of polyamides [85], gradient and patterned polymer brushes [86], and binder agents for the modification of diamond powder and polyethylene terephthalate films [87].

$$\begin{aligned}
&(Pn\text{-})(\text{-}Pm)C(\text{-}R)^\bullet \rightarrow (Pn\text{-})(\text{-}Pm)C(\text{-}R)\text{-}(M)_n &&(1)\\
&(Pn\text{-})(\text{-}Pm)C(\text{-}R)\text{-}A + Pm \rightarrow (Pn\text{-})(\text{-}Pm)C(\text{-}R)\text{-}Pm &&(2)\\
&(Pn\text{-})(\text{-}Pm)C(\text{-}R)\text{-}M + (Pn\text{-})(\text{-}Pm)C(\text{-}R)\text{-}M\\
&\quad\rightarrow (Pn\text{-})(\text{-}Pm)C(\text{-}R)\text{-}M\text{-}M\text{-}(\text{-}R)C(Pn)(\text{-}Pm) &&(3)
\end{aligned} \qquad (4.1)$$

11) Gradient photopolymerization. This can be achieved by the creation of a double bond conversion gradient using the attenuation of the excitation light that penetrates into the depth of the acrylate film. As a consequence, the polymerized material exhibits a difference in structure on the top and at the bottom (e.g., for a 5 mm thick sample [88]).
12) Frontal photopolymerization [89–91]. In contrast to most photopolymerization reactions that spatially and temporally occur in a homogeneous way, frontal photopolymerization is confined within a localized narrow area, proceeds in a heterogeneous manner as a function of time on the local stimulus produced by heat (as the reaction is highly exothermic), and propagates through the sample. This finds applications in the production of polymers with graded optical and mechanical properties, and IPNs.
13) Controlled polymerization. This occurs using photoiniferters [92, 93] or other strategies (see details in Section 4.2.11) and photo-living polymerization [94, 95].
14) Photoassisted oxypolymerization. This is encountered in the drying of alkyd resins. The thermal oxypolymerization involves molecular oxygen and a catalyst dryer [96]. The introduction of a photochemical process allows acceleration of the drying of the resin [97, 98].
15) Template photopolymerization. It can be used for the development of nanostructured organic–inorganic composite materials that are reliably and controllably synthesized [99, 100]. For example, photopolymerizable lyotropic liquid crystals allow to template specific structures onto the organic and inorganic phases [101].

16) Photopolymerization in multilayers. This allows the preparation of extended and highly organized ultrathin layers of polymers (by solid-state photopolymerization of multilayers built from monomolecular films of monomers using the Langmuir–Blodgett technique [102–105]), growing nanofruit structures (on polymer surfaces through layer-by-layer grafting [106]), and self-assembled monolayers [107].
17) Spatially controlled photopolymerization, for example, (i) two-photon photopolymerization [108–110], (ii) photopolymerization with evanescent waves [111], (iii) manufacture of optical elements such as microlenses [112], (iv) holographic recording, (v) photo-oriented polymerization (e.g., a self-focusing laser excitation controls the polymerization in a hybrid SiO_2 glass waveguide by a high degree of spatial localization, thereby leading to optical fibers on chips [113]).
18) *In situ* photopolymerization of (i) microfluidic devices (allowing the production of polymer microstructures such as fibers, tubes, spheres, and membrane-type microvalves for biomedical applications [114–117]), (ii) polymer particles [118], (iii) biodegradable robust hydrogels [119], (iv) macroporous sol–gel monoliths [120], and (v) pyrrole in mesoporous titanium dioxide [121].
19) Pinpoint and self-guiding photopolymerization of micro-/nanostructures [122, 123]. For example, the photopolymerization of a monomer at the end of an optical fiber leads to an increase in the refractive index that induces a self-guiding effect, counterbalancing the natural divergence of the light beam along hundreds of micrometers. This results in the fabrication of a polymer tip.
20) *In vivo* light-induced polymerization [124].
21) Photopolymerization under magnetic field. The reaction shows a faster rate of monomer conversion that is connected with the role of the radical pair formed after excitation of the PI [125] (Section 5.7.4).
22) True photo-cross-linking reactions. They lead to a structured material and occur in:
    a. prepolymers or polymers, for example, having photosensitive pendant groups [126–128], photo-cross-linkable termini [129], or photodimerizable units such as cinnamoyl moieties [130–132], iminopyridinium ylides [133], coumarins [134], and chalcones [135];
    b. organized media [136] such as in the case of the topochemical photopolymerization that takes place within the solid state of organic crystals made up by packing molecules into a three-dimensional regular lattice. Stereospecifically regulated polymers are thus formed [137]. The understanding of the topochemical solid-state photopolymerization [138] as well as the application to, for example, polydiacetylenes (which combine highly ordered and conjugated backbones with tailorable pendant side groups and terminal functionalities) [139] or thin polycrystalline diacetylene films [140] has been reported;
    c. photodimerizable polyethyleneglycol (PEG) derivatives [141].

4.2
Typical Characteristics of Selected Photopolymerization Reactions

4.2.1
Film Radical Photopolymerization of Acrylates

The radical photopolymerization of multifunctional monomers and oligomers in film is very fast (in the 1 s time range under Hg lamps in industrial conditions) and generally sensitive to the presence of oxygen. It presents a relatively slow posteffect (Section 7.1.1). Under continuous laser sources, the exposure time can drop to 1 ms or less.

Usual monomers and oligomers are based on acrylates, for example, **1–5** and **6–9**. Their viscosities range from 10 (hexanediol diacrylate (HDDA)), 15 (tripropyleneglycol diacrylate (TPGDA)), 115 (trimethylolpropane triacrylate (TMPTA)) to 500/50 000 (polyester acrylate), 7500/25 000 (epoxy acrylate), 3500/45 000 mPa s^{-1} (polyurethane acrylate). Sometimes, they possess a strong odor and exhibit skin-irritating properties. Unsaturated polyester resins dissolved in styrene are well known in the curing of glass-reinforced materials.

New structures include

1) Highly reactive acrylate monomers (HRAMs). They correspond to labile hydrogen-containing monofunctional monomers (carrying, e.g., an oxazolidone, a carbamate, a carbonate, an ester, or an ether group). An example concerns the lactone-derived acrylate where the labile hydrogen is located at the α-carbon of the lactone carbonyl group [142] (**10–12**). The enhanced performances were discussed in terms of hydrogen abstraction/chain transfer, hydrogen bonding, electronic, and resonance effects [143, 144].

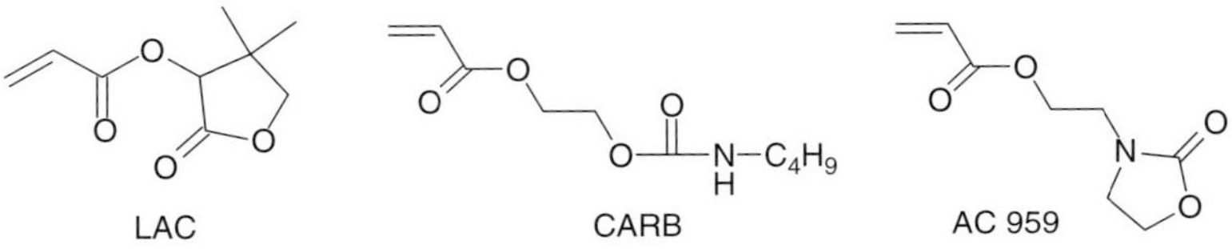

LAC   CARB   AC 959

10–12

2) Starlike or treelike hyperbranched monomers (exhibiting many functionalities, lower viscosity, and enhanced pigment wetting) [145], thioether linkage containing hyperbranched polyester acrylates (for a less sensitivity to oxygen) [146].
3) Liquid crystalline acrylates [147, 148] and macroscopically oriented liquid crystalline elastomers [149].
4) Ionic liquid monomers (Eq. (4.2)) (see a review in [150]).

$$[R_3NH^+][CH_2=CHCOO^-] \quad [R_3NH^+][CH_2=C(CH_3)-COO^-]$$
$$[R_2(CH_2=CRCOO)NH^+][CH_2=CHCOO^-] \qquad (4.2)$$

5) Phosphonated [151] or fluorinated [152] methacrylates, acrylate containing cyclic acetals [153], multifunctional pyrrolidone methacrylates [154], bicomponent acrylates [155], acrylated telechelic polysulfones [156], sulfide group containing acrylated oligomers [157], silicone acrylates [1, 2, 5], methacryloxy-functionalized benzoxazines [158], and alkali-soluble photosensitive polysiloxane urethane acrylates [159].

A growing interest is noted in the use of new starting monomer structures for getting novel end-use properties:

1) Low-toxicity monomers based on ethoxylated TMPTA, propoxylated neopentylglycol diacrylate, and glyceryl propoxy triacrylate [5].
2) Renewable monomers (acrylated vegetable soybean oils [160]), natural or naturally derived products (photo-cross-linkable polylactides [161], ε-caprolactone (CL) leading to a polycaprolactone (PCL) structure (**13 and 14**)), biocompatible [162] and biodegradable materials (e.g., poly (lactide-co-ethylene oxide-co-fumarate) for tissue regeneration [163], methacrylate-based gelatin derivatives [164], acrylate-modified starch [165], and itaconic-acid-based

photocurable polyesters (for biomedical applications) [166].

$$\underset{\textbf{13}}{\underset{\text{CL}}{\text{[cyclic caprolactone]}}} \qquad \underset{\textbf{14}}{\underset{\text{PCL}}{+\!\!\text{O}-(\text{CH}_2)_5-\overset{\text{O}}{\overset{\|}{\text{C}}}\!\!+_{\!n}}}$$

3) Reworkable materials [167, 168]. As photo-cross-linked polymers consist of infusible and insoluble networks, their removal from strong scratching or chemical treatments usually leads to damage of the underlying substrate. Therefore, the preparation of a starting material containing cross-linkable units and thermally degradable moieties allows a further destruction of the coating on baking.

Oxygen inhibition significantly affects the top surface layer (over ~1–10 μm) and the network structure in terms of physical properties as confirmed by a recent investigation using advanced surface characterization techniques [169]. In films up to 100 μm, oxygen leads to a decrease in the reaction rate and a spatial anisotropy of the polymer yield [170]. Many efforts have been made to overcome the O_2 inhibition (Section 7.5) by using various strategies [171–173]. The introduction of less sensitive monomers such as allyl functionalized derivatives has also been proposed.

Photopolymerization of thick clear coatings (centimeter range) can be achieved provided a suitable photoinitiating system is used. Indeed, the penetration of the light in the depth of the film is favored if an important photobleaching occurs. On the other hand, the photopolymerization of thick pigmented coatings appears as a real challenge owing to the strong absorption of the pigment. High-performance suitable photoinitiating systems have been, however, proposed for the curing of heavily pigmented paints (a few hundred micrometers) usable on industrial lines (Section 10.2.10).

The addition of UV absorbers (UVAs) and sterically hindered amine light stabilizers (HALSs), useful for the further photostabilization of the coating under outdoor exposure, reduces the curing speed to a small extent (Section 7.6).

The substrate can have a specific influence on the cure speed (Section 7.7; for example, phenolic compounds that are inherently present in cellulosic materials sharply decrease the photopolymerization efficiency in wood coatings [174]).

Outstanding properties can be achieved if the matrix is properly selected [1, 2, 5, 175], for example: high Tg and low shrinkage for isobornyl acrylates, high flexibility in ethoxylated TMPTAs, high tensile strengths in polyether acrylates, hardness and flexibility in carbamate acrylate monomer/epoxy acrylates, and good thermomechanical properties [7]. Adhesion is improved [2, 5, 175] by a surface treatment with chemicals or plasmas, the use of adhesion promoters, the chemical modification of monomers and oligomers, the adequate formulation, and selection of existing compounds.

4.2.2
Film Cationic Photopolymerization

The cationic photopolymerization in film is fast, usually insensitive to oxygen (except in free-radical-promoted cationic photopolymerization (FRPCP); Section 12.5), sensitive to moisture and water. An important dark posteffect is noted (Section 7.1.1). The range of available monomers and oligomers, for example, **15–25** has been largely expanded. Epoxides and VEs are very widely used and are known to give coatings with high thermal capability, excellent adhesion, good chemical resistance, and environmentally friendly character [176]. VEs (e.g., diethyleneglycol divinyl ether) polymerize faster than epoxides (e.g., cycloaliphatic diepoxide) [4]. Cyclic ethers such as oxetanes [177–179] can be alternatives to epoxides for getting faster curing speeds in industrial lines.

15–25

The reaction rate can be increased through the use of additional initiation processes originating from the aryl free radical formed in the iodonium salt photodecomposition [180] (Section 12.2.1.5). A proper selection of cationic monomers possessing readily abstractable hydrogen atoms helps in the design of high-performance systems. Recent developments include hybrid monomers (epoxide-vinyl ethers, epoxide-propenyl ethers, epoxide-acetals) and monomers bearing functional groups [181].

Recent works have explored many aspects of the cationic photopolymerization reaction:

1) Reactivity of various monomers: alkyl VEs [182], acetal-group-bearing epoxides, phosphonate-containing divinyl ethers [183], thiophenes [184], vinyl cyclopropanes [185], carbazole monomers [186], hyperbranched monomers [187], oxiranes [188], mono- and bifunctional epoxides [189], and bisphenol-A-based VEs [190].
2) Role of thick media [191].

3) Manufacture of metal containing epoxy nanocomposites (e.g., Zr [192]).
4) Design of new monomers such as benzoxazines, thiocarbonates, functionalized epoxides, and VEs, expanding monomers (when incorporated into a classical formulation, polycyclic monomers such as spiro orthocarbonates lead to a linear polymer chain on irradiation, which results in a molecular expansion and reduces the shrinkage of the coating; see, e.g., in [193]), oxetane end-capped hyperbranched polyesters [194], perfluoropolyethers [195], liquid fullerene derivatives [196], and trioxaspiro [4,4] nonanes [197].
5) Presence of pigments (e.g., carbon black nanoparticles [198]).

Epoxy-modified silicone monomers can be polymerized using cationic PIs [1, 5, 199]. High cure speeds are attained as in acrylate-modified silicones. The final properties correspond to those of a silicone polymer (excellent temperature, chemical and environmental resistance, and good electrical characteristics).

Organic–inorganic hybrid resins based on a siloxane structure bearing pendant epoxy groups display a high reactivity in photoinitiated cationic ring-opening polymerization. The mechanical properties are related to the length and type of the spacer group between the epoxy group and the siloxane chain [200].

Efforts toward the use of renewable monomers have to be outlined. Promising monomer candidates are epoxidized sunflower for oil/organo clay nanocomposite coatings [201] and hydroxystearic acid nanofibers [202], epoxidized soybean oil (ESO) (**26** and **27**), epoxidized linseed oil, epoxidized vernonia oil, epoxidized castor oil [203] limonene dioxide (LDO) [204], epoxidized natural rubbers [205], and epoxidized vegetable oils [206]. Biodegradable polymers also receive more and more important attention.

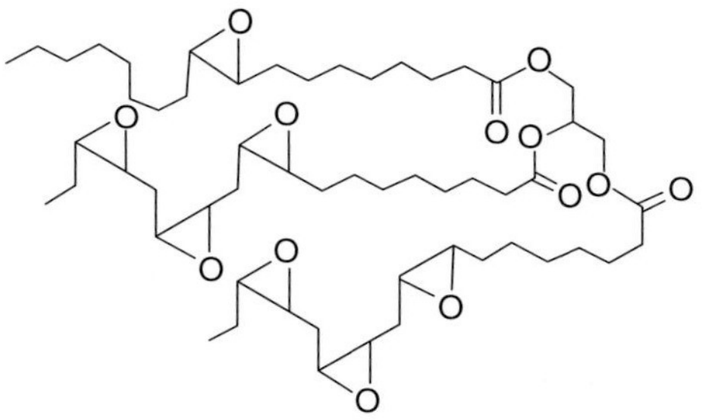

ESO
26

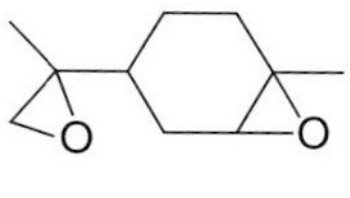

LDO
27

4.2.3
Thiol-ene Photopolymerization

As reviewed in [207], thiol-ene photopolymerization shows some interesting features: very fast process, low or even no oxygen inhibition effect, and formation of highly cross-linked networks with good adhesion, reduced shrinkage and stress, and improved physical and mechanical properties. In the past years, it has known a noticeably new revival of interest. Thiol-ene photopolymerization is based on a stoichiometric reaction of multifunctional olefins and thiols, for example, thiol-vinyl ether, thiol-allyl ether, thiol-acrylate, thiol-yne, or thiol-allyl ether-acrylate [51, 208–218].

A lot of olefins (allyl ethers, VEs, allylic urethanes, ureas and phosphazenes, vinyl-functionalized silicones, and norbornenes (NORs)) and di-, tri-, and tetrafunctional thiols have been proposed. Examples are shown in structures **28–36**. Introducing a tetrafunctional thiol and a difunctional alkyne in thiol-yne photopolymerizations [219] allows formation of an inherently higher cross-link density network displaying a higher Tg and rubbery modulus.

The reaction refers to the step-growth addition of a thiol to the double bond (vinyl, allyl, acrylate, and methacrylate). The invoked mechanism is recalled in Eq. (4.3). The polymerization reaction in PI-free conditions was recently revisited. This last reaction is helpful when a non-UV-absorbing coating must be obtained because, in that case, no PI decomposition products are obviously present in the cured coating.

In the absence of a UV PI, the reaction proceeds, however, more slowly.

Initiation: PI $\to$ PI* $(h\nu)$ and PI* + RSH $\to$ PIH$^\bullet$ + RS$^\bullet$
Propagation: RS$^\bullet$ + –CH=CH$_2$ $\to$ –CH$^\bullet$–CH$_2$SR
and –CH$^\bullet$–CH$_2$SR + RSH $\to$ –CH$_2$–CH$_2$SR + RS$^\bullet$
Termination: 2 RS$^\bullet$ $\to$ RS–SR;
2 –CH$^\bullet$–CH$_2$SR $\to$ RSCH$_2$CH–CH–CH$_2$SR;
–CH$^\bullet$–CH$_2$SR + RS$^\bullet$ $\to$ RS(–)CH–CH$_2$SR   (4.3)

Many papers are currently devoted to the study of the following points:

1) Propagation mechanism of the polymerization reaction and the formation of the polymer network [220], behavior of liquid crystalline thiol-enes [221], and stress relaxation via an addition-fragmentation chain transfer in thiol-ene systems [222].
2) Photopolymerization of pigmented thiol-enes [223], thiol-acrylates (for the synthesis of PEG-peptide hydrogels) [224], linear polymers attached to silica nanoparticles [225], and thiol-acrylate nanocomposites [226].
3) Oxygen inhibition in thiol-acrylate photopolymerization [227].
4) Mechanisms and kinetics of the thiol-vinyl, thiol-acrylate, and thiol-yne photopolymerization [228–231].
5) Analysis of the network formed from ternary thiol-vinyl-acrylates [232–235], thiol-isocyanate-enes [236], or thiol-enes [237]; effect of the chain length distribution [238]; influence of the thiol structure [239–241].
6) Synthesis of allyl-ether-functionalized dendrimers [242], novel second-generation dendrimers [243], and hyperbranched dendrimers containing thio-enes [244].
7) Photorheology and photoplasticity of thio-ene allylic sulfide networks [245, 246]; solvent dependence morphologies in thiol-ene systems [247].
8) Rapid prototyping of multilayer thiol-ene microfluidic chips [248].
9) Development of thiol-ene click functionalization [249]; cleavage of surface-bound linear polymers formed using thiol-ene photopolymerization [250]; formation of shape memory polymers from thiol-enes [251]; fabrication of networks for membrane carbon dioxide separation [252].
10) Aging of thiol-ene networks [253].

All these systems present an interest in the field of coatings, adhesives, laser imaging, and optics (the thiol moieties allow a high refractive index of the cured system). The problem of the smelling of thiols is nowadays solved by using new high-molecular-weight compounds and is not important in the case of a grafted system. The thiol-isocyanate (or urea)-acrylate ternary systems have a higher glass transition temperature and refractive index [254].

On the basis of the same principle, the first approach of silane-ene and silane-acrylate photopolymerization has been proposed [255] (Section 9.1.2.7).

4.2.4
Photopolymerization of Water-Borne Light Curable Systems

Owing to the current pressure of new environmental regulations [5, 256], the interest in solventless systems increases. Water-based formulations are less viscous than conventional acrylate mixtures. When used on porous substrates such as wood, the formulation swells the wood fibers and increases the wood–resin interface resulting in better adhesion. These systems are available, for example, **37** as water-soluble media (urethane acrylates with low molecular weights, water-soluble polyether acrylates, diethylamino ethyl acrylate, and vinyl caprolactam), emulsions (ethylene-vinyl acetate-methacrylate or methacrylate-butylacrylate-glycidylmethacrylate), dispersions (polyethylene glycol diacrylate, butanediol monoacrylate, polypropylene glycol monoacrylate; urethane acrylates, epoxy acrylates, or polyester acrylates stabilized with an added surfactant, or functionalized with a surfactant moiety). Reactive monomer/oligomer cross-linkers can be added.

37

Tack-free clear coatings are obtained using, for example, the photopolymerization of an aqueous polyurethane dispersion in the presence of a radical PI. Desiring, however, good surface properties (hardness and chemical resistance) requires the drying of the coating before light exposure. Clear water-borne latexes for outdoor applications cross-link under sunlight via an oxidative mechanism. Sunlight curing of pigmented water-borne paints under air has been successfully achieved using a latex and a cross-linker [257]. Works are in progress, for example, on the rapid prototyping of cellular materials [258] and the design of silane-monomer-modified styrene/acrylate microemulsion coatings [259].

4.2.5
Photopolymerization of Powder Formulations

Powder coatings are very attractive compared to liquid finishes since they are VOC-free and easier to transport and handle [5]. The thermal cross-linking of conventional systems occurs at high temperature (200 °C). The interest in using UV powder coatings are (i) easy handling, (ii) no use of solvent, (iii) low-space equipment, (iv) fast drying, and (v) two-step procedure. The powder is applied onto the substrate and exposed to an IR source to get the coalescence of the solid particles. Then, the cross-linking reaction is carried out under UV (or visible) lights in a very short time and at a lower temperature (100–120 °C). This allows use of these coatings on heat-sensitive substrates such as wood (e.g., for medium-density fiber (MDF) board). Applications are discussed in [260, 261].

The nature of the resin governs the final properties of the cured coating: alkydes (adhesion, brightness, flexibility), epoxides (mechanical properties, resistance to solvents and corrosion, insulating properties), polyurethanes (light fastness), and unsaturated polyesters (hardness). Largely encountered systems are based on polyesters (with maleate or fumarate unsaturations) and polyurethane/VEs, which copolymerize according to a radical process. A combination of amorphous and functionalized polyesters with allyl ethers has been proposed. Allyl ethers are cheaper, available in a wide range of functionalities and structures, thereby allowing a tuning of the final properties [262].

4.2.6
Charge-Transfer Photopolymerization

A CT photopolymerization refers to a reaction between a monomer containing an electron poor moiety (acceptor molecule A) and a monomer bearing an electron-rich moiety (donor molecule D). A charge-transfer complex (CTC) whose absorption is red-shifted (Sections 5.2 and 9.3.2) is formed between A and D (Eq. (4.4)). A CTC can either (i) exist in its ground state and then be excited (1) or (ii) be generated in the excited state after reaction between, for example, an unexcited donor and the excited acceptor (2).

$$D + A \rightarrow [D^{\delta+} \ldots A^{\delta-}] \quad \text{and} \quad [D^{\delta+} \ldots A^{\delta-}] \rightarrow [D^{\delta+} \ldots A^{\delta-}]^* \ (h\nu) \quad (1)$$
$$A \rightarrow A^* \ (h\nu) \quad \text{and} \quad D + A^* \rightarrow [D^{\delta+} \ldots A^{\delta-}]^* \quad (2) \quad (4.4)$$

For example, a maleimide (absorption around 300–350) and a VE (absorption below 250 nm) can act as an acceptor and a donor, respectively (**38–40**). The CTC absorbs above 350 nm.

$$\text{HBVE} \quad \text{PMI} \quad \text{TMPAE}$$

38   **39**   **40**

The photopolymerization reactions of maleimide-VEs exhibit two interesting features [263–267]: (i) very high cure speeds (as those obtained with acrylates) and (ii) no or low sensitivity to oxygen. Using VEs appears as an environment-friendly alternative to acrylate monomers (∼no smell and no skin and eye irritation).

In a PI-free formulation, the initiation is thought to occur through a CTC as explained above. After excitation, the generated intermediate might be a biradical •D–A•, a zwitterion D^+–A^-, or a radical ion pair $D^{•+}$–$A^{•-}$. The polymerization efficiency sharply increases, however, in the presence of a UV or visible radical PI. Homopolymerization and cross propagation are invoked to explain the formation of alternate copolymers in VE/acceptor monomer M (Scheme 4.1). In

Homopolymerization

$$PI \xrightarrow{h\nu} R^{\bullet} \xrightarrow{[VE \ldots M]} R-VE-M^{\bullet} \xrightarrow{[VE \ldots M]} R-VE-M-VE-M^{\bullet}$$

Cross propagation

$$PI \xrightarrow{h\nu} R^{\bullet} \xrightarrow{VE} R-VE^{\bullet} \xrightarrow{M} R-VE-M^{\bullet} \xrightarrow{VE} R-VE-M-VE^{\bullet}$$

Scheme 4.1

maleate-VEs, the presence of an alternate copolymer is explained on the basis of a homopolymerization through a CTC rather than a cross propagation. A primary hydrogen abstraction process with the maleate has also been demonstrated to directly generate free radicals that can react with the monomer or participate in the propagation step of the polymerization through a chain transfer reaction. Papers are currently devoted to the study of the propagation of the polymerization reaction and the formation of the polymer network, the kinetics of the reaction, the primary processes (hydrogen abstraction), and the possible applications [268].

One drawback of such CT photopolymerizable systems is the high cost and the availability of the polyfunctional monomers (e.g., maleimides). PI-free and UV/visible PI-containing maleimide-allyl-ether systems exhibit a quite good efficiency; they are less expensive and less irritating than maleimide-VEs. The fast curing of a maleimide-allyl or -VE film formulation was achieved within 1 s in the presence of a four-component photoinitiating system under low light intensity [269]. Other systems include donor/acceptor vinyl monomers [270], silicon-containing maleimides [271], and bismaleimides [272].

4.2.7
Dual Cure Photopolymerization

The dual cure technique is a two-step process [1, 2, 27, 30, 273] corresponding to four possible strategies.

1) Two consecutive UV exposures under either (i) a low- and a high-intensity source (pregelification followed by complete curing) or (ii) a short- and a high-wavelength Hg lamp (surface cure followed by body cure).
2) Two consecutive processes (UV irradiation and thermal drying). This can be helpful in curing shadow areas. In that case, a typical dual cure system consists in a two-pack material, for example, based on a polyisocyanate and an acrylate. The acrylate is first cured on light exposure and the subsequent dark reaction of the isocyanates with the hydroxyl groups of the acrylates allows full curing of the coating, thereby producing a tack-free film.
   A recent example is based on a PI/benzoxazine combination that starts the photopolymerization reaction and introduces a benzoxazine moiety in the polymer chains [274]. A further thermal ring-opening process of the benzoxazine allows an additional cross-linking of the material.
3) UV curing followed by (room temperature) air drying.

4) UV curing and catalyzed moisture curing.

Examples of dual cure also include the cationic photopolymerization of renewable monomers (e.g., partially epoxidized soilbean oil [275]).

4.2.8
Hybrid Cure Photopolymerization

The hybrid cure technique [276, 277] corresponds to the photopolymerization of a blend of two monomers/oligomers Mc and Mr exhibiting a different chemistry (here cationic for Mc and radical for Mr). Mc and Mr obviously react through two kinds of mechanisms on two different light exposures ν_c and ν_r [278–282] according to either a simultaneous one-step or a consecutive two-step irradiation procedure of a blend of Mc, Mr, radical PI, and cationic PI (Section 12.8). Cross-linked copolymers can thus be formed as an IPN combining the properties of the two polymer backbones.

Such a blend of an acrylate and a vinyl monomer in the presence of a radical and a cationic PI will lead to a different network compared to those obtained in separate radical and cationic formulations. The network structure also depends on the one- or two-step procedure. The starting point of the reaction is driven by a selective and separated excitation of the two PIs depending on the wavelengths used (see, e.g., in [205, 283–286] and references therein). A hybrid cure has been already applied to various epoxides (or epoxidized materials, VEs, and cycloepoxides)/acrylates and acrylate/hexamethoxymethyl melamines.

The decrease in the oxygen inhibition [287] has also been pointed out. The first example of an IPN formed in a thiol-ene radical/-ene cationic two-step hybrid photopolymerization using a trithiol and a trivinyl ether has been recently reported [288, 289].

4.2.9
Anionic Photopolymerization

The use of anionic photopolymerization is almost unknown in the industrial applications of the radiation curing area. Indeed, owing to the high reactivity of the propagating species (Eq. (4.5)), the anionic reaction has to be carried out in the absence of impurities and compounds sensitive to nucleophiles. Polycyanoacrylates are obtained according to this process [290, 291]. Metallocenophanes such as silicon-bridged [1]ferrocenophanes [292] have also been reported to be candidates for such a reaction. They lead to a class of metallopolymers (exhibiting a controlled response to redox stimuli, an etch resistance to plasma, high refractive indices, a behavior of precursor to nanostructured magnetic ceramics, or catalysts of carbon nanotube growth). Anionic PIs are discussed in Chapter 13.

$$\text{Photoinitiator (AC)} + M \rightarrow AM^- \ C^+ \ (h\nu)$$
$$AM^-C^+ + M \rightarrow AM_2^- \ C^+$$
$$A(M)_n^- \ C^+ + M \rightarrow A(M)_{n+1}^- \ C^+ \tag{4.5}$$

4.2.10
Metathesis Photopolymerization

Few examples have been reported in the past. They concern (Eq. (4.6)), for example, the photopolymerization of (i) alkynes by W(CO)$_6$ (the irradiation leads to an unstable (CO)$_5$W(η2-alkyne) complex from which the polymerization proceeds; the mechanism is rather complicated) [293] and (ii) cyclopentene using amine or carbene complexes of pentacarbonyltungstene/TiCl$_4$ [294].

$$W(CO)_6 + RCCH \rightarrow W(CO)_5-(RCCH) + CO \; (h\nu)$$
$$W(CO)_5-(RCCH) + nRCCH \rightarrow \rightarrow \rightarrow -(RC=CH)_x- \quad (4.6)$$

The ethylene photopolymerization initiated by tetra-neopentyltitanium (Neo)$_4$Ti most certainly occurs through a coordinated radical mechanism (Eq. (4.7)) [295] where the photolysis of the catalyst is assumed to lead to homolysis.

$$(Neo)_4Ti \rightarrow (Neo)_3Ti + Neo \; (h\nu)$$
$$(Neo)_3Ti + M \rightarrow (Neo)_3TiM$$
$$(Neo)_3TiM \leftrightarrow [(Neo)_3Ti + M^\bullet]$$
$$[(Neo)_3Ti + M^\bullet] + M \rightarrow [(Neo)_3Ti + (M)_2^\bullet] \leftrightarrow (Neo)_3Ti\,(M_2) \quad (4.7)$$

Ring-opening metathesis photopolymerization (ROMP) of cyclic monomers (e.g., NOR) has been observed in the presence of bis (arene) Ru (II) complexes Ar-Ru-Ar [296] (Eq. (4.8)) or Ru (II)-N-heterocyclic carbene complexes [297].

$$Ar\text{-}Ru\text{-}Ar \rightarrow Ar\text{-}Ru\text{-}(Solv)_3 + Ar \; (\text{light, solvent})$$
$$Ar\text{-}Ru\text{-}(Solv)_3 \rightarrow Ru\text{-}(Solv)_6 + Ar$$
$$Ru\text{-}(Solv)_6 + NOR \rightarrow ROMP \; (\text{fast})$$
$$Ar\text{-}Ru\text{-}(Solv)_3 + NOR \rightarrow ROMP \; (\text{slow}) \quad (4.8)$$

4.2.11
Controlled Photopolymerization Reactions

Controlled or living radical photopolymerizations are the most encountered reactions compared to cationic or anionic controlled/living reactions.

4.2.11.1 Radical Photopolymerization Reactions
A conventional radical polymerization terminates when the reactive intermediates are destroyed or rendered inactive through bimolecular reactions. The radical generation is irreversible without any possibility of controlling the final properties of the polymer formed such as the number average molecular weight M_n and the molecular weight distribution (MWD). A highly uniform and well-defined product is therefore difficult to achieve. The same holds true in specialized applications (where a control is necessary) such as for the design of block copolymers or star copolymers, the preparation of end functional polymers, the development of multilayer systems, or microfluidic devices.

Polymer — X ⇌ Polymer$^\bullet$ + X$^\bullet$ ⟶ Polymerization
(Dormant chain)   (Living chain)

Scheme 4.2

In thermal-controlled (or living) radical polymerization [298, 299], the strategy consists in avoiding the simultaneous initiation, propagation, and termination steps by temporarily blocking (and in a reversible way) the growing polymer chain (Scheme 4.2). The radicals are reversibly generated according to three largely used methodologies [300]: atom transfer radical polymerization (ATRP), reversible addition-fragmentation transfer (RAFT), and nitroxide-mediated polymerization (NMP). The use of iniferters is also encountered.

Few things are relatively known in the control of photopolymerization reactions although strong interests are noted, for example, for the applications in photografting or the integrated surface modification of microfluidic devices [301].

For example, in a light-induced reaction, a photoiniferter [302–307] that behaves as a PI, a transfer agent, and a chain terminator can be used. Both the initiation and the reversible termination are photoinduced. The mechanism is recalled in Eq. (4.9): cleavage of the photoiniferter structure R–DC to produce an initiating R$^\bullet$ and a terminator DC$^\bullet$ radical (usually a dithiocarbamyl radical), usual addition of the initiating radical to the first monomer unit and propagation of the polymerization. The crucial difference from a noncontrolled radical polymerization process lies in the preferred termination of the reaction by the DC$^\bullet$ radical. This reaction is photochemically reversible and brings the living character of the polymerization (providing that the exchange between the reactive and the dormant species is fast in comparison with the propagation). The molecular weight increases with the conversion, in contrast with classical radical photopolymerizations. Experimental as well as theoretical works will probably help in the design of high-performance structures (Sections 8.30 and 18.19). The insertion of the dormant species (R-M$_n$-DC) in a polymer matrix is worthwhile for many reactions. The synthesis of copolymers (e.g., PMMA-polystyrene) becomes achievable. Applications in viscous monomer/oligomer media, such as films, might also be of interest.

$$R\text{-}DC \rightarrow R^\bullet + DC^\bullet \ (h\nu)$$
$$R^\bullet + M \rightarrow RM^\bullet \quad \text{and} \quad RM^\bullet + M \rightarrow \rightarrow R(M)_n^\bullet$$
$$R(M)_n^\bullet + DC^\bullet \rightleftarrows R(M)_n\text{-}DC \ (h\nu) \qquad (4.9)$$

Photo NMP or NMP^2 is almost unknown [308–310]. The most recent real way toward NMP^2, as described in Section 8.30, is satisfactory and promising [311–313]. It consists in starting with a cleavable PI being able to form an initiating radical R$^\bullet$ and a nitroxide radical for the control (Scheme 4.3). The nitroxide plays the same role as DC$^\bullet$ in the photoiniferter.

The nitroxide-mediated photo-living radical polymerization of acrylates is discussed in [314, 315] but the mechanisms are unclear. In particular, the nature of the initiating and controlling agents remain the subject of discussion [316].

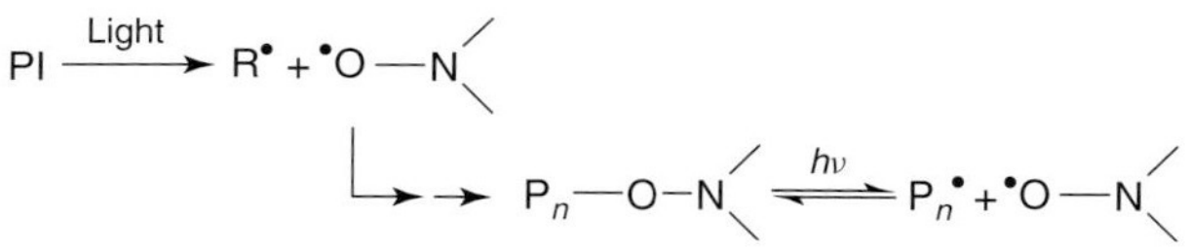

$$\text{PI} \xrightarrow{\text{Light}} R^\bullet + {}^\bullet O-N\begin{matrix}/\\\backslash\end{matrix}$$

$$\hookrightarrow P_n-O-N\begin{matrix}/\\\backslash\end{matrix} \xrightleftharpoons{h\nu} P_n^\bullet + {}^\bullet O-N\begin{matrix}/\\\backslash\end{matrix}$$

Scheme 4.3

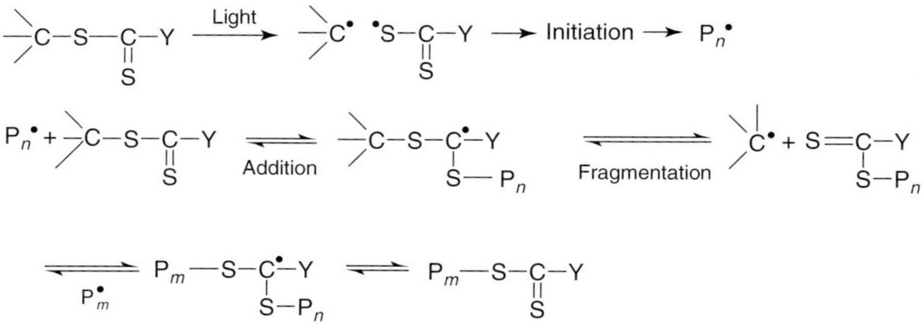

$$\begin{matrix}\backslash\\/\end{matrix}C-S-\underset{\underset{S}{\|}}{C}-Y \xrightarrow{\text{Light}} \begin{matrix}\backslash\\/\end{matrix}C^\bullet \quad {}^\bullet S-\underset{\underset{S}{\|}}{C}-Y \longrightarrow \text{Initiation} \longrightarrow P_n^\bullet$$

$$P_n^\bullet + \begin{matrix}\backslash\\/\end{matrix}C-S-\underset{\underset{S}{\|}}{C}-Y \xrightleftharpoons[\text{Addition}]{} \begin{matrix}\backslash\\/\end{matrix}C-S-\underset{\underset{S-P_n}{|}}{C^\bullet}-Y \xrightleftharpoons[\text{Fragmentation}]{} \begin{matrix}\backslash\\/\end{matrix}C^\bullet + S=\underset{\underset{S-P_n}{|}}{C}-Y$$

$$\xrightleftharpoons[P_m^\bullet]{} P_m-S-\underset{\underset{S-P_n}{|}}{C^\bullet}-Y \xrightleftharpoons{} P_m-S-\underset{\underset{S}{\|}}{C}-Y$$

Scheme 4.4

$$\text{Met} \xrightarrow[RX]{\text{Light}} R^\bullet + \text{Met}^+ X^- \xrightleftharpoons{} R\underset{Z}{\diagdown\diagup} {}^\bullet + \text{Met}^+ X^- \xrightleftharpoons{} R\underset{Z}{\diagdown\diagup}{}^X + \text{Met}$$

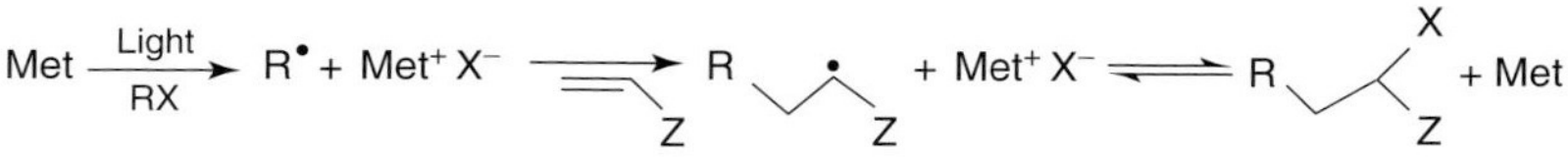

Scheme 4.5

In photo RAFT, a suitable cleavable PI leads to two radicals $R^\bullet$; at least, one of them should initiate the polymerization. The control is achieved through an addition/fragmentation process involving the growing polymer chain $P_n^\bullet$ (see the principle in Scheme 4.4). Examples are shown in [317–325].

Basically, in photo ATRP, a transition-metal derivative Met both acts as a PI and reacts with an alkyl halide RX through a redox process (Scheme 4.5). The control is owing to the repetition of this process. Examples of applications are shown in [161, 326–337]. Incorporation of a trithiocarbonate moiety into a main polymer chain allows mediation of the controlled insertion photopolymerization of styrene [338].

4.2.11.2 Cationic Photopolymerization Reactions

A few examples of photoinitiated controlled/living polymerization have been reported (Sections 12.7 and 13.4), for example, in (i) cyclic monomers [339] (oxolane, tert-butylazirdine, ethers) in the presence of iodonium or sulfonium salts (the living polymerization occurs photochemically until the salts are consumed or the light is switched off and then can continue through a thermal process) and (ii) isobutyl VE using iodonium salts and zinc halides [340] (the growing carbocations are stabilized because of the high nucleophilicity of the salt anion/$ZnCl_2$ association that prevents the chain-breaking processes). These reactions exhibit a living character rather than a control.

4.2.11.3 Anionic Photopolymerization Reactions

Rare examples are available. Photocontrolled living anionic ring-opening polymerization of silicon (or phosphorus)-bridged ferrocenophanes was reported [341–343] (Chapter 3). The anionic photopolymerization becomes a living reaction provided that a mild nucleophile is added as an initiator, the reaction being controlled by simply switching on/off the light.

4.2.12
Hybrid Sol–Gel Photopolymerization

Interesting alternative materials based on cross-linked glasslike matrices bearing Si-O-Si linkages can be synthesized from a photosensitive hybrid sol gel (HSG) formulation. They combine the properties of glasses (high refractive index, heat, and mechanical properties) and polymers (flexibility and versatility) [344, 345]. The incorporation of organic, organometallic, or bio-organic molecules still extends the scope of applications. Such a HSG system consists, for example, in an acrylated silicon alkoxide derivative $Si(-OR)_3$-M and a PI that allows the free radical photopolymerization of the acrylate M (the requirements for the choice of the PI in such a matrix are crucial toward solubility, thermal stability, pH compatibility, and polarity). The reaction can be carried out, for example, according to a consecutive two-step procedure. First, it involves a conventional acid- or base-catalyzed hydrolysis of the alkoxysilane moieties leading to M–Si(–OR_2)–O–Si(–OR)$_2$–M. Then, the presence of the acrylate monomer M allows a further photoinduced densification of the gel [346].

A single-step procedure was recently proposed. It is based on the concomitant UV curing of a suitable organic–inorganic matrix (e.g., an epoxy-modified alkoxysilane) where the protic acid generated from a cationic PI both catalyzes the sol–gel reaction and initiates the epoxy cationic polymerization [347]. This concept has been extended to the simultaneous use of a radical and a cationic PI for the one-step curing of a methacrylate trimethoxysilane matrix [348].

Applications to wood–inorganic composites [349], acrylic resin/titania systems [350], bridged polysilsesquioxane films [351], nanocomposite coatings [352], and clay/polyurethane nanocomposites [353] as well as the introduction of new properties (anisotropy, electrical, optoelectronic, or mechanical properties) [354] have been mentioned.

4.2.13
Photo-Cross-Linking Reactions in the Presence of Photobases or Photoacids

UV deblockable acid- or base-releasing systems (known as *photolatent systems*) allow starting of a cross-linking reaction on demand as the generation of the acid or the base is triggered by the light exposure. This was encountered, for example, in the curing of aminoplast resins, which is catalyzed by acids [355, 356]. In photoresist applications, for example, iminosulfonates (also referred to as photoacid generators (PAGs)) yield a sulfonic acid (Chapter 14) that induces the acid-catalyzed cross-linking of epoxy-group-bearing polymers (Scheme 4.6a) [357].

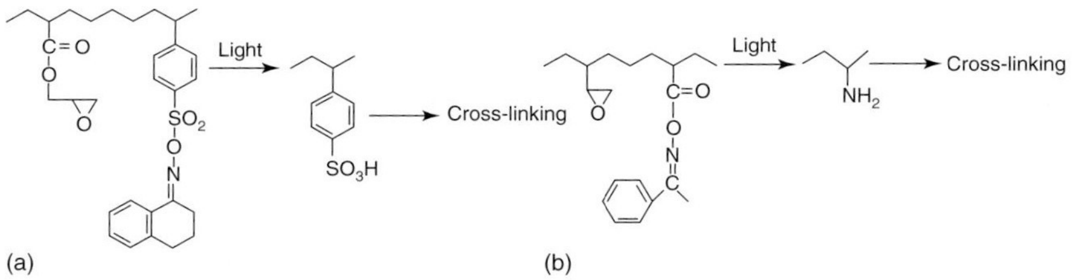

Scheme 4.6

In the same way, acyloxime derivatives (defined as photobase generators (PBGs)) generate an amine on irradiation (Chapter 15), which can also cross-link the epoxy resins (Scheme 4.6b).

More recently, the photolatent technology was applied to the curing-on-demand of two-component polyurethane (2K-PUR), for example, for the curing of coatings on a complete car body at a low temperature [358]. This new chemistry involving a cross-linking between an isocyanate and a polyol [359] or an isocyanate and a thiol-functionalized polymer [360] in the presence of an amine was shown to produce a thio-urethane network. The amine is formed on light exposure of a suitable compound acting as a photolatent catalyst. The system has a low sensitivity to oxygen and a complete cure occurs in the shadowed areas. The curing of acetonate (or malonate)-acrylate system (working through a Michael-type addition base-catalyzed reaction) or epoxide-thiol-based resins was also reported [361, 362].

4.3
Two-Photon Absorption-Induced Polymerization

The sequential absorption of two photons is governed by the same one-photon selection rules. On the contrary, a two-photon absorption (TPA) process relates to the simultaneous absorption (via a virtual state) of two in-phase photons with a total energy close to the energy of the electronic transition of the molecule that is usually accessible through the classical one-photon absorption (Section 5.2). The TPA-induced polymerization is based on this nonlinear optical property of the absorbing materials and can generate polymeric microstructures and nanorods [363] (Section 3.11). Extensive works are carried out on the design of suitable TPA compounds (Section 18.20). Recent papers review some of the applications [364].

4.4
Remote Curing: Photopolymerization without Light

In a remote curing process, no light is used. One interesting application is in the polymerization of coatings in the hidden parts of a substrate or the inside of a tube or a container. This can be exemplified through two particular situations. The first one refers to a thermal process in suitable systems that produce an electronically excited molecule from which initiating species are formed. In such a case [365], a reaction between an oxalate ester and hydrogen peroxide is able to produce a high-energy-content molecule, which decomposes into radicals (similar to those resulting from a direct light excitation) that further add to an acrylate monomer. The second one involves a trialkyl borane (which is released *in situ* from an amine–borane complex with a stream of a carrier gas) that further reacts with oxygen and generates alkyl radicals usable for the initiation of the acrylate polymerization (Eq. (4.10)) [366].

$$\text{Amine borane} \rightarrow \rightarrow \rightarrow R_3B$$
$$R_3B + O_2 \rightarrow \rightarrow \rightarrow R^\bullet + R_2\text{-BOO}^\bullet \tag{4.10}$$

4.5
Photoactivated Hydrosilylation Reactions

Hydrosilylation reactions of silicone polymers containing Si-H/Si-vinyl and Si-H/Si-epoxide groups have been often carried out in the past by homogeneous catalysis using β-dicarbonyl platinum complexes [367–370].

This allows, for example, the preparation of hydrophilic/hydrophobic-patterned surfaces [371]. A hydrophobic region is created in the irradiated area through hydrosilylation of the vinyl-terminated poly(dimethylsiloxane) with the Si-H groups present on the surface. Then, a hydrophilic region is obtained in the unexposed areas through the conversion of the remaining Si-H into Si-OH groups.

Recent radical-based hydrosilylation reactions involving silanes and double and triple bonds can be initiated through the primary formation of a silyl radical (Section 18.20) [372, 373].

References

1. Fouassier, J.P. (1995) *Photoinitiation, Photopolymerization, Photocuring*, Hanser, Münich.
2. Davidson, S. (1999) *Exploring the Science, Technology and Application of UV and EB Curing*, Sita Technology Ltd, London.
3. Neckers, D.C. (1999) *UV and EB at the Millenium*, Sita Technology Ltd, London.
4. Belfield, K.D. and Crivello, J.V. (eds) (2003) *Photoinitiated Polymerization*, ACS Symposium Series, Vol. 847, American Chemical Society, Washington, DC.
5. Schwalm, R. (2007) *UV Coatings: Basics, Recent Developments and New Applications*, Elsevier, Oxford.
6. Allen, N.S. (ed.) (2010) *Photochemistry and Photophysics of Polymer Materials*, John Wiley & Sons, Inc., Hoboken, NJ.
7. Ortega, A.M., Kasprzak, S.E., Yackacki, C.M., Diani, J., Greenberg, A.R., and Gall, K. (2008) *J. Appl. Polym. Sci.*, **110**, 1559–1572.
8. Mougharbel, A., Mallegol, J., and Coqueret, X. (2009) *Langmuir*, **17**, 9831–9839.
9. Dworak, D.P. and Soucek, M.D. (2006) *Surf. Coat. Int., B: Coat. Trans.*, **89**, 169–175.

10. Esen, H., Kusefoglu, S., and Wool, R. (2006) *J. Appl. Polym. Sci.*, **103**, 626–633.
11. Wei, H., Lee, T.Y., Miao, W., Fortenberry, R., Magers, D.H., Hait, S., Guymon, A.C., Joensson, S.E., and Hoyle, C.E. (2007) *Macromolecules*, **40**, 6172–6178.
12. Liu, T.-M., Zheng, Y.-S., Zhang, G.-C., and Chen, T. (2009) *J. Appl. Polym. Sci.*, **112**, 3041–3047.
13. Momotake, A. and Arai, T. (2004) *Polymer*, **45** (16), 5369–5390.
14. Asif, A. and Shi, W.F. (2004) *Polym. Adv. Technol.*, **15**, 669–675.
15. Guymon, C.A. and Norton, J.H. (2002) Proceedings of RadTech, pp. 338–348.
16. Decker, C. (2006) *Polym. Nanocompos.*, **188**, 205–210.
17. Kim, H.G., Han, D.H., Lim, J.C., Oh, D.H., and Min, K.E. (2006) *J. Appl. Polym. Sci.*, **6**, 3609–3615.
18. Tan, H. and Nie, J. (2007) *J. Appl. Polym. Sci.*, **106**, 2656–2660.
19. Owusu Adom, K. and Guymon, C.A. (2008) *Polymer*, **49**, 2636–2643.
20. Owusu-Adom, K. and Guymon, C.A. (2009) *Macromolecules*, **42**, 180–187.
21. Li, F., Zhou, S., You, B., and Wu, L. (2006) *J. Appl. Polym. Sci.*, **99**, 3281–3287.
22. Jonsson, S., Kalyanaraman, V., Lindgren, K., Swami, S., and Ng, L.T. (2003) *PSME Polym. Prepr.*, **44** (1), 7–8.
23. Monier, M., Wei, Y., Sarhan, A.A., and Ayad, D.M. (2010) *Polymer*, **51**, 1002–1009.
24. Bader, R.A. (2008) *Acta Biomater.*, **4**, 967–975.
25. Haraguchi, K. and Takada, T. (2010) *Macromolecules*, **43**, 4294–4299.
26. Xiong, H.M., Wang, Z.D., and Xia, Y.Y. (2006) *Adv. Mater.*, **18**, 748–751.
27. Studer, K., Decker, C., Beck, E., Schwalm, R., and Gruber, N. (2005) *Prog. Org. Coat.*, **53**, 126–133.
28. Zhou, H., Ly, Q., Lee, T.Y., Guymon, C.A., Jonsson, E.S., and Hoyle, C.E. (2006) *Macromolecules*, **39**, 8269–8273.
29. Zhou, H., Li, Q., Lee, T.Y., Guymon, C.A., Joensson, E.S., and Hoyle, C.E. (2006) *Macromolecules*, **39**, 8269–8273.
30. Broer, D.J. (1993) in *Radiation Curing in Polymer Science and Technology*, vol. 3 (eds J.P. Fouassier and J.F. Rabek), Elsevier, Barking, pp. 383–444.
31. Degirmenci, M., Hizal, G., and Yagci, Y. (2002) *Macromolecules*, **35** (22), 8265–8270.
32. Cakmak, I. and Ozturk, T. (2005) *J. Polym. Res.*, **12**, 121–126.
33. Durmaz, Y.Y., Kukut, M., Moszner, R., and Yagci, Y. (2009) *J. Polym. Sci., A: Polym. Chem.*, **47**, 4793–4799.
34. Temel, G., Aydogan, B., Arsu, N., and Yagci, Y. (2009) *J. Polym. Sci., A: Polym. Chem.*, **47**, 2938–2947.
35. Muftuoglu, A.E., Tasdelen, M.A., and Yagci, Y. (2006) in *Photochemistry and UV Curing* (ed. J.P. Fouassier), Research Signpost, Trivandrum, pp. 343–354.
36. Liu, L. and Yang, W. (2004) *J. Polym. Sci., B: Polym. Phys.*, **42**, 846–852.
37. Oh, J.K., Wu, J., Minnik, M.A., Craun, G.P., Rademacher, J., and Farwaha, R. (2002) *J. Polym. Sci., B: Polym. Chem.*, **40**, 1594–1607.
38. Shim, S.E., Jung, H., Lee, H., Biswas, J., and Choe, S. (2003) *Polymer*, **44**, 5563–5572.
39. Jain, K., Klier, J., and Scranton, A.B. (2005) *Polymer*, **46**, 11273–11278.
40. Peinado, C., Bosch, P., Martin, V., and Corrales, T. (2006) *J. Polym. Sci., A: Polym. Chem.*, **44**, 5291–5303.
41. Wan, T., Wang, Y.-C., and Feng, F. (2006) *J. Appl. Polym. Sci.*, **102**, 5105–5112.
42. Gao, F., Ho, C.-C., and Co, C.C. (2006) *Macromolecules*, **39**, 9467–9472.
43. Chemtob, A., Kunstler, B., Crouxte-Barghorn, C., and Fouchard, S. (2010) *Colloid Polym. Sci.*, **288**, 579–587.
44. Tan, J., Wu, B., Yang, J., Zhu, Y., and Zeng, Z. (2010) *Polymer*, **51**, 3394–3401.
45. Huang, Z., Sun, F., Liang, S., Wang, H., Shi, G., Tao, L., and Tan, J. (2010) *Macromol. Chem. Phys.*, **211**, 1868–1878.
46. Pioge, S., Nesterenko, A., Brotons, G., Pascual, S., Fontaine, L., Gaillard, C., and Nicol, E. (2011) *Macromolecules*, **44**, 594–603.

47. Lime, F. and Irgum, K. (2007) *Macromolecules*, **40**, 1962–1968.
48. Chen, J., Zeng, Z., Yang, J., and Chen, Y. (2008) *J. Polym. Sci., A: Polym. Chem.*, **46**, 1329–2338.
49. Deng, Y., Li, N., He, Y., and Wang, X. (2007) *Macromolecules*, **40**, 6669–6678.
50. Tomatsu, I., Hashidzume, A., and Harada, A. (2009) *Macromol. Chem. Phys.*, **210**, 1640–1646.
51. Trey, S.M., Nilsson, C., Malmstroem, E., and Johansson, M. (2010) *Prog. Org. Coat.*, **67**, 348–355.
52. Baker, M.V., Brown, D.H., Casadio, Y.S., and Chirila, T.V. (2009) *Polymer*, **50**, 5918–5927.
53. Jiang, Y., Liu, Y., Ma, Y., Liu, X., Qin, Z., Zhu, M., and Adler, H.J. (2008) *Macromol. Symp.*, **264**, 95–99.
54. Croute-Barghorn, C., Calixto, S., and Lougnot, D.J. (1997) *Proc. SPIE*, **2998**, 222–231.
55. Chen, Z. and Webster, D.C. (2007) *J. Photochem. Photobiol. A: Chem.*, **185**, 115–126.
56. Urech, L. and Lippert, T. (2010) in *Photochemistry and Photophysics of Polymer Materials* (ed. N.S. Allen), John Wiley & Sons, Inc., Hoboken, NJ, pp. 541–568.
57. Higashi, M. and Niwa, M. (1993) in *Radiation Curing in Polymer Science and Technology*, vol. 3 (eds J.P. Fouassier and J.F. Rabek), Elsevier, Barking, pp. 367–382.
58. Paul, R., Schmidt, R., Feng, J., and Dyer, D.J. (2002) *J. Polym. Sci., A: Polym. Chem.*, **40**, 3284–3291.
59. Janorkar, A.V., Proulx, S.E., Metters, A.T., and Hirt, D.E. (2006) *J. Polym. Sci., A: Polym. Chem.*, **44**, 6534–6543.
60. Chen, F., Jiang, X., Liu, R., and Yin, J. (2010) *ACS Appl. Mater. Interfaces*, **2**, 3628–3635.
61. Chen, F., Jiang, X., Liu, R., and Yin, J. (2011) *Polym. Chem.*, **2**, 614–618.
62. Rajaheyaganthan, R., Kessler, F., Leal, P., Kuehn, S., and Weibel, D.E. (2011) *Macromol. Symp.*, **299**, 175–182.
63. Chen, F.C. and Lackritz, H.S. (1997) *Macromolecules*, **30**, 5986–5996.
64. Esen, C., Kaiser, T., Borchers, M.A., and Schweiger, G. (1997) *Colloid Polym. Sci.*, **275**, 131–137.
65. Harrisson, S., Mackenzie, S.R., and Haddleton, D.M. (2003) *Macromolecules*, **36**, 5072–5075.
66. Andrzejewska, E., Podgoska-Golubska, M., Stepniak, I., and Andrzejewski, M. (2009) *Polymer*, **50**, 2040–2047.
67. Jimenez, Z., Bounds, C., Hoyle, C.E., Lowe, A.B., Zhou, H., and Pojman, J.A. (2007) *J. Polym. Sci., A: Polym. Chem.*, **45**, 3009–3021.
68. Chesnokov, S.A., Zakharina, M.Y., Shaplov, A.S., Chechet, Y.V., Lozinskaya, E.I., Mel'nik, O.A., Vygodskii, Y.S., and Abakumov, G.A. (2008) *Polym. Int.*, **57**, 538–545.
69. Zhou, H., Jimenez, Z., Pojman, J.A., Paley, M.S., and Hoyle, C.E. (2008) *J. Polym. Sci., A: Polym. Chem.*, **46**, 3766–3773.
70. Nonoguchi, Y., Nakashima, T., Sakashita, M., and Kawai, T. (2008) *Jpn. J. Appl. Phys.*, **47**, 1385–1388.
71. Batra, D., Hay, D.N.T., and Firestone, M.A. (2007) *Chem. Mater.*, **19**, 4423–4431.
72. Chesnokov, S.A., Zakharina, M.Y., Shaplov, A.S., Lozinskaya, E.I., Malyshkina, I.A., Abakumov, G.A., Vidal, F., and Vygodskii, Y.S. (2010) *J. Polym. Sci., A: Polym. Chem.*, **48**, 2388–2409.
73. Bongartz, N.A. and Goodby, J.W. (2010) *Chem. Commun.*, **46**, 6452–6454.
74. Dworjanyn, P.A. and Garnett, J.L. (1993) in *Radiation Curing in Polymer Science and Technology*, vol. 1 (eds J.P. Fouassier and J.F. Rabek), Elsevier, Barking, pp. 263–328.
75. Muftuogli, A.E., Tasdelen, M.A., and Yagci, Y. (2010) in *Photochemistry and Photophysics of Polymer Materials* (ed. N.S. Allen), John Wiley & Sons, Inc., Hoboken, NJ, pp. 509–540.
76. Luo, N., Hutchison, J.B., Anseth, K.S., and Bowman, C.N. (2002) *J. Polym. Sci., A: Polym. Chem.*, **40**, 1885–1891.
77. Susanto, H. and Ulbricht, M. (2007) *Langmuir*, **23**, 7818–7830.
78. Yang, H., Lazos, D., and Ulbricht, M. (2005) *J. Appl. Polym. Sci.*, **97**, 158–164.

79. Vicini, S., Princi, E., and Pedemonto, E. (2002) *Recent Res. Dev. Macromol.*, **6**, 59–79.
80. Wang, X., Song, X., Lin, M., Wang, H., Zhao, Y., Zhong, W., and Du, Q. (2007) *Polymer*, **48**, 5834–5838.
81. Derouet, D., Tran, Q.N., and Thuc, H.H. (2009) *J. Appl. Polym. Sci.*, **114**, 2149–2160.
82. Yamamoto, Y., Suksawad, P., Pukkate, N., Horimai, T., Wakisaka, O., and Kawahara, S. (2010) *J. Polym. Sci., A: Polym. Chem.*, **48**, 2418–2424.
83. Gao, J., Huang, B., Lei, J., and Zheng, Z. (2010) *J. Appl. Polym. Sci.*, **115**, 2156–2161.
84. He, D. and Ulbricht, M. (2009) *Macromol. Chem. Phys.*, **210**, 1149–1158.
85. Liu, N., Sun, G., Gaan, S., and Rupper, P. (2010) *J. Appl. Polym. Sci.*, **116**, 3629–3637.
86. Enright, T.P., Hagaman, D., Kokoruz, M., Coleman, N., and Sidorenko, A. (2010) *J. Polym. Sci., B: Polym. Phys.*, **48**, 1616–1622.
87. Hsu, Y.G., Su, X.W., Lin, K.H., Wan, Y.S., Chen, J.C., Guu, J.A., and Tsai, P.C. (2010) *J. Appl. Polym. Sci.*, **119**, 693–701.
88. Desilles, N., Lecamp, L., Lebaudy, P., and Bunel, C. (2003) *Polymer*, **44**, 6159–6165.
89. Falk, B., Zonca, M.R. Jr., and Crivello, J.V. (2005) *Macromol. Symp.*, **226**, 97–107.
90. Cui, Y., Yang, J., Zhan, Y., Zeng, Z., and Chen, Y. (2008) *Colloid Polym. Sci.*, **286**, 97–106.
91. Yayki, N., Lecamp, L., Desilles, N., and Lebaudy, P. (2009) *Macromolecules*, **43**, 177–184.
92. Otsu, T. (2000) *J. Polym. Sci., Polym. Chem.*, **38**, 2121–2127.
93. Kwak, J., Lacroix-Desmazes, P., Robin, J.J., Boutevin, B., and Torres, N. (2003) *Polymer*, **44**, 5119–5125.
94. Yoshida, E. (2009) *Colloid Polym. Sci.*, **287**, 767–772.
95. Yoshida, E. (2010) *Colloid Polym. Sci.*, **288**, 341–345.
96. Micciche, F., Long, G.J., Shahin, A.M., Grandjean, F., Ming, W., Van Haveren, J., and van der Linde, R. (2007) *Inorg. Chim. Acta*, **163**, 535–545.
97. Hubert, J.C., Venderbosch, R.A.M., Muizebelt, W.J., Klaasen, R.P., and Zabel, K.H. (1997) *J. Coat. Technol.*, **69**, 59–64.
98. Ye, G., Courtecuisse, F., Allonas, X., Ley, C., Crouxte-Barghorn, C., Raja, P., and Taylor, P. (2012) Gwenaelle Bescond *Progress in Organic Coatings*, **73**, 366–373.
99. Clapper, J.D., Sievens-Figueroa, L., and Guymon, C.A. (2008) *Chem. Mater.*, **20**, 768–781.
100. Forney, B. and Guymon, C.A. (2010) *Macromolecules*, **43**, 8502–8510.
101. Zhang, J., Xie, Z., Hua, S., Hoang, M., Hill, A.J., Min gao, W., and Kong, L.X. (2011) *J. Appl. Polym. Sci.*, **120**, 1817–1821.
102. Seki, T., Tanaka, K., and Ichimura, K. (1998) *Polym. J. Tokyo*, **30**, 646–652.
103. Schmelmer, U., Paul, A., Kueller, A., Jordan, R., Goelzhaeuser, A., Grunze, M., and Ulman, A. (2004) *Macromol. Symp.*, **217**, 223–230.
104. Dultsev, F.N. and Badmaeva, I.A. (2009) *Thin Solid Films*, **518**, 328–332.
105. Badmaeva, I.A., Surovtsev, N.V., Malinovskii, V.K., and Sveshnikova, L.L. (2010) *J. Struct. Chem.*, **51**, 244–250.
106. Ford, J., Marder, S.R., and Yang, S. (2009) *Chem. Mater.*, **21**, 476–483.
107. Li, F., Shishkin, E., Mastro, M.A., Hite, J.K., Eddy, C.R. Jr., Edgar, J.H., and Ito, T. (2010) *Langmuir*, **26**, 10725–10730.
108. Belfield, K.D., Ren, X., Van Stryland, E.W., Hagan, D.J., Dubikovsky, V., and Miesak, E.J. (2000) *J. Am. Chem. Soc.*, **122**, 1217–1218.
109. Sun, H.B. and Kawata, S. (2004) *Adv. Polym. Sci.*, **170**, 169–273.
110. Yan, Y.X., Tao, X.T., Sun, Y.H., Wang, C.K., Xu, G.B., Yu, W.T., Zhao, H.P., Yang, J.X., Yu, X.Q., Wu, Y.Z., Zhao, X., and Jiang, M.H. (2005) *New J. Chem.*, **29**, 479–484.
111. Ecoffet, C., Espanet, A., and Lougnot, D.J. (1998) *Adv. Mater.*, **10**, 411–414.
112. Croutxe-Barghorn, C., Soppera, O., and Lougnot, D.J. (2000) *Appl. Surf. Sci.*, **168**, 89–91.

113. Saravanamuttu, K. and Andrews, M.P. (1999) *Polym. Mater. Sci. Eng.*, **81**, 477–478.
114. Xu, S., Nie, Z., Seo, M., Lewis, P., Kumacheva, E., Stone, H.A., Garstecki, P., Weibel, D.B., Gitlin, I., and Whitesides, G.M. (2004) *Angew. Chem. Int. Ed.*, **43**, 2–2.
115. Yeh, C. and Lin, Y.-C. (2009) *Microfluid. Nanofluidics*, **6**, 277–283.
116. Turri, S., Levi, M., Emilitri, E., Suriano, R., and Bongiovanni, R. (2010) *Macromol. Chem. Phys.*, **211**, 879–887.
117. Park, W., Han, S., and Kwon, S. (2010) *Lab Chip*, **10**, 2814–2817.
118. Nie, Z., Xu, S., Seo, M., Lewis, P.C., and Kumacheva, E. (2005) *J. Am. Chem. Soc.*, **127**, 8058–8063.
119. Hiemstra, C., Zhou, W., Zhong, Z., Wouters, M., and Feijen, J. (2007) *J. Am. Chem. Soc.*, **129**, 9918–9926.
120. Dulay, M.T., Choi, H.N., and Zare, R.N. (2007) *J. Sep. Sci.*, **30**, 2979–2985.
121. Strandwitz, N.C., Nonoguchi, Y., Boettcher, S.W., and Stucky, G.D. (2010) *Langmuir*, **26**, 5319–5322.
122. Takada, K., Kaneko, K., Li, Y.D., Kawata, S., Chen, Q.D., and Sun, H.B. (2008) *Appl. Phys. Lett.*, **92**, 1–3.
123. Soppera, O., Turck, C., and Lougnot, D.J. (2009) *Opt. Lett.*, **34**, 461–464.
124. de Groot, J.H., Dillingham, K., Deuring, H., Haitjema, H.J., van Beijma, F.J., Hodd, K., and Norrby, S. (2001) *Biomacromolecules*, **2**, 1271–1278.
125. Khudyakov, I.V., Arsu, N., Jockusch, S., and Turro, N.J. (2003) *Des. Monom. Polym.*, **6**, 91–95.
126. Chae, H.S. and Park, Y.H. (2007) *Mol. Cryst. Liq. Cryst.*, **472**, 263–270.
127. Yoshida, T., Kanaoka, S., and Aoshima, S. (2005) *J. Polym. Sci., A: Polym. Chem.*, **43**, 4292–4297.
128. Mahy, R., Bouammali, B., Oulmidi, A., Challioui, A., Derouet, D., and Brosse, J.C. (2006) *Eur. Polym. J.*, **42**, 2389–2397.
129. Chen, Y. and Kennedy, J.P. (2008) *J. Polym. Sci., A: Polym. Chem.*, **46**, 174–185.
130. Fan, H., Zhu, X., Gao, L., Li, Z., and Huang, J. (2008) *J. Phys. Chem. B*, **112**, 10165–10170.
131. Kumar, R.M., Balamurugan, R., and Kannan, P. (2007) *Polym. Int.*, **56**, 1230–1239.
132. Jiao, Y., Guo, J., Dong, X., Li, R., and Wei, J. (2010) *J. Appl. Polym. Sci.*, **116**, 3569–3580.
133. Klinger, D., Nilles, K., and Theato, P. (2010) *J. Polym. Sci., A: Polym. Chem.*, **48**, 832–844.
134. He, J., Tremblay, L., Lacelle, S., and Zhao, Y. (2011) *Soft Matter.*, **7**, 2380–2386.
135. Wen, P., Wang, L., Zhang, A., Li, X.D., and Lee, M.H. (2011) *Mater. Chem. Phys.*, **126**, 832–835.
136. Hasegawa, M., Hashimoto, Y. and Chung, C.M. (1993) in *Radiation Curing in Polymer Science and Technology*, vol. 3 (eds J.P. Fouassier and J.F. Rabek), Elsevier, Barking, pp. 341–366.
137. Hasegawa, M., Chung, C.M., and Kinbara, K. (1997) *J. Photosci.*, **4**, 147–159.
138. Guo, F., Marti-Rujas, J., Pan, Z., Hugues, C.E., and Harris, K.D.M. (2008) *J. Phys. Chem. C*, **112**, 19793–19796.
139. Wang, Y., Li, L., Yang, K., Samuelson, L.A., and Kumar, J. (2007) *J. Am. Chem. Soc.*, **129**, 7238–7239.
140. Spagnoli, S., Fave, J.L., and Schott, M. (2011) *Macromolecules*, **44**, 2613–2625.
141. Nakayama, Y., Ishikawa, A., Sato, R., Uchida, K., and Kambe, N. (2008) *Polym. J.*, **40**, 1060–1066.
142. Lalevée, J., Allonas, X., and Fouassier, J.P. (2006) *Chem. Phys. Lett.*, **429**, 282–285.
143. Lee, T.Y., Roper, T.M., Guymon, C.A., Jonnson, E.S., and Hoyle, C.E. (2004) *PSME, Polym. Prepr.*, **45**, 49–50.
144. Deepak, V.D., Rajan, J., and Asha, S.K. (2006) *J. Polym. Sci., A: Polym. Chem.*, **44**, 4384–4395.
145. Sangermano, M., Di Gianni, A., Bongiovanni, R., Priola, A., Voit, B., Pospiech, D., and Appelhans, D. (2005) *Macromol. Mater. Eng.*, **290**, 721–725.
146. Zhang, Y., Miao, H., and Shi, W. (2011) *Prog. Org. Coat.*, **71**, 48–55.
147. Duran, H., Meng, S., Kim, N., Hu, J., Kyu, T., Natarajan, L.V., Tondiglia,

V.P., and Bunning, T.J. (2008) *Polymer*, **49**, 534–545.

148. Sievens-Figueroa, L. and Allan Guymon, C. (2008) *Polymer*, **49**, 2260–2267.
149. Beyer, P., Terentjev, E.M., and Zentel, R. (2007) *Macromol. Rapid Commun.*, **28**, 1485–1490.
150. Andrzejewska, E. (2010) in *Basics of Photopolymerization Reactions*, vol. 2 (eds J.P. Fouassier and X. Allonas), Research Signpost, Trivandrum, pp. 245–257.
151. Sahin, G., Avci, D., Karahan, O., and Moszner, N. (2009) *J. Appl. Polym. Sci.*, **114**, 97–106.
152. Shim, S.Y. and Kim, E.S. (2009) *Macromol. Symp.*, **277**, 201–206.
153. Berchtold, K.A., Hacioglu, B., Nie, J., Cramer, N.B., Stansbury, J.W., and Bowman, C.N. (2009) *Macromolecules*, **42**, 2433–2437.
154. Morizur, J.-F., Zhou, H., Hoyle, C.E., and Mathias, L.J. (2010) *Polymer*, **51**, 848–853.
155. Lapsa, K., Marcinkowska, A., Rachocki, A., Andrzejewska, E., and Tritt-Goc, J. (2010) *J. Polym. Sci., B: Polym. Phys.*, **48**, 1336–1348.
156. Dizman, C., Ates, S., Torun, L., and Yagci, Y. (2010) *Beilstein J. Org. Chem.*, **6**, 1–7.
157. Aerykssen, J.H. and Khudyakov, I.V. (2011) *Ind. Eng. Chem. Res.*, **50**, 1523–1529.
158. Jin, L., Agag, T., Yagci, Y., and Ishida, H. (2011) *Macromolecules*, **44**, 767–772.
159. sun, F., Liao, B., Zhang, L., Du, H.G., and Huang, Y.D. (2011) *J. Appl. Polym. Sci.*, **120**, 3604–3612.
160. Pelletier, H., Belgacem, N., and Gandini, A. (2006) *J. Appl. Polym. Sci.*, **99**, 3218–3221.
161. Durmaz, Y.Y., Karagoz, B., Bicak, N., and Yagci, Y. (2008) *Polym. Int.*, **57**, 1182–1187.
162. Dworak, C., Koch, T., Varga, F., and Liska, R. (2010) *J. Polym. Sci., A: Polym. Chem.*, **48**, 2916–2924.
163. Jabbari, E. and He, X. (2008) *J. Mater. Sci., Mater. Med.*, **19**, 311–316.
164. Schuster, M., Turecek, C., Weigel, G., Saf, R., Stampfl, J., Varga, F., and Liska, R. (2009) *J. Polym. Sci., A: Polym. Chem.*, **47**, 7078–7089.
165. Hedin, J., Oestlund, A., and Nyden, M. (2010) *Carbohydr. Polym.*, **79**, 606–613.
166. Barrett, D.G., Merkel, T.J., Luft, J.C., and Yousaf, M.N. (2010) *Macromolecules*, **43**, 9660–9667.
167. Matsukawa, D., Okamura, H., and Shirai, M. (2010) *Polym. Int.*, **59**, 263–268.
168. Okamura, H., Terakawa, T., and Shirai, M. (2009) *Res. Chem. Intermed.*, **35**, 865–878.
169. Cao, H., Currie, E., Tilley, M., and Jean, Y.C. (2003) in *Photoinitiated Polymerization*, ACS Symposium Series, Vol. 847 (eds K.D. Belfield and J. Crivello), pp. 152–159.
170. Krongauz, V.V., Chawla, C.P., and Dupré, J. (2003) in *Photoinitiated Polymerization*, ACS Symposium Series, Vol. 847 (eds K.D. Belfield and J. Crivello), American Chemical Society, pp. 165–175.
171. Sato, R., Kurihara, T., and Takeishi, M. (1998) *Polym. J.*, **30** (2), 158–160.
172. Kuang, W., Hoyle, C.E., Viswanathan, K., and Jonsson, S. (2002) RadTech Europe, pp. 292–299.
173. Studer, K., Decker, C., Beck, E., and Schwalm, R. (2003) *Prog. Org. Coat.*, **48**, 92–100.
174. Yin, S., Merlin, A., Pizzi, A., Deglise, X., George, B., and Sylla, M. (2004) *J. Appl. Polym. Sci.*, **92** (6), 3499–3507.
175. Drobny, J.G. (2003 and 2010) *Radiation Technology for Polymers*, CRC Press, Boca Raton, FL.
176. Crivello, J.V. and Macromol, J. (2008) *Sci., A: Pure Appl. Chem.*, **45** (8), 591–598.
177. Kato, H. and Sasaki, H. (2003) in *Photoinitiated Polymerization*, ACS Symposium Series, Vol. 847, Chapter 25 (eds K.D. Belfield and J. Crivello), American Chemical Society, pp. 285–305.
178. Sangermano, M., Di Gianni, A., Malucelli, G., Roncuzzi, C., Priola, A., and Voit, B. (2005) *J. Appl. Polym. Sci.*, **97**, 293–299.
179. Bulut, U. and Crivello, J.V. (2005) *J. Polym. Sci. A: Polym. Chem.*, **43** (15), 3205–3220.

180. Crivello, J.V. (2003) in *Photoinitiated Polymerization*, ACS Symposium Series, Vol. 847 (eds K.D. Belfield and J. Crivello), American Chemical Society, pp. 178–206.
181. Crivello, J.V. and Ortiz, R.A. (2001) *J. Polym. Sci., A: Polym. Chem.*, **39**, 2385–2395.
182. Cho, J.D. and Hong, J.W. (2005) *J. Appl. Polym. Sci.*, **97**, 1345–1351.
183. Minegishi, S., Otsuka, T., Kameyama, A., and Nishikubo, T. (2005) *J. Polym. Sci., A: Polym. Chem.*, **43**, 3105–3115.
184. Yagci, Y., Yilmaz, F., Kiralp, S., and Toppare, L. (2005) *Macromol. Chem. Phys.*, **206**, 1178–1182.
185. Abdo Al-Doaiss, A., Guenther, W., Klemm, E., and Stadermann, D. (2005) *Macromol. Chem. Phys.*, **206**, 2348–2353.
186. Lengvinaite, S., Sangermano, M., Malucelli, G., Priola, A., Grigalevicius, S., Grazulevicius, J.V., and Getautis, V. (2007) *Eur. Polym. J.*, **43**, 380–387.
187. Sangermano, M., Amerio, E., Di Gianni, A., Priola, A., Pospiech, D., and Voit, B. (2007) *Macromol. Symp.*, **254**, 9–15.
188. Chappelow, C.C., Pinzino, C.S., Chen, S.-S., Li, J., and Eick, J.D. (2006) *J. Appl. Polym. Sci.*, **103**, 336–344.
189. Sangermano, M., Tasdelen, M.A., and Yagci, Y. (2007) *J. Polym. Sci., A: Polym. Chem.*, **45**, 4914–4920.
190. Stanislovaityte, E., Priola, A., Sangermano, M., Malucelli, G., Simokaitiene, J., Lazauskaite, R., and Grazulevicius, J.V. (2009) *Prog. Org. Coat.*, **65**, 337–340.
191. Ficek, B.A., Thiesen, A.M., and Scranton, A.B. (2008) *Eur. Polym. J.*, **44**, 98–105.
192. Bongiovanni, R., Casciola, M., Di Gianni, A., Donnadio, A., and Malucelli, G. (2009) *Eur. Polym. J.*, **45**, 2487–2493.
193. Kahveci, M.U., Gilmaz, A.G., and Yagci, Y. (2010) in *Photochemistry and Photophysics of Polymer Materials* (ed. N.S. Allen), John Wiley & Sons, Inc., Hoboken, NJ, pp. 421–478.
194. Zhan, F., Asif, A., Liu, J., Wang, H., and Shi, W. (2010) *Polymer*, **51**, 3402–3409.
195. Sangermano, M., Messori, M., Rizzoli, A., and Grassini, S. (2010) *Prog. Org. Coat.*, **68**, 323–327.
196. Lintinen, K., Efimov, A., Hietala, S., Nagao, S., Jalkanen, P., Tkachenko, N., and Lemmetyinen, H. (2008) *J. Polym. Sci., A: Polym. Chem.*, **46**, 5194–5201.
197. Hsu, Y.G., Wan, Y.S., Lin, W.Y., and Hsieh, W.L. (2010) *Macromolecules*, **43**, 8430–8435.
198. Hoppe, C.C., Ficek, B.A., Eom, H.S., and Scranton, A.B. (2010) *Polymer*, **51**, 6151–6160.
199. Fouassier, J.P. and Rabek, J.F. (eds) (1993) *Radiation Curing in Polymer Science and Technology*, Chapman & Hall, London.
200. Song, K.Y., Ghoshal, R., and Crivello, J.V. (2003) in *Photoinitiated Polymerization*, ACS Symposium Series, Vol. 847 (eds K.D. Belfield and J. Crivello), American Chemical Society, p. 253.
201. Jiratumnukul, N. and Itarat, R. (2008) *J. Appl. Polym. Sci.*, **110**, 2164–2167.
202. Shibata, M., Teramoto, N., Someya, Y., and Suzuki, S. (2009) *J. Polym. Sci., B: Polym. Phys.*, **47**, 669–673.
203. Gupta, M.K. and Singh, R.P. (2010) in *Basics of Photopolymerization Reactions*, vol. 1 (eds J.P. Fouassier and X. Allonas), Research Signpost, Trivandrum, pp. 23–35.
204. Crivello, J.V. (2009) *J. Macromol. Sci.*, **46**, 1535–1541.
205. Decker, C., Nguyen Thi Viet, T., and Le Xuan, H. (1996) *Eur. Polym. J.*, **32**, 1319–1326.
206. Crivello, J.V., Narajan, R., and Sternstein, S.S. (1997) *J. Appl. Polym. Sci.*, **64**, 2073–2087.
207. Jacobine, A. (1993) in *Radiation Curing in Polymer Science and Technology*, vol. 3 (eds J.P. Fouassier and J.F. Rabek), Elsevier, Barking, pp. 219–268.
208. Carlsson, A., Harden, S., Lundmark, A., Manea, N., Rehnberg, L., and Svensson, L. (2003) in *Photoinitiated Polymerization*, ACS Symposium Series, Vol. 847 (eds K.D. Belfield and J. Crivello), American Chemical Society, pp. 65–75.
209. Reddy, S.K., Cramer, N.B., O'Brien, A.K., Cross, T., Raj, R., and Bowman,

C.N. (2004) *Macromol. Symp.*, **206**, 361–374.
210. Hoyle, C.E., Lee, T.Y., and Roper, T. (2004) *J. Polym. Sci., A: Polym. Chem.*, **42**, 5301–5338.
211. Shipp, D.A., McQuinn, C.W., Rutherglen, B.G., and McBath, R.A. (2009) *Chem. Commun.*, **42**, 6415–6417.
212. Li, Q., Zhou, H., and Hoyle, C.E. (2009) *Polymer*, **50**, 2237–2245.
213. Kim, Y.B., Kim, H.K., Choi, H.C., and Hong, J.W. (2005) *J. Appl. Polym. Sci.*, **95**, 342–350.
214. Reddy, S.K., Sebra, R.P., Anseth, K.S., and Bowman, C.N. (2005) *J. Polym. Sci., A: Polym. Chem.*, **43**, 2134–2144.
215. Chan, J.W., Wei, H., Zhou, H., and Hoyle, C.E. (2009) *Eur. Polym. J.*, **45** (9), 2717–2725.
216. Owusu-Adom, K., Schall, J., and Guymon, C.A. (2009) *Macromolecules*, **42**, 3275–3284.
217. Chan, J.W., Zhou, H., Hoyle, C.E., and Lowe, A.B. (2009) *Chem. Mater.*, **21**, 1579–1585.
218. Chan, J.W., Shin, J., Hoyle, C.E., Bowman, C.N., and Lowe, A.B. (2010) *Macromolecules*, **43**, 4937–4942.
219. Fairbanks, B.D., Scott, T.F., Kloxin, C.J., Anseth, K.S., and Bowman, C.N. (2008) *Macromolecules*, **41**, 6019–6026.
220. Shin, J., Nazarenko, S., and Hoyle, C.E. (2008) *Polym. Prepr.*, **49**, 170–171.
221. Wilderbeek, H.T.A., Van der Meer, M.G.M., Bastiaansen, C.W.M., and Broer, D.J. (2002) *J. Phys. Chem. B*, **106** (50), 12874–12883.
222. Kloxin, C.J., Scott, T.F., and Bowman, C.N. (2009) *Macromolecules*, **42**, 211–217.
223. Roper, T.M., Kwee, T., Lee, T.Y., Guymon, C.A., and Hoyle, C.E. (2004) *Polymer*, **45**, 2921–2929.
224. Salinas, C.N. and Anseth, K.S. (2008) *Macromolecules*, **41**, 6019–6026.
225. Khire, V.S., Kloxin, A.M., Couch, C.L., Anseth, K.S., and Bowman, C.N. (2008) *J. Polym. Sci., A: Polym. Chem.*, **46**, 6896–6906.
226. Kim, S.K. and Guymon, C.A. (2010) *J. Polym. Sci., A: Polym. Chem.*, **49**, 465–475.
227. O'Brien, A.K., Cramer, N.B., and Bowman, C.N. (2006) *J. Polym. Sci., A: Polym. Chem.*, **44**, 2007–2014.
228. Reddy, S.K., Cramer, N.B., and Bowman, C.N. (2006) *Macromolecules*, **39**, 3673–3680.
229. Scott, T.F., Kloxin, C.J., Draughon, R.B., and Bowman, C.N. (2008) *Macromolecules*, **41**, 2987–2989.
230. Johnson, P.M., Stansbury, J.W., and Bowman, C.N. (2008) *J. Polym. Sci., A: Polym. Chem.*, **46**, 1502–1509.
231. Fairbanks, B.D., Sims, E.A., Anseth, K.S., and Bowman, C.N. (2010) *Macromolecules*, **43**, 4113–4119.
232. Reddy, S.K., Cramer, N.B., Kalvaitas, M., Lee, T.Y., and Bowman, C.N. (2006) *New J. Chem.*, **59**, 586–593.
233. Wei, H., Senyurt, A.F., Jonsson, S., and Hoyle, C.E. (2007) *J. Polym. Sci., A: Polym. Chem.*, **45**, 822–829.
234. Senyurt, A.F., Wei, H., Hoyle, C.E., Piland, S.G., and Gould, T.E. (2007) *Macromolecules*, **40**, 4901–4909.
235. Senyurt, A.F., Wei, H., Phillips, B., Cole, M., Nazarenko, S., Hoyle, C.E., Piland, S.G., and Gould, T.E. (2006) *Macromolecules*, **39**, 6315–6317.
236. Shin, J., Matsushima, H., Comer, C.M., Bowman, C.N., and Hoyle, C.E. (2010) *Chem. Mater.*, **22**, 2616–2625.
237. Trey, S.M., Nilsson, C., Malmstroem, E., and Johansson, M. (2010) *Prog. Org. Coat.*, **68**, 151–158.
238. Rydholm, A.E., Held, N.L., Bowman, C.N., and Anseth, K.S. (2006) *Macromolecules*, **39**, 7882–7888.
239. Wutticharoenwong, K. and Soucek, M.D. (2008) *Macromol. Mater. Eng.*, **293**, 45–56.
240. Wutticharoenwong, K. and Soucek, M.D. (2009) *J. Appl. Polym. Sci.*, **113**, 2173–2185.
241. Wutticharoenwong, K. and Soucek, M.D. (2010) in *Basics of Photopolymerization Reactions*, vol. 2 (eds J.P. Fouassier and X. Allonas), Research Signpost, Trivandrum, pp. 163–178.
242. Nilsson, C., Simpson, N., Malkoch, M., Johansson, M., Malkoch, E., Johansson, M., and Malmstroem, E. (2008) *J. Polym. Sci., A: Polym. Chem.*, **46** (4), 1339–1348.

243. Ortiz, R.A., Flores, R.V.G., Garcia Valdez, A.E., and Duarte, M.L.B. (2010) *Prog. Org. Coat.*, **69**, 463–469.
244. Fu, Q., Liu, J., and Shi, W. (2008) *Prog. Org. Coat.*, **63**, 100–109.
245. Cook, W.D., Chausson, S., Chen, F., Le Pluart, L., Bowman, C.N., and Scott, T.F. (2008) *Polym. Int.*, **57**, 469–478.
246. Cook, W.D., Chen, F., Nghiem, Q.D., Scott, T.F., Bowman, C.N., Chausson, S., and Le Pluart, L. (2010) *Macromol. Symp.*, **291**, 50–65.
247. Balasubramanian, R., Kalaitzis, Z.M., and Cao, W. (2010) *J. Mater. Chem.*, **20**, 6539–6543.
248. Natali, M., Begolo, S., Carofiglio, T., and Mistura, G. (2008) *Lab Chip*, **8**, 492–494.
249. Campos, L.M., Killops, K.L., Sakai, R., Paulusse, J.M.J., Damiron, D., Drockenmuller, E., Messmore, B.W., and Hawker, C.J. (2008) *Macromolecules*, **41**, 7063–7070.
250. Khire, V.S., Lee, T.Y., and Bowman, C.N. (2008) *Macromolecules*, **41**, 7440–7447.
251. Nair, D.P., Cramer, N.B., Scott, T.F., Bowman, C.N., and Shandas, R. (2010) *Polymer*, **51** (19), 4383–4389.
252. Kwisnek, L. and Nazarenko, S. (2010) *Polym. Prepr.*, **51**, 695–696.
253. Li, Q., Zhou, H., Hoyle, C.E., and Wicks, D.A. (2010) in *Basics of Photopolymerization Reactions*, vol. 2 (eds J.P. Fouassier and X. Allonas), Research Signpost, Trivandrum India, pp. 225–234.
254. Matsushima, H., Shin, J., Chan, J.W., Shirai, M., and Hoyle, C.E. (2010) *J. Photopolym. Sci. Technol.*, **23**, 121–124.
255. El Roz, M., Lalevée, J., Allonas, X., and Fouassier, J.P. (2008) *Macromol. Rapid Commun.*, **29**, 804–808.
256. Decker, C., Masson, F., and Schwalm, R. (2004) *J. Coat. Technol.*, **1**, 127–136.
257. Bibaut-Renaud, C., Burget, D., Fouassier, J.P., Varelas, C.G., Thomatos, J., Tsagaropoulos, G., Ryrfors, L.O., and Karlsonn, O.J. (2002) *J. Polym. Sci., A: Polym. Chem.*, **40**, 3171–3181.
258. Liska, R., Schwager, F., Maier, C., Cano-Vives, R., and Stampfl, J. (2005) *J. Appl. Polym. Sci.*, **97**, 2286–2298.
259. Wan, T., Hu, Z.W., Ma, X.L., Yao, J., and Lu, K. (2008) *Prog. Org. Coat.*, **62**, 219–225.
260. Bastani, S. and Moradian, S. (2006) *Prog. Org. Coat.*, **56**, 248–251.
261. Boncza-Tomaszewski, Z., Penczek, P., and Bankowska, A. (2006) *Surf. Coat. Int., B: Coat. Trans.*, **89**, 157–161.
262. Manea, M., Ogemark, K., and Svensson, L.S. (2006) in *Photochemistry and UV Curing* (ed. J.P. Fouassier), Research Signpost, Trivandrum India, pp. 445–460.
263. Hoyle, C.E., Clark, S.C., Viswanathan, K., and Jonsson, S. (2003) *Photochem. Photobiol. Sci.*, **2** (11), 1074–1079.
264. Nguyen, C.K., Cavitt, T.B., Hoyle, C.E., Kalyanaraman, V., and Jonsson, S. (2003) in *Photoinitiated Polymerization*, ACS Symposium Series, Vol. 847, American Chemical Society, pp. 27–40.
265. Decker, C., Bianchi, C., and Jönsson, S. (2004) *Polymer*, **45**, 5803–5811.
266. Cramer, N.B., Reddy, S.K., Cole, M., Hoyle, C., and Bowman, C.N. (2004) *J. Polym. Sci., B: Polym. Chem.*, **22**, 5817–5826.
267. Zhang, X., Li, Z.C., Wang, Z.M., Sun, H.L., He, Z., Li, K.B., Wei, L.H., Lin, S., Du, F.S., and Li, F.M. (2006) *J. Polym. Sci., A: Polym. Chem.*, **44**, 304–313.
268. Haraldsson, T., Johansson, M., and Hult, A. (2010) *J. Polym. Sci., A: Polym. Chem.*, **48**, 2810–2816.
269. Burget, D., Mallein, C., and Fouassier, J.P. (2003) *Polymer*, **44**, 7671–7678.
270. Zhang, X., Li, Z.-C., Li, K.B., Lin, S., Du, F.-S., and Li, F.-M. (2006) *Prog. Polym. Sci.*, **31**, 893–948.
271. Ahn, K.D., Kang, J.H., Yoo, K.W., and Choo, D.J. (2007) *Macromol. Symp.*, **254**, 46–53.
272. Vazquez, C.P., Joly-Duhamel, C., and Boutevin, B. (2009) *Macromol. Chem. Phys.*, **210**, 269–278.
273. Decker, C., Masson, F., and Schwalm, R. (2003) *Macromol. Mater. Eng.*, **288** (1), 17–28.
274. Tasdelen, M.A., Kiskan, B., and Yagci, Y. (2006) *Macromol. Rapid Commun.*, **27**, 1539–1544.

275. Black, S.M., Whittemore, J.H.I.V., and Rawlins, J.W. (2007) *Polym. Prepr.*, **96**, 909–910.
276. Peeters, S. (1993) in *Radiation Curing in Polymer Science and Technology*, vol. 3 (eds J.P. Fouassier and J.F. Rabek), Elsevier, Barking, pp. 177–218.
277. Magwood, L., Ficek, B.A., Coretsopoulos, C.N., and Scranton, A.B. (2010) in *Basics of Photopolymerization Reactions*, vol. 2 (eds J.P. Fouassier and X. Allonas), Research Signpost, Trivandrum, pp. 213–224.
278. Decker, C. (2003) in *Photoinitiated Polymerization*, ACS Symposium Series, Vol. 847, Chapter 23 (eds K.D. Belfield and J. Crivello), American Chemical Society, p. 266.
279. Choe, J.D. and Hong, J.W. (2004) *J. Appl. Polym. Sci.*, **93**, 1473–1483.
280. Crivello, J.V. (2007) *J. Polym. Sci., A: Polym. Chem.*, **45**, 3759–3769.
281. He, Y., Xiao, M., Wu, F., and Nie, J. (2007) *Polym. Int.*, **56**, 1292–1297.
282. De Ruiter, B., El Ghayoury, A., Hofmeir, H., Schubert, U.S., and Manea, M. (2006) *Prog. Org. Coat.*, **55**, 154–159.
283. Sangermano, M., Carbonaro, W., Mallucelli, G., and Priola, A. (2008) *Macromol. Mater. Eng.*, **293**, 515–520.
284. Lecamp, L., Pavillon, C., Lebaudy, P., and Bunuel, C. (2005) *Eur. Polym. J.*, **41**, 169–175.
285. Rajaraman, C., Mowers, W.A., and Crivello, J.V. (1999) *Macromolecules*, **32**, 36–41.
286. Decker, C. and Bendaikha, T. (1998) *J. Appl. Polym. Sci.*, **70**, 2269–2282.
287. Cai, Y. and Jessop, J.L.P. (2006) *Polymer*, **47**, 6560–6566.
288. Wei, H., Li, Q., Ojelade, M., Madbouly, S., Otaigbe, J.U., and Hoyle, C.E. (2007) *Macromolecules*, **40**, 8788–8793.
289. Ortiz, R.A., Urbina, B.A.P., Valdez, L.V.C., Duarte, L.B., Santos, R.G., Valdez, A.I.G., Santos, R.G., Valdez, A.E.G., and Soucek, M.D. (2007) *J. Polym. Sci., A: Polym. Chem.*, **45**, 4829–4843.
290. Kutal, C., Grutsch, P.A., and Yang, D.B. (1991) *Macromolecules*, **24**, 6872–6873.
291. Palmer, B.J., Kutal, C., Billing, R., and Henning, H. (1995) *Macromolecules*, **28**, 1328–1331.
292. Chang, W.Y., Lough, A.J., and Manners, I. (2007) *Chem. A Eur. J.*, **13**, 8867–8876.
293. Landon, S., Shulman, P.M., and Geoffroy, G.L. (1985) *J. Am. Chem. Soc.*, **107**, 6739–6744.
294. Chauvin, Y., Commereuc, D., and Cruypelinck, D. (1976) *Makromol. Chem.*, **177**, 2637–2642.
295. Chien, J.C.W., Wu, J.C., and Rausch, M.D. (1981) *J. Am. Chem. Soc.*, **103**, 1180–1185.
296. Hafner, A., van der Schaaf, P.A., Muhlebach, A., Bernhard, P., Schaedeli, U., Karlen, T., and Ludi, A. (1997) *Prog. Org. Coat.*, **32**, 89–96.
297. Wang, D., Wurst, K., and Buchmeiser, R.M. (2010) *Chem. A Eur. J.*, **16**, 12928–12934.
298. Greszta, D., Mardare, D., and Matyjaszewski, K. (1994) *Macromolecules*, **27**, 638–644.
299. Braunecker, A. and Matyjaszewski, K. (2007) *Prog. Polym. Sci.*, **32**, 93–146.
300. Matyjaszewski, K., Gnanou, Y., and Leibler, L. (eds) (2007) *Macromolecular Engineering: from Precise Macromolecular Synthesis to Macroscopic Materials Properties and Applications*, vol. 1, Wiley-VCH Verlag GmbH, Weinheim, pp. 643–672.
301. Sebra, R.P., Anseth, K.S., and Bowman, C.N. (2006) *J. Polym. Sci., A: Polym. Chem.*, **44**, 1404–1413.
302. Di, J. and Sogah, D.Y. (2006) *Macromolecules*, **39**, 1020–1028.
303. Kilambi, H., Reddy, S.K., and Bowman, C.N. (2007) *Macromolecules*, **40**, 6131–6135.
304. Lalevée, J., Allonas, X., and Fouassier, J.P. (2006) *Macromolecules*, **39**, 8216–8220.
305. Ran, R., Wan, T., Gao, T., Gao, J., and Chen, Z. (2008) *Polym. Int.*, **57**, 28–34.
306. Nemoto, Y. and Nakayama, Y. (2008) *J. Polym. Sci., A: Polym. Chem.*, **46**, 4505–4512.
307. Nordborg, A. and Irgum, K. (2010) *J. Appl. Polym. Sci.*, **117**, 2781–2789.
308. Tanabe, M., Vandermeulen, G.W.M., Chan, W.Y., Cyr, P.W., Vanderark, L.,

Rider, D.A., and Manners, I. (2006) *Nat. Mater.*, **5**, 467–470.
309. Goto, A., Scaiano, J.C., and Maretti, L. (2007) *Photochem. Photobiol. Sci.*, **6** (8), 833–835.
310. Guillaneuf, Y., Couturier, J.L., Gigmes, D., Marque, S.R.A., Tordo, P., and Bertin, D. (2008) *J. Org. Chem.*, **73** (12), 4728–4731.
311. Guillaneuf, Y., Bertin, D., Gigmes, D., Versace, D.L., Lalevée, J., and Fouassier, J.P. (2010) *Macromolecules*, **43**, 2204–2212.
312. Versace, D.L., Lalevée, J., Fouassier, J.-P., Guillaneuf, Y., Bertin, D., and Gigmes, D. (2010) *Macromol. Rapid Commun.*, **31**, 1383–1388.
313. Guillaneuf, Y., Versace, D.L., Bertin, D., Lalevée, J., Gigmes, D., and Fouassier, J.P. (2010) *Macromol. Rapid Commun.*, **31**, 1909–1913.
314. Yoshida, E. (2010) *Colloid Polym. Sci.*, **288**, 901–905.
315. Yoshida, E. (2010) *Colloid Polym. Sci.*, **288**, 1027–1030.
316. Yagci, Y., Jockusch, S., and Turro, N.J. (2010) *Macromolecules*, **65**, 6245–6260.
317. Omrane, K., Partch, R.E., and Shipp, D.A. (2005) *PMSE Prepr.*, **93**, 566–567.
318. Buback, M., Junkers, T., and Vana, P. (2005) *Polymer Preprints*, **46**, 362–363.
319. Muthukrishnan, S., Pan, E.H., Stenzel, M.H., Barner-Kowollik, C., Davis, T.P., Lewis, D., and Barner, L. (2007) *Macromolecules*, **40**, 2978–2980.
320. Ran, R., Yu, Y., and Wan, T. (2007) *J. Appl. Polym. Sci.*, **105**, 398–404.
321. Lu, L., Zhang, H., Yang, N., and Cai, Y. (2006) *Macromolecules*, **39**, 3770–3776.
322. Kitchin, A.D., Velate, S., Chen, M., Ghiggino, K.P., Smith, T.A., and Steer, R.P. (2007) *Photochem. Photobiol. Sci.*, **6**, 853–856.
323. Bouchekif, H. and Narain, R. (2007) *Polym. Prepr.*, **48**, 403–404.
324. Zhang, H., Deng, J., Lu, L., and Cai, Y. (2007) *Macromolecules*, **40**, 9252–9261.
325. Luo, Q., Zheng, H., Peng, Y., Gao, H., Lu, L., and Cai, Y. (2009) *J. Polym. Sci., A: Polym. Chem.*, **47**, 6668–6681.
326. Muftuoglu, A.E., Cianga, I., Colak, D., and Yagci, Y. (2004) *Des. Monom. Polym.*, **7**, 563–582.
327. Roof, A.C., Tillman, E.S., Malik, R.E., Roland, A.M., Miller, D.J., and Sarry, L.R. (2006) *Polymer*, **47**, 3325–3335.
328. Johnson, J.A., Finn, M.G., Koberstein, J.T., and Turro, N.J. (2007) *Macromolecules*, **40**, 3589–3359.
329. Cummins, D., Wyman, P., Duxbury, C.J., Thies, J., Koning, C.E., and Heise, A. (2007) *Chem. Mater.*, **19**, 5285–5292.
330. Ishizu, K., Kobayakawa, N., Takano, S., Tokuno, Y., and Ozawa, M. (2007) *J. Polym. Sci. A: Polym. Chem.*, **45**, 1771–1777.
331. Ishizu, K., Murakami, T., and Takano, S. (2008) *J. Colloid Interface Sci.*, **322**, 59–64.
332. Cohen, N.A., Tillman, E.S., Thakur, S., Smith, J.R., Eckenhoff, W.T., and Pintauer, T. (2009) *Macromol. Chem. Phys.*, **3**, 263–268.
333. Durmaz, Y.Y., Kumbaraci, V., Demirel, L.A., Talinli, N., and Yagci, Y. (2009) *Macromolecules*, **42**, 2899–2902.
334. Acik, G., Kahveci, M.U., and Yagci, Y. (2010) *Macromolecules*, **43**, 9198–9201.
335. Tasdelen, M.A., Uygun, M., and Yagci, Y. (2010) *Macromol. Chem. Phys.*, **211**, 2271–2275.
336. Tasdelen, M.A., Uygun, M., and Yagci, Y. (2011) *Macromol. Rapid*, **32**, 58–62.
337. Kwak, Y. and Matyjaszewski, K. (2010) *Macromolecules*, **43**, 5180–5183.
338. Motokucho, S., Sudo, A., and Endo, T. (2006) *J. Polym. Sci. B: Polym. Chem.*, **44**, 6324–6331.
339. Rempp, P. and Merryl, E.W. (1991) *Polymer Synthesis*, Huthig and Wepf, Basel. pp. 120–134.
340. Kwon, S., Lee, Y., Jeon, H., Han, K., and Mah, S. (2006) *J. Appl. Polym. Sci.*, **101**, 3581–3586.
341. Herbert, D.E., Gilroy, J.B., Chan, W.Y., Chabanne, L., Staubitz, A., Lough, A.J., and Manners, I. (2009) *J. Am. Chem. Soc.*, **131**, 14958–14968.
342. Patra, S.K., Whittell, G.R., Nagiah, S., Ho, C.L., Wong, W.-Y., and Manners, I. (2010) *Chem. A Eur. J.*, **16** (3240), 1–5.
343. Smith, G.S., Patra, S.K., Vanderark, L., Saithong, S., Charmant, J.P.H., and Manners, I. (2010) *Macromol. Chem. Phys.*, **211**, 303–312.

344. Soppera, O. and Croutxe Barghorn, C. (2003) *J. Polym. Sci., A: Polym. Chem.*, **41**, 831–838.
345. Chemtob, A., Belon, C., Mouazen, M., and Crouxte-Barghorn, C. (2010) in *Basics of Photopolymerization Reactions* (eds J.P. Fouassier and X. Allonas), Research Signpost, Trivandrum. pp. 104–115.
346. Kim, W.-S., Houbertz, R., Lee, T.H., and Bae, B.-S. (2004) *J. Polym. Sci. B: Polym. Chem.*, **42**, 1979–1986.
347. Chemtob, A., Versace, D.L., Belon, C., Croutxé-Barghorn, C., and Rigolet, S. (2008) *Macromolecules*, **41**, 7390–7398.
348. Croutxé-Barghorn, C., Chemtob, A., and Belon, C. (2010) *J. Photopolym. Sci. Technol.*, **23**, 129–131.
349. Miyafuji, H., Kokaji, H., and Saka, S. (2004) *J. Wood Sci.*, **50**, 130–135.
350. Sangermano, M., Amerio, E., Priola, A., Di Gianni, A., and Voit, B. (2006) *J. Appl. Polym. Sci.*, **102**, 4659–4664.
351. Chemtob, A., Belon, C., Croutxé-Barghorn, C., Brendle, J., Soulard, M., Rigolet, S., Le Houerou, V., and Gauthier, C. (2010) *New J. Chem.*, **34**, 1068–1072.
352. Belon, C., Chemtob, A., Croutxé-Barghorn, C., Rigolet, S., Schmitt, M., Bistac, S., Le Houerou, V., and Gauthier, C. (2010) *Polym. Int.*, **59**, 1175–1186.
353. Tan, H., Yang, D., Han, J., Xiao, M., and Nie, J. (2008) *Appl. Clay Sci.*, **42**, 25–31.
354. Wingfield, C., Baski, A., Bertino, M.F., Leventis, N., Mohite, D.P., and Lu, H. (2009) *Chem. Mater.*, **21**, 2108–2114.
355. Berner, G., Rist, G., Rutsch, W., and Kirchmayer, R. (1985) *Radcure Basel*, Technical Paper FC85-446, SME Ed., Dearborn, MI.
356. Li Bassi, G., Cadona, L., and Broggi, F. (1986) *Radcure 86*, Technical Paper 4-27, SME Ed., Dearborn, MI.
357. Shirai, M., Suyama, K., and Tsunooka, M. (1999) in *Photosensitive Systems for Photopolymerization Reactions*, Trends in Photochemistry and Photobiology, Vol. 5 (ed. J.P. Fouassier), Research Trends, Trivandrum. pp. 188–209.
358. Dietliker, K., Braig, A., and Ricci, A. (2010) *Photochemistry*, **38**, 344–368.
359. Dietliker, K., Misteli, K., Carroy, A., Peregi, E., Blickenstorfer, B., and Sitzmann, E. (2010) in *Basics of Photopolymerization Reactions*, vol. 3 (eds J.P. Fouassier and X. Allonas), Research Signpost, Trivandrum, pp. 1–22.
360. Dogan, N., Klinkenberg, H., Reinerie, L., Ruigrok, D., and Wijnands, P. (2006) Radtech News, 1, pp. 6–9.
361. Jensen, K.H. and Hanson, J.E. (2002) *Chem. Mater.*, **14**, 918–923.
362. Dietliker, K., Hüsler, R., Birbaum, J.L., Ilg, S., Villeneuve, S., Studer, K., Jung, T., Benkhoff, J., Kura, H., Matsumoto, A., and Oka, H. (2007) *Prog. Org. Coat.*, **58**, 146–157.
363. Li, Y., Cui, H., Qi, F., Yang, H., and Gong, Q. (2008) *Nanotechnology*, **19**, 375304/1–375304/4.
364. Baldeck, P., Stephan, O., and Andraud, C. (2010) in *Basics of Photopolymerization Reactions*, vol. 3 (eds J.P. Fouassier and X. Allonas), Research Signpost, Trivandrum, pp. 200–220.
365. Ermoshkin, A.A., Neckers, D.C., and Fedorov, A.V. (2006) *Macromolecules*, **39**, 5669–5674.
366. Fedorov, A.V., Ermoshkin, A.A., and Neckers, D.C. (2008) *J. Appl. Polym. Sci.*, **107**, 147–152.
367. Boardman, L. (1993) *Organometallics*, **446**, 15–19.
368. Lewis, L.W. and Salvi, G.D. (1995) *Inorg. Chem.*, **34**, 3182–3187.
369. Fry, B.E. and Neckers, D.C. (1996) *Macromolecules*, **29**, 5306–5311.
370. Mayer, T., Burget, D., Mignani, G., and Fouassier, J.P. (1996) *J. Polym. Sci., Polym. Chem.*, **34**, 3141–3147.
371. Furukawa, Y., Hoshino, T., and Morizawa, Y. (2009) *J. Appl. Polym. Sci.*, **112**, 910–916.
372. Naga, N., Kihara, Y., Miyanaga, T., and Furukawa, H. (2009) *Macromolecules*, **42**, 3454–3462.
373. Postigo, A., Kopsov, S., Zlotsky, S.S., Ferreri, C., and Chatgilialoglu, C. (2009) *Organometallics*, **28**, 3282–3287.

5
Photosensitive Systems

Chapters 3 and 4 clearly show the various applications of light-induced photopolymerizations and the different experimental conditions that can be encountered. The characteristics that must be exhibited by the photosensitive system are now detailed.

5.1
General Properties

The properties that should fulfill a photosensitive photoinitiating system (PIS) fall into two categories:

1) Practical properties oriented to the end uses. They concern the following: (i) no effect of PIS on the final properties of the polymerized material, (ii) no formation (or as low as possible) of yellowing photolysis products, (iii) low sensitivity to air and moisture, (iv) convenient handling of the formulation, (v) good solubility or compatibility in the monomer/oligomer matrix, (vi) good shelf stability, (vii) good acceptance of the marketing considerations (low price), (viii) excellent safety (absence of odor, toxicity, and extractable detrimental compounds), and (ix) biocompatibility. Certain properties that are important in a given area are not decisive in another one. For example, the price could be crucial in some applications of radiation curing but presumably has no importance in the imaging technology.
2) Intrinsic properties that govern the reactivity of PIS in terms of quantum yields (see below). They include (i) excellent light absorption properties, (ii) a reduced sensitivity to oxygen, (iii) an excellent photochemical reactivity, and (iv) a high chemical reactivity of the initiating species. For a given matrix, the reactivity of the photosensitive system governs the practical efficiency, that is, the possibility of having a fast (or ultrafast) polymerization and a high final conversion (Section 7.3).

Photoinitiators for Polymer Synthesis: Scope, Reactivity and Efficiency, First Edition.
Jean-Pierre Fouassier and Jacques Lalevée.
© 2012 Wiley-VCH Verlag GmbH & Co. KGaA. Published 2012 by Wiley-VCH Verlag GmbH & Co. KGaA.

5.2
Absorption of Light by a Molecule

5.2.1
Absorption

A molecule exhibits an absorption spectrum in the UV, visible, or very near-infrared wavelength range as a function of its skeleton [1–4]. An absorption spectrum reflects the wavelengths of the photons that can be absorbed. The absorbance or the optical density (OD) at a given wavelength is defined by Beer's law (OD $= \varepsilon l c$ where l is the optical path length in centimeter, c is the molar concentration in moles per liter, and ε is the molar extinction coefficient in mol^{-1} l cm^{-1}); ε is a function of the wavelength.

5.2.2
Molecular Orbitals and Energy Levels

A molecule is described in terms of molecular orbitals (MOs) and state energy levels [2]. The MOs in a chemical bond between two atoms can be constructed from the atomic orbitals (AOs) through the LCAO concept (linear combination of atomic orbitals). The energy of a given state corresponds to the sum of the energy of all the electrons located in the MOs. In singlet states S_n, the molecule only possesses paired electrons (according to the Pauli principle, all MOs are filled with two electrons having opposite spins +1/2 and −1/2). In triplet states T_n, spins are parallel (e.g., +1/2 and +1/2). In usual organic molecules, n (nonbonding), π or σ (bonding), and π* or σ* (antibonding) MOs are encountered (e.g., see the OMs of the carbonyl chromophore in Section 8.1).

5.2.3
Absorption of Light and Optical Transitions

When the AO energy is the same (e.g., in an A–A bond), the coupling is symmetric and strong. When the AO have a different energy (e.g., in an A–B bond), the coupling becomes asymmetric and weaker. In that case, the bond is said to be polarized and the electronic density on each atom is different. Partial charges δ appear and the molecule has a permanent dipole moment μ (Eq. (5.1) where q is the electron charge and d is the bond length).

$$\mu = \delta q \; d \tag{5.1}$$

The absorption of light by a substrate results from the light/matter interaction. The electric field component E of the electromagnetic radiation interacts with the molecule dipole moment (any molecule possesses a transitory dipole moment μ^* induced by the interactions of E and the electrons). The *polarizability* α is defined as the magnitude of this induced dipole moment μ^* over the E intensity (Eq. (5.2)). When the applied field is very strong such as in laser beams, the induced dipole moment contains other terms in E^n (Eq. (5.3) where β is the first-order hyperpolarizability, the following terms including the second-order hyperpolarizability, etc.).

$$\alpha = \mu^*/E \qquad (5.2)$$
$$\mu^* = \alpha E + \tfrac{1}{2}\beta E^2 + \ldots \qquad (5.3)$$

The absorption of a photon by a molecule causes an electronic transition between a ground state (singlet state S_0) and an excited state (singlet state S_1) of the molecule and corresponds to a jump of an electron from an occupied MO to a vacant MO (usually, the lowest energy transition corresponds to transition from the highest occupied molecular orbital (HOMO) to the lowest unoccupied molecular orbital (LUMO)).

In the usual UV/visible wavelength range considered for photopolymerization reactions, three main transitions generally occur in the traditionally used organic photoinitiator (PI) molecules ($\sigma\sigma^*$ transitions are seldom encountered). In organometallic compounds, the situation is different as metal–ligand transitions occur.

1) **nπ^* transitions**. They occur between symmetry different orbitals and are therefore symmetry forbidden. For example, in the RR'C=O group, n is the lone pair on the oxygen atom; the absorption wavelengths are generally located around 300–380 nm with low absorption coefficients. These transitions usually correspond to localized excitations.
2) **$\pi\pi^*$ transitions**. They arise between the π orbitals and are symmetry-allowed transitions. The absorption coefficients are consequently very high. The absorption often occurs at short wavelength (e.g., in the RR'C=CR''R''' group). In addition, these transitions can also arise between strongly delocalized orbitals (e.g., in conjugated π systems) resulting in a delocalization of the excitation energy.
3) **Charge-transfer (CT) transitions**. These transitions are generally observed in strongly delocalized and polarized substituted molecules. The resulting excitation energy is delocalized over the whole molecular system. The absorption coefficients are high and the transition, compared to that of the parent unsubstituted molecule, is significantly shifted to the red (e.g., in 4-phenyl benzophenone compared to benzophenone). CT transitions also occur in donor/acceptor couples (one electron is promoted from the HOMO of the donor to the LUMO of the acceptor).

Any transition energy is dependent on the solvent. A blue-shift (or hypsochromic shift) or a red-shift (or bathochromic shift) of the absorption maximum wavelength is observed when changing the solvent or the polarity of the medium according to the involved transitions. For example, the nπ^* transition of the carbonyl group is blue-shifted and the $\pi\pi^*$ transition is red-shifted when the polarity increases. H-bonding is also very important and leads to enhanced shifts. A polar solvent is characterized by a dipole moment (a polarized bond is present) and a dielectric constant (as encountered in the coulombic strength). A protic solvent is able to share its hydrogen atom with a nucleophilic-site-possessing molecule. Solute–solvent interactions are classified into dipolar and specific (such as the hydrogen bonding) interactions. The former involves the dipole moment and the polarizability properties of both the solute and the solvent and is described by the dielectric continuum model [5]. The latter [6] requires an empirical polarity scale for the solvents and is very often treated according either to the Dimroth's empirical parameter $E_T(30)$ or the solvatochromic comparison method (SCM). The SCM method is based on an empirical solvent polarity scale where the solvent is characterized by three parameters: (i) an index of solvent polarity/polarizability that measures the ability of the solvent, through its dielectric properties, to stabilize a dipole (π^*), (ii) the ability of the solvent to form the hydrogen bond with the solute (α), and (iii) the ability of the solvent to accept a hydrogen bond from the solute (β). These models allow to explain the solvent effects on the electronic transitions.

The intensity I at a given wavelength is driven by the molar extinction coefficient ε (Eq. (5.4) where I_0 is the incident light intensity in photons s^{-1} cm^{-2}); sometimes, an absorption cross section σ (in square centimeters per molecule) that is proportional to ε is used ($\varepsilon = 6 \times 10^{23}\sigma/2300$). These ε or σ terms experimentally reflect the theoretical transition probability: they depend on the molecular structure. An electronic transition obeys selection rules, which states whether a transition is theoretically allowed or not. For the absorption process there are two rules. One refers to the total spin change, which must be equal to zero (only S–S or T–T transitions are allowed), the other to the symmetry properties of the molecule.

$$dI/dx = -2.3\,\varepsilon\,c\,I_0 \text{ and } I = I_0 \exp(-2.3\varepsilon lc) \qquad (5.4)$$

The amount of light absorbed I_{abs} by a molecule irradiated by a light beam having an intensity I_0 at a given wavelength is written as Eq. (5.5). As a consequence, a high absorption is achieved by using molecules having high extinction coefficients at the emission lines of the light source.

$$I_{abs} = I_0\,[1 - \exp(-2.3\varepsilon lc)] \qquad (5.5)$$

The rate of absorption defined in Eq. (5.6) (where I_0 has to be introduced in photons s^{-1} l^{-1}) governs the concentration of the excited states and then the concentration of the initiating species (in mol l^{-1}).

$$-d\,[S_0]/dt = d\,[S_1]/dt = I_{abs} \qquad (5.6)$$

5.2.4
Reciprocity Law

The exposure reciprocity law (ERL) states that a given photochemical reaction depends on the energy dose E_d, which is the product of the incident light intensity I_0 and the irradiation time t_{irr} [7]. This means that the same result should be obtained by varying either I_0 or t_{irr} while keeping E_d constant. ERL holds true in a lot of cases when at least I_0 and t_{irr} remain in a reasonable range of values (such as, e.g., in dental applications) although the nature of the formulation can introduce some deviations. An ERL failure is, however, often observed in specific experimental conditions (e.g., under high intensities and short exposure times in laser curing) or in particular curable systems (such as fluorinated hyperbranched acrylate nanocomposites).

5.2.5
Multiphotonic Absorption

Multiphotonic absorption refers to the absorption of several photons (Eq. (5.7)). For example, compared to a one-photon absorption (1), two cases (2), and (3) are encountered for the absorption of two photons.

$$\text{PI}(S_0) \rightarrow \text{PI}(S_1)(1h\nu) \text{ and PI}(S_1) \rightarrow \text{PI}(T_1)(\text{ISC}) \quad (1)$$
$$\text{and} \quad \text{PI}(T_1) \rightarrow \text{PI}(T_2)(1h\nu \text{ or } 1h\nu') \quad (2)$$
$$\text{PI}(S_0) \rightarrow \text{PI}(S_1)(2h\nu'' \text{with}: \nu'' = \nu/2) \quad (3a)$$
$$\text{PI}(S_0) \rightarrow \text{PI}^+(2h\nu''') \quad (3b) \quad (5.7)$$

1) Under conventional excitation (lamps and ambient light), the absorption of light by a molecule PI is a one-photon process, that is, one photon ν is used to generate a given excited state, for example, T_1 (1), this process being characterized by the cross section σ. When using lasers (or even very high-intensity nonlaser light sources), an excited state, for example, T_2 (2) can be formed through a biphotonic process that requires a stepwise photon absorption where the second photon (that may have a different wavelength ν') is absorbed by another excited state (here T_1). In that case, a different photochemistry might be observed if directly originating from T_2.

2) On the contrary, a simultaneous two-photon absorption (TPA) allows the formation of the same S_1 state (3a) using a photon ν'' (with $\nu'' = \nu/2$). In other cases, a high-energy state PI* can be reached (3b) only by the addition of two photons (e.g., in some photoionization processes). The TPA process is characterized by a TPA cross section σ_2, which is different from the usual one-photon absorption cross section σ. It depends on the second-order hyperpolarizability and could be enhanced by using D-π-D, D-π-A, A-π-A, or D-π-A-π-D structures where D, π, A stand for a donor, a π CT system and an acceptor. In usual compounds, σ_2 have very low values, typically 5–30 GM (1GM = 10^{-50} cm^4 s per photon). In suitable compounds, σ_2 can reach, for example, 500–6800 GM. In a TPA process, the number of excited molecules per square centimeter (which is equal to the number n_p of photon absorbed per square centimeter) per unit time $d[S_1]/dt$ will depend on the square of the light intensity (Eq. (5.8)) where N is the density of absorbing species; the first term that corresponds to the one-photon absorption becomes negligible.

$$d[S_1]/dt = dn_p/dt = \sigma N I_0 + \sigma_2 N I_0^2 \quad (5.8)$$

5.3
Jablonski's Diagram

The evolution of the excited states after light absorption is depicted by the familiar Jablonski's diagram (Figure 5.1) [2]. The molecular states are composed of electronic (see above) and vibrational and rotational energy levels due to the motions of the atoms in the molecule (the energy of these levels is quantified by the v and J numbers that are introduced in the (an)harmonic oscillator and rotator models, respectively). Transitions are classified as radiative (a photon is involved) and nonradiative or radiationless (no photon is involved). After the S_0 excitation, a triplet state T_1 is generated from S_1 through intersystem crossing (ISC). The $S_1 - T_1$ transition is in competition with other deactivation pathways of S_1 to S_0: fluorescence (emission of a photon) and internal conversion (the energy excess is dissipated as heat). In T_1, deactivation occurs through ISC and phosphorescence (emission of a photon). Triplet–triplet annihilation ($T_1 + T_1$ reaction) and self-quenching ($T_1 + S_0$ reaction) can also occur.

All these processes are defined as photophysical processes as no chemical modification of the starting molecule occurs. On the other hand, photochemical processes refer to processes that originate from S_1 or T_1, undergo the generation of chemical intermediates (radicals and ionic species) and lead to new molecules defined as the photolysis products. This is exemplified by the cleavage reaction shown in Eq. (5.9). In that case, the photolysis products are very often formed through bimolecular recombination or hydrogen abstraction from a hydrogen donor molecule SH. In the case of a PI, these chemical intermediates, for example, radicals in Eq. (5.9) correspond to the initiating species of the polymerization reaction. This initiation reaction (radical/monomer addition) obviously competes with the formation of the photolysis products.

$$R_1 - R_2 \rightarrow R_1^\bullet + R_2^\bullet \quad (h\nu)$$
$$R_1^\bullet + R_2^\bullet \rightarrow R_1 - R_1, \quad R_2 - R_2, \quad R_1H, R_2H \tag{5.9}$$

Excited S_1 or T_1 state can also be indirectly populated using an excited suitable molecule, which is called a photosensitizer (PS) (Section 5.5).

5.4
Kinetics of the Excited State Processes

Each of the processes of Figure 5.1 corresponds to a reaction that obeys to a first-order kinetics, characterized by a rate constant k, for example, k_{isc} for the ISC process. The disappearance of the excited singlet state through ISC is given in Eq. (5.10).

$$d[S_1]/dt = -k_{isc}[S_1] \tag{5.10}$$

5.4 Kinetics of the Excited State Processes

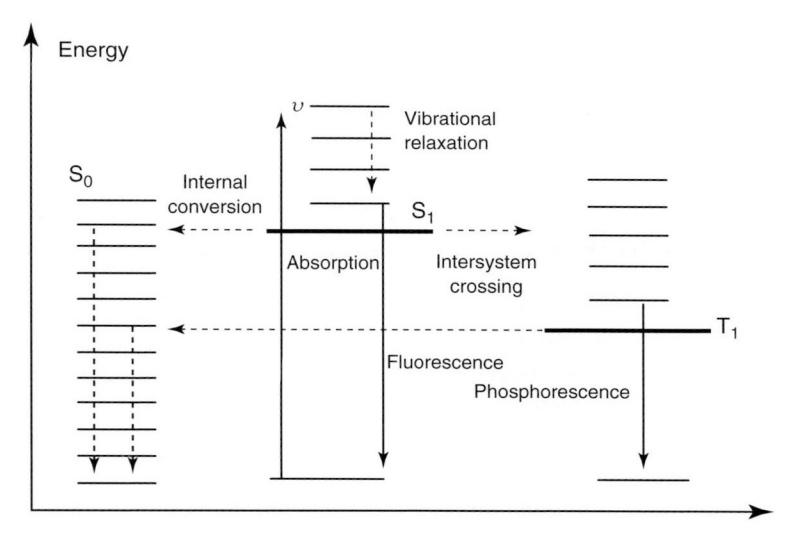

Figure 5.1 Photophysical processes involved in an electronically excited molecule.

The same holds true for the two other processes: fluorescence (k_r) and internal conversion (k_{ci}). The lifetime of the S_1 state is thus defined by $\tau^0{}_S$ in Eq. (5.11).

$$1/\tau^0{}_S = k_{isc} + k_r + k_{ci} \tag{5.11}$$

The quantum yield ϕ of a photophysical process is expressed by the product of the deactivation rate constant and the lifetime of the considered excited state, for example, Eq. (5.12). Such a quantum yield represents the efficiency of a given process over all the processes. The sum of the ISC, fluorescence, and internal conversion quantum yields ($\phi_{isc} + \phi_f + \phi_{ci}$) will obviously be equal to 1. In a similar way, one can define a triplet state lifetime $\tau^0{}_T$ as well as the ISC ($T_1 \rightarrow S_0$) and phosphorescence quantum yields.

$$\phi_{isc} = k_{isc}\tau^0{}_S \tag{5.12}$$

A concomitant photochemical reaction that generates a chemical intermediate (k_{react}) can also arise, for example, in the singlet state. The excited state lifetime τ'_S defined by Eq. (5.13) is therefore shortened. It is obvious from Eqs. (5.11) and (5.13) that the shorter the lifetime, the more efficient the photophysical/photochemical processes.

$$1/\tau'_S = k_{isc} + k_r + k_{ci} + k_{react} \tag{5.13}$$

The addition of another molecule (denoted as a quencher Q) can lead to an interaction (denoted as a quenching) that deactivates the excited state or even regenerates the starting molecule ground state. The reaction has a pseudo first-order rate constant k_q [Q] corresponding to the product of the bimolecular quenching rate constant k_q and the concentration in Q. The lifetime in the absence of Q ($\tau^0{}_S$) will be reduced to τ_S in the presence of Q. Equation (5.14), known as the *Stern–Volmer relationship*, is easily derived.

$$1/\tau_S = 1/\tau^0{}_S + k_q [Q] \qquad (5.14)$$

5.5
Photoinitiator and Photosensitizer

A PI is employed alone, absorbs the light, and leads to initiating species, for example, radicals in Eq. (5.15). It can also be used in the presence of a co-initiator (co-I) which does not absorb the light but reacts with PI to produce the initiating species (1). Both cases constitute what could be truly called a *(direct) photoinitiation process*.

$$\text{PI} \rightarrow \text{PI}^* \rightarrow \text{radicals } (h\nu) \text{ or PI}^* + \text{co-I} \rightarrow \text{radicals } (h\nu) \qquad (1)$$
$$\text{PS} \rightarrow \text{PS}^* (h\nu) \text{ and PS}^* + \text{PI} \rightarrow \text{radicals} \qquad (2) \qquad (5.15)$$

A PS absorbs the light and transfers its energy excess to PI (Eq. (5.15)): this way, (2) is denoted as a photosensitized initiation process. The PI and/or the PS are the basic components of a photosensitive system PIS and are responsible for the absorption properties. The term *photoinitiation* is commonly used whatever the presence of PS or not.

Photosensitization can be described as resulting from the two following main processes largely encountered in photochemistry: energy transfer and electron transfer. Basically, photosensitization relates to an energy transfer between a suitable PS and a PI so that the energy is moving from the excited PS (e.g., the PS triplet state ^{3}PS) to the unexcited PI (in its ground state). This generates the excited PI (in its triplet state ^{3}PI in that case) which will then react as if it would have been directly produced (Scheme 5.1). Singlet–singlet energy transfer also exists in suitable systems. In fluid-like media, this triplet–triplet energy transfer occurs by electron exchange through a collisional mechanism, that is, the two partners have to move to each other. As a consequence, the reaction is diffusion controlled: the viscosity of the matrix will play a decisive role.

Energetic considerations are required. An efficient process only occurs if the energy level E of the donor (here, ^{3}PS) is higher than that of the acceptor (^{3}PI). In that case, the energy transfer is said to be exothermic. The efficiency of this energy transfer rapidly decreases with the energy difference: the rate constant is

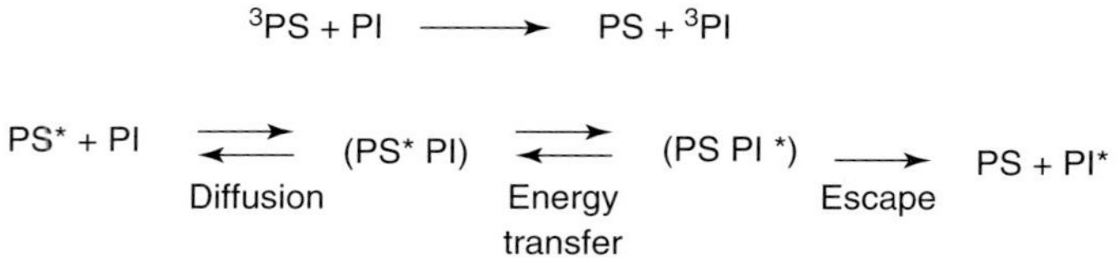

$$^3\text{PS} + \text{PI} \longrightarrow \text{PS} + {}^3\text{PI}$$

$$\text{PS}^* + \text{PI} \underset{\text{Diffusion}}{\rightleftarrows} (\text{PS}^* \text{ PI}) \underset{\text{Energy transfer}}{\rightleftarrows} (\text{PS PI}^*) \underset{\text{Escape}}{\longrightarrow} \text{PS} + \text{PI}^*$$

Scheme 5.1

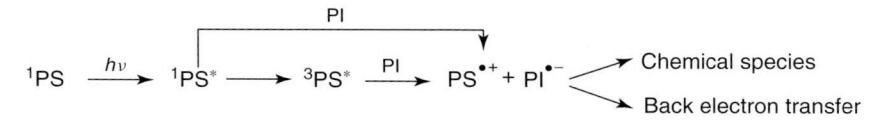

Scheme 5.2

$$^1PS \xrightarrow{h\nu} {}^1PS^* \longrightarrow {}^3PS^* \xrightarrow{PI} PS^{\bullet+} + PI^{\bullet-} \longrightarrow \text{Chemical species / Back electron transfer}$$

(with PI also acting directly on ${}^1PS^*$)

10–100 times lower for a few kilocalories per mole change (rendering the process endothermic). A long-distance process (through a coulombic mechanism) can sometimes occur in singlet–singlet energy transfer. In polymer-like media, energy transfer can be also observed even in the absence of diffusion [2].

An electron transfer process from PS to PI (e.g., PS acts as a donor and PI as an acceptor) is very often encountered by extension, it is also depicted as a photosensitization (Scheme 5.2). A radical ion pair is usually formed (of course, if one of the PS or PI molecule has a charge, the pair consists of a radical and a radical ion) either from the singlet or the triplet state. It undergoes subsequent reactions leading to reactive species. Back electron transfer can occur. The nature of the precursor state affects the yield of the different chemical species. Electron transfer from PI to PS also exists depending of the PS/PI couple. In fluid media, the electron transfer is a diffusion-controlled process. The rate constant is affected when the viscosity increases.

The efficiency of this process is driven by the free energy change ΔG_{et} of the reaction (Section 6.4).

5.6 Absorption of a Photosensitive System

A good matching of the absorption spectrum of PI (and/or PS) and the emission spectrum of the light source must be achieved in order to (i) recover the maximum of photons emitted in the visible region, (ii) eventually use a well-defined and monochromatic excitation (e.g., when using lasers), and (iii) polymerize a pigmented medium. These requirements have moved the design of systems sensitive to UV lights toward that of systems exhibiting red-shifted or visible light absorption. Figure 5.2 shows that, very often, but not always, various compounds allow to get such a good matching (in this example, one of them is able to absorb the green line of an argon ion laser at 514 nm).

A part of the incident light is very often not available for the absorption of the photosensitive systems as if the light was filtered. This phenomenon is called an *inner filter effect*. These competitive light absorptions can arise from

1) pigments incorporated in the formulations to give an opaque film, for example, in paints (Section 7.4);
2) molecules that have been added to further allow a new property, for example, a light-stabilizing effect during the aging of the coating (Section 7.6);
3) photolysis products that are, in most cases, colored (Sections 8.1.2–8.1.4);

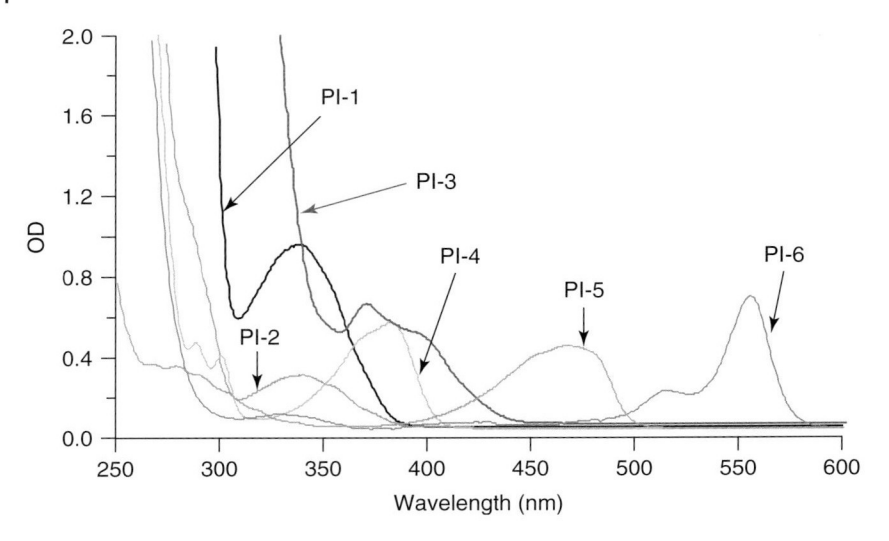

Figure 5.2 Typical absorption spectra of various compounds used as PI.

4) compounds that could interfere with PIS, for example, the phenols in the wood coating curing (Section 8.6.4).

5.7
Initiation Step of a Photoinduced Polymerization

5.7.1
Production of Initiating Species

The initiating species generated through the different kinds of excited state processes react with the monomer/oligomer and form the first reactive monomer species. This corresponds to the initiation step. A further chain reaction allows the buildup of the macromolecular chain (Eq. (1.4)).

The photosensitive system in a photoinduced polymerization reaction can be based on a PI alone (sometimes called *Type I photoinitiator*) or consists of various PI/co-I (*Type II photoinitiator*), PI/PS, PS/PI/co-I, eventually PS/PI/co-I/additive combinations.

In a general way, radical PIs work according to two main types of processes:

1) **Homolytic cleavage**. This leads to two radicals (see (1) in Eq. (5.16)).
2) **Hydrogen transfer**. Two processes can occur (Eq. (5.16)). In the first one (2), a PI/DH hydrogen abstraction (DH being a hydrogen donor) generates two radicals. In the second one (3), an electron/proton transfer PI/AH (where AH is an electron donor bearing a labile hydrogen) corresponds to an electron transfer (that yields a charge transfer complex (CTC)) followed by a proton

transfer that creates the radicals.

$$PI \rightarrow {}^1PI(h\nu) \text{ and } {}^1PI \rightarrow {}^3PI$$
$$^3PI \rightarrow R'^\bullet + R''^\bullet \quad (1)$$
$$^3PI \rightarrow DH \rightarrow PI - H^\bullet + D^\bullet \quad (2)$$
$$^3PI + AH \rightarrow CTC \text{ and } CTC[PI^{\bullet-}AH^{\bullet+}] \rightarrow PI - H^\bullet + D^\bullet \quad (3) \quad (5.16)$$

Usual onium salts used as cationic PIs undergo a homolytic or/and a heterolytic cleavage followed by hydrogen transfer reactions that lead to a protic species (Section 7.1.1).

The rate of initiation R_i (Eq. (5.17)) which is connected with the rate of polymerization Rp (Section 7.2) is a function of the amount of absorbed light I_{abs} and the initiation quantum yield ϕ_i (see below).

$$R_i = \phi_i I_{abs} \quad (5.17)$$

5.7.2
Competitive Reactions in the Excited States

The initiation quantum yield ϕ_i depends on the relative efficiency of the initiation route over all the relaxation and quenching processes of the excited states (by the monomer, oxygen, and additives) and the by-side reactions of the initiating species.

For example, let us consider a Type I PI that cleaves in its triplet state (rate constant: k_c, lifetime: $\tau^0{}_T$) in a deaerated monomer medium (bimolecular quenching rate constant of the triplet state by the monomer: k_q). The initiation quantum yield ϕ_i is expressed as a product of the ISC quantum yield ϕ_{isc} with ϕ_c and ϕ_{RM} (Eq. (5.18)), where ϕ_c is the yield in initiating radicals, ϕ_{RM} the yield in the first monomer radical, k_i the initiation rate constant that refers to the initiating species/monomer interaction, k_r^0 the sum of the pseudo first-order rate constants for the decay of the radical including true first-order and bimolecular reactions. The measured cleavage or dissociation quantum yield corresponds to ϕ_{diss} (Section 6.1.3).

$$\phi_i = \phi_{isc} \phi_c \phi_{RM} \quad \text{with} \quad \phi_{diss} = \phi_{isc} \phi_c$$
$$\text{and}: \quad \phi_{RM} = k_i[M]/(k_r^0 + k_i[M]) \quad \text{and} \quad \phi_c = k_c/(1/\tau^0{}_T + k_q[M]) \quad (5.18)$$

Similarly, for a Type II PI (PI/hydrogen donor DH or AH; see Eq. (5.16)) working in its triplet state, ϕ_i is defined by Eq. (5.19), where ϕ_H is the yield in hydrogen transfer (direct H-transfer with DH) or electron transfer (with AH) and ϕ_R the yield in initiating radicals, respectively; ϕ_R is usually equal to 1 in the direct H-transfer or <1 in the electron/proton transfer sequence). The measured initiating radical quantum yield corresponds to $\phi_{ir} = \phi_{isc} \phi_H \phi_R$.

$$\phi_i = \phi_{isc} \phi_H \phi_R \phi_{RM} \quad \text{with:} \quad \phi_H = k_H[DH]/(1/\tau^0{}_T + k_H[DH] + k_q[M])$$
$$(5.19)$$

5.7.3
Reactivity in Bulk versus Solution: Role of the Diffusion

Bulk monomer or oligomer media are characterized by the presence of a high concentration of monomer double bonds [M] and a viscosity higher than that of monomer solutions. These two factors can be expected to have two opposite effects in bimolecular reactions: (i) an increase in k_q [M] because of the concentration effect and (ii) a decrease in k_q due to the decrease of the diffusion rate constant k_{diff} when the viscosity increases (Eq. (5.20)), where R is the gas constant, T is the temperature, and η is the viscosity. In cleavable PIs where the bond scission is not dependent on the macroscopic viscosity, the cleavage rate constant is not affected but the back reaction can be more efficient in highly viscous media (see the next paragraph).

$$k_{diff} = 8RT/3000\, \eta \tag{5.20}$$

For example, the viscosity of multifunctional monomers is typically 10–110 cP at 25 °C (1 cP = 1 mPa s). The viscosity of a monomer/oligomer matrix can be in the range of 3000 cP at 60 °C. By comparison, the viscosity of usual organic solvents is a few centipoises, that of a toluene/13.7% PMMA solution (molecular weight: 120 000) is 700 cP. The estimated diffusion rate constant k_{diff} is about 1.1×10^{10}, 8.4×10^6, and 4×10^5 mol^{-1} l s^{-1} in media having a viscosity of 4.0, 700 and >1000 cP. In the two last media, all bimolecular rate constants will level off (Section 16.1). This is qualitatively exemplified in Figure 5.3 for a radical photopolymerization: at low viscosity, an addition rate constant k_{add} of an initiating

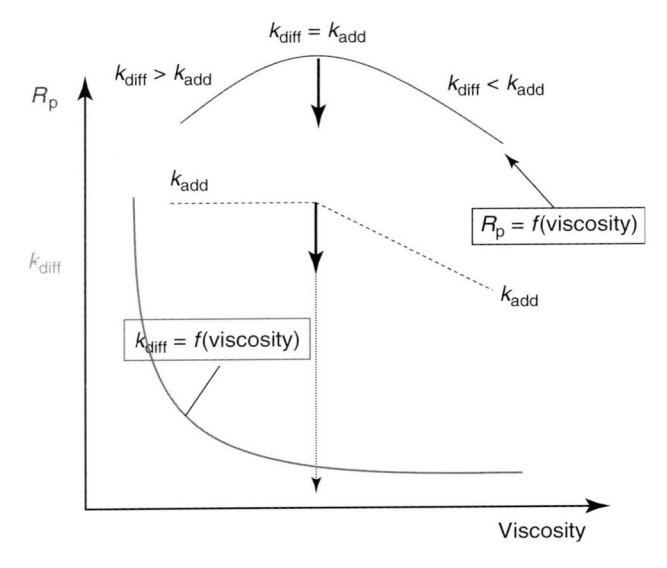

Figure 5.3 Schematic representation of the evolution of R_p, k_{diff}, and k_{add} as a function of the viscosity of the medium. See text.

species to a monomer lower than k_{diff} keeps the same value as in solution until $k_{add} = k_{diff}$. For $k_{add} > k_{diff}$, k_{add} decreases on an increase in the viscosity. As a consequence, for a given monomer, the polymerization rate R_p (which is a function of k_{add}) will increase and then decrease with the viscosity of the medium. For example, two radicals having $k_{add} = 10^6$ and $10^7\,\text{mol}^{-1}\,\text{l}\,\text{s}^{-1}$ in an organic solvent will exhibit an almost similar $k_{add} \sim 4 \times 10^5\,\text{mol}^{-1}\,\text{l}\,\text{s}^{-1}$ in an epoxy acrylate matrix ($\eta > 1000$ cP). This means that the relative reactivity of two given structures in a bimolecular process observed in solution does not necessarily parallel that encountered in a high-viscosity medium. In some particular cases, the quenching rate constants of a PI in its triplet state by, for example, a monomer or an amine can also be affected by the polarity of the medium (Section 9.1.4).

5.7.4
Cage Effects

From a qualitative and simplified point of view, the solvent cage effect corresponds to a situation where the two primary generated species A and B (e.g., a pair of radicals RP arising from a PI bond cleavage or a pair of radical-ions formed through an electron transfer in a PI/amine system) can remain in close proximity for a finite period. This is due to the presence of the surrounding solvent molecules that play the role of a wall (Scheme 5.3) and confine both A and B (these two partners collide with each other inside the cage). The recombination reactions are, therefore, favored and, as a consequence, the yield in escape reactions decreases [8]. The initiation quantum yield can be expressed as $\phi_i = \phi_i^0 f$ (where f is the fraction of initiating radicals that escape out of the cage and ϕ_i^0 the initiation quantum yield in a medium when no cage effect occurs).

In the common simple exponential model, cage recombination and cage escape are described as first-order reactions. In the more refined contact model, the reactive species undergo a series of contacts/separations before their final escape. In viscous media, cage effects involving radical pairs can dramatically decrease the yield in escaping free radicals [9]. A combination of both models has to be introduced to account for the RP decay as a function of the medium viscosity (see a review in [10]). Cage effect dynamics, kinetics of geminate recombination, evolution of the cage effect during the course of a photoinduced polymerization reaction (where the viscosity of the medium increases) have been discussed in several papers [11–15].

In fluid media and for PIs that react in their triplet states (Scheme 5.4), the radical pair formed after light excitation has a triplet character ^{3}RP as the radical

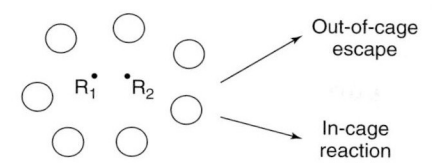

Scheme 5.3

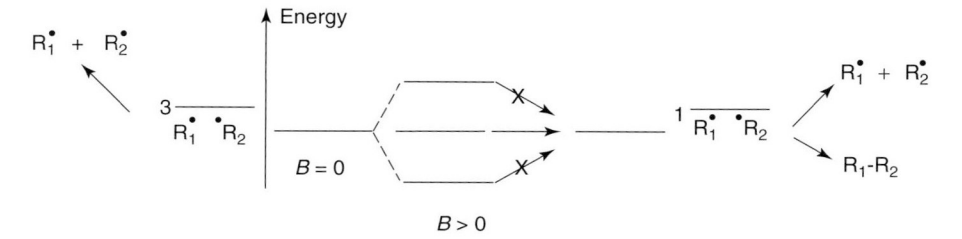

Scheme 5.4

pair possesses the spin memory of the precursor state. It quasi instantaneously dissociates into free radicals as in ^{3}RP, the radicals cannot recombine because of the two parallel spins. As a consequence, almost no cage effect occurs. On the contrary, in viscous media, ^{3}RP has enough time to undergo ISC (because of the reduced diffusion) and form a singlet radical pair ^{1}RP. Recombination or disproportionation of the radicals is highly favorable in ^{1}RP and the yield in escaping free radicals decreases. This results in an important cage effect.

The application of an external (low, moderate, or high) magnetic field B affects the free radical escape in (viscous) solvent cages or micelles [16]. For example, a low B (~10 mT) suppresses the degeneracy of a triplet pair and changes the ^{3}RP/^{1}RP interconversion yield (Scheme 5.4). As a consequence, under such conditions, the amount of the escaping free radicals increases as well as the rates of polymerization. Magnetic field effects on photochemical reactions have been reviewed in, for example, [17–20].

5.8
Reactivity of a Photosensitive System

A general discussion of the PI reactivity is nowadays based on both experimental and theoretical considerations. For example, the overall diagram of evolution of the excited states and reactive intermediates of a radical PI shown in Scheme 5.5 summarizes the different kinetic and thermodynamical parameters that are by now accessible: (i) quantum yields of ISC ϕ_{isc} and dissociation of cleavable PIs ϕ_{diss}, (ii) yields in initiating radicals ϕ_R and monomer radicals ϕ_{RM}, (iii) triplet-state energy levels E_T, (iv) excited-state lifetimes τ_S and τ_T, (v) interaction rate constants of triplet states with hydrogen donors k_e, oxygen, monomer or other additives, cleavage rate constants k_c, (vi) PS/PI energy (k_{ET}) or electron (k_{eT}) transfer, (vii) bond dissociation energies (BDEs), reaction enthalpies ΔH_R, and (vii) addition rate constants of the initiating species to the monomer k_i.

By using both these data and the polymerization results, it becomes now possible to derive very interesting (photo)chemical reactivity/practical efficiency relationships. The huge progress made in the theoretical approach carried out in the recent years really improves the interpretation of these relationships compared to what was done in the past. Today, the design of more or less sophisticated PIS

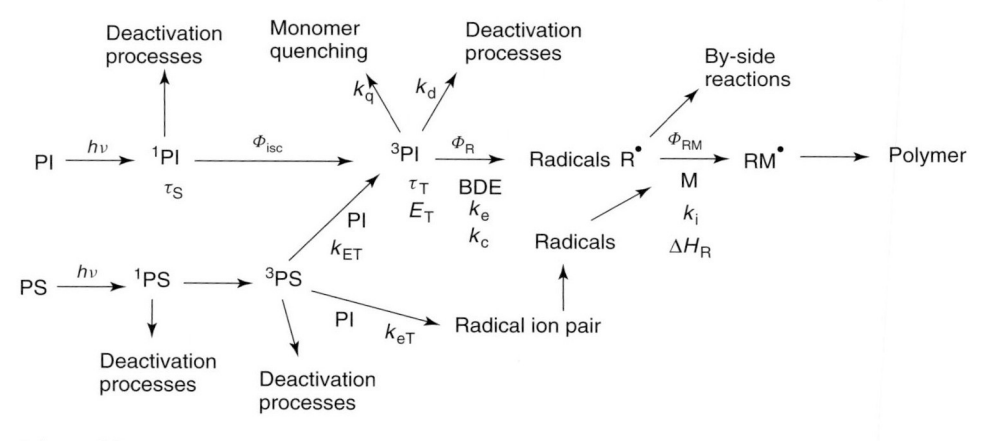

Scheme 5.5

has induced a large variety of processes occurring in the excited states and the overall mechanisms become often complicated. It is therefore of prime importance to investigate these processes in order to propose some basic guidelines useful for the development of novel photoinitiating systems.

References

1. Wayne, R.P. (1988) *Principles and Applications of Photochemistry*, Oxford Science and Publications, Oxford.
2. Turro, N.J. (1978 and 1990) *Modern Molecular Photochemistry*, Benjamin, New York.
3. Gilbert, A. and Baggot, J. (1991) *Essentials of Molecular Photochemistry*, Blackwell Scientific Publications, Oxford.
4. Suppan, P. (1994) *Chemistry and Light*, The Royal Society of Chemistry.
5. Bötcher, C.J.F. and Bordewijk, P. (1977) *Theory of Electric Polarization*, Elsevier, Amsterdam.
6. Kamlet, M.J., Abboud, J.M.L., Abraham, M.H., and Taft, R.W. (1983) *J. Org. Chem.*, **48**, 2877–2882.
7. Feng, L. and Suh, B.I. (2007) *Macromol. Chem. Phys.*, **208**, 295–306.
8. Noyes, R.M. (1961) *Prog. React. Kinet.*, **1**, 129–135.
9. Savitsky, A.N., Paul, H., and Shushin, A.I. (2006) *Helv. Chim. Acta*, **89**, 2533–2543.
10. Khudyakov, I.V. and Turro, N.J. (2010) *Des. Monomer Polym.*, **13**, 487–496.
11. Khudyakov, I.V. and Turro, N.J. (2010) in *Carbon-Centered Free Radicals and Radical Cations*, Chapter 12 (ed. M.D.E. Forbes), John Wiley & Sons, Inc., New York, pp. 249–279.
12. Turro, N.J., Ramamurthy, V., and Scaiano, J.C. (2009) *Principles of Molecular Photochemistry*, Chapter 7, University Science Books, Sausalito.
13. Khudyakov, I., Levin, P., and Kuzmin, V. (2008) *Photochem. Photobiol. Sci.*, **7**, 1540–1546.
14. Levin, P.P., Khudyakov, I.V., and Kuzmin, V.A. (1989) *J. Phys. Chem.*, **93**, 208–213.
15. Khudyakov, I.V., Zharikov, A., and Burshtein, A. (2010) *J. Chem. Phys.*, **132**, 14104–14109.
16. Khudyakov, I.V., Serebrennikov, Y., and Turro, N.J. (1993) *Chem. Rev.*, **93**, 537–542.
17. Murai, H. and Hayashi, H. (1993) in *Radiation Curing in Polymer Science and Technology*, vol. **2** (eds J.P. Fouassier

and J.F. Rabek), Elsevier, Barking, pp. 63–154.
18. Steiner, U.E. and Wolff, H.J. (1991) in *Photochemistry and Photophysics*, vol. IV (ed. J.F. Rabek), CRC Press, Boca Raton, FL, pp. 1–130.
19. Hayashi, H. (1990) in *Photochemistry and Photophysics*, vol. I (ed. J.F. Rabek), CRC Press, Boca Raton, FL, pp. 59–136.
20. Khudyakov, I.V., Arsu, N., Jockusch, S., and Turro, N.J. (2003) *Des. Monomers Polym.*, **6**, 91–101.

6
Approach of the Photochemical and Chemical Reactivity

The investigation of the mechanisms involved in photosensitive systems is mainly carried out using steady-state and time-resolved techniques, for example, (i) time-resolved laser absorption (sometimes fluorescence) spectroscopies on the nanosecond and picosecond timescale ([1–4]; see also references in Parts 2–4), (ii) time-resolved photothermal techniques [5], (iii) time-resolved photoconductivity (for the direct evidence of the ions or radical ions generation) [6], (iv) time-resolved Fourier transform infrared (FTIR) vibrational spectroscopy [7, 8], (v) electron spin resonance techniques [9, 10] (ESR) (steady-state ESR, time-resolved ESR, and ESR spin trapping), and (vi) (time-resolved) chemically induced nuclear polarization (CIDNP) [11] and chemically induced electron polarization (CIDEP) [12]. Steady-state UV photolysis techniques commonly employed in photochemistry [13–17] can provide some complementary information such as the nature of the photoinitiator (PI) decomposition products in an organic solvent or a given medium (e.g., the photoproduct analysis by analytical techniques such as chromatography, NMR, and MS). Polymer end group analysis is particularly interesting as it gives an access to the efficiency of the PI-derived radical toward initiation and chain-growth termination; recent progress has been achieved using soft-ionization mass spectrometry techniques [18]. Apart from the experimental studies, MO or quantum mechanical calculations (or molecular modeling) performed at high level of theory give interesting thermodynamic (or even kinetic) data useful to propose a final reaction scheme.

6.1
Analysis of the Excited-State Processes

We mainly focus on the techniques that were (and still are) able to provide a large set of data.

6.1.1
Nanosecond Laser Flash Photolysis

Nanosecond laser flash photolysis (LFP) is certainly the most popular technique in this area [1–3]. It is usually based on a Q-switched nanosecond Nd/YAG laser (λ_{exc} = 266, 355, 532, and 1064 nm), which delivers an exciting pulse (approximately a few

Photoinitiators for Polymer Synthesis: Scope, Reactivity and Efficiency, First Edition.
Jean-Pierre Fouassier and Jacques Lalevée.
© 2012 Wiley-VCH Verlag GmbH & Co. KGaA. Published 2012 by Wiley-VCH Verlag GmbH & Co. KGaA.

nanoseconds duration; energy reduced down to a few tens of millijoules). An optical parametric oscillator (OPO) crystal pumped by the Nd/YAG laser provides excitation laser wavelengths tunable from the UV to the near IR. The laser pulse allows to populate the spectroscopic excited states through the primary $S_0 - S_1$ transition (or even $S_0 - S_2$ usually followed by internal conversion to S_1) and the subsequent intersystem crossing process. Energy levels of transient species (e.g., radicals) can also be reached. The analyzing system lies on a time-resolved colorimetric device consisting of a white light emitted by a pulsed xenon lamp, a monochromator, a fast photomultiplier, and a transient digitizer (the time resolution is limited by the rise time of the photomultiplier). The information $OD = f(t)$ at a given λ is first recorded. Changing the wavelength allows to obtain the transient absorption spectra $OD = f(\lambda)$ at a given time. Using a CCD camera instead of a photomultiplier offers the possibility to record the absorption spectrum in one shot. Addition of a quencher shortens the transient lifetime: the bimolecular quenching reaction rate constant is obtained through a Stern Volmer treatment (Eq. (5.14)) [15].

6.1.2
Picosecond Pump-Probe Spectroscopy

The picosecond setup is based on a usual pump-probe arrangement [4]. The principle consists in generating a transient species on excitation with a short laser pulse (pump pulse) and passing a white light pulse (probe pulse) through the sample so that an absorption can be optically detected. The picosecond pulses can be delivered by a passively/actively mode-locked Nd/YAG laser operating in a repetitive mode [19]. The third harmonic (355 nm) and the fundamental (1064 nm) laser emissions are used to excite the sample and generate the white light continuum probe beam (by focusing the laser pulse in a D_2O/H_2O mixture), respectively. The white light beam illuminates both the sample and the reference cells. The transmitted light is sent out to the dispersive element of a double diode array multichannel analyzer. Using a given amount of laser shots (~400) allows the recording of one transient absorption spectrum at a given time that corresponds to the delay between the excitation pulse and the analyzing pulse (a delay of up to 6 ns could be achieved using a computer-controlled micrometer translation stage). Therefore, the absorption of the transient is directly recorded as a function of wavelength and time $OD = f(t, \lambda)$. The time resolution of such an experimental setup is ~5–10 ps. Pump-probe spectroscopy has also been extended to the femtosecond timescale [20, 21].

Picosecond or femtosecond transient grating spectroscopy as well as fluorescence up-conversion and time-correlated single-photon counting are possible complementary techniques [22].

6.1.3
Photothermal Techniques

Photothermal techniques are based on the presence of nonradiative transitions [23]. After excitation of a molecule, the radiationless deactivation of a metastable species releases thermal energy and causes local heating in the medium. This results in (i) a thermal wave (giving rise to a temperature jump ΔT) leading to a pressure variation ΔP and formation of an acoustic wave or (ii) a change of the refractive index Δn that generates an optical element. In thermal lens spectroscopy (TLS) [24], the growing of this optical element is monitored by a CW probe laser: the recorded signal corresponds to the defocusing and the refocusing of the analytic laser beam corresponding to the formation and disappearance of the metastable species. The time window of the phenomena that can be studied lies between 100 ns and 0.1 ms. In laser-induced photoacoustic calorimetry (LIPAC) [23], the acoustic wave is detected by an ultrasonic transducer. Signals ranging between 20 ns and a few microseconds are time resolved (signals owing to excited states deactivating in <20 ns are too fast; signals higher than a few microseconds are too slow). The energy balance is a function of the fluorescence quantum yield ϕ_F, the energy of the first excited-singlet-state E_F, and the thermal energy αE_a. The last term represents a fraction α of the absorbed photon energy E_a released in the medium either in a time shorter than the available time windows (it includes internal conversion, intersystem crossing ϕ_{isc}, and heat of the photoreaction following the excitation) or stored in the long-lived photoproducts that do not recombine in the apparatus time window. A simple energy balance consideration leads to Eq. (6.1). The fraction of energy released αE_a is divided into two terms corresponding to the fast ($\alpha_{Fast} E_a$) and slow ($\alpha_{Slow} E_a$) heat deposit. Only $\alpha_{Fast} E_a$ is detected in LIPAC, whereas both the fast and slow signals are observed in TLS.

$$E_a = \phi_F E_F + \alpha E_a = \phi_F E_F + \alpha_{Fast} E_a + \alpha_{Slow} E_a \tag{6.1}$$

In the recent years, this technique appears as very powerful in the application to PIs [24]: it allows to determine intersystem crossing quantum yields, triplet-state energy levels, formation enthalpies, bond dissociation energies of coinitiators (amines and thiols), reaction enthalpies, initiation rate constants, and dissociation quantum yields ϕ_{diss}.

6.1.4
Time-Resolved FTIR Spectroscopy

Laser-induced step-scan FTIR spectroscopy (TR-S^{2}FTIR) and time-resolved Fourier transform infra red (TR-FTIR) are powerful tools to characterize the reactivity of transient intermediates exhibiting a weak absorption in the UV–visible region and being relatively difficult to detect and, in more general manner, to assign a chemical structure to the transient (as usually done for the chemical analysis of a compound by steady-state IR or FTIR). In TR-S^{2}FTIR, which is rather hard to handle, a pulsed Nd/YAG laser excites the sample and a step-scan FTIR spectrometer yields the IR spectrum of the transient. Both spectral and time resolution (in a few ten nanoseconds) are simultaneously obtained. Few data are available [7, 8]. Time-resolved Raman spectroscopy techniques can also provide a spectral and time resolution [25–27]. Diode laser-based TR-IR is very similar to LFP. The detected signal corresponds to the evolution of the IR optical density at a given wave number after a pulsed laser excitation and using a suitable diode laser for the analyzing wavelength [7]. The decay of the benzoyl radical produced from the cleavage of a suitable ketone has been monitored at 1825 cm^{-1} [28].

6.1.5
Direct Detection of Radicals

Real progresses in the investigation of the radical reactivity have been recently made, thanks to their direct detection. In fact, the direct observation of a radical is rather difficult compared to the usual detection of other transient states because of an absorption located in the UV range [29–33]. Few examples of a direct optical detection have been provided. On the other hand, indirect methods were used but they are very often complicated. In the past few years, LIPAC was shown as an interesting technique. A procedure was recently developed [34]. It consists (Eq. (6.2)) in (i) generating a particular radical R$^•$ by LFP, as already proposed in [35], through the photolysis of di-*tert*-butylperoxide and reaction of the produced *tert*-butoxyl radical with a suitable compound AH (1), for example, triethylamine (TEA), leading to the A$^•$ radical (here TEA$^•$) through hydrogen abstraction (2), (ii) detecting the optical absorption of A$^•$, and (iii) investigating the subsequent reactions of A$^•$ (3).

$$t\text{-BuOO}t\text{-Bu} \rightarrow 2\ t\text{-BuO}^• \ (h\nu) \quad (1)$$
$$t\text{-BuO}^• + \text{AH} \rightarrow \text{A}^• + t\text{-BuOH} \quad (2)$$
$$\text{A}^• + \text{M} \rightarrow \text{AM}^• \quad (3) \quad\quad\quad (6.2)$$

This has been successfully applied to the detection of a lot of radicals (Section 17.2.1), for example, A$^•$ as well as the first monomer radical A–M$^•$ formed by addition of A$^•$ to a monomer double bond M. The recombination of two AM$^•$ radicals, the addition of AM$^•$ to a new monomer unit, and the interaction of AM$^•$ with various additives were easily followed.

6.1.6
CIDNP, CIDEP, and ESR Spectroscopy

CIDNP and CIDEP techniques are sophisticated analytical methods that allow to get interesting information on a process because of an excellent characterization of the generated transients or photolysis products [12, 36, 37]. These phenomena are encountered in chemical reactions occurring through radical pairs and biradicals where anomalous intensities (arising from radicals and reaction products) are found in the ESR and NMR spectra. Most of them are explained on the basis of the singlet–triplet conversion of radical pairs and biradicals (radical pair mechanism (RPM)), the spin polarization of the precursor triplet state (triplet mechanism (TM)), and the spin-correlated radical pair mechanism (SCRP). CIDNP and CIDEP can also be time resolved. For example, the reversible addition reaction of a radical to an acrylate at initial stages (this phenomenon is usually rather weak) has been studied using ^{1}H CIDNP [38]. Kinetic data are accessible by time-resolved ESR (TR-ESR); the term kinetic ESR (K-ESR) being preferably used for a lamp excitation

[9, 39, 40]. Electron spin resonance spin trapping (ESR-ST) [10] appears as a powerful technique for the investigation of the mechanisms involved in PIs [41]. In ESR-ST, the ESR spectrum of an adduct is recorded. This adduct is formed by the addition of a radical R$^\bullet$ to the N=C double bond of a spin trap YY'–C=N$^+$(–O–)R'R'' (see, e.g., phenyl-N-tbutylnitrone (PBN) or 5,5'-dimethyl-1-pyrroline N-oxide (DMPO)). The spin trap is linked to R$^\bullet$ by a C–R bond and exhibits a characteristic ESR spectrum due to the nitroxide. The coupling of this radical with the nuclear spin of the nitrogen and the hydrogen atom of ST (a_N and a_H hyperfine splitting HFS constants) is dependent on the size and nature of the substituents and allows an unambiguous assignment of the adduct and R$^\bullet$ structure. When the trapped radical contains atoms with a high gyromagnetic ratio such as P or B, an additional coupling can occur.

6.2
Quantum Mechanical Calculations

Nowadays, quantum mechanical or MO calculations have become largely accessible to everybody, but they require a careful attention to the hypothesis used and the levels of theory and a thorough analysis of the results. Basically, the problem consists in solving the Schröndinger's equation, which has no exact solution in the case of large systems. Depending on the approximations used, several methods are employed: the semiempirical methods (they are fast but the results are not accurate and only allow a trend), the density functional theory (DFT) methods (good results are obtained together with reasonable computing times), and *ab initio* methods (they lead to good results but require time-consuming calculations). The Gaussian 03 or 09 suite of programs [42] is well known and DFT (e.g. B3LYP) or time-dependent-DFT methods are very often used. Other functionals exist (e.g., PBE1PBE and MPW1PW91). Various procedures and basis sets can be considered depending on the level of theory and the desired parameter calculations such as UQCISD(T), CCSD(T), CBS-RAD, CBS-QB3, G2, B3LYP/6-311++G**, B3LYP/6-31G*. For thermodynamic or spectroscopic parameters, different appropriated methods or levels of theory must be used.

These methods allow to calculate a lot of interesting molecular characteristics (Parts 2–4): MOs, energy levels of excited states and transient species (singlet, triplet, and radicals), geometry of the molecules, spectral properties (ground state, triplet state, and radical absorption spectra), potential energy surface (PES) (these curves connect the reactants with the products), thermodynamical properties (bond dissociation energy (BDE), ionization potential (IP), electronic affinity (EA), reaction enthalpy ΔH_r, formation enthalpy ΔH_f, transition state (TS), activation energy or barrier E_a^{TS}, distance d(C–C) between an attacked carbon and a radical center in the TS), kinetic parameters (reaction rate constants), and charge density (in a triplet state, on a radical site). The ability of the different computational methods for a given application has to be evaluated. For example, a reliable description of the barrier of a chemical reaction (e.g., for the addition reactions to double bonds) is not always straightforward.

6.3
Cleavage Process

In photochemistry, a qualitative theory of photoreactions is based on the MO interactions and the state correlation diagrams [15]. The photodissociation favorably occurs for an excited molecule having a localized excitation energy. Therefore, most systems dissociate from initial $n\pi^*$ states. On the contrary, $\pi\pi^*$ states hardly dissociate. The requirements for an efficient dissociation are the following: (i) the molecule should possess a bond characterized by a rather low BDE (however, the BDE must be high enough to avoid a thermal degradation of the starting molecule), (ii) the energy of the excited state from which the dissociation arises must be higher (at least similar) than the BDE of the considered bond, and (iii) the energy barrier E_a which is connected with the reaction rate constant k and the pre-exponential factor A_{exp} according to the usual Arrhenius' plot (Eq. (6.3)) must be as low as possible. Assuming a pre-exponential factor of 10^{13} s^{-1}, a lifetime of 1 ns and 1 μs for the singlet and triplet states, respectively, the E_a values should be <9.5 kcal mol^{-1} and <5.5 kcal mol^{-1} for getting an efficient cleavage in the corresponding states. The crossing between the $\pi\pi^*$ and the $n\pi^*$ states occurs, however, at a much higher energy than the corresponding minima of the PESs. The height of the barrier is expected to strongly affect ϕ_{diss}. The photodissociation is more favorable in triplet states because of their longer lifetimes.

$$k = A_{exp} \exp\left(\frac{-E_a}{RT}\right) \tag{6.3}$$

For example, the Norrish I cleavage at the α-position of a carbonyl group that has been studied a long time ago [15, 43] is the basic mechanism encountered in many PIs. State correlation diagrams have been proposed [15, 44–48] for small ketones and extended to more complex ketones based on a benzoyl moiety [49–52]. Knowing the orbital symmetries with respect to the symmetry plane of the reaction allows building up the MO correlation diagram from which the state correlation diagram is deduced [15]. The evolution of the energy of the reactant along the coordinate for the reaction (or the energy profile of the reaction) is referred, as usual, as the PES.

The triplet-state reactivity was usually explained on the basis of the thermal activation theory [53, 54] and the tunneling effect [55]. In the thermal activation hypothesis, an activation energy is needed to cleave the C–C bond in the excited $n\pi^*$ triplet state. The reaction rate constants are not correlated with the stability of the radicals produced after the cleavage process, indicating a TS geometry close to that of the triplet. The reactivity is controlled by the steric effects, the electronegativity of the substituents on the benzoyl group, and the negative partial charge on the carbon atom of the carbonyl group in the TS. In the tunneling theory, no activation energy is needed; the system reaches the PES of the radicals from the triplet surface as if there was a tunnel in the barrier. The cleavage reaction mainly depends on the radical stability and the relative position of the triplet and the radical energy surfaces. New investigations (rather limited to small ketones in the gas phase) have been carried out using molecular modeling in order to have an access to the reaction rate constants and to identify the MOs involved in the cleavage reaction [56].

Calculation of the reaction rate constant k_{TST} can be achieved using the transition state theory (TST) as done for a unimolecular reaction [57]. Quite good trends between k_{TST} and the experimental values k_{exp} are obtained but, generally, k_{exp} does not exactly correspond to the calculated value. A 10-fold factor is even considered as excellent, the discrepancy being due to the measurement uncertainty and the calculation approximations (partition functions and calculations in gas phase).

In a recent work that considered a series of aryl alkyl ketones [58], it was shown that (i) the tunneling effect is weak, (ii) the reaction enthalpy ΔH_r influences the activation energy, (iii) the polar effects in terms of partial charges δ in the TS play a significant role (a rough trend between the rate constants and δ was noted), and (iv) the change of the pre-exponential factor of the Arrhenius' law might be an important parameter (this change was ascribed to the entropy variations probably because of the difference in the geometry relaxation occurring during the cleavage reaction).

6.4
Hydrogen Transfer Processes

As stated above (Eq. (5.19)), hydrogen transfer can be divided into two processes. In the first one, the direct or pure hydrogen transfer (PHT) (or H-abstraction) directly generates two radicals. In the second one, the electron/proton hydrogen transfer (EPHT) corresponds to an electron transfer (that yields a CTC) followed by a proton transfer, which creates the radicals [59–62].

The PHT process between a ketone in its triplet state and a H-donor is also described by a state correlation diagram [15]. Only $n\pi^*$ states are reactive. The BDE is a driving factor.

The EPHT process can involve singlet or triplet states having an $n\pi^*$ or a $\pi\pi^*$ spectroscopic character. The electron transfer step is explained on the basis of a well-known approach in photochemistry (Marcus and Rehm–Weller theories) [16]. The proton transfer ability is not fully understood yet. The efficiency of the proton transfer step is affected by a competitive back electron transfer reaction [17].

The overall free energy change ΔG^0 is given by Eq. (6.4) where E_{red} is the reduction potential of the electron acceptor, E_{ox} and $E_{0,0}$ are the oxidation potential and the excitation energy of the electron donor. The E_{ox} value is connected with the IP of the donor (IP $= a + b\, E_{ox}$). When ionic species are present, a coulombic term C is added (Eq. (6.5)).

$$\Delta G^0 = E_{ox} - E_{red} - E_{0,0} \tag{6.4}$$
$$\Delta G_{eT} = \Delta G^0 + C \tag{6.5}$$

In the Marcus theory [16], the nuclear configurations of the reactant and product encounter pair fluctuate and are different for the reactant and the product except at the crossing point of the PES, where they are identical. The electron transfer rate constant k is a function of the free energy of activation ΔG^* (Eq. (6.6)), which contains a term corresponding to the reorganization energy. An inverted region is predicted (i.e., for very exothermic processes, the rate constants decrease [63]).

$$k = Z \exp\left(\frac{-\Delta G^*}{RT}\right) \tag{6.6}$$

In the Rehm–Weller approach [16], a dependence of the experimental electron transfer rate constant k_q on ΔG_{eT} (similar to that given by Eq. (6.6) considering that ΔG^* is replaced by a function of ΔG_{eT}) is observed. In the exothermic region, k_q is governed by the diffusion, and in the endothermic region, k_q dramatically decreases. The inverted region has not been observed in compounds used as photoinitiating systems.

6.5
Energy Transfer

In photoinitiating systems, energy transfer usually occurs through a collisional (or an overlap) mechanism that is operative at short distance; it involves triplet states. The energy requirement (Section 5.5) makes decisive the evaluation of the triplet-state energy E_T. The E_Ts can be experimentally determined through

1) Phosphorescence measurements [15] (a spectroscopic energy E_T^{sp} is determined).
2) Energy-transfer measurements [15] with known triplet energy donors or acceptors (a relaxed energy E_T^{rel} is usually expected but this is not necessarily true in any case). They are largely used (Figure 6.1).
3) LIPAC measurements (a relaxed energy E_T^{rel} is obtained).

In the usual energy transfer (vertical energy transfer (VET)), the curves (Figure 6.1) are generally well fitted by the Sandros' model [64] where the energy transfer rate constant k is linked to the energy difference between the unknown triplet energy and the donor energies in photosensitized experiment or the acceptor energies in quenching experiment (e.g., as in Eq. (6.7) where E_D is the triplet donor energy and E_A the unknown acceptor energy). The determined triplet energy corresponds to a value between E_T^{sp} and E_T^{rel}.

$$k = \frac{k_{diff}}{1 + \exp\{-(E_D - E_A)/RT\}} \tag{6.7}$$

Sometimes, the decrease of the energy transfer efficiency is less important than expected in the endothermic region. This behavior is called a nonvertical energy transfer (NVET) and appears when a change of geometry arises between the relaxed S_0 and T_1 states and decreases the S_0–T_1 energy difference [65–68]. For example, the NVET character of flexible molecules could be related to a bond torsion (see [69] and references therein). Thermally activated molecules can find configurations from which the energy required to reach the triplet PES is lower than that necessary when starting from the ground state. This occurs, for example, in

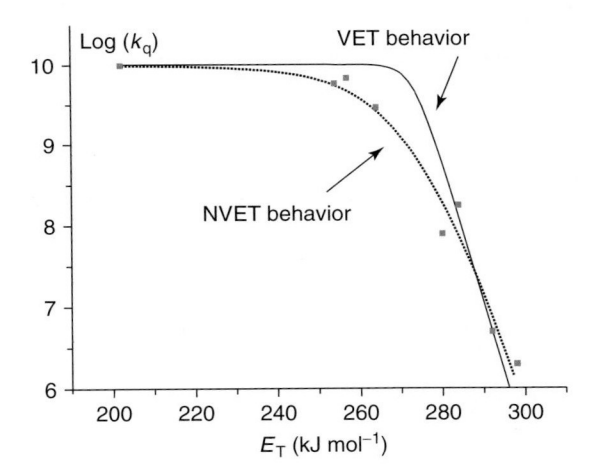

Figure 6.1 Typical VET and NVET energy transfer curves for the quenching of a molecule by various energy acceptors.

the NVET involving oxime or stilbene derivatives where a thermal bond activation model allowed to correctly fit the curves and gave an access to E_T^{sp}.

When NVET occurs, the curves are often experimentally fitted by the Balzani's model [70] (derived from the Agmon–Levine's model used in electron transfer reactions), which proposes the following relationship (Eq. (6.8)) (where k_{ET} is the energy transfer rate constant), which is qualitatively similar to Eq. (6.6) and contains a term of reorganization energy. The E_T value refers here to E_T^{rel}.

$$k = \frac{k_{diff}}{1 + (k_{-diff}/k_{ET}) \exp\{-(E_D - E_A)/RT\}} \quad (6.8)$$

For molecules that exhibit the usual VET character and a low relaxation energy (e.g., rigid molecules), E_T^{rel} and E_T^{sp} are almost the same. When they present a high relaxation energy, E_T^{rel} and E_T^{sp} are different, for example, 51 vs. 60 kcal mol^{-1} for styrene where an important relaxation occurs along a coordinate (the double-bond torsion here) that cannot be thermally activated. MO calculations allow to get E_T^{rel} and E_T^{S0-T1}.

Apparent discrepancies between the values determined by the various experiments and the calculations are obviously often noted. In most cases, in PIs, there is no dramatic geometry change and the discrepancies are not so crucial. On the contrary, the use of a triplet energy value in a flexible molecule (in which a substantial geometry change can occur) is not trivial. Indeed, deducing the occurrence of a photosensitized reaction on the basis of a triplet energy value found somewhere in the literature should be done with a special care. A combination of phosphorescence, triplet–triplet energy transfer, and LIPAC experiments is worthwhile to determine the triplet-state energy levels (relaxed, spectroscopic).

6.6
Reactivity of Radicals

According to the Arrhenius's law, the reaction rate constant for the addition of a radical to a monomer is a function of the activation energy (the barrier height). The reaction is usually described by a state correlation diagram (e.g., [71]), which depicts the reaction profile (expressed as the evolution of the energy when going from the reactants to the products) as a function of the reaction coordinate (Figure 6.2). It involves the reactant ground-state configuration $R^{\bullet}/M$ (where M is the double bond) and the excited reactant configuration $R^{\bullet}/^3M$. The interaction between the singly occupied molecular orbital (SOMO) of the radical and the lowest unoccupied molecular (LUMO) of the double bond induces a charge transfer from the radical to the monomer (R^+/M^-) in the TS and changes the TS energy. On the contrary, the interaction between the SOMO and the highest occupied molecular (HOMO) leads to a charge transfer from the monomer to the radical (R^-/M^+). These two polar charge transfer configurations R^+/M^- and R^-/M^+ mix with the other two ($R^{\bullet}/M$ and $R^{\bullet}/^3M$) and lead to a change of the crossing region. In addition to possible repulsive steric effects, the exothermicity of the reaction (i.e., the reaction enthalpy ΔH_r) as well as the polar effects strongly affect the barrier height.

The barrier is expressed by Eq. (6.9) where E_{enth} is the enthalpy term and ΔE_{pol} the polar energy change. The separation of the relative contributions of the enthalpy and polar effect of the activation energies allows explaining the reactivity.

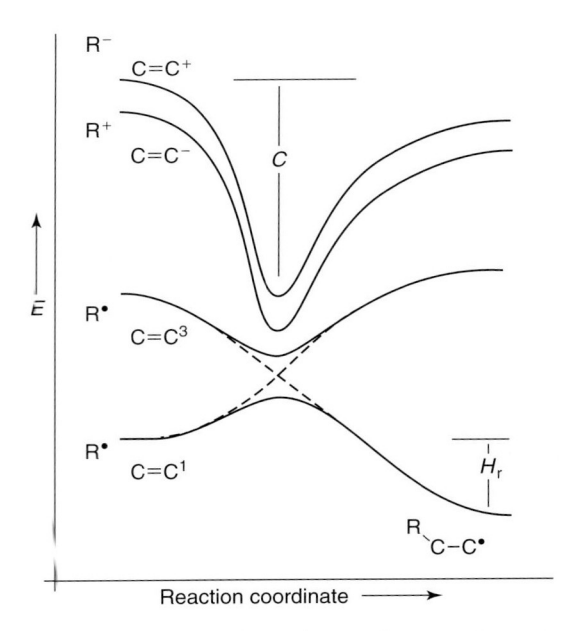

Figure 6.2 State correlation diagram for the addition of a radical to a double bond [71]. (Source: Reproduced with permission of Wiley.)

The interpretation of free radical processes, for example, the addition or H-transfer reaction is a fascinating challenge. For example (Section 17.4), in addition reactions that are only governed by the exothermicity, the barrier E_a is usually described by a linear relationship with ΔH_r (Eq. (6.10)).

$$E_a = E_{enth} - \Delta E_{pol} \tag{6.9}$$

$$E_a = a + b\Delta H_r \tag{6.10}$$

The electron deficient or electron rich character of a double bond or a radical is represented by the absolute electronegativity χ and the hardness η, which are connected with the IP and the EA (Eq. (6.11)). The polar energy change ΔE_{pol} is expressed from the χ and η values of the monomer and the radical.

$$\Delta E_{pol} = \frac{(\chi_M - \chi_R)^2}{4(\eta_M - \eta_R)} \quad \text{with} \quad \chi = \frac{IP + EA}{2} \quad \text{and} \quad \eta = \frac{IP - EA}{2} \tag{6.11}$$

Calculating E_a and ΔE_{pol} yields E_{enth} (Eq. (6.9)). Linear relationships between E_{enth} and ΔH_r are generally found (Eq. (6.12)). The E_{enth} term is also expressed as a function of E_a^0 (barrier at $\Delta H_r = 0$ in the absence of any polar effect) and ΔE_{enth} (contribution of the enthalpy change to the barrier). The enthalpy change ΔE_{enth} is therefore a function of the addition reaction enthalpy $(-c\Delta H_r)$. The calculated barrier is now expressed by Eq. (6.13). From this description, the barrier can be calculated from the exothermicity of the process and from the absolute electronegativity and hardness of the reactants.

$$E_{enth} = C + c\Delta H_r = E_a^0 - \Delta E_{enth} \tag{6.12}$$

$$E_a = C - \Delta E_{pol} - \Delta E_{enth} \tag{6.13}$$

References

1. Fouassier, J.P. (1995) *Photoinitiation, Photopolymerization, Photocuring*, Hanser, Münich.
2. Fouassier, J.P. (ed.) (2006) *Photochemistry and UV Curing*, Research Signpost, Trivandrum.
3. Schnabel, W. (1990) in *Lasers in Polymer Science and Technology*, vol. II (eds J.P. Fouassier and J.F. Rabek), CRC Press, Boca Raton, FL, pp. 95–143.
4. Phillips, D. and Rumbles, G. (1990) in *Lasers in Polymer Science and Technology*, vol. I (eds J.P. Fouassier and J.F. Rabek), CRC Press, Boca Raton, FL, pp. 91–146.
5. Allonas, X., Lalevée, J., and Fouassier, J.P. (2003) in *Photoinitiated Polymerization*, ACS Symposium Series, vol. 847 (eds K.D. Belfield and J. Crivello), American Chemical Society, pp. 140–157.
6. Allonas, X., Ley, C., Gaume, R., Jacques, P., and Fouassier, J.P. (1999) *Trends Photochem. Photobiol.*, **5**, 93–102.
7. Colley, C.S., Grills, D.C., Besley, N.A., Jockusch, S., Matousek, P., Parker, A.W., Towrie, M., Turro, N.J., Gill, P.M.W., and George, M.W. (2002) *J. Am. Chem. Soc.*, **124**, 14952–14958.
8. Fedorov, A.V., Danilov, E.O., Merzlikine, A.G., Rodgers, M.A.J., and Neckers, D.C. (2003) *J. Phys. Chem.*, **107**, 3208–3214.
9. Gatlik, I., Rzadek, P., Gescheidt, G., Rist, G., Hellrung, B., Wirz, J., Dietliker, K., Hug, G., Kunz, M., and Wolf, J.-P. (1999) *J. Am. Chem. Soc.*, **121**, 8332–8336.
10. Gigmes, D., Berchadsky, Y., Finet, J.-P., Siri, D., and Tordo, P. (2003) *J. Phys. Chem. A*, **107**, 9652–9657.

11. Hayashi, H. (1990) in *Photochemistry and Photophysics*, vol. I (ed. J.F. Rabek), CRC Press, Boca Raton, FL, pp. 59–136.
12. Dreeskamp, H. and Palm, W.U. (1990) in *Lasers in Polymer Science and Technology*, vol. II (eds J.P. Fouassier and J.F. Rabek), CRC Press, Boca Raton, FL, pp. 47–56.
13. Braun, A., Maurette, M.T., and Oliveros, E. (1986) *Technologie Photochimique*, Presses Polytechniques Romandes, Lausanne.
14. Wayne, R.P. (1988) *Principles and Applications of Photochemistry*, Oxford Science and Publications, Oxford.
15. Turro, N.J. (1978 and 1990) *Modern Molecular Photochemistry*, Benjamin, New York.
16. Gilbert, A. and Baggot, J. (1991) *Essentials of Molecular Photochemistry*, Blackwell Scientific Publications, Oxford.
17. Suppan, P. (1994) *Chemistry and Light*, The Royal Society of Chemistry.
18. Carr, S.A., Baeza-Romero, M.T., Blitz, M.A., Price, B.J.S., and Seakins, P.W. (2008) *Int. J. Chem. Kinet.*, **40**, 504–514.
19. Morlet-Savary, F., Ley, C., Jacques, P., and Fouassier, J.P. (2001) *J. Phys. Chem.*, **105**, 11026–11033.
20. Baskin, J.S., Bañares, L., Pedersen, S., and Zewail, A.H. (1996) *J. Phys. Chem.*, **100**, 11920–11933.
21. Oelkers, A.B., Scatene, L.F., and Tyler, D.R. (2007) *J. Phys. Chem. A*, **111**, 5353–5360.
22. Morandeira, A., Furstenberg, A., and Vauthey, E. (2004) *J. Phys. Chem.*, **108**, 8190–8200.
23. Braslavsky, S.E. and Heibel, G.E. (1992) *Chem. Rev*, **92** (6), 1381–1410.
24. Lalevée, J., Allonas, X., and Fouassier, J.P. (2006) in *Photochemistry and UV Curing: New Trends* (ed. J.P. Fouassier), Research Signpost, Trivandrum, pp. 229–240.
25. Ong, S.Y., Chan, P.Y., Zhu, P., Leung, K.H., and Phillips, D.L. (2003) *J. Phys. Chem. A*, **107**, 3858–3865.
26. Mc Camant, D.W., Kukura, P., and Mathies, R.A. (2005) *J. Phys. Chem. A*, **109**, 10449–10457.
27. Mohapatra, H. and Umapathy, S. (2009) *J. Phys. Chem. A*, **113**, 6904–6909.
28. Allonas, X., Lalevée, J., Morlet-Savary, F., and Fouassier, J.P. (2006) *Polimeri*, **51**, 491–498.
29. Knolle, W., Muller, U., and Mehnert, R. (2000) *Phys. Chem. Chem. Phys.*, **2**, 1425–1430.
30. Knolle, W. and Mehnert, R. (1995) *Nucl. Instrum. Methods Phys. Res. B*, **105**, 154–158.
31. Takacs, E. and Wojnarovits, L. (1995) *Nucl. Instrum. Methods Phys. Res. B*, **105**, 282–284.
32. Kujawa, P., Mohid, N., Zaman, K., Manshol, W., Ulanski, P., and Rosiak, J.M. (1998) *Rad. Phys. Chem.*, **53**, 403–409.
33. Martschke, R., Farley, R.D., and Fischer, H. (1997) *Helv. Chim. Acta*, **80**, 1363–1374.
34. Lalevée, J., Allonas, X., and Fouassier, J.P. (2006) in *Photochemistry and UV Curing: new trends* (ed. J.P. Fouassier), Research Signpost, Trivandrum, pp. 117–126.
35. Scaiano, J.C. (1981) *J. Phys. Chem.*, **85**, 2851.
36. Murai, H. and Hayashi, H. (1993) in *Radiation Curing in Polymer Science and Technology*, vol. 2 (eds J.P. Fouassier and J.F. Rabek), Elsevier, Barking, pp. 63–154.
37. Mc Lauchlan, K.A. (1990) in *Lasers in Polymer Science and Technology*, vol. I (eds J.P. Fouassier and J.F. Rabek), CRC Press, Boca Raton, FL, pp. 259–304.
38. Griesser, M., Neshchadin, D., Dietliker, K., Moszner, R., Liska, R., and Gescheidt, G. (2009) *Angew. Chem. Int. Ed.*, **48**, 9359–9361.
39. Sartori, E., Khudyakov, I.V., Lei, X., and Turro, N.J. (2007) *J. Am. Chem. Soc.*, **129**, 7785–7792.
40. Weber, M., Turro, N.J., and Beckert, D. (2002) *Phys. Chem. Chem. Phys.*, **4**, 168–172.
41. Weber, M., Khudyakov, I.V., and Turro, N.J. (2002) *J. Phys. Chem. A*, **106**, 1938–1945.
42. (a) Frisch, M.J., Trucks, G.W., Schlegel, H.B., Scuseria, G.E., Robb, M.A., Cheeseman, J.R., Zakrzewski, V.G., Montgomery, J.A. Jr., Stratmann, R.E., Burant, J.C., Dapprich, S., Millam, J.M., Daniels, A.D., Kudin, K.N., Strain,

M.C., Farkas, O., Tomasi, J., Barone, V., Cossi, M., Cammi, R., Mennucci, B., Pomelli, C., Adamo, C., Clifford, S., Ochterski, J., Petersson, G.A., Ayala, P.Y., Cui, Q., Morokuma, K., Salvador, P., Dannenberg, J.J., Malick, D.K., Rabuck, A.D., Raghavachari, K., Foresman, J.B., Cioslowski, J., Ortiz, J.V., Baboul, A.G., Stefanov, B.B., Liu, G., Liashenko, A., Piskorz, P., Komaromi, I., Gomperts, R., Martin, R.L., Fox, D.J., Keith, T., Al-Laham, M.A., Peng, C.Y., Nanayakkara, A., Challacombe, M., Gill, P.M.W., Johnson, B., Chen, W., Wong, M.W., Andres, J.L., Gonzalez, C., Head-Gordon, M., Replogle, E.S., and Pople, J.A. (2001) *Gaussian 03, Revision B2*, Gaussian, Inc., Pittsburgh; (b)Foresman, J.B. and Frisch, A. (1996) *Exploring Chemistry with Electronic Structure Methods*, 2nd edn, Gaussian, Inc.

43. Bamford, C.H. and Norrish, R.G.W. (1935) *J. Chem. Soc.*, 1504–1511.
44. Sevin, A., Bigot, B., and Devaquet, A. (1978) *Tetrahedron*, **34**, 3275–3280.
45. Salem, L. (1974) *J. Am. Chem. Soc.*, **96**, 3486–3501.
46. Rauk, A. (2000) *Orbital Interaction Theory of Organic Chemistry*, 2nd edn, John Wiley & Sons, Inc., pp. 212–216.
47. Chandra, A.K. and Sumathi, K. (1990) *J. Photochem. Photobiol., A: Chem.*, **52**, 213–234.
48. Dauben, W.G., Salem, L., and Turro, N.J. (1975) *Acc. Chem. Res.*, **8**, 41–54.
49. Heine, H.-G. (1972) *Tetrahedron Lett.*, **13**, 3411–3414.
50. Heine, H.-G. (1972) *Tetrahedron Lett.*, **13**, 4755–4758.
51. Heine, H.-G., Hartmann, W., Kory, D.R., Magyar, J.G., Hoyle, C.E., McVey, J.K., and Lewis, F.D. (1974) *J. Org. Chem.*, **39**, 691–698.
52. Lewis, F.D., Hoyle, C.H., Magyar, J.G., Heine, H.-G., and Hartmann, W. (1975) *J. Org. Chem.*, **40**, 488–492.
53. Encina, M.V., Lissi, E.A., Lemp, E., Zanocco, A., and Scaiano, J.C. (1983) *J. Am. Chem. Soc.*, **105**, 1856–1860.
54. Heine, H.-G., Hartmann, W., Lewis, F.D., and Lauterbach, R.T. (1976) *J. Org. Chem.*, **41**, 1907–1912.
55. Arnaut, L.G. and Formosinho, S.J. (1985) *J. Photochem.*, **31**, 315–332.
56. Bigot, B., Devaquet, A., and Turro, N.J. (1981) *J. Am. Chem. Soc.*, **103**, 6–12.
57. Cramer, C.J. (2003) *Essentials of Computational Chemistry, Theory and Models*, John Wiley & Sons, Ltd.
58. Dietlin, C., Allonas, X., Fouassier, J.P., and Defoin, A. (2008) *Photochem. Photobiol. Sci.*, **7**, 558–565.
59. Roth, H.D. (1990) *Photoinduced Electron Transfer I*, Springer-Verlag, Heidelberg.
60. Vauthey, E., Pilloud, D., Haselbach, E., Suppan, P., and Jacques, P. (1993) *Chem. Phys. Lett.*, **215**, 264–268.
61. Mohammed, O.F., Adamczyk, K., Banerji, N., Dreyer, J., Lang, B., Nibbering, E.T.J., and Vauthey, E. (2008) *Angew. Chem. Int. Ed.*, **47**, 9044–9048.
62. Mohammed, O.F., Banerji, N., Lang, B., Nibbering, E.T.J., and Vauthey, E. (2006) *J. Phys. Chem. A*, **110**, 13676–13680.
63. Could, I.R., Ege, D., Moser, J.E., and Farid, S. (1990) *J. Am. Chem. Soc.*, **112**, 4290–4301.
64. Sandros, K. (1964) *Acta Chem. Scand.*, **18**, 2355–2374.
65. Saltiel, J., Charlton, J.L., and Mueller, W.B. (1979) *J. Am. Chem. Soc.*, **101**, 1347–1352.
66. Ramamurthy, V. and Liu, R.S.H. (1976) *J. Am. Chem. Soc.*, **98**, 2935–2941.
67. Hammond, G.S. and Saltiel, J. (1963) *J. Am. Chem. Soc.*, **85**, 2516–2521.
68. Catalan, J. and Saltiel, J. (2001) *J. Phys. Chem. A.*, **105**, 6273.
69. Lalevée, J., Allonas, X., Louerat, F., Fouassier, J.P., Tachi, H., Izumitani, A., Shirai, M., and Tsuooka, M. (2001) *Phys. Chem. Chem. Phys.*, **3**, 2721–2722.
70. Balzani, V., Bolletta, F., and Scandola, F. (1980) *J. Am. Chem. Soc.*, **102**, 2152–2153.
71. Fischer, H. and Radom, L. (2001) *Angew. Chem. Int. Ed.*, **40**, 1340–1349.

7
Efficiency of a Photopolymerization Reaction

7.1
Kinetic Laws

7.1.1
Radical Photopolymerization

The different steps of a radical photopolymerization reaction [1] are shown in Eq. (7.1). Under light irradiation of the photoinitiator (PI), initiating radicals R$^\bullet$ are created. Then they react with the monomer M to give the first macroradical. The propagation of the reaction takes place through the subsequent addition of monomer units to the growing macroradical. The last step is the mono- or bimolecular termination reaction, depending on the experimental conditions. The bimolecular termination occurs through the recombination of two polymer chains (coupling is preferred for radicals presenting small steric effect) and disproportionation (that is more important with, e.g., disubstituted radicals). The monomolecular termination arises when the photopolymerizable resin becomes so viscous or rigid that the diffusion of the remaining monomer close to the reactive centers is prevented. Therefore, the reactive centers are possibly trapped within the polymeric medium, and as a consequence, the conversion of the monomer is lower than 100%.

$$\begin{aligned}
&\text{PI} \rightarrow \text{R}^\bullet \quad (h\nu) &&\text{formation of initiating radicals} \\
&\text{R}^\bullet + \text{M} \rightarrow \text{RM}^\bullet &&\text{initiation} \\
&\text{RM}^\bullet + n\text{M} \rightarrow \text{RM}_{n+1}^\bullet &&\text{propagation} \\
&\text{R(M)}_n^\bullet + \text{R(M)}_m^\bullet \rightarrow \text{R(M)}_n\text{-(M)}_m\text{R} &&\text{termination (coupling)} \\
&\text{R(M)}_n^\bullet + \text{R(M)}_m^\bullet \rightarrow \text{R(M)}_n\text{-H} + \text{R(M)}_{m-1}\text{M}_{-\text{H}} &&\text{termination (disproportionation)}
\end{aligned} \quad (7.1)$$

During the reaction, a consumption of the monomer double bonds occurs. The photopolymerization profile (or the conversion–time curve) has a characteristic S-shape (Figure 7.1) and can be divided into three different regimes: (i) in the early stages after the light illumination, the excited states and the reactive species react with the inhibitors such as oxygen and stabilizers (see below) leading to an induction period during which the polymerization hardly starts; (ii) when all the

Photoinitiators for Polymer Synthesis: Scope, Reactivity and Efficiency, First Edition.
Jean-Pierre Fouassier and Jacques Lalevée.
© 2012 Wiley-VCH Verlag GmbH & Co. KGaA. Published 2012 by Wiley-VCH Verlag GmbH & Co. KGaA.

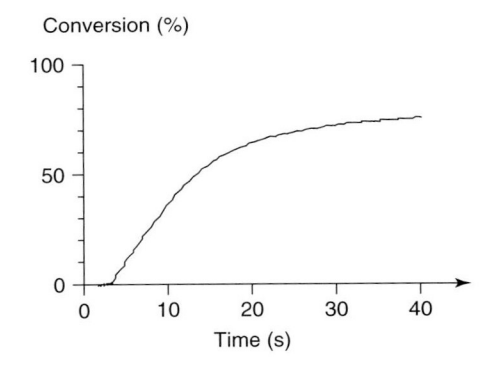

Figure 7.1 Typical time evolution of the radical photopolymerization reaction of an acrylate.

inhibitors are consumed, the reactive species react with the monomer and lead to the formation of macroradicals that propagate (at low conversion, change of refractive index already occurs); thus, the polymerization rate rapidly increases to reach a maximum value R_p^{max} (the medium becomes tack free); and (iii) the medium becomes more and more solid and the polymerization reaction slows down until a plateau is reached for the conversion. These processes have led to numerous modelization studies and are now well understood [2–6]. An important problem encountered in free radical polymerizations is the volume shrinkage that is attributable to the contraction of the material following the single- to double-bond conversion [7, 8]. Film photopolymerization of multifunctional acrylates are fast (within the 10 s timescale).

Once the light irradiation that has induced the photopolymerization has been turned off, a dark polymerization (also referred to as *posteffect*) can take place and contributes to the further consumption of the residual monomer. A radical photopolymerization exhibits a relatively low posteffect. For example, after switching off the light at 50% conversion obtained in 0.3 s for a polyurethane acrylate [9], the increase in conversion is ~10/15% (after 1 s in the dark). This posteffect is due to the presence of long-lived radicals trapped in the polymer network: for example, ESR spectra can be recorded after 3 h of storage (in air and room temperature) of the cured film [10]. When the monomer conversion increases, termination mostly occurs through radical occlusion.

In the early part of a radical photopolymerization in film, where the termination occurs through a bimolecular reaction, the rate of polymerization R_p and the initiation rate R_i are given, as in solution, by the usual relationship (Eq. (7.2)) where k_p is the propagation rate constant, k_t is the termination rate constant, [M] is the monomer concentration, I_{abs} is the amount of energy absorbed at the considered wavelength, and ϕ_i is the initiation quantum yield (number of starting chains per photon absorbed). The polymerization quantum yield ϕ_m relates to the number of monomer units polymerized per photon absorbed (Eq. (7.3)). The kinetic chain length l is expressed by Eq. (7.4).

$$R_p = \left(\frac{k_p}{k_t^{0.5}}\right) R_i^{0.5}[M] \text{ with } R_i = \phi_i I_{abs} = \phi_i I_0(\lambda_m)\left[1 - \exp(-2.3\varepsilon_m lc)\right] \quad (7.2)$$

$$\phi_m = \frac{R_p}{I_{abs}} \quad (7.3)$$

$$l = \frac{\phi_m}{\phi_i} \quad (7.4)$$

The intrinsic reactivity of a PI is expressed by the initiation quantum yield ϕ_i. The absolute evaluation of ϕ_i from R_p through Eq. (7.2) is not easy as it requires a knowledge of k_p and k_t and a precise measurement of I_{abs} in a monochromatic illumination. In the literature, this has led to a lot of discrepancies in the proposed values.

Therefore a relative value, defined by Eq. (7.5), is preferably accessible using a compound selected as a reference. Equation (7.5) can also be used in a polychromatic illumination; indeed, the expression for the amount of absorbed energy defined in Eq. (7.2) is still valid provided that the term I_{abs} is replaced by an integral over the whole wavelength range. The trivial effect of I_0, ε, and concentration are thus avoided. In this way, a reactivity scale ϕ_i/ϕ_i (ref) for the photoinitiating systems (PISs) can be defined.

$$\frac{\phi_i}{\phi_i(\text{ref})} = \left\{\frac{R_p^2}{R_p^2(\text{ref})}\right\} \frac{I_{abs}(\text{ref})}{I_{abs}} \quad (7.5)$$

In the photoresist and imaging technology fields, the photosensitivity S of a formulation (Eq. (7.6)) corresponds to the amount of energy (often expressed in mJ cm^{-2}) required to polymerize half of the reactive functions (50% conversion in the time $t_{1/2}$).

$$S = t_{1/2} I_{abs} h\nu \quad (7.6)$$

The polymerization quantum yield ϕ_m observed with multifunctional acrylates in film can be very high (e.g., in the range of 200–13 000 under air or 20 000 in laminated conditions). The kinetic chain length is typically in the range of 10 000–50 000 for the most reactive formulations [11]. Photosensitivities in the imaging area can be as low as 0.1 mJ cm^{-2}.

7.1.2
Cationic Photopolymerization

The kinetic scheme for a cationic polymerization is more complicated [1]. The Brönsted H$^+$ or the Lewis acid LA$^+$ created on light exposure of the cationic PI reacts with a monomer unit. The propagation is ensured by the further reaction of the macrocation with the monomer. The termination of the cationic polymerization is generally due to the presence of nucleophilic species such as water-, amino-, or hydroxyl-containing compounds. Therefore, the cationic polymerization is sensitive to moisture. In the case of the polymerization of epoxides or vinyl ethers, the polymerization proceeds by a ring opening or a double-bond addition, respectively.

The polymerization rates are generally lower than those obtained for the acrylate monomers because of the low propagation rate constants: for example, the rates of the polymerization reaction (carried out under a laser beam) of a cycloaliphatic epoxide and a polyurethane acrylate are in a 10/16 ratio [11].

The nucleophilic character of the counteranion is of prime importance. Indeed, a strongly nucleophilic anion prevents an efficient reaction of the cationic center with the monomer. The initiation rate R_i is defined as in Eq. (7.2). Kinetic rate constants of epoxides are accessible [12]. The polymerization quantum yield of a diethyleneglycol divinylether (DVE) can reach 200 and the kinetic chain length ~400 [13]. The posteffect is important. For example, it can account for over 80% of the final conversion if the light is turned off at 10% conversion.

7.1.3
Dependence of the Photopolymerization Rate

The rates of polymerization R_p of a photopolymerizable formulation (e.g., 3–100 µm or more) are a function of both the intrinsic parameters that depend on the PI and the monomer molecular properties (1, 2) and the experimental conditions (3, 4).

1) The monomer (propagation and termination rate constants, chain transfer [14]).
2) The chemical structure of the PI (Parts 2–4) governs the photochemical and chemical reactivity (ϕ_i) and the absorption properties (λ, ε).
3) The PI and monomer concentrations, sample thickness, light intensity I_0, temperature [15], and spectral and time distribution of light [16].
4) The presence of air as oxygen quenches the excited states and reacts with radicals (see below).

All the R_p's dependencies were originally studied in many papers [11, 13, 17–19] and other papers mentioned in Part 4) dealing with radical and cationic photopolymerizations. They outline the role of

1) the PI on R_p, the residual monomer concentration, the inhibition time, the light penetration, the remaining PI concentration;
2) the film thickness on the polymerization profile;
3) the presence of oxygen on R_p, conversion, and inhibition period;
4) the acrylate monomer (acrylate structures can possess mono-, di-, tri-, or tetrafunctionality) in the conversion–time profiles, the amount of residual unsaturations, the coating hardness, and flexibility;
5) the cationic monomer (epoxides, vinyl ethers, etc.) in the polymerization profiles and post effects;
6) the other photopolymerizable monomers (self-initiating monomers, charge transfer monomer couples, thiol-enes, or thiol-ynes).

These points were largely and continuously explored in the more recent years (see e.g., in [20–27] and papers cited throughout Part 4). Examples are provided in Parts 2 and 3.

The conversion observed during a photopolymerization depends on the film thickness. For example, in a polyurethane acrylate matrix [11], the double-bond conversions after 1.5 s of exposure are 5, 20, 40, 60, and 70% for thicknesses of 3, 6, 10, 24, and 30 μm, respectively. This phenomenon is due to the fast diffusion of oxygen into the film where the oxygen molecules that are depleted by the radical oxidation reactions are replaced by new ones. This phenomenon is obviously more important in a thin film and leads to a decrease of the polymerization rate. On the other hand, the core of a thick film can be considered as an oxygen-free medium once oxygen has been consumed. To overcome the continuous reoxygenation of the 1–3 μm top layer of a film in radical photopolymerization experiments under air, a high PI concentration (i.e., a high OD) should be used.

The penetration profile of the light in a film depends on the optical density, that is, on the PI concentration and the wavelength (through ε). For example, the fraction of light absorbed f by a 30 μm film in a given condition changes from ~10 to 60% on a 10-fold increase of the concentration or the ε. On a 100-fold increase, ~100% of the light is absorbed within 10 μm.

In the fast photopolymerization reactions for coating applications, less than 10–15% of the PI is consumed during the light exposure. When the photolysis of PI leads to colored products, an inner filter effect appears, that is, the penetration of the light decreases and less amount of photons is absorbed by PI (as the other photons are consumed by these photolysis products). Interesting PIs might be those that bleach under irradiation (Section 8.6). When increasing the concentration, the absorption properties might be modified (e.g., in a dye where aggregation processes can occur: both the concentration and the ε values are changed at the same time).

In a radical film polymerization, R_p is very often proportional to $I_0^{0.5}$. Other dependencies occur, such as (i) in the presence of a monomer exhibiting unimolecular termination when a tight network is formed and decreases the chain mobility or (ii) when a more complicated kinetic scheme is involved in the system, for example, biphotonic processes in which the consecutive absorption of two photons is required can sometimes arise in PIs in conventional curing (Section 8.11).

Recently studied or revisited problems include the effect of the coating thickness [28], the temperature and penetration depth [29], the depth profile [30, 31], and the light intensity [32].

7.1.4
Laser-Induced Photopolymerization

The laser-induced polymerization exhibits some particular features with respect to the conventional irradiation by usual lamps. The polymerization develops very rapidly within the 10 μs range when using 800 W cm^{-2} delivered by a focused CW Ar$^+$ laser [33–35] compared to the 0.5, 5, and 50 ms time range for a 400 W cm^{-2} CW Ar$^+$ laser, 10 W cm^{-2} UV source, and 1 W cm^{-2} UV source, respectively [36]. After an intense laser irradiation, dark polymerization can take place even for a radical polymerization. The ultrafast photopolymerization reaction is particularly effective in applications such as laser direct imaging or stereolithography (Sections 3.6 and 3.7) and allows the formation of a solid polymer by a short and intense pulse after which the material is expanded within few seconds. The intense laser pulse creates a high concentration of initiating species that affects the polymerization in two different ways. First, it drastically increases R_p [37]. Second, the dissolved oxygen is completely consumed at the early stage after the light exposure. The corresponding time t_{th} (gel point) is used (Eq. (7.7)) to define a threshold energy E_{th} at which an insoluble polymer is formed.

$$E_{th} = I_{abs} t_{th} \tag{7.7}$$

With pulsed repetitive lasers, a single UV pulse is sufficient to produce a tack-free surface or to allow a complete insolubilization of oxygen-free resins. The repetition rate has a marked effect both on the polymer yield and on the molecular weight distribution, presumably because the polymerization develops to a high extent in the dark period between two consecutive pulses separated by a few seconds or more [38].

It was shown that for the monofunctional monomers studied in [39], the polymerization process requires the same energy whatever be the light intensity. In such a case, this obeys the reciprocity law (Section 5.6). Interestingly, the formation of the polymeric network was found to linearly depend on the light intensity. Indeed, increasing the laser intensity leads to the generation of more radicals and therefore, the kinetic chain length decreases. This could affect the final properties of the materials [40, 41].

7.1.5
Kinetics of the Photopolymerization in Bulk

The kinetics of a polymerization reaction corresponds to the analysis of the evolution of the monomer conversion as a function of time. This evolution is accompanied by a change of the physical properties of the medium (a treatment of the photomechanics of light-activated polymers is reported in [42]). The complete investigation of the overall kinetics of the fast-film photopolymerization of multifunctional monomers is a rather hard and complex task [4]. In a quasi-stationary regime, it is admitted that the usual R_p expression for a free radical chain process can be successfully applied only at the early stages of such a polymerization (see above). A network is rapidly formed, even at low conversion, and the viscosity increases: as a consequence, the propagation k_p and termination k_t rate constants will change as a function of the conversion C as well as the initial quantum yield, which is dependent on the cage-escape reactions of the radicals. On the basis of the reaction diffusion concept [43], a relationship (Eq. (7.8)) for the photopolymerization of difunctional/monofunctional acrylate coatings having $T_g = -5\,°C$ and $+63\,°C$ has been recently proposed [4] in the range $C = 20-100\%$. This means that for $C > 20\%$, k_p and k_t are not constant. For monofunctional monomers, these rate constants can be considered as almost not affected until $C < 80\%$.

$$\frac{k_t}{[k_p(1-C)]} = \text{constant} \tag{7.8}$$

Various aspects of the polymerization kinetics have been studied, namely, kinetics and structural evolution [44] or characterization of nanoheterogeneities [45] in crosslinking polymerizations, gel time prediction in multiacrylates [46], comparison of the temperature dependence of siloxanes and dimethacrylates [47], viscosity effects [48] and thickness effects [49], time-intensity superposition laws [50], nonstationary polymerization kinetics [51, 52], stationary kinetics of a linear polymerization in the high conversion domain [53], sequential photoinduced living graft polymerizations [54], chain length dependence [55, 56], modeling of the thermal and optical effects [57], modeling of the polymerization reaction during the formation of holographic gratings [58], autoacceleration effects [59], kinetic model for radical trapping [60], determination of copolymerization ratios [61], heat and mass transfer effects in thick films [62], bulk photopolymerization at high conversions [63], propagation and termination kinetics [64], monomer segregation [65], dark photopolymerization [66], photoinitiation rate in thick coatings [67] and its spatial and temporal evolution [68], use of light intensity gradients in high throughput conversion analysis [69], role of the persistent radicals, and transfer reactions in the acrylate postpolymerization [70]. Polymerization kinetics have been explored in thiol-enes and thiol-acrylates [71–74], acrylate/diacrylate copolymers [75, 76], cationic monomers [77], acrylamide in inverse emulsion [78], novel methacrylic monomers [79], and highly reactive acrylate monomers (HRAMs) [80, 81] (in HRAM, a direct correlation between the hydrogen bonding/dipole moment and the rate of polymerization has been proposed; the hydrogen bonding affects the tacticity of the formed polymer; the propagation and termination steps are influenced by the hydrogen bonding and the dipole moment, respectively). Frontal photopolymerization [82–84], cure depth in photopolymerization [85–87], pulsed laser polymerization (PLP) [88–90], and potential effects of oxygen in the presence of high-power lasers [91] were also studied.

7.2
Monitoring the Photopolymerization Reaction

The investigation of the photopolymerization reactions is based on different kinds of analytical methods reported in the relevant literature or in books [8, 92–95].

7.2.1
FTIR Analysis

The most popular method is the FTIR analysis of the polymerizable medium. For example, in a radical photopolymerization, the monomer double bonds disappear as a function of the light exposure time t (Figure 7.1). Following the decrease of a suitable IR band allows to calculate (i) the percentage conversion using Eq. (7.9) (where A_0 and A_t represent the IR band area at time $= 0$ and t, respectively) and (ii) R_p at the origin of the % conv $= f(t)$ curve.

$$\% \text{ conv} = 100 \frac{(A_0 - A_t)}{A_0} \tag{7.9}$$

This can be achieved either point by point (by measuring $A(t)$ at different exposure times) or continuously by using, as originally proposed in [96], a real-time Fourier transform infrared technique (RT-FTIR, operating in transmission in the 500–4000 cm^{-1} range) where the sample is simultaneously exposed to a light source for excitation and an analyzing IR beam at a given wave number.

7.2.2
Photocalorimetry

Photocalorimetry measurements (photo differential scanning calorimetry (DSC)) are also often carried out [97]. They require the use of a differential scanning calorimeter in which the sample is irradiated. The heat flow dH/dt (which is related to R_p) is measured [98] and the total amount of evolved heat ΔH (related to the conversion degree) is calculated using Eq. (7.10) where ΔH_0 stands for the standard heat of polymerization (with a and b = constants).

$$R_p = a \left(\frac{1}{\Delta H_0} \right) \frac{dH}{dt}; \; \Delta H = b \int \left(\frac{dH}{dt} \right) dt; \; \% \text{ conv} = 100 \frac{\Delta H}{\Delta H_0} \tag{7.10}$$

7.2.3
Optical Pyrometry

In optical pyrometry [99], the temperature of the sample during the course of the photopolymerization is recorded by measuring the infrared emission of the medium using an IR camera. The temperature rise reflects the monomer conversion; the slope of the temperature curve is clearly correlated with the rate of polymerization for the first time of the reaction.

7.2.4
Other Methods

Other methods include (see, e.g., in [8, 92, 99] and references therein) the following:

1) Raman spectroscopy [100].
2) Attenuated reflectance (ATR), multiple internal reflectance (MIR), near-IR reflectance (NIR) IR spectroscopy [101].
3) Nuclear magnetic resonance (NMR) spectroscopy [102].
4) Photoacoustic spectroscopy. The photoacoustic signal is connected with the intensity of the IR bands [103].
5) Microwave dielectric measurements. The loss factor $\Delta \varepsilon'$ (i.e., the effective dielectric constant variation) of the polymerizing mixture placed in a microwave cavity is related to the monomer concentration [104].

6) Dilatometry and shrinkage [105] measurements. The volume contraction (typically a small percentage) or the thickness decrease of the film observed during the polymerization can be directly connected with the degree of conversion.
7) Holographic methods [106]. They rely on the fact that the refractive index n of a photopolymerizable film increases with the monomer conversion. The refractive index modulation $(n - n_0)$ is calculated from the diffraction efficiency η (ratio of the intensity of the diffracted beam to that of the incident beam when the sample is exposed to a reading laser beam). In a general way, the changes of η reflect the changes of the monomer conversion.
8) Gel point determination [99]. The gel time measurement gives the characteristic increase in viscosity of the liquid sample as it changes during the photopolymerization reaction.
9) Fluorescence emission of a probe sensitive to the local viscosity change. The fluorescence intensity is connected with the conversion [107–111].
10) Change of the dynamic viscosity, rheological analysis, IR radiometry, laser nephelometry (change of turbidity), gravimetry for solution polymerization, and so on [112].

7.3
Efficiency versus Reactivity

One has to make a distinction between the practical efficiency and the reactivity of a PI in the sense of organic chemistry. The practical efficiency is truly expressed by the rate of polymerization R_p. It corresponds to the result obtained in the selected experimental conditions and is of interest in practical applications. The reactivity is expressed by ϕ_i. It refers to the driving role of the chemical and photochemical processes and is not dependent on the experimental conditions (in the absence of concentration and light-intensity-dependent reactions).

Usually, in the given experimental conditions, the R_p values are determined relative to that of a PI chosen as a reference R_p (ref) and the relative efficiency is calculated from R_p/R_p (ref). As stated above, the R_ps of a monomer in the presence of various PIs on a given light source depend on the absorbed energy I_{abs}, which is a function of the light intensity, PI molar extinction coefficients, and concentration. R_ps are usually evaluated using the same PI concentration, and the relative efficiency does not take into account the change in the absorbed energy due to the fact that the PI absorption spectra are different. As the initiation quantum yield, ϕ_i is rather hard to determine (Section 7.1.3), a better and easier way consists in using Eq. (7.5) to define the relative reactivity by ϕ_i/ϕ_i (ref). In this expression, the total absorbed energy I_{abs} is taken into account.

In order to illustrate this important fact, let us consider a situation where two PIs A and B are used, B having a better matching of its absorption spectrum with the emission spectrum of the light source and compound A exhibiting a better reactivity, for example, $\phi_i(A) = 2\phi_i(B)$. If $I_{abs}(A)/I_{abs}(B) = 1/8$ at the same [A] and [B], for example, under a polychromatic light excitation, the $R_p(A)/R_p(B)$ ratio is equal to 1/2. PI A will be said to be less efficient than B. On the other

Table 7.1 Effect of various photoinitiating systems (at the same w/w concentration) on the photopolymerization of a polyurethane acrylate resin.

Photoinitiating system	f (%)	R_p (mol kg^{-1} s^{-1})	$R_p/f^{1/2}$ (mol kg^{-1} s^{-1})
BP/MDEA	17	20	4.8
CTX/MDEA	60	70	9
HAP	13	110	30.5
DMPA	30	120	21.9
HCAP	12	140	40.4
TPMK	43	180	27.4

BP, benzophenone; CTX, chlorothioxanthone; DMPA, 2,2-dimethoxy -2 phenyl-acetophenone; HAP, 2-hydroxy-2- methyl-1- phenyl-1- propanone; HCAP, 1-hydroxy- cyclohexyl-1- phenyl ketone; MDEA, methyldiethanolamine; TPMK, 2-methyl-1-(4-methylthiobenzoyl)-2-morpholino-propan-1-one.
Fraction of light absorbed f by the formulation and measured rate of polymerization R_p. Calculated corrected rate of polymerization $R_p/f^{1/2}$. See text. The formulas of BP, CTX, MDEA, HAP, DMPA, HCAP, and TPMK can be found in Part 2.
Source: From Ref. [113].

hand, in other irradiation conditions, using $I_{abs}(A)/I_{abs}(B) = 1$ (by adjusting the concentrations of A and B), for example, under a monochromatic light, excitation leads to $R_p(A)/R_p(B) = \sqrt{2}$: PI A will be claimed more efficient than B. This outlines how the term *efficiency* should be cautiously considered.

The experimental results shown in Table 7.1 exemplify this point. The measured R_ps (from [113]) reflect the relative practical efficiency of the PISs, whereas the corrected values $R_p/f^{1/2}$ reflect the relative reactivity of these PIS. Therefore, the most efficient system is TPMK (>HCAP) but the most reactive is HCAP (>TPMK).

In common language, the term *efficiency* instead of reactivity and vice versa is, unfortunately, often used, which could lead to a misunderstanding.

7.4
Absorption of Light by a Pigment

As shown in Figure 7.2, competitive light absorption by a pigment can be detrimental to the light absorption by the PI. As a consequence, the design of a system containing a photosensitizer (PS) that absorbs in the spectral window offered by the pigment is obviously necessary [114].

The pigments used in the radiation curing technologies consist, for example, in (i) white pigments such as TiO_2 (rutile or anatase), oxides (e.g., ZnO or ZnS), sulfates (e.g., $BaSO_4$), titanates (e.g., Mg TiO_3); (ii) mineral pigments such as Fe oxides, Cd sulfides, Pb chromates, Cr oxides, silico aluminates, and mixed metallic oxides; (iii) organic pigments based on various structures such as azoic groups, phthalocyanine, anthraquinone, oxazine, azomethine, and thioindigo; and (iv) carbon black.

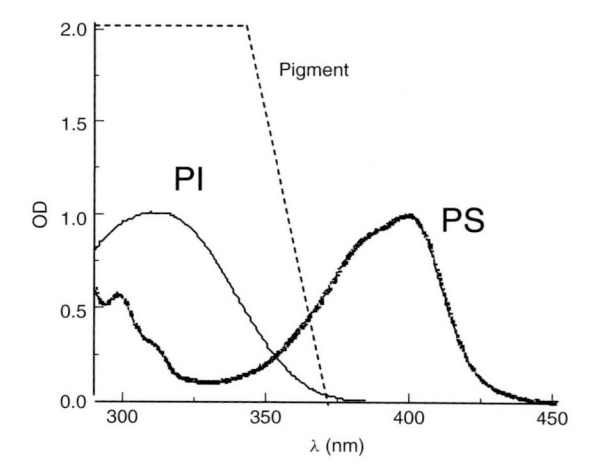

Figure 7.2 Typical absorption of a white pigment, a photoinitiator (PI), and a photosensitizer (PS).

When incorporated into a photocurable coating, a pigment has several effects such as (i) direct light absorption that decreases the amount of light absorbed by the photosensitive system, (ii) light reflection or diffusion capability that affects the transmission of the light through the film (a part of the incident light is reflected by the surface and the remaining light is absorbed by the PI and the pigment or diffused by the pigment particles through many internal reflections), and (iii) a possible photocatalytic or photosensitizing effect (in fact, the light absorption by the pigment can also result in the generation of initiating species – see Section 11.2).

The UV curing efficiency of pigmented media (in terms of the polymerized thickness) depends on the color and nature of the pigment (transparent platelet-shaped pigments, pearl luster pigments, and metallic pigments). The photopolymerization of inks (that can be considered as thin pigmented coatings) under air require to use a high PI concentration and a highly reactive PI [115] to ensure a complete curing of the aerated surface. In the case of paints (thick pigmented coatings), the situation is complicated. The PI concentration should be adjusted to allow penetration of the light in the coating together with an efficient surface cure [116, 117]. These two antagonistic requirements are almost impossible to fulfill. A blend of two PIs (with suitable concentrations) can be used: one for the surface cure and one for the body cure. A multicomponent PIS (where additional thermal reactions occur) can also achieve both a surface and a body cure [118, 119].

7.5
Oxygen Inhibition

Oxygen is a paramagnetic species that possesses two unpaired electrons (each of them are located in the degenerate antibonding π^* orbitals. It is highly reactive toward triplet states, and singlet oxygen is generated (Eq. (7.11) where ^{1}PI stands for the PI ground state).

$$^3\text{PI} + {}^3\text{O}_2 \left({}^3\sum\right) \rightarrow {}^1\text{PI} + {}^1\text{O}_2 \, ({}^1\Delta)$$
$$^3\text{PI} + {}^3\text{O}_2 \left({}^3\sum\right) \rightarrow {}^1\text{PI} + {}^1\text{O}_2 \left({}^1\sum\right) \quad (7.11)$$

It also increases the intersystem crossing pathway (Eq. (7.12)). An electron transfer yielding the oxygen radical anion (the superoxide anion) as well as the formation of an addition biradical can also occur in the presence of a PS.

$$^1\text{PI}^* + {}^3\text{O}_2 \left({}^3\sum\right) \rightarrow {}^3\text{PI} + {}^3\text{O}_2 \left({}^3\sum\right) \quad (7.12)$$

As known, oxygen is also an efficient radical scavenger as exemplified by the usual photooxidation sequence leading to peroxyl and oxyl radicals (Scheme 7.1).

Oxygen leads to a number of detrimental effects in radical or radical-sensitized cationic photopolymerization, for example, (i) the presence of an induction period, (ii) the decrease of the polymerization rate and the final conversion, (iii) the reduction of the polymer chain length, and (iv) the formation of a tacky surface on the coating. These effects have been observed through the recording of the photopolymerization profiles or in FTIR imaging [120]. The impact on the kinetics has also been discussed [121, 122].

In photoinitiated polymerization, the excited singlet and triplet states of PI are quenched by O_2, which decreases the yield in initiating species. Both the initiating and propagating radicals (P$^\bullet$ in Scheme 7.1) are scavenged by O_2 and yield highly stable peroxyl radicals that cannot participate in any further polymerization initiation reactions. The oxygen/radical interaction is a nearly diffusion-controlled reaction. The polymerization only starts when oxygen is consumed. In highly viscous or thick samples, the reoxygenation process is quite slow, leading to an easier polymerization after the inhibition period. On the contrary, in very low-viscosity or thin samples, the reoxygenation remains efficient. Therefore, strongly reduced monomer conversions are achieved. Moreover, the lower the light intensity is, the lower the initial O_2 consumption.

To overcome the inhibitory effects of oxygen, the following methods have been proposed (e.g., see a review in [123]):

1) The purge with an inert gas such as nitrogen, usually, [124] or, in rare cases until now, carbon dioxide [95].
2) The consumption of the dissolved oxygen in the film. This can be achieved through the use of additives containing easily extractable hydrogen atoms. This has been claimed [125] in the case of amines (Scheme 7.2; see also in Section 18.5), thiols, boranes, and particular monomers (e.g., N-vinylamides).

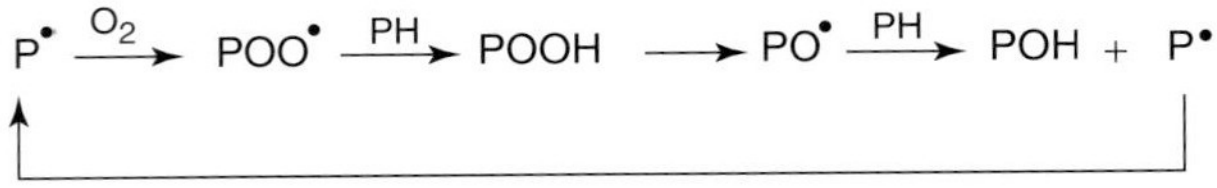

$$\text{P}^\bullet \xrightarrow{O_2} \text{POO}^\bullet \xrightarrow{PH} \text{POOH} \longrightarrow \text{PO}^\bullet \xrightarrow{PH} \text{POH} + \text{P}^\bullet$$

Scheme 7.1

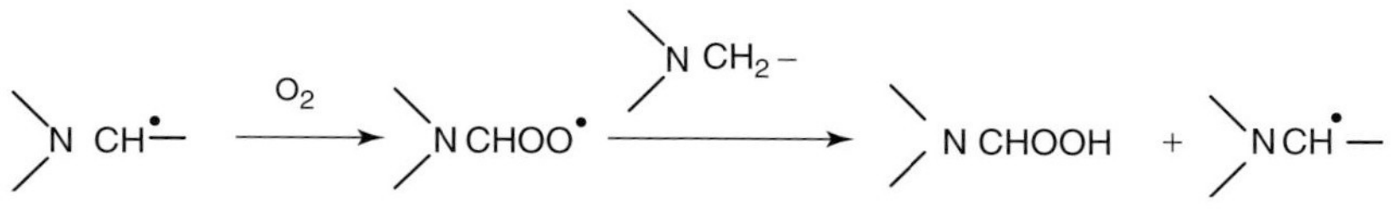

$$\diagdown\!\!\!\!_{/}\!\text{N}-\text{CH}{-}\overset{\bullet}{} \xrightarrow{O_2} \diagdown\!\!\!\!_{/}\!\text{N}-\text{CHOO}^{\bullet} \xrightarrow{\diagdown\!\!\!\!_{/}\!\text{N}-\text{CH}_2-} \diagdown\!\!\!\!_{/}\!\text{N}-\text{CHOOH} + \diagdown\!\!\!\!_{/}\!\text{N}-\overset{\bullet}{\text{CH}}{-}$$

Scheme 7.2

3) The use of trivalent phosphorus compounds, for example, triphenylphosphine PAr$_3$ [126, 127] (Section 18.6) or phosphates that scavenge the peroxyls and alkoxyls.
4) The use of a physical barrier such as paraffin waxes or a protective film (polyethylene and polyol).
5) The incorporation of a surface-active PI (carrying a fluorinated substituent) [128] or the addition of a lithium alkyl sulfonate [129].
6) The use of oxygen scavengers and sensitizers. Reports have suggested the use of a photochemical method in which two compounds G and T are added to the PI-containing system. Compound G acts as a singlet oxygen generator (through the G excited state/O$_2$ interaction) and T as a singlet oxygen trap. For example, in [130], G is methylene blue and T is 1,3 diphenyl isobenzofuran. The procedure requires two kinds of photons: a visible one to excite G and a UV one to excite PI. To avoid this drawback, the principle has just been reconsidered using compounds being able to act both as a Type II PI and G (e.g., 1,2 diones), T being a 9,10-disubstituted anthracene [131]. Other systems have been proposed [132].
7) The design of an oxygen-sensitive PI. Incorporation of an anthracene moiety into a thioxanthone skeleton (Section 9.1.13.5) yielded an interesting improvement of the radical polymerization under air [133].
8) The introduction of silanes R$_3$Si-H or boranes (L $\rightarrow$ BH$_3$). The recent introduction of the silyl chemistry allowed to enhance radical- and free-radical-promoted cationic photopolymerization reactions in aerated conditions [134, 135] (Sections 8.20, 9.1.2.7, 9.3.15, 9.3, and 12.5.2.2). Peroxyls are quenched by the silane (hydrogen abstraction) and new initiating silyls are formed.

7.6
Absorption of Light Stabilizers

The basic idea is to incorporate in the photopolymerizable formulation, before the UV curing step, one or several compounds known as *photostabilizers* or light stabilizers (LSs). Indeed, on further exposure of the cured coating, photooxidation of the polymer matrix occurs as a consequence of the influence of the UV–visible light, oxygen, humidity, and toxic agents of the atmosphere. This leads to a damage of the cured material (aging), which results in loss of gloss, cracking, color changes, yellowing, blistering, and loss of adhesion [87, 136–138].

LSs must be capable of limiting these detrimental effects. For practical use, LSs have to exhibit many properties, especially the absence of interaction with the formulation. The most usual LSs are the UV absorbers (UVAs) (based on 2-(2-hydroxyphenyl)-benzotriazoles, 2-hydroxy-benzophenone, and 2-hydroxyphenyl-triazines) and the sterically hindered amines HALS (derived from 2′,2′,6,6′−tetramethyl-piperidine) [8, 92, 139–141]. Two examples are shown in **1** and **2**.

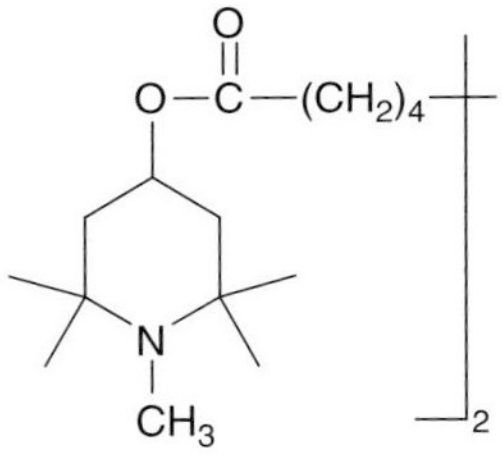

HALS
1

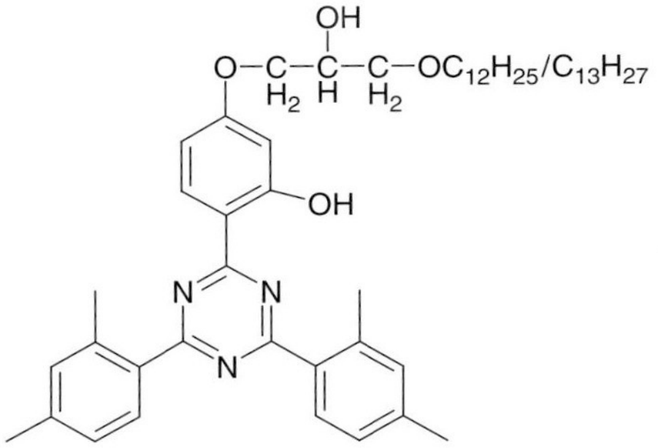

UVA
2

A UVA has to absorb the light received by the coating during its aging and must not induce any photochemical process. The deactivation of the electronically excited UVA where the H atom of the OH group has moved from oxygen to nitrogen in the case of **1** and **2** occurs according to a nonradiative mechanism (associated to a back movement of the H atom) without generation of any radical species (Eq. (7.13)).

$$\text{UVA} \rightarrow \text{UVA}^* \text{ (light) and UVA}^* \rightarrow \text{UVA} \tag{7.13}$$

A HALS predominantly behaves (Scheme 7.3) as a chain-breaking electron acceptor. The absorption of light by HALS in the presence of oxygen generates a nitroxide, which forms an alkoxyamine on combination with a radical. This alkoxyamine further reacts with peroxyl radicals, forms a peroxide, and regenerates the nitroxide.

Typical absorption spectra of a UVA and a HALS compound are shown in Figure 7.3. The polymerization efficiency is generally weakly affected by the presence of UVA and HALS. Interactions between PI excited states and LS are discussed elsewhere (Section 9.1.2.1).

The photo-oxidation and the photodegradation of polymer materials under light exposure (see e.g., in [142–144] and references therein) as well as their photostabilization are largely investigated in the relevant literature [140]. The aging of the coating results from the degradation of the starting macromolecular chains

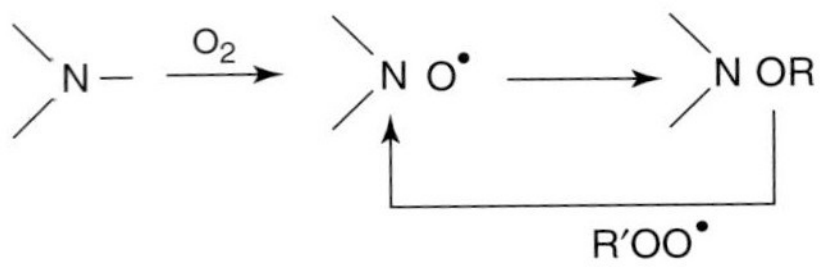

Scheme 7.3

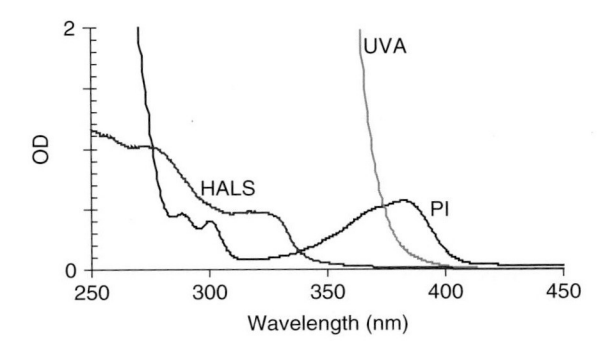

Figure 7.3 Typical ground-state absorption spectra of a UVA, a HALS, and a given photoinitiator.

through chain scission and cross-linking under daylight or sunlight exposure, which leads to a loss of the mechanical, physical, optical, and surface properties. The initial presence of residual PI molecules and photolysis products as well as the long-term formation of colored oxidation moieties on the polymer backbone on aging is responsible for the increased light absorption of the coating (and as a consequence, the increased destruction of the material) as a function of time.

7.7 Role of the Environment

The environment, that is to say everything other than the monomer/oligomer and the PIS, plays a role through the various compounds that can be present in the formulation and lead to different possible interactions, for example:

1) UVA and HALS. They exhibit a weakly important inner filter effect (see above).
2) Phenolic-type radical inhibitors. They are present in the starting monomer (for ensuring a good shell life), for example, hydroquinone methyl ether (HQME). As far as the initiation step is concerned, the excited states of a PI and the produced initiating radicals usually react with phenolic compounds. The basic mechanism (electron transfer/proton transfer vs. direct hydrogen transfer) is a subject of debate (Chapter 18); an example of the PI/phenolic compound interactions is described in Section 8.6.4.
3) Phenols. Tannins and phenolic species especially encountered in exotic woods are known [145] as strong inhibitors in the polymerization of coatings in the wood finishing industry. In fact, these species are present at the UV curable resin–substrate interface and can migrate from the wood surface into the resin. Colored phenols can directly absorb the light, thus leading to a more or less important inner filter effect. In addition, as in thermal polymerization, the presence of phenols obviously affects the propagation and termination steps of the reaction. Recent examples include the interaction of thioxanthone with

indolic and phenolic compounds [146], the near-UV photolysis of substituted phenols [147], and the role of phenols in Type I and Type II PI [148, 149].

4) Oxyl and peroxyl radicals formed during the photopolymerization in aerated media. They usually react (Sections 8.10 and 8.11) with a lot of substrates, here, especially with the polymer chains (in chain branching or chain termination reactions).
5) Moisture and water in cationic photopolymerization reactions.
6) Micelles in photopolymerization in heterogeneous media. The localization of the PI in the surfactant assembly, the effect of ionic versus nonionic PIs, the exit efficiency of the initiating radicals from the micelle, and the relative efficiency of the monomolecular versus bimolecular reactions are some of the interesting topics to be considered (Section 17.3.2).

References

1. Matyjaszewski, K., Gnanou, Y., and Leibler L. (eds) (2007) *Macromolecular Engineering: from Precise Macromolecular Synthesis to Macroscopic Materials Properties and Applications*, vol. 1, Wiley-VCH Verlag GmbH, Weinheim, pp. 643–672.
2. Lovestead, T.M., O'Brien, A.K., and Bowman, C.N. (2003) *J. Photochem. Photobiol. A: Chem.*, **159**, 135–143.
3. O'Brien, A.K. and Bowman, C.N. (2006) *Macromol. Theor. Simul.*, **15**, 176–183.
4. Khudyakov, I.V., Purvis, M.B., and Turro, N.J. (2003) in *Photoinitiated Polymerization*, ACS Symposium Series, Vol. 847 (eds K.D. Belfield and J. Crivello), American Chemical Society, pp. 113–126,
5. Khudyakov, I.V., Legg, J.C., Purvis, M.B., and Overton, B.J. (1999) *Ind. Eng. Chem. Res.*, **38**, 3353–3359.
6. Khudyakov, I.V., Fox, W.S., and Purvis, M.B. (2001) *Ind. Eng. Chem. Res.*, **40**, 3092–3097.
7. Roffey, C.G. (1982) *Photopolymerization of Surface Coatings*, John Wiley & Sons, Inc., New York.
8. Davidson, S. (1999) *Exploring the Science, Technology and Application of UV and EB Curing*, Sita Technology, Ltd, London.
9. Decker, C. and Moussa, K. (1988) *Macromol. Chem.*, **189**, 2381–2386.
10. Selli, E., Oliva, C., Galbiata, M., and Bellobono, I.R. (1992) *J. Chem. Soc., Perkin Trans. 2*, 1391–1398.
11. Decker, C. (1990) *Macromolecules*, **23**, 5217–5220.
12. Sipani, V., Kirsch, A., and Scranton, A.B. (2004) *Polym. Prepr.*, **45**, 3–4.
13. Decker, C. and Moussa, K. (1990) *J. Polym. Sci., A: Polym. Chem.*, **28**, 3429–3443.
14. Encinas, M.V. and Lissi, E. (2010) in *Basics of Photopolymerization Reactions*, vol. 2 (eds J.P. Fouassier and X. Allonas), Research Signpost, Trivandrum, pp. 1–18.
15. Corcione, C.E., Greco, A., and Maffezzoli, A. (2005) *Polymer*, **46**, 8018–8027.
16. Sun, X., Yin, D., Dai, H., Liu, J., Lu, R., and Wu, S.T. (2008) *Appl. Phys. B: Lasers Opt.*, **92**, 93–98.
17. Decker, C. and Moussa, K. (1989) *Macromolecules*, **22**, 4455–4462.
18. Decker, C. (1992) *J. Polym. Sci., A: Polym. Chem.*, **30**, 913–920.
19. Decker, C., Moussa, K., and Bendaikha, T. (1991) *J. Polym. Sci. Part A: Polym. Chem.*, **29**, 739.
20. Decker, C. (1999) in *Macromolecules*, vol. 143 (ed. K.P. Ghiggino), Wiley-VCH Verlag GmbH, Weinheim, pp. 45–63.
21. Decker, C. and Hoang Ngoc, T. (1998) in *Functional Polymers, Modern Synthetic Methods and Novel Structures*, ACS Symposium Series (eds A.O. Patil, D.N.

Schultz, and B.M. Novak), American Chemical Society, pp. 286–302,
22. Decker, C. (1998) *Des. Monomers Polym.*, **1**, 47–64.
23. Decker, C. and Elzaouk, B. (1997) in *Current Trends in Polymer Photochemistry* (eds N.S. Allen, M. Edge, I. Bellobono, and E. Selli), Ellis Horwood, New York, pp. 131–148.
24. Decker, C. (1997) in *Materials Science and Technology*, vol. 18 (ed. H.E.H. Meijer), Wiley-VCH Verlag GmbH, Weinheim, pp. 615–657.
25. Decker, C. (1996) in *Polymeric Materials Encyclopedia*, vol. 7 (ed. J.C. Salamone), CRC Press, Boca Raton, FL, pp. 5181–5190.
26. Decker, C. (2001) in *Specialty Polymer Additives* (eds S. Al Malaika, A. Golovoy, and C.A. Wilkic), Backwell Science, Oxford, pp. 139–154.
27. Decker, C. (2003) in *Photoinitiated Polymerization*, ACS Symposium Series, Vol. 847, Chapter 23 (eds K.D. Belfield and J. Crivello), American Chemical Society, p. 266.
28. Mirschel, G., Heymann, K., Scherzer, T., and Buchmeiser, M.R. (2009) *Polymer*, **50**, 1895–1900.
29. Nekkanti, V. and Jachuck, R.J.J. (2010) *J. Appl. Polym. Sci.*, **116**, 1940–1947.
30. Bao, R. and Jonsson, S. (2008) *Prog. Org. Coat.*, **61**, 176–180.
31. Bao, R. (2010) in *Basics of Photopolymerization Reactions*, vol. 1 (eds J.P. Fouassier and X. Allonas), Research Signpost, Trivandrum, pp. 175–188.
32. Baker, C.C. and De Meter, E.C. (2008) *J. Adhesion Sci. Technol.*, **22**, 1105–1121.
33. Decker, C. and Fouassier, J.P. (1993) in *Radiation Curing in Polymer Science and Technology*, vol. 3 (eds J.P. Fouassier and J.F. Rabek), Elsevier, Barking, pp. 1–36.
34. Decker, C. (1996) *Prog. Polym. Sci.*, **21**, 593–650.
35. He, B., Wang, F., Sun, X., Dai, H., and Liu, J. (2007) *J. Korean Phys. Soc.*, **51**, 1587–1592.
36. Decker, C. and Moussa, K. (1990) in *Radiation Curing of Polymeric Materials*, ACS Symposium Series, Vol. 417 (eds C.E. Hoyle and J.F. Kinstle), American Chemical Society, pp. 439–451.
37. Decker, C. (1999) *Nucl. Instrum. Methods Phys. Res. B.*, **151**, 22–28.
38. Hoyle, C.E., Trapp, M.A., Chang, C.H., Latham, D.D., and Mc Laughin, K.W. (1993) *Macromolecules*, **26**, 844–849.
39. Feng, L. and Suh, B.I. (2007) *Macromol. Chem. Phys.*, **208**, 295–306.
40. Hoyle, C.E., Hansel, R.D., and Grubb, M.B. (1984) *Polym. Photochem.*, **4**, 69.
41. Hoyle, C.E., Chang, C.H., and Trapp, M.A. (1989) *Macromolecules*, **22**, 3607–3611.
42. Long, K.N., Scott, T.F., Qi, H.J., Bowman, C.N., and Dunn, M.L. (2009) *J. Mech. Phys. Solids*, **57**, 1103–1121.
43. Schultz, G.V. (1956) *Z. Phys. Chem.*, **8**, 290–293.
44. Bowman, C.N. and Anseth, K.S. (1997) *Polym. Mater. Sci. Eng.*, **77**, 375–376.
45. Krzeminski, M., Molinari, M., Troyon, M., and Coqueret, X. (2010) *Macromolecules*, **43**, 8121–8127.
46. Boddapati, A., Rahane, S.B., Slopek, R.P., Breedveld, V., Henderson, C.L., and Grower, M.A. (2011) *Polymer*, **52**, 866–873.
47. Uygun, M., Cook, W.D., Moorhoff, C., Chen, F., Vallo, C., Yagci, Y., and Sangermano, M. (2011) *Macromolecules*, **44**, 1792–1800.
48. Andrzejewska, E. and Marcinkowska, A. (2008) *J. Appl. Polym. Sci.*, **110**, 2780–2786.
49. Zhang, Y., Kranbuehl, D.E., Sautereau, H., Seytre, G., and Dupuy, J. (2009) *Macromolecules*, **42**, 203–210.
50. Dalle Vacche, S., Geiser, V., Leterrier, Y., and Manson, J.-A.E. (2010) *Polymer*, **51**, 334–341.
51. Medvedevskikh, Y.G., Kytsya, A.R., Bazylyak, L.I., Turovsky, A.A., and Zaikov, G.E. (2006) in *Chemical Reactions in Condensed Phase* (eds. G.E. Zaikov, V.G. Zaikov, and A.K. Mikitaev), Nova Science Publishers, New York, pp. 97–137.
52. Kytsya, A.R., Bazylyak, L.I., Turovsky, A.A., and Zaikov, G.E. (2005) in *Trends in Molecular and High Molecular Science* (eds. G.E. Zaikov, Yu.B. Monakov, and A. Jimenez), Nova Science Publishers, New York, pp. 239–274.

53. Medvedevskikh, Y.G., Kytsya, A.R., Bazylyak, L.I., Turovsky, A.A., and Zaikov, G.E. (2006) *Divers. Chem. React.*, 227–249.
54. Ma, H., Davis, R.H., and Bowman, C.N. (2000) *Macromolecules*, **33**, 331–335.
55. Berchtold, K.A., Hacioglu, B., Lovell, L., Nie, J., and Bowman, C.N. (2001) *Macromolecules*, **34**, 5103–5111.
56. Lovestead, T.M. and Bowman, C.N. (2005) *Macromolecules*, **38**, 4913–4919.
57. O'Brien, A.K. and Bowman, C.N. (2003) *Macromolecules*, **36**, 7777–7782.
58. Gleeson, M.R. and Sheridan, J.T. (2009) *J. Optics. A: Pure Appl. Opt.*, **11** (24008), 1–12.
59. Goodner, M.D. and Bowman, C.N. (1999) *Macromolecules*, **32**, 6552–6559.
60. Wen, M. and Mc-Cormick, A.V. (2000) *Macromolecules*, **33**, 9247–9254.
61. Jansen, J.F.G.A., Houben, E.E.J.E., Tummers, P.H.G., Wienke, D., and Hoffman, J. (2004) *Macromolecules*, **37**, 2275–2286.
62. Goodner, M.D. and Bowman, C.N. (2002) *Chem. Eng. Sci.*, **57**, 887–900.
63. Schmelmer, U., Paul, A., Kueller, A., Jordan, R., Goelzhaeuser, A., Grunze, M., and Ulman, A. (2004) *Macromol. Symp.*, **217**, 223–230.
64. Berchtold, K.A., Randolph, T.W., and Bowman, C.N. (2005) *Macromolecules*, **38**, 6954–6964.
65. Depierro, M.A. and Guymon, C.A. (2006) *Macromolecules*, **39**, 617–626.
66. Kilambi, H., Reddy, S.K., Schneidewind, L., Stansbury, J.W., and Bowman, C.N. (2007) *Polymer*, **48**, 2014–2021.
67. Kenning, N.S., Kriks, D., El-Maazawi, M., and Scranton, A. (2006) *Polym. Int.*, **55**, 994–1006.
68. Kenning, N.S., Ficek, B.A., Hoppe, C.C., and Scranton, A.B. (2008) *Polym. Int.*, **57**, 1134–1140.
69. Johnson, P.M., Stansbury, J.W., and Bowman, C.N. (2007) *Polymer*, **48**, 6319–6324.
70. Garcia, N., Tiemblo, P., Hermosilla, L., Calle, P., Sieiro, C., and Guzman, J. (2007) *Macromolecules*, **40**, 8168–8177.
71. Scott, T.F., Kloxin, C.J., Draughon, R.B., and Bowman, C.N. (2008) *Macromolecules*, **41**, 2987–2989.
72. Johnson, P.M., Stansbury, J.W., and Bowman, C.N. (2008) *J. Polym. Sci., Part A: Polym. Chem.*, 1502–1509.
73. Cramer, N.B. and Bowman, C.N. (2001) *J. Polym. Sci., Part A: Polym. Chem.*, **39** (19), 3311–3319.
74. Reddy, S.K., Sebra, R.P., Anseth, K.S., and Bowman, C.N. (2005) *J. Polym. Sci., Part A: Polym. Chem.*, **43**, 2134–2144.
75. Johnson, P.M., Stansbury, J.W., and Bowman, C.N. (2007) *J. Comb. Chem.*, **9**, 1149–1156.
76. Johnson, P.M., Stansbury, J.W., and Bowman, C.N. (2007) *Macromolecules*, **40**, 6112–6118.
77. Crivello, J.V., Ma, J., Jiang, F., Hua, H., Ahn, J., and Ortiz, R.A. (2004) *Macromol. Symp.*, 165–177.
78. Liu, L. and Yang, W. (2004) *J. Polym. Sci., Part B: Polym. Phys.*, **42**, 846–852.
79. Ng, L.-T., Swami, S., and Jonsson, S. (2004) *Rad. Phys. Chem.*, **69** (4), 321–328.
80. Jansen, J.F.G.A., Dias, A.A., Dorschu, M., and Coussens, B. (2003) *Macromolecules*, **36**, 3861–3873.
81. Jansen, J.F.G.A., Dias, A.A., Dorschu, M., and Coussens, B. (2002) *Macromolecules*, **35**, 7529–7531.
82. Belk, M., Kostarev, K.G., Volpert, V., and Yudina, T.M. (2003) *J. Phys. Chem.*, **107** (37), 10292–10299.
83. Tao, Y., Yang, J., Zeng, Z., Cui, Y., and Chen, Y. (2006) *Polym. Int.*, **55**, 418–425.
84. Crivello, J.V. (2007) *J. Polym. Sci., Part A: Polym. Chem.*, **45**, 4331–4340.
85. Lee, J.H., Prud'Homme, R.K., and Aksay, I.A. (2001) *J. Mater. Res.*, **16** (12), 3536–3544.
86. Stephenson, N., Kriks, D., El-Maazawi, M., and Scranton, A. (2005) *Polym. Int.*, **54**, 1429–1439.
87. Aloui, F., Ahajji, A., Irmouli, Y., George, B., Charrier, B., and Merlin, M. (2007) *Appl. Surf. Sci.*, **253**, 3737–3745.
88. Buback, M., Egorov, M., and Feldermann, A. (2004) *Macromolecules*, **37**, 1768–1776.

89. Asua, J.M., Bauermann, S., Buback, M., Castignolles, P., Charleux, B., Gilbert, R.G., Hutchinson, R.A., Leiza, J.R., Nikitin, A.N., Vairon, J.P., and van Herk, A.M. (2004) *Macromol. Chem. Phys.*, **205**, 2151–2156.
90. Willemse, R.X.E. and van Herk, A.M. (2010) *Macromol. Chem. Phys.*, **211**, 539–545.
91. Castignolles, P., Nikitin, A.N., Couvreur, L., Mouraret, G., Charleux, B., and Vairon, J.P. (2006) *Macromol. Chem. Phys.*, **207**, 81–89.
92. Fouassier, J.P. (1995) *Photoinitiation, Photopolymerization, Photocuring*, Hanser, Münich.
93. Drobny, J.G. (2003 and 2010) *Radiation Technology for Polymers*, CRC Press, Boca Raton, FL.
94. Belfield, K.D. and Crivello, J.V. (eds) (2003) *Photoinitiated Polymerization*, ACS Symposium Series, Vol. 847, American Chemical Society, Washington, DC.
95. Schwalm, R. (2007) *UV Coatings: Basics, Recent Developments and New Applications*, Elsevier, Oxford.
96. Decker, C. and Moussa, K. (1988) *Eur. Pol. J.*, **135**, 57–61.
97. Ropper, T.M., Hoyle, C.E., and Magers, D.H. (2006) in *Photochemistry and UV Curing: New Trends* (ed. J.P. Fouassier), Research Signpost, Trivandrum, pp. 254–264.
98. Chen, S., Cook, W.D., and Chen, F. (2009) *Macromolecules*, **42**, 5965–5975.
99. Crivello, J.V. (2006) in *Photochemistry and UV Curing: new trends* (ed. J.P. Fouassier), Research Signpost, Trivandrum, pp. 265–277.
100. Mathur, A.M., Drescher, B., and Scranton, A.B. (2000) *J. Spectrosc.*, **15**, 36–37.
101. Scherzer, T. and Decker, U. (1999) *Nucl. Instrum. Methods Phys. Res., Sect. B*, **151**, 306–312.
102. Schweri, R. (1991) *J. Radiat. Curing*, **1**, 36–39.
103. Davidson, R.S. and Lowe, C. (1989) *Eur. Polym. J.*, **25**, 159–165.
104. Xu, K., Zhou, S., and Wu, L. (2009) *Prog. Org. Coat.*, **65**, 237–245.
105. Neo, W.K. and Chan-Park, M.B. (2005) *Macromol. Rapid Commun.*, **26**, 1008–1013.
106. Carré, C., Lougnot, D.J., and Fouassier, J.P. (1989) *Macromolecules*, **22**, 791–795.
107. Nelson, E.W. and Scranton, A.B. (1996 and 2004) *J. Polym. Sci., Part B: Polym. Chem.*, **34**, 403–408.
108. Hotta, Y., Komatsu, K., and Wang, F.W. (2009) *J. Appl. Polym. Sci.*, **112**, 2441–2444.
109. Ortyl, J., Sawicz, K., and Popielarz, R. (2010) *J. Polym. Sci., Part A: Polym. Chem.*, **48**, 4522–4528.
110. Kabatc, J., Bajorek, A., and Dobosz, R. (2010) *J. Mol. Struct.*, **985**, 95–104.
111. Peinado, C., Catalina, F., Bosch, P., Abrusci, C., and Corrales, T. (2010) in *Basics of Photopolymerization Reactions*, vol. 1 (eds J.P. Fouassier and X. Allonas), Research Signpost, Trivandrum, pp. 189–216.
112. Schall, J.D., Jacobine, A.F., Woods, J.G., and Coffey, R.N. (2007) *Polym. Prepr.*, **97**, 941–942.
113. Decker, C. and Moussa, K. (1991) *Eur. Polym. J.*, **27**, 881–889.
114. Lowe, C. (1993) in *Radiation Curing in Polymer Science and Technology*, vol. 4 (ed. J.P. Fouassier), Elsevier, Barking, pp. 87–104.
115. Valis, J. and Jasurek, B. (2010) *Basics of Photopolymerization Reactions*, vol. 3 (eds J.P. Fouassier and X. Allonas), Research Signpost, Trivandrum, pp. 133–146.
116. Dietliker, K. (2002) *A Compilation of Photoinitiators Commercially Available for UV Today*, Sita Technology, Ltd, London.
117. Dietliker, K., Hüsler, R., Birbaum, J.L., Ilg, S., Villeneuve, S., Studer, K., Jung, T., Benkhoff, J., Kura, H., Matsumoto, A., and Oka, H. (2007) *Progr. Org. Coat.*, **58**, 146–157.
118. Catilaz, L. and Fouassier, J.P. (1995) *Eur. Coat. J.*, **4**, 272.
119. Catilaz, L. and Fouassier, J.P. (2001) *J. Appl. Polym. Sci.*, **79**, 1911.
120. Biswal, D. and Hilt, J.Z. (2009) *Macromolecules*, **42**, 973–979.
121. O'Brien, A.K. and Bowman, C.N. (2006) *Macromolecules*, **39**, 2501–2506.

122. Feng, L. and Suh, B.I. (2009) *J. Appl. Polym. Sci.*, **112**, 1565–1571.
123. Gou, L., Opheim, B., and Scranton, A. (2006) in *Photochemistry and UV Curing: New Trends* (ed. J.P. Fouassier), Research Signpost, Trivandrum, pp. 301–310.
124. Crivello, J.V. and Dietliker, K. (1999) *Photoinitiators for Free Radical, Cationic and Anionic Photopolymerization*, of the Surface Coatings Technology Series, vol. 3, (ed. G. Bradley), John Wiley & Sons, Inc, NY and Sita Techn., London.
125. Decker, C. and Jenkins, A.D. (1985) *Macromolecules*, **18**, 1241–1244.
126. Hageman, H.J. (1985) *Prog. Org. Coat.*, **13** (2), 123–150.
127. Belon, C., Allonas, X., Crouxte-Barghorn, C., and Lalevée, J. (2010) *J. Polym. Sci., Part A: Polym. Chem.*, **48**, 2462–2469.
128. Hageman, J. (1984) European Patent 0, 037, 152.
129. Hageman, J.H. and Jansen, L.G.J. (1988) *Makromol. Chem.*, **189**, 2781–2788.
130. Decker, C. (1979) *Macromol. Chem.*, **180**, 2027–2030.
131. Hoefer, M., Moszner, N., and Liska, R. (2008) *J. Polym. Sci., Part A: Polym. Chem.*, **46**, 6916–6927.
132. Shenoy, R. and Bowman, C.N. (2010) *Macromolecules*, **43**, 7964–7970.
133. Balta, D.K., Arsu, N., Yagci, Y., Jockusch, S., and Turro, N.J. (2007) *Macromolecules*, **40**, 4138–4141.
134. Lalevée, J., El Roz, M., Allonas, X., and Fouassier, J.P. (2009) in *Organosilanes: Properties, Performance and Applications*, Chapter 6 (eds E. Wyman and M.C. Skief), Nova Science Publishers, Hauppauge, pp. 174–198.
135. Lalevée, J., Tehfe, M.A., Allonas, X., and Fouassier, J.P. (2010) in *Polymer Initiators*, Chapter 8 (ed. W.J. Ackrine), Nova Science Publishers, Hauppauge, pp. 201–234.
136. Fluegge, A.P., Waiblinger, F., Stein, M., Keck, J., Kramer, H.E.A., Fischer, P., Wood, M.G., DeBellis, A.D., Ravichandran, R., and Leppard, D. (2007) *J. Phys. Chem. A*, **111**, 9733–9744.
137. Musto, P., Ragosta, G., Abbate, M., and Scarinzi, G. (2008) *Macromolecules*, **41**, 5729–5743.
138. Sarvestani, A.S., Xu, W., He, X., and Jabbari, E. (2007) *Polymer*, **48**, 7113–7120.
139. Bortolus, P. (1993) in *Radiation Curing in Polymer Science and Technology*, vol. 2 (eds J.P. Fouassier and J.F. Rabek), Elsevier, Barking, pp. 603–636.
140. Gisjman, P. (2010) in *Photochemistry and Photophysics of Polymer Materials* (ed. N.S. Allen), John Wiley & Sons, Inc., Hoboken, pp. 627–680.
141. Schaller, C., Rogez, D., and Braig, A. (2008) Proceedings of the 29th FATIPEC Congress, pp. 174–183.
142. Ranby, B. and Rabek, J.F. (1975) *Photodegradation, Photooxidation, and Photostabilization of Polymers: Principles and Applications*, John Wiley & Sons, Inc., New York.
143. Rabek, J.F. (1995) *Polymer Photodegradation: Mechanisms and Experimental Methods*, Chapman & Hall.
144. Gardette, J.L., Rivaton, A., and Therias, S. (2010) in *Photochemistry and Photophysics of Polymer Materials* (ed. N.S. Allen), John Wiley & Sons, Inc., Hoboken, pp. 569–602.
145. Stoye, D. and Freitag, W. (eds) (1996) *Resins for Coatings*, Hanser Publishers, Munich.
146. Das, D. and Nath, D.N. (2007) *J. Phys. Chem. B*, **111**, 11009–11015.
147. King, G.A., Devine, A.L., Nix, M.G.D., Kelly, D.E., and Ashfold, M.N.R. (2008) *Phys. Chem. Chem. Phys.*, **10**, 6417–6429.
148. Allonas, X., Dossot, M., Merlin, A., Sylla, M., Jacques, P., and Fouassier, J.P. (2000) *J. Appl. Polym. Sci.*, **78**, 2061–2069.
149. Dossot, M., Obeid, H., Allonas, X., Jacques, P., Fouassier, J.P., and Merlin, A. (2004) *J. Appl. Polym. Sci.*, **92**, 1154–1164.

Part II
Radical Photoinitiating Systems

On light excitation, a radical photoinitiator (PI) produces different kinds of reactive species according to its chemical structure, as already explained in Section 5.7, through cleavage, hydrogen abstraction, or electron transfer. Radicals are the most widely used initiating species in photopolymerization reactions both in academic and industrial applications. This chapter is devoted to a broad presentation of the available radical PIs and their reactivity. All along the past 20 years, review chapters or books have given some general information on one or the other points [1–35]. More recent advances in this field appear in specific reviews [36–49]. The reactivity of radicals has been discussed in Part 4.

The development of radical photoinitiating systems (PISs) has induced a lot of works in many directions:

1) Increase in the chemical reactivity through the design of new or modified structures.
    a. Search for highly efficient compounds photosensitive to UV and visible lights. Since the originally and largely explored triplet-state C–C bond cleavage, this has led to new works on the ability of, for example, C–S, C–P, C–Ge, C–Si, Si–Si, S–Si, S–S, C–O, N–O bond to dissociate. In the same way, singlet-state cleavable PIs might be a new opportunity. Hydrogen-transfer reactions have been extensively studied. Although initiating carbon-centered radicals and amino alkyl radicals have been (and still are) usually encountered for a long time, other radicals are of interest currently such as the phosphinoyl, thiyl, silyl, germyl, boryl, dithiocarbamyl, and nitroxyl radicals. New radical chemistries (based on silyls and boryls) can be highly useful to overcome the oxygen inhibition in the polymerization (see below).
    b. Synthesis of improved existing structures (e.g., with triplet states exhibiting a low deactivation by the monomer or generating new efficient initiating radicals) or macrophotoinitiators for achieving new properties and enhanced performances (absorption, reactivity, efficiency, and handling).
    c. Incorporation of various chemical groups (substituents, chromophores, and conjugated chains) to change the absorption properties for recovering more

visible photons, avoiding the competitive pigment absorption, and shifting the absorption to specific visible laser lines.
2) Introduction of new properties into the starting structures in order to enhance the following:
   a. The handling of PIs in the photopolymerizable medium (better compatibility in the matrix using polymeric PI or oligomeric PI).
   b. The reduction of the migration and extractability (after light exposure) of the remaining PIs, photolysis products, and volatile or odorous by-products using copolymerizable PIs.
   c. The development of acrylated PIs for getting a high-quality surface modification of the polymer or smart priming, where the surface property of a polymer has to be modified without altering the bulk properties (the acrylated PI is deposited onto a plasma- or corona-treated polymer surface, where a grafted reaction is thus achieved; then, the surface is coated with a monomer/oligomer film and the photopolymerization can start again on light exposure).
   d. The change of the hydrophobicity (e.g., by introducing long alkyl chains) for a better solubility in solvent-borne media.
   e. The change of the hydrophilicity (by incorporating, e.g., hydroxylic chains) or the water solubility (using water-solubilizing groups) for working in water-borne media such as water-thinnable or water-soluble resins, and aqueous dispersions or emulsions of oligomers or prepolymers.
3) Development of PI systems less sensitive to oxygen and usable in free radical photopolymerization and free-radical-promoted cationic photopolymerization, especially for low-viscosity media under low light exposure. Different strategies include the use of PIs that trap oxygen to generate an initiating radical, co-initiators that scavenge the peroxyls and create a new initiating radical, and compounds that decompose the hydroperoxides.
4) Search for synergistic effects (a synergy is observed when the effect on a given property produced by a mixture of two compounds is greater than that expected from a simple addition). In other words, this research consists here in the design of the following:
   a. Efficient photosensitizer/photoinitiator (PS/PI) couples where PS should exhibit a suitable light absorption spectrum and an efficient energy or electron transfer to PIs.
   b. Multicomponent combinations to further enhance the photoinitiation ability through a judicious choice of the different compounds.

The excited-state processes and photodecomposition reactions of PIs on light excitation can be investigated (Chapter 6) through (i) steady-state photolysis and time-resolved spectroscopy techniques such as analytical chemistry methods, ESR, NMR, CIDEP, CIDNP, FTIR, Raman spectroscopy, (ii) laser flash photolysis (LFP), and (iii) molecular orbitals (MOs) calculations. This approach (i) helps in the understanding of the mechanisms involved, (ii) makes it possible to identify the degradation products that are playing a role in the further modification or the properties of the cured polymer material, and (iii) allows proposals for the design of new systems.

The practical efficiency of all these compounds has been largely evaluated in many works. In the literature, the reactivity of some particular structures was more extensively investigated than others. In the following, selected studies of reactivity are used as examples. Experimental or theoretical data available for radical PIS are given in several tables: absorption properties (wavelengths and molar extinction coefficients), reaction rate constants (cleavage, triplet-state interaction with monomer, hydrogen donor, amines, and oxygen), addition rate constants of radicals to monomer, quantum yields (dissociation and intersystem crossing (ISC)), calculated thermodynamical data (bond dissociation energy (BDE) and reaction enthalpy), transition state (TS) characteristics (energy barrier and amount of charge transfer) and are considered for discussion.

In this Part, the photoinitiating systems arbitrarily classified according to the number of components (one-component or Type I systems, two-component or Type II systems, three- or four-component systems) are presented in detail; PI-free systems as well as some other PISs are also included. The PIS reactivity has been also largely discussed in Part 4.

8
One-Component Photoinitiating Systems

8.1
Benzoyl-Chromophore-Based Photoinitiators

Many efficient PIs (that have been primarily developed for applications in the UV curing area and have driven the large use of photopolymerization reactions in other fields) are based on the benzoyl chromophore and, as a consequence, absorb in the UV or near-UV range (Figure 8.1). Their absorption spectra match quite well the 313 and 365 nm (and, of course, 254 nm) Hg lines.

The carbonyl group [50] possesses three important MOs (Scheme 8.1): n (non-bonding MO on the oxygen atom), π (bonding MO), and π* (antibonding MO); π and π* are located on the carbonyl double bond.

Different distributions of the electrons (configurations) in the three available MOs appear in Scheme 8.2. They correspond to one singlet ground state S_0 and two excited states (nπ* and ππ*), each of them having either a singlet (S_1 or S_2) or a triplet (T_1 or T_2) multiplicity. Five molecular states are thus possible: S_0, S_1 (nπ*), S_2 (ππ*), T_1 (nπ*), and T_2 (ππ*).

According to these excited-state energy levels, two electronic transitions (having a theoretically different probability to occur) can be observed: one is allowed (ππ* transition from S_0 to S_2), the other is forbidden (nπ* transition from S_0 to S_1). The maximum molar extinction coefficients reflect this transition probability: they are high ($\sim$10 000 mol^{-1} l cm^{-1}) for the ππ* transition and low ($\sim$100 mol^{-1} l cm^{-1}) for the nπ* transition. After excitation in the S_2 state, internal conversion leads to the S_1 state. The UV absorption exhibits two absorption bands: an intense one at a short wavelength (ππ* transition) around 280–300 nm and a weak one at a longer wavelength in the 320–360 nm range (nπ* transition) as qualitatively represented in Figure 8.1. A shift of the absorption maximum wavelength toward the blue (nπ* transition) or the red (ππ* transition) is usually observed on increasing the polarity of the medium.

The X and Y groups on the benzoyl chromophore (Figure 8.1) affect the absorption properties through a delocalization/interaction of the benzoyl MOs with those of the X and Y substituents. For example, the molar extinction coefficients ε of the n-π* transition are typically 24, 71, 44, 325 mol^{-1}l cm^{-1} at 366 nm for Y=H, CH$_3$ S, (CH$_3$)$_2$N, and the morpholyl group Mor, respectively, X being

Photoinitiators for Polymer Synthesis: Scope, Reactivity and Efficiency, First Edition.
Jean-Pierre Fouassier and Jacques Lalevée.
© 2012 Wiley-VCH Verlag GmbH & Co. KGaA. Published 2012 by Wiley-VCH Verlag GmbH & Co. KGaA.

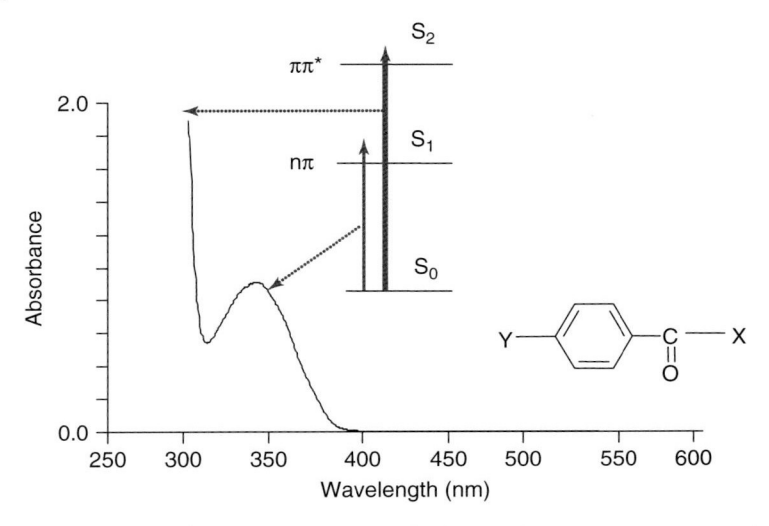

Figure 8.1 Typical absorption spectrum of a benzoyl-chromophore-containing photoinitiator (S_0–S_1 and S_0–S_2 transitions). X=Y=H.

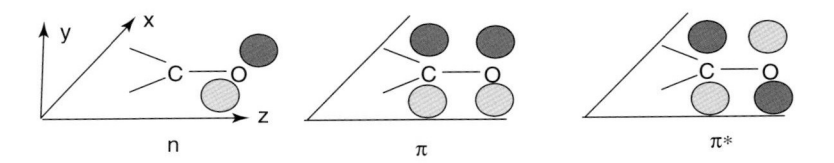

Scheme 8.1

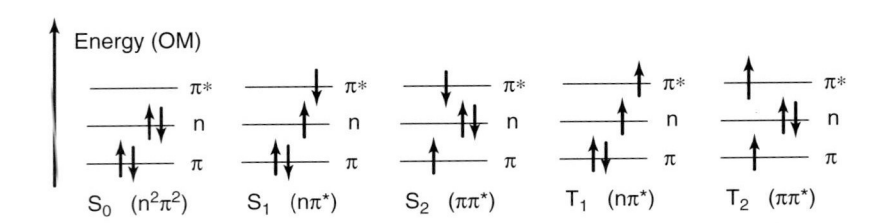

Scheme 8.2

C[(CH$_3$)$_2$]-Mor. A significant participation of the substituent in the highest occupied molecular orbital (HOMO) is found for Y=Mor and X=C[(CH$_3$)$_2$]-Mor leading to a partial π character for this MO (Figure 8.2) and explaining the increase of ε.

The benzoyl-chromophore-based PIs undergo a cleavage that can occur at the α- (Norrish I process), β-, or γ-position of the carbonyl group. Very few examples relate to an intramolecular hydrogen abstraction (Norrish II process) or to a bimolecular

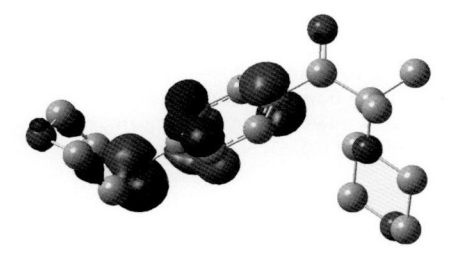

Figure 8.2 HOMO of the benzoyl chromophore for Y=Mor and X=C[(CH$_3$)$_2$]-Mor at density functional theory (DFT) level (H atoms are missing). (Source: From J. Lalevée, unpublished data.)

process (e.g., in aldehydes). In every case, the aryloyl radical has been recognized as one of the initiating radicals. The reactivity of various substituted benzoyl radicals has been elegantly studied by time-resolved infrared spectroscopy [51] (Section 18.3).

8.1.1
Benzoin Derivatives

The photochemistry of aromatic ketones has been largely investigated in the 1960s and 1970s. Acetophenone derivatives produce radicals according to a Norrish I process [50]. In former works, benzoin (BZ) and its substituted analogs [52], deoxybenzoin (DB) derivatives, and benzoin esters (BEs) (1–3) were tested as PIs ([19, 53, 54]).

A rather slow triplet cleavage is observed in DB (together with low photopolymerization ability). The addition rate constant of the benzoyl radical to a monomer double bond is ~100-fold higher than that of the benzyl radical (Section 18.3). BZ (and to a lesser 3) more efficiently cleaves on the nanosecond timescale (Table 8.1) and intrinsically appears as good PIs. The TS is stabilized by polar effects thereby leading to a Ph–C(=O)$^{\delta-}$–C$^{\delta+}$ structure. Substituents that stabilize the partial positive charge at the α-carbon lead to a lowering of the energy barrier [55]. In the same way, polar solvents stabilize TS as revealed by the 50-fold increase of the cleavage rate constant when going from hexane to methanol [56]. The cleavage of BZ and the evolution of the radicals have been deeply investigated through CIDNP-detected LFP [57] and the polymerization initiation by the benzoyl and mesitoyl radicals through size-exclusion chromatography/electrospray ionization mass spectrometry

Table 8.1 Examples of available data concerning cleavable photoinitiators based on the benzoyl chromophore: cleavage rate constant k_c (upper value taken as the reciprocal value of the triplet-state lifetimes τ_T^0), triplet-state energy levels, rate constants of interaction k_q with oxygen, monomer, and hydrogen donor (most data are taken from the references cited in the text, [19], authors' unpublished work or revisited values), intersystem crossing quantum yield ϕ_{isc}, measured dissociation quantum yield ϕ_{diss} (from Refs. [60, 61]; values in parentheses were reported in [9]).

Compound	$10^{-6} k_c$ (s^{-1})	E_T (kcal mol^{-1})	$10^{-6} k_q$ (mol^{-1} l s^{-1}) Monomer	$10^{-6} k_q$ (mol^{-1} l s^{-1}) H-donor	ϕ_{diss}
Ph–C(=O)–CH(OH)–Ph	1200	–	–	–	0.5[a]
Ph–C(=O)–C(CH$_2$OH)(OH)–Ph	–	–	–	–	0.5
Ph–C(=O)–CH$_2$–Ph	4	70	6	4 (THF)	–
Ph–C(=O)–CH(OCOCH$_3$)–Ph BAc	53	–	800 (MMA) 4800 (styrene) 60 (VA) 1300 (AN)	–	–
Ph–C(=O)–C(OCH$_3$)$_2$–Ph DMPA	4000	66 (70)	–	–	0.95; 1[a]
Ph–C(=O)–C(OPh)(CH$_2$OPh)–	100	–	–	–	–

Table 8.1 *(continued)*

Compound	$10^{-6} k_c$ (s^{-1})	E_T (kcal mol^{-1})	$10^{-6} k_q$ $(mol^{-1} l\, s^{-1})$ Monomer	$10^{-6} k_q$ $(mol^{-1} l\, s^{-1})$ H-donor	ϕ_{diss}
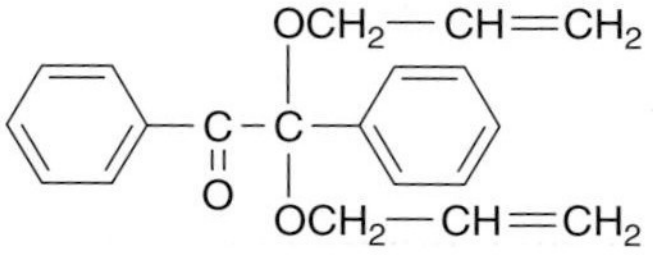 Ph–C(=O)–C(OCH$_2$–CH=CH$_2$)(OCH$_2$–CH=CH$_2$)–Ph	>1000	–	–	–	–
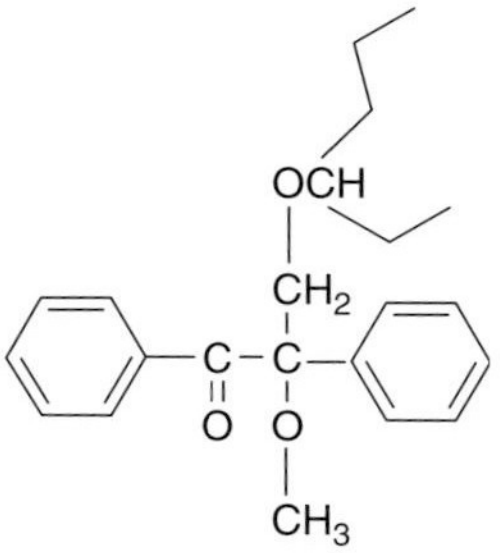 Ph–C(=O)–C(OCH(CH$_2$CH$_2$CH$_2$CH$_3$))(O-CH$_3$)–Ph	1600	–	470	–	–
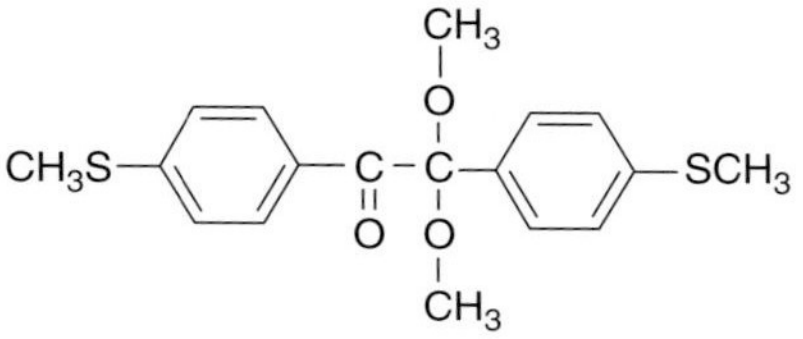 CH$_3$S–C$_6$H$_4$–C(=O)–C(OCH$_3$)(OCH$_3$)–C$_6$H$_4$–SCH$_3$	1	–	1.3 (MMA)	300 (MDEA)	–
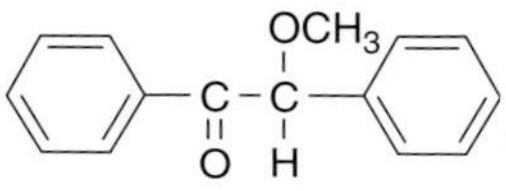 Ph–C(=O)–C(OCH$_3$)(H)–Ph BME	–	–	200	–	0.5
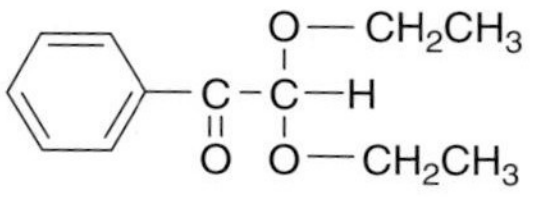 Ph–C(=O)–C(O–CH$_2$CH$_3$)(H)(O–CH$_2$CH$_3$) DEAP	>1000	–	–	–	–

(continued overleaf)

Table 8.1 (continued)

Compound	$10^{-6} k_c$ (s^{-1})	E_T (kcal mol^{-1})	$10^{-6} k_q$ (mol^{-1} l s^{-1}) Monomer	$10^{-6} k_q$ (mol^{-1} l s^{-1}) H-donor	ϕ_{diss}
DPAP	100	–	100	–	

MDEA, methyldiethanolamine; MMA, methylmethacrylate; AN, acrylonitrile; VA, vinylacetate; VE, vinylether; THF, tetrahydrofuran. See text.
[a] From [42].

[58, 59]. Interaction rate constants with monomers and hydrogen donors are given in Table 8.1 (see references in [9, 19, 60, 61]). The triplet-state energy levels of acrylates and styrene are >70 and $=62$ kcal mol^{-1}, respectively.

8.1.2
Benzoin Ether Derivatives

Benzoin ethers, among them DMPA or BME (Table 8.1) are more convenient compounds and have been the most widely used UV PIs for a long time. Their photochemistry is described in [62–67]. The UV spectrum of DMPA is characterized by a maximum absorption at 343 nm ($\varepsilon = 230$ mol^{-1} l cm^{-1}). The absorption at 366 nm (Hg line) remains important ($\varepsilon = 120$ mol^{-1} l cm^{-1}). The absorption properties are given in Table 8.2. These compounds lead to a fast curing of multifunctional acrylates (Figure 8.3).

Recent MO calculations bring some new information [68]. In the DMPA ground state, the benzoyl group is slightly twisted and the second phenyl group is clearly out of the benzoyl pseudoplane. In the triplet state, the geometry does not drastically change. The geometry plays a role in the reactivity of a PI with possible through-space interaction (Section 8.2.3.2).

The oscillator strength of the S_0–S_1 transition (mostly HOMO to lowest unoccupied molecular orbital (LUMO)) is low (0.0012). The HOMO orbital (Figure 8.4) has an important contribution of the in-plane lone pair of the oxygen atom of the carbonyl group, together with a noticeable hyperconjugation on the α-phenyl. The LUMO π^* orbital is delocalized on the whole benzoyl group, resulting in a typical $n\pi^*$ transition.

Table 8.2 Absorption properties (at $\lambda > 300$ nm) of different photoinitiators in *n*-hexane. The extinction coefficients at 366 nm (a well-known Hg line) are also given. See formulas of HAP and HCAP in Section 8.1.6, TPMK and MPPK in Section 8.2.2.

	λ_{max} (nm)	ε_{max} (mol^{-1} l cm^{1})	ε_{366} (mol^{-1} l cm^{-1})
DMPA	343	230	117
HCAP	311	93	9
HAP	305	260	5
TPMK	300	13 000	70
MPPK	308	20 900	270

From J. Lalevée (unpublished data).

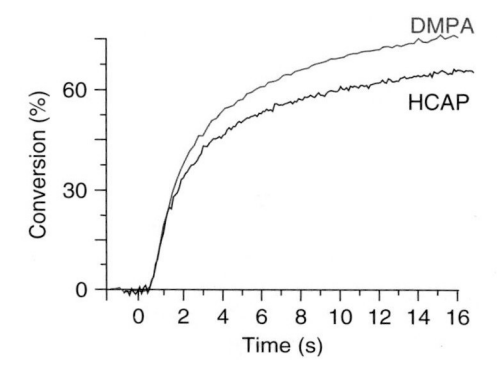

Figure 8.3 Typical photopolymerization profiles of an epoxy acrylate in the presence of DMPA and HCAP in laminate. Under polychromatic light (Hg lamp, ~11 mW cm^{-2} in the 300–450 nm range). See text. (Source: From J. Lalevée, unpublished data.)

The photolysis of DMPA in benzene [69] (Scheme 8.3) leads to benzaldehyde (quantum yield $\phi = 0.19$), benzyl ($\phi = 0.09$), and acetophenone ($\phi = 0.05$); the benzaldehyde quantum yield increases in hydrogen-donating solvents such as isopropanol. A yellowing of the photolyzed medium is noted because of benzyl and other secondary photolysis products having a quinoid structure [9]. Benzaldehyde is produced as in any benzoyl-chromophore-based PIs: this is a serious drawback for UV-curing applications in food packaging.

The cleavage process of DMPA at the C–C bond in the triplet state is very fast compared to that of some related structures (for a long time, the rate constant was estimated to be $> 10^{10}$ s^{-1} [9]). A secondary photochemical (and thermal to a lesser extent) cleavage can arise in the dimethoxy benzyl radical and leads to a methyl radical and methyl benzoate. The three radicals produced were till recently detected by ESR experiments [70].

Recent time-resolved laser spectroscopy of DMPA in acetonitrile on the picosecond timescale [68] leads to a transient absorption spectrum attributed to the triplet

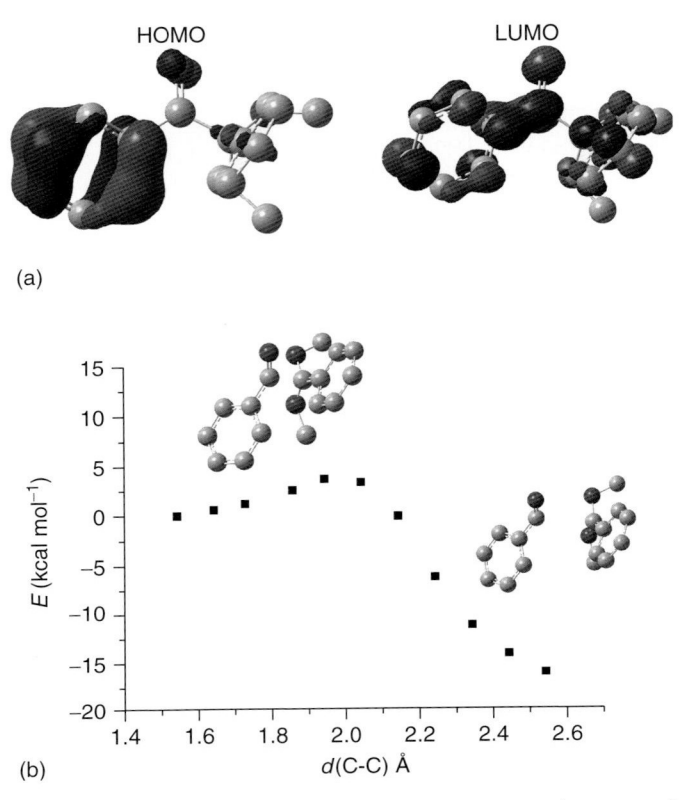

Figure 8.4 (a) HOMO and LUMOs of DMPA. (b) Potential energy surface (PES) for the cleavage of the triplet state. (Source: J. Lalevée, unpublished data.)

state that rises up within 15 ps and whose lifetime is 250 ps. The upper value of the cleavage rate constant k_c is thus 4×10^9 mol^{-1} l s^{-1}. The lifetimes are 300 ps in cyclohexane, 120 ps in acetonitrile, and 75 ps in methanol. Monomer and oxygen quenching cannot compete.

As obtained by laser-induced photocalorimetry, the cleavage quantum yield (Table 8.1) is almost unity for DMPA (0.95) [71], in line with the very short triplet state. For BME, the cleavage quantum yield is lower (0.5). The BDE of the involved carbon–carbon bond is 51 kcal mol^{-1} and the cleavage reaction enthalpy ΔH_r is -15.5 kcal mol^{-1}, rendering this reaction very exothermic. The relaxed triplet energy level is located at 68.2 kcal mol^{-1} in very good agreement with the experimental value (66.3 kcal mol^{-1} [9]) and the spectroscopic energy (69.4 kcal mol^{-1}) in the gas phase.

The geometry of the TS is close to that of the relaxed triplet state (Figure 8.4), except the length of the cleavable C–C bond that increases from 1.570 Å (triplet state) to 1.994 Å (TS). The energy barrier ΔE for the cleavage process is small as expected for a very efficient process (4.2 kcal mol^{-1}).

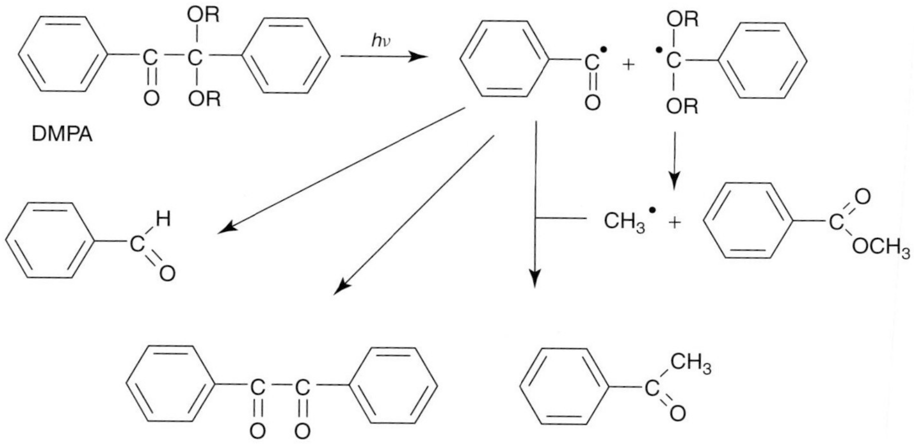

Scheme 8.3

The initiating species are the benzoyl radical, the methyl radical and, to a lesser extent, the dimethoxy benzyl radical that rather behaves as a chain-terminating agent [72] (Sections 18.1 and 18.3). In the presence of solvents (such as THF in solution photopolymerization) or substrate containing a labile hydrogen, the benzoyl radical can abstract a hydrogen atom and create a new initiating radical on the solvent or the substrate. Both, the methyl and the benzoyl radical, efficiently [73] react with monomer double bonds (Section 18.3).

The addition reaction enthalpy ΔH_r of the benzoyl radical to methylacrylate (MA) is -23.1 kcal mol^{-1}. The benzoyl radical behaves as a nucleophilic radical for the addition to an electron-deficient monomer such as MA (indeed, a high charge transfer occurs between the benzoyl and MA in TS). The polar effect is rather weak so that its contribution to the decrease in the barrier is not important. As a consequence, the barrier remains quite high and the addition rate constant is expected to exhibit a rather low value (experimental value: 2.7×10^5 mol^{-1} l s^{-1}). These calculations are currently extended to other radicals and monomers (Chapter 18).

A tentative study was done to increase the secondary cleavage route shown in Scheme 8.3 through the synthesis of derivatives containing a benzyloxybenzyl group such as in BE-OP (**4**), which was expected to undergo the formation of a benzyl radical with a lower activation energy than that of other alkoxybenzyl radicals [74]. The polymerization ability and the cleavage efficiency, however, remain lower than those of DMPA [75]. Structural modifications through the introduction (**5–6**) of a copolymerizable group in BE-cop or a polymer chain in BE-pol [31] have been studied. The reactivity is not significantly changed [76, 77] (Table 8.1) but the handling is better.

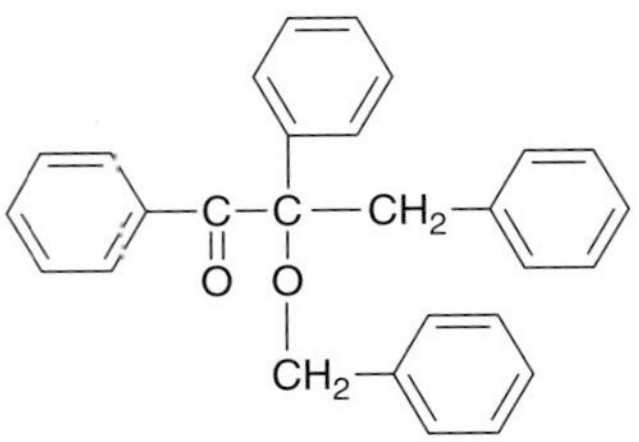

BE-OP	BE-cop	BE-pol
4	5	6

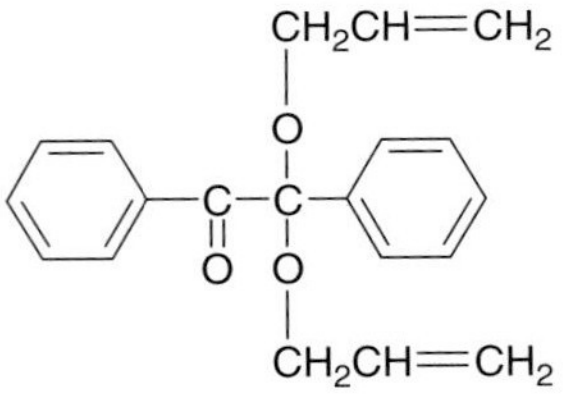

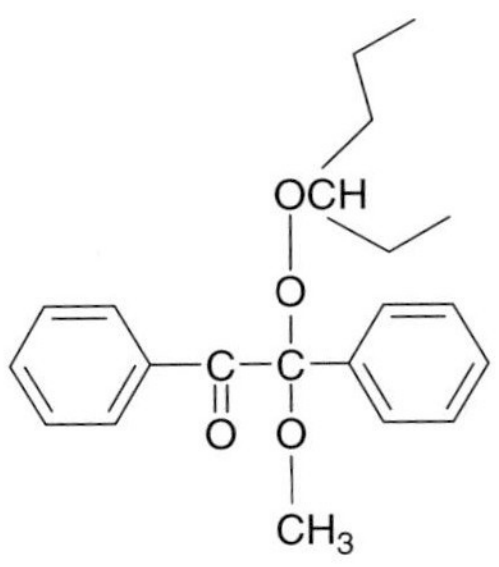

8.1.3
Halogenated Ketones

It is known for a long time that, on light exposure, α-halogenoaryl ketones (8.1) liberate a chlorine atom and a phenacyl radical through a β cleavage process [78].

$$Ph-C(O)-CH_2Cl \longrightarrow Cl^{\bullet} + Ph-C(O)-CH_2^{\bullet} \; (h\nu) \qquad (8.1)$$

8.1.4
Dialkoxyacetophenones and Diphenylacetophenones

Dialkoxyacetophenones that have been proposed in the past [79] exhibit a fast cleavage ($\geq 10^9$ s^{-1}) and a polymerization quantum yield close to that of DMPA [80]. A Norrish I (main process) and a Norrish II intramolecular hydrogen abstraction reaction (minor process) occur in the triplet state (Scheme 8.4 where R=R' for DEAP). The 1,4 biradical leads to a four-membered ring cyclization or a C_2–C_3 cleavage (with formation of an ethylenic and a ketone). On the other hand, α, α diphenylacetophenone derivatives have longer lived triplet states. The unsubstituted compound cleaves, however, within 10 ns (Table 8.1).

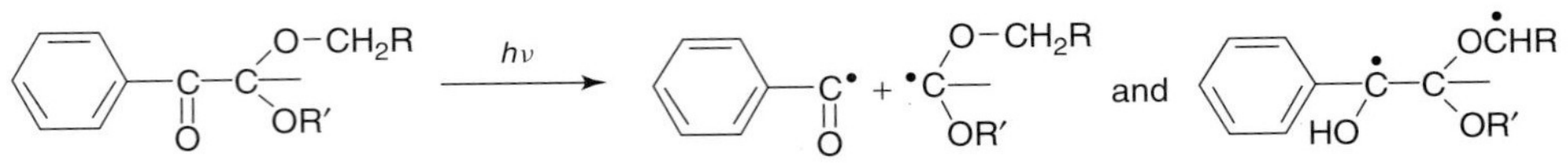

DEAP

Scheme 8.4

8.1.5
Morpholino and Amino Ketones

The search for a structure partially different from that of benzoin ethers and being able to generate another efficient initiating radical has led to the change of the dialkoxy benzyl group for a dimethylamino group or a morpholinoalkyl moiety. Excellent starting representatives [9] are PMK and PPK (**7** and **8**). They exhibit the usual absorption bands of aryl ketones (Figure 8.1), for example, for PMK: $\pi\pi^*$ transition centered at 276 nm ($\varepsilon = 3700$ mol^{-1}l cm^{-1}) with an $n\pi^*$ small shoulder around 340 nm ($\varepsilon = 100-200$ mol^{-1}l cm^{-1}).

<pre>
       PMK   PPK
        7   8
</pre>

In PMK [81], an S_0–S_1 absorption band centered at 383 nm is predicted to be a pure HOMO–LUMO transition (MO calculations) both with a moderate oscillator strength ($f = 0.0025$) and an expected significant charge transfer from the morpholino moiety to the phenyl group. The LUMO orbital is typically a π^* MO delocalized on the whole phenyl ring. By contrast, the HOMO is an n MO mainly centered on the nitrogen atom of the morpholino moiety; a significant contribution of the morpholino C–H sigma bonds occurs through hyperconjugation. Higher electronic transitions arise at 339 nm ($f = 0.0024$) and 246 nm.

PMK and PPK behave as DMPA and lead to a well-documented Norrish I cleavage in the nanosecond timescale. At high concentration and in a hydrogen-donating solvent, photoreduction can occur as revealed by ESR [82]. On light exposure, PMK and PPK present a good activity. The cleavage mechanisms and the reactivity of the excited states (Table 8.3) have been reported [4, 83]. Efficient initiating radicals are generated.

8.1.6
Hydroxy Alkyl Acetophenones

The basic structures HAP and HCAP (where the 2-hydroxyl-propyl group is changed for 2-hydroxy cyclohexyl) are also well known (Scheme 8.5 where R=CH$_3$ for HAP) [84] and still largely used and studied [19, 22, 85]. Their absorption spectra are rather similar to that of DMPA (Table 8.2). From their triplet states, HAP and HCAP yield an efficient benzoyl and 2-hydroxyl-propyl (or -cyclohexyl) radical. This last radical quickly adds to acrylates [86] (Section 18.1). Benzaldehyde and cyclohexanone can be also produced to a minor extent through an in-cage process in the HCAP radical pair. HAP and HCAP are highly efficient PIs.

Table 8.3 Examples of available data concerning cleavable photoinitiators derived from aminoketones: cleavage rate constant k_c (upper value taken as the reciprocal value of the triplet-state lifetimes τ_T^0), triplet-state energy levels, and interaction rate constants k_q with oxygen, monomers, and hydrogen donors.

Compound	$10^{-6} k_c$ (s^{-1})	E_T (kcal mol^{-1})	$10^{-6} k_q$ $(l\,mol^{-1}\,s^{-1})$ Monomer	$10^{-6} k_q$ $(l\,mol^{-1}\,s^{-1})$ H-donor	ϕ_c
TMPK ($H_3C-S-C_6H_4-CO-C(CH_3)_2-N(morpholine)$)	100	61 (63)	2 (MMA)	–	0.3; 0.13[a]
($CH_3O-C_6H_4-CO-C(CH_3)_2-N(morpholine)$)	2500	65	–	–	–
($(CH_3)_2N-C_6H_4-CO-C(CH_3)_2-N(morpholine)$)	0.5	63	5 (MMA)	<0.1 (MDEA)	–
PMK ($C_6H_5-CO-C(CH_3)_2-N(morpholine)$)	1000	71	200 (MMA)	–	0.3
MPPK (morpholino-$C_6H_4-CO-C(CH_3)(C_6H_5)-N$)	0.6	–	3.5 (MMA) 0.22 (MA)	<0.1 (MDEA)	0.9 (0.24)
(morpholino-$C_6H_4-CO-C(CH_3)(CHCH_2)-N$)	0.7	–	<0.01 (MMA)	<0.1 (MDEA)	–

Calculated cleavage yields ϕ_c. See legend of Table 8.1.
[a] Measured dissociation quantum yields from [42].
From Refs. [4, 22, 23].

8.1 Benzoyl-Chromophore-Based Photoinitiators

[Scheme 8.5 structure: HAP cleavage showing Ph-C(=O)-C(R)(R)-OH → Ph-C•(=O) + •C(R)(R)-OH]

HAP

Scheme 8.5

Different conformations for the ground- and triplet-state geometry of HAP were obtained in the gas phase [87]. Three absorption bands are observed in HAP, for example, in acetonitrile: (i) S_0–$S_3(\pi\pi^*$ at 244 nm; high extinction coefficients $\varepsilon = 10^4$ mol^{-1} l cm^{-1}), (ii) S_0–$S_2(\pi\pi^*$ at 279 nm, $\varepsilon = 900$ mol^{-1} l cm^{-1}), (iii) S_0–$S_1(n\pi^*$ at 319 nm, HOMO$_{-2}$–LUMO, $f = 0.0004$–0.0007, $\varepsilon = 85$ mol^{-1} l cm^{-1}). The HOMO$_{-2}$ orbital exhibits a noticeable hyperconjugation on the two C(C=O)–C σ bonds. The LUMO π^* orbital is delocalized on the whole phenyl ring and the carbonyl group.

HAP efficiently cleaves as the triplet energy is 6–8.4 kcal mol^{-1} higher than the BDE of the cleavable C–C bond ($\sim$65 kcal mol^{-1}). Calculations of the potential energy surfaces (PESs) show that (i) the triplet state near the equilibrium geometry has an $n\pi^*$ character, (ii) when the C–C cleavable bond length increases, a partial-state mixing occurs with an increased participation of a σ orbital (and a decreased involvement of the π^* MO) to the singly highest occupied molecular

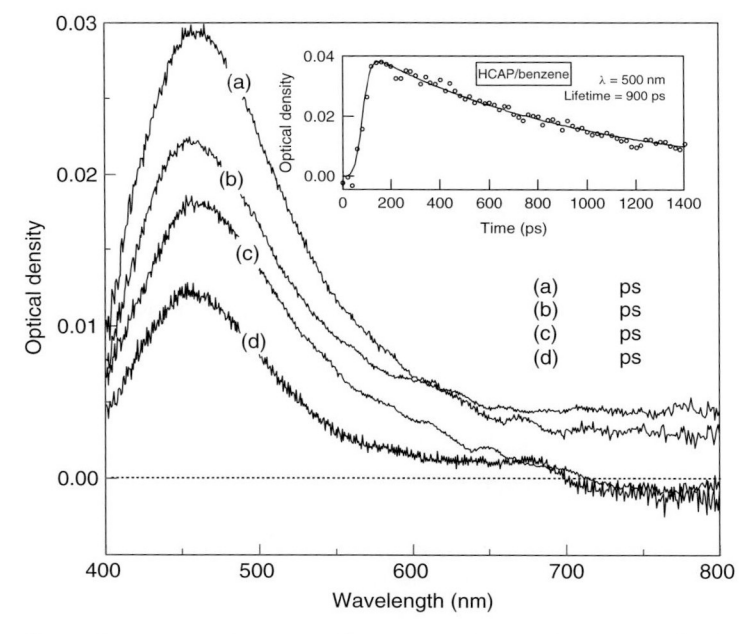

Figure 8.5 Picosecond pump-probe spectroscopy of 1-hydroxy-cyclohexyl-1-phenyl ketone (HCAP): triplet-state absorption spectra at various time delays. Inset: decay trace as a function of time. (Source: From Ref. [87].)

orbital (SHOMO), and (iii) the triplet energy slightly increases. In the TS, the spin density is mainly localized at the carbonyl group with a partial contribution of the σ bond; the contribution of the phenyl group is low. Such a high-spin localization has been expected for a long time to favor the Norrish I process in ketones [88].

The HAP and HCAP triplet states (e.g., in Figure 8.5) can be directly observed by using picosecond absorption spectroscopy [87]. ISC occurs within 10–20 ps. The triplet-state lifetime τ_T^0 of HAP is 450 ps in benzene; it decreases as the polarity of the medium increases. A sevenfold decrease of τ_T^0 is observed when going from hexane (380 ps) to methanol (55 ps). In ethylene glycol, the lifetime is nearly twice the value in methanol (120 ps). This was well reproduced by the calculations and ascribed to a decrease in the TS energy barrier, possibly through the stabilization of the polar TS [52, 89] as in BZ and benzoin ether derivatives (see above). The very low energy barrier is in line with the high value of the reported dissociation quantum yield (0.8). The presence of the hydroxy group near the carbonyl moiety leads to the formation of an intramolecular hydrogen bond that competes, in polar solvents, with the ketone/solvent hydrogen bond. The quite good linear relationship found between τ_T^0 and the β-parameter of the SCM model (Section 5.7) confirms that the ketone/solvent hydrogen bond plays a significant role.

Interaction rate constants between ^{3}HAP and various additives have been often obtained through indirect measurements (Table 8.4). Substitution on the O atom of the OH group (e.g., with C(=O)CH=CH$_2$) decreases the reactivity [90], whereas a modification on the benzoyl chromophore appears interesting (Section 8.2).

8.1.7
Ketone Sulfonic Esters

A careful change of the substituent at the β carbon atom of the carbonyl group leads to an efficient radical PI family (Scheme 8.6) [63]. It allows favoring of the α-cleavage process and yields a benzoyl radical and a benzyl-type radical, which liberates, in a second step, an oxysulfonyl radical through a subsequent cleavage. The reactivity of sulfamic esters of benzoin ethers [91] was also discussed.

8.1.8
Thiobenzoate Derivatives

Thiobenzoates exhibit an α-cleavage process (Scheme 8.7) whose efficiency is dependent on the presence of a substituted benzoyl moiety or a benzophenone (BP) moiety as the absorbing chromophore [92, 93]. The usual benzoyl-skeleton-based derivatives work according to an efficient cleavage (short triplet state ~2 ns) that yields a benzoyl and a thiyl radical. In a general way, thiyl radicals are very interesting initiating structures as they are not selective toward the addition to monomers, exhibit a low sensitivity to oxygen, and are therefore highly efficient (Section 8.2).

Table 8.4 Examples of available data concerning cleavable photoinitiators derived from hydroxyalkylacetophenones: cleavage rate constant k_c (upper value taken as the reciprocal value of the triplet-state lifetimes τ_T^0), triplet-state energy levels, and interaction rate constants k_q with oxygen, monomers, and hydrogen donors.

Compound	$10^{-6} k_c$ (s^{-1})	E_T (kcal mol^{-1})	$10^{-6} k_q$ (mol^{-1} l s^{-1}) Monomer	$10^{-6} k_q$ (mol^{-1} l s^{-1}) H-donor	ϕ_{diss}
HCAP	1100	67	—	—	0.8
HAP	2100	72	250 (MMA) 330 (MA)	—	0.8 (0.38)
CH$_3$S— (compound)	0.25	—	8 (MMA)	80 (MDEA)	—
CH$_3$O— (compound)	87	65	20 (MMA)	260 (MDEA)	0.38[a]
(CH$_3$)$_2$N— (compound)	0.3	63	4.5 (MMA)	<1 (MDEA)	0.03[a]

(continued overleaf)

Table 8.4 (continued)

Compound	$10^{-6} k_c$ (s^{-1})	E_T (kcal mol^{-1})	$10^{-6} k_q$ (mol^{-1} l s^{-1}) Monomer	$10^{-6} k_q$ (mol^{-1} l s^{-1}) H-donor	ϕ_{diss}
HO(CH$_2$)$_2$O—C$_6$H$_4$—C(=O)—C(CH$_3$)$_2$—OH	83	–	22 (MMA)	200 (MDEA)	–
CH$_2$CHOCO(CH$_2$)$_2$O—C$_6$H$_4$—C(=O)—C(CH$_3$)$_2$—OH	120	–	20 (MMA)	150 (MDEA)	–
CH$_2$CHOCO(CH$_2$)$_2$O—C$_6$H$_4$—C(=O)—C(CH$_3$)$_2$—OCOCHCH$_2$	80	–	20 (MMA)	50 (MDEA)	–
C$_6$H$_5$—C(=O)—C(CH$_3$)$_2$—OCOCHCH$_2$	>300	–	–	–	–

Measured dissociation quantum yields ϕ_{diss}. See legend of Table 8.1.
[a]From [42].
From [19, 22].

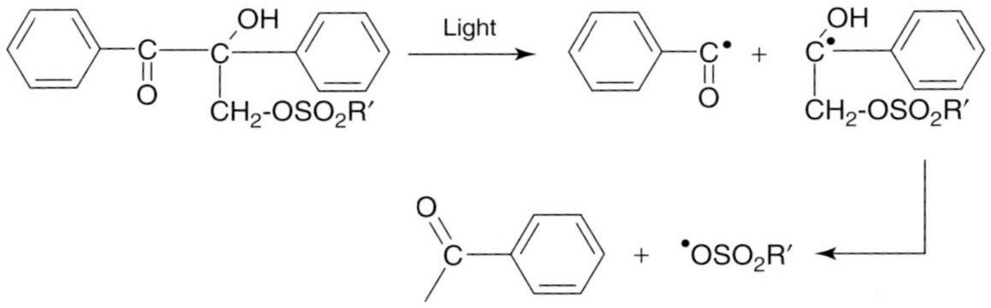

Scheme 8.6

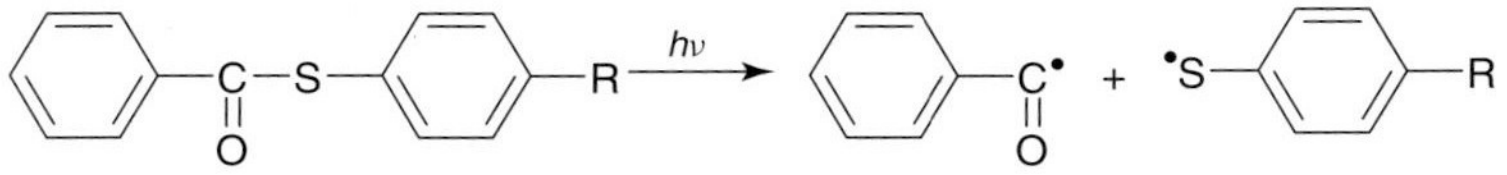

Scheme 8.7

8.1.9
Sulfonyl Ketones

A large variety of sulfonyl ketones (Scheme 8.8) has been developed [32, 94]. They undergo a β-cleavage process that occurs in the triplet state [22, 95, 96] and yields a (substituted) phenacyl radical and an aryl sulfonyl radical. In the presence of a hydrogen donor, the corresponding sulfinic acid is formed, which then liberates a sulfonic acid. The sulfinic acid is also released (together with an unsaturated ketone) through an in-cage hydrogen abstraction reaction (Section 18.8).

The absorption properties are strongly dependent on the substitution at the para position of the benzoyl chromophore. The unsubstituted compound has a low absorption in the near UV–visible region (owing to the $n\pi^*$ transition) and a quite reactive triplet state (Table 8.5) [97]. Their ability to initiate a polymerization is correct.

Different sulfonyl ketones substituted on the benzoyl group and at the sulfur atom [98] have been studied (Section 8.2.5). The triplet energies vary from 60 to 74 kcal mol^{-1} and the BDE (β C–S) from 37.7 to 42.1 kcal mol^{-1}. The dissociation reaction enthalpy from the triplet state ΔH_r ($\Delta H_r = $ BDE $- E_T$) lies from -36.3 to -18.9 kcal mol^{-1}. The reaction is therefore strongly exothermic, in contrast with the low dissociation observed for some compounds. As a consequence, ΔH_r is not the main parameter governing the cleavage.

If the triplet state is delocalized, the spin density being distributed over all the chromophoric groups should affect the bond dissociation. Calculations show that the $C_{C=O}$ spin density strongly ranges from 0.54 to 0.01 depending on the chemical structure (it decreases when the delocalization increases). As revealed by

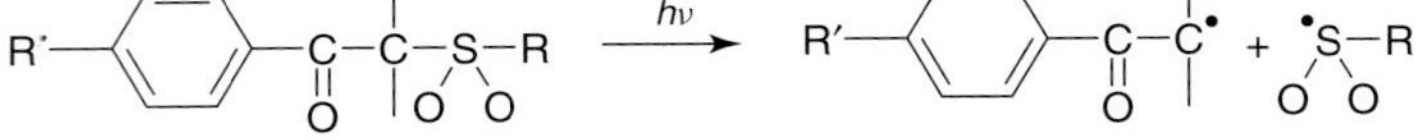

Scheme 8.8

8.1 Benzoyl-Chromophore-Based Photoinitiators

Table 8.5 Examples of available data concerning cleavable photoinitiators derived from various ketone structures: cleavage rate constant k_c (upper value taken as the reciprocal value of the triplet-state lifetimes τ_T^0), triplet-state energy levels, and interaction rate constants k_q with oxygen, monomers, and hydrogen donors.

Compound	$10^{-6} k_c$ (s^{-1})	E_T (kcal mol^{-1})	$10^{-6} k_q$ (mol^{-1} l s^{-1}) Monomer	$10^{-6} k_q$ (mol^{-1} l s^{-1}) H-donor	ϕ_{diss}
TPO	~10 000	63	—	—	0.7
$CH_3-S-\underset{\underset{O}{\parallel}}{\overset{\overset{O}{\parallel}}{C}}-\underset{\underset{O}{\parallel}}{\overset{\overset{O}{\parallel}}{C}}-S-\text{(p-tolyl)}$	40	—	100 (MMA)	700 (MDEA)	—
$CH_3-S-\underset{\underset{O}{\parallel}}{\overset{\overset{O}{\parallel}}{C}}-\underset{\underset{O}{\parallel}}{\overset{\overset{O}{\parallel}}{C}}-\text{(p-tolyl)}$	13	—	40 (MMA)	—	—
R–S–R (Where R is p-tolyl-SO$_2$-C(=O)-C$_6$H$_4$–)	0.15	—	0.8 (MMA)	140 (MDEA)	—
R with R = p-tolyl-SO$_2$-C(=O)-C$_6$H$_4$–	0.17	—	320 (MDEA)	1 (MMA)	—
Morpholino-substituted analog	1	—	130 (MDEA)	1.2 (MMA)	—

(continued overleaf)

Table 8.5 (continued)

Compound	$10^{-6} k_c$ (s^{-1})	E_T (kcal mol^{-1})	$10^{-6} k_q$ (mol^{-1} l s^{-1}) Monomer	$10^{-6} k_q$ (mol^{-1} l s^{-1}) H-donor	ϕ_{diss}
C(=O)Ph–C–OSO$_2$Ph	5	–	250 (MMA)	3300 (MDEA)	–
PhC(=O)–C(OH)(Ph')–CH$_2$OSO$_2$R'	150	–	–	–	–
PhC(=O)–C$_6$H$_4$–S–Ph	200	–	400 (MMA)	–	–
PhC(=O)–C$_6$H$_4$–C(=O)Ph	1.6	–	2.3 (MMA)	–	–
PhC(=O)–C$_6$H$_4$–S–C(CH$_3$)$_2$–C(=O)–C$_6$H$_4$–CH$_3$ with SO$_2$	0.15	65	0.18 (MMA)	390 (MDEA)	–
PhC(=O)–C$_6$H$_4$–S–Ph	0.6	–	0.8 (MMA)	1800 (MDEA)	–
PhC(=O)–C$_6$H$_4$–S–R with S–O	150	–	–	–	–

See legend of Table 8.1.
From [97].

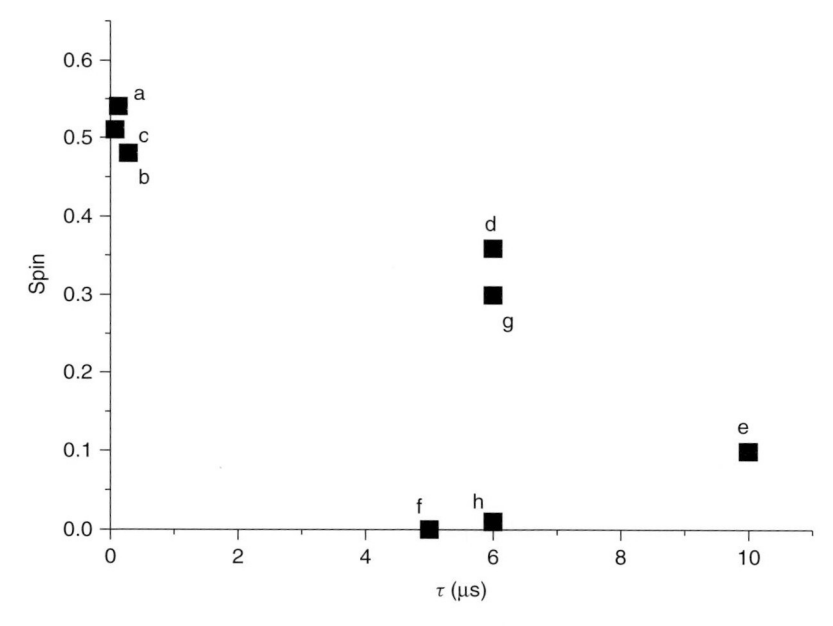

Figure 8.6 Spin density versus the triplet-state lifetime in a series of substituted sulfonyl ketones. See Scheme 8.8 where a: R′ = H, R=CH$_3$; b: R′=CH$_3$O, R=Ph − CH$_3$; c: R′=CH$_3$, R=Ph−CH$_3$; d: R′=CH$_3$ S, R=Ph − CH$_3$; e: R′=PhS, R=Ph−CH$_3$; f: R′=Ph−C(=O)−PhS, R=Ph−CH$_3$; g: R′=CH$_3$ S, R=CH$_3$; h: R′=Ph, R=Ph−CH$_3$.

the triplet-state lifetimes, the dissociation experimentally occurs for compounds exhibiting a spin density >0.4 (Figure 8.6). Therefore, the dissociation is favored in localized triplet states. A large delocalization of the chromophoric group decreases the spin density near the C–S bond, thereby rendering the dissociation less efficient and leading to a relative long lifetime of the corresponding triplet state. Here, the spin density on the carbon of the cleavable C(=O)–C bond is likely a decisive factor. This behavior cannot be generalized (Section 8.3).

8.1.10
Oxysulfonyl Ketones

The triplet cleavage of oxysulfonyl ketones (Scheme 8.9) occurs at the C–O bond [99, 100]. The cleavage is slow ($\sim 10^7$ s^{-1}). Their radical polymerization ability is poor [101]. Interestingly, under UV light exposure, they are able to act as a photolatent acid catalyst through the generation of a sulfonic acid by a hydrogen abstraction

Scheme 8.9

reaction either between the oxy sulfonyl radical and a hydrogen donor or involving both radicals in the radical pair. The benzene sulfonyl radical reacts with an amine (e.g., methyldiethanolamine (MDEA)) and THF with rate constants of $\sim 10^9$ and 10^8 mol^{-1} l s^{-1}, respectively.

8.1.11
Oxime Esters

The cleavage of 1-phenyl 2-propanedione-2 (ethoxycarbonyl) oxime PDO primarily occurs at the γ-position of the carbonyl (N–O bond) [19, 102–104]. A further cleavage process leads to a benzoyl radical, a nitrile, and carbon dioxide (Scheme 8.10). The cleavage is presumably fast (nanosecond timescale). The photochemistry of other keto-oxime derivatives is rather complex. A direct α-cleavage has been proposed in O–acetyl-diacetyl mono-oxime esters. The addition rate constant of the acetyl radical to styrene and methylmethacrylate (MMA) is one order of magnitude higher than that of the benzoyl radical [19]. A suitable substitution on the phenyl ring allows a shift of the ground-state absorption.

A tailor-made keto-oxime ester derivative (**9**) was proposed [106]: it exhibits an interesting red-shifted absorption, a high solubility, a reduced volatility, and an improved reactivity. The cleavage process is claimed analog to that of PDO.

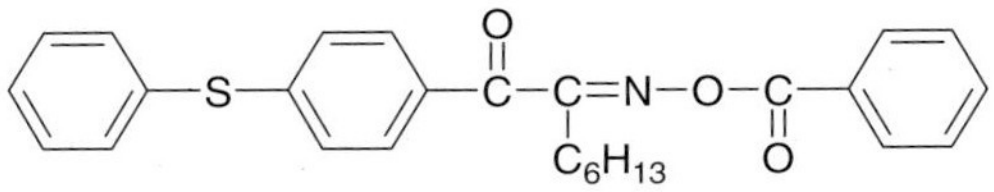

9

8.2
Substituted Benzoyl-Chromophore-Based Photoinitiators

The search of structures exhibiting a red-shifted absorption has driven a lot of works devoted to the introduction of a suitable substituent on the aromatic ring of the benzoyl chromophore. Thioether, ether, and amino groups were largely used. The observed effect is strongly dependent on the starting skeleton. The reactivity is

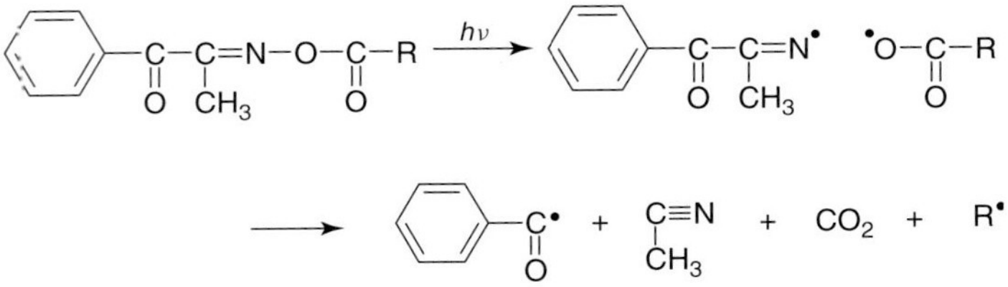

Scheme 8.10

often modified. The same holds true when changing the counterpart of the benzoyl moiety.

8.2.1
Benzoin Ether Series

In the benzoin ether series, the introduction of a thioether at the para position of both the phenyl rings of DMPA was not successful as shown in Table 8.1. The cleavage rate constant is more than 1000 times lower than that of DMPA [107].

8.2.2
Morpholino Ketone and Amino Ketone Series

An excellent representative of these new morpholino ketones is TMPK (Scheme 8.11). On light exposure [9], TPMK in benzene yields methylthiobenzaldehyde and morpholino propene through disproportionation in the radical pair. An escape reaction pathway is supported by the formation of morpholino propane through a bimolecular reaction between two morpholino isopropyl radicals. Photoreduction of TPMK by alcohols in concentrated solutions was observed by ESR: the ketyl intermediate rearranges to an α-keto radical by amine elimination [82].

In amino ketones, the introduction of a morpholino moiety on a benzoyl group such as in MPPK (Scheme 8.12) also leads to a red-shifted absorption (Table 8.2). The photolysis of MPPK gave evidence for an α-cleavage process as the main deactivation pathway of the excited triplet state [9]. Careful examination of the photolysis products, however, suggests that a β-cleavage and/or a Norrish II hydrogen abstraction should also contribute to a minor extent. At high concentration, an ^{3}MPPK/MPPK quenching through electron transfer can occur as shown by ESR [109].

TPMK and MPPK present a good activity [9]. The cleavage mechanisms and the reactivity of the excited states have been fully reported (Table 8.3) [4]. Efficient

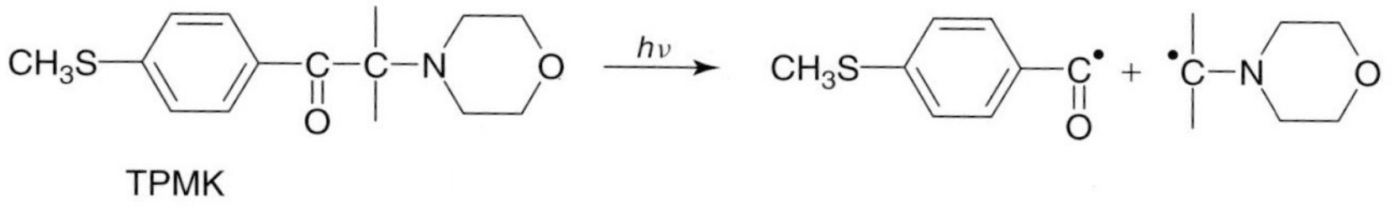

TPMK

Scheme 8.11

MPPK

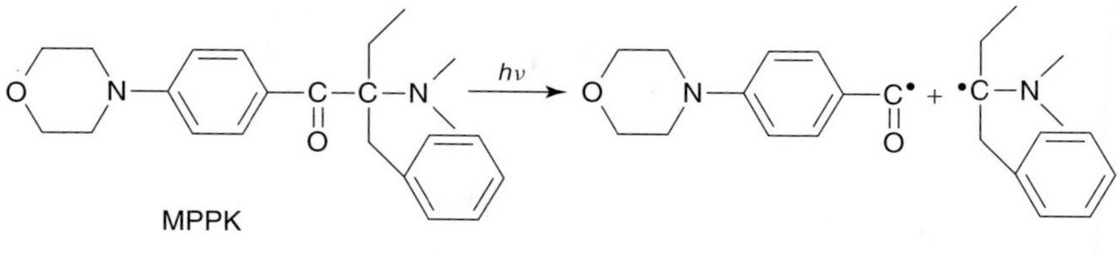

Scheme 8.12

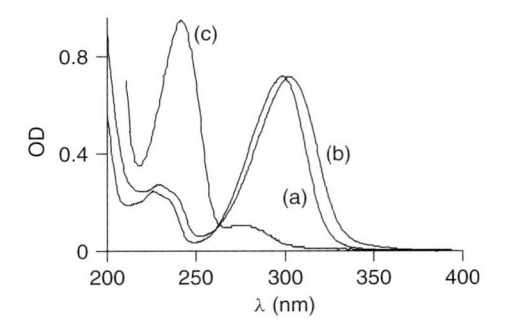

Figure 8.7 Absorption spectra of (a) TPK, (b) TPMK, and (c) PMK in acetonitrile. OD, optical density.

initiating radicals are generated. The methyl thio benzoyl and morpholino benzoyl radicals exhibit the same reactivity as the benzoyl radical: both of them are σ radicals and the para substituent has a minor influence on the aryloyl radical reactivity (Section 18.3).

The striking feature of the TPMK and MPPK compounds is the possibility of enhancing the light absorption (Table 8.2). Ground-state absorption spectra are reported in Figure 8.7 for PMK (7), TPMK, and a parent compound (TPK: 1-[4-(methylthio) phenyl]-ethanone). The UV spectra of TPK and TPMK are very similar and present intense and broad absorption bands centered around 300 nm with high molar extinction coefficients ($\varepsilon = 22000\, \text{mol}^{-1}\text{l}\,\text{cm}^{-1}$). At 366 nm, $\varepsilon = 70\, \text{mol}^{-1}\text{l}\,\text{cm}^{-1}$ for TPMK in acetonitrile. For MPPK, $\varepsilon = 25000, 325$, and $50\, \text{mol}^{-1}\text{l}\,\text{cm}^{-1}$ at 312, 366, and 400 nm, respectively.

The calculated absorption bands of TPMK are in line with the experimental results [81]: (i) S_0–S_1 transition (HOMO–LUMO transition predicted at 418 nm; $n \rightarrow \pi^*$ character; $f = 0.001$; the LUMO orbital is delocalized on the whole phenyl ring; the HOMO is centered on the nitrogen atom with a significant contribution of the morpholino C–H σ bonds through hyperconjugation) and (ii) S_0–S_2 and S_0–S_3 transitions (predominantly $\pi \rightarrow \pi^*$) at 311 and 304 nm ($f = 0.20$ and 0.25). The MOs are displayed in Figure 8.8.

As in TPMK, the p-substitution on the PMK structure using an ether or an amino group completely changes the ground-state absorption and red-shifts the

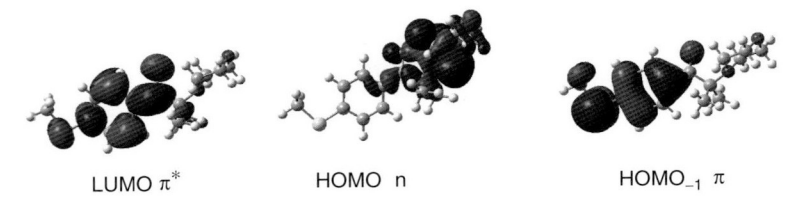

Figure 8.8 Molecular orbitals involved in the S_0–S_1 and S_0–S_2 absorption of TPMK. (Source: From [81].)

absorption maxima: 242, 301, 269, and 323 nm for PMK, TPMK, CH$_3$O-PMK, and (CH$_3$)$_2$N-PMK, respectively. The reactivity is dramatically affected (Table 8.3).

The first excited singlet-state S$_1$ is hardly produced by direct excitation but rather formed through internal conversion from the S$_2$ or S$_3$ excited state. Owing to its low absorption intensity, the S$_0$–S$_1$ band cannot be observed. This is confirmed by picosecond laser spectroscopy [81]: (i) the S$_0$ ground-state absorption leads to the second excited singlet $\pi\pi^*$ state S$_2$, (ii) a nonradiative relaxation process to the first excited singlet $n\pi^*$ state S$_1$ occurs, (iii) the $\pi\pi^*$ T$_2$ is formed within 700 ps from S$_2$, (iv) the $n\pi^*$ T$_1$ triplet state is generated in <10 ps from S$_1$, and (v) a very fast internal conversion of T$_2$ to T$_1$ arises. The T$_1$ state has a lifetime of 10 ns, which yields an upper limit of $10^8 \times s^{-1}$ for the cleavage rate constant.

The introduction of an ether (EPMK) or an amino group (APMK) at the para position of the benzoyl moiety of PMK has also been achieved [9]. The qualitative effect on the absorption properties are almost the same as in TPMK: presence of a new intense band at $\lambda_{max} = 269$ ($\varepsilon = 14\,500$ mol^{-1} l cm^{-1}) and 323 nm (25 500 mol^{-1} l cm^{-1}) for EPMK and APMK, respectively, in cyclohexane. The reactivity is affected (Table 8.3) [111]. The same holds true in the case of MPPK, where a slow cleavage in the microsecond range and a low monomer quenching are observed. The change of the dialkyl amino group for a morpholino group, however, enhances both the cleavage efficiency ($\sim 10^7$ s^{-1}) and (unfortunately) the monomer quenching so that the yield in radicals in the presence of a monomer remains low (Chapter 17).

One interesting feature in the thioether-modified benzoyl chromophore is the lowering of the triplet state energy (Tables 8.3 and 8.4) which makes feasible a photosensitized route for the cleavage process (Section 9.3.5).

Oligo α-aminoketones, based on a PMK structure (where the morpholine has been changed for a piperazine), have been synthesized [112]. Introducing a phenyl group instead of a naphthyl group in PMK yields 2-morpholino acetonaphthone [113].

Changing the thioether for a mercaptopropylthioether in the TPMK structure has been also achieved [114]. Time-resolved ESR and ESR spin-trapping experiments show that (i) an α-cleavage takes place and (ii) a sulfur-centered radical is readily produced by a hydrogen abstraction between the primary radicals and the thiol group.

8.2.3
Hydroxy Alkyl Acetophenone Series

8.2.3.1 Oil- or Water-Soluble Compounds

The para substitution of the benzoyl group of HAP (Scheme 8.5) with an amino (AHAP), an ether (EHAP), or a thioether (THAP) group modifies the electronic transitions. The absorption maxima are 244, 305, 271, and 332 nm for HAP, THAP, EHAP, and AHAP, respectively. This substitution drastically affects the triplet-state reactivity (Table 8.4). While TPMK and PMK exhibit a rather close reactivity, THAP is noticeably less reactive than HAP (the cleavage ability dramatically decreases).

However, this para substitution does not significantly change the reactivity of the derived benzoyl-initiating radical (Section 18.3).

Owing to their polar structure, hydrophilic- or water-soluble PIs derived from the corresponding oil-soluble structures are usable in water-soluble monomers or aqueous dispersions. This has been elegantly achieved with hydroxyl alkyl ketone derivatives [115] whose reactivity has been studied in [116, 117]. Compound **10** behaves as an EHAP derivative (Table 8.4). Both the cleavage and the monomer quenching are less efficient. The cleavage yield remains close to that of HAP itself. This compound behaves as an excellent PI. The comparative reactivity of several derivatives where the $O(CH_2)_2OH$ group is replaced by $S(CH_2)_2OH$, OCH_2COOH, $O(CH_2)_2ONa$ has been discussed.

10

Owing to its outstanding features (water compatibility, low volatility, no benzaldehyde release, easy chemical modifications of the OH group), compound **10** has opened new opportunities in water-borne coatings, UV-curable lacquers, biomedical applications, development of tailor-made mono- or multifunctional PI [9]. Other derivatives of HAP modified, for example, by a carbohydrate residue [118] have been proposed: a glucose-modified PI leads to the best results in terms of compatibility, efficiency, and gel content.

8.2.3.2 Difunctional Compounds

Multifunctional photoinitiators (mPIs) where several PI units are bound in a single molecule are attractive (**11**). As reported in a recent review [119], difunctional photoinitiators (dPIs) containing two covalently linked cleavable moieties exhibit a practical efficiency, which is sometimes better than that of the mono derivatives.

11

A fascinating question refers to the description of the MOs in such dPIs. This has been achieved using two selected compounds (**11**) consisting of two HAP moieties linked by either an oxygen atom (DHAP-O) or a methylene group (DHAP-C) [105]. The experimental absorption bands of DHAP-O (Figure 8.9) correspond to a $\pi\pi^*$ ($\sim$280 nm, $\varepsilon = \sim$25 000 $mol^{-1} 1 cm^{-1}$) and an $n\pi^*$ transition (shoulder above 320 nm). The maxima of the $\pi\pi^*$ transitions is red-shifted compared to that of the parent HAP compound ($\sim$240 nm).

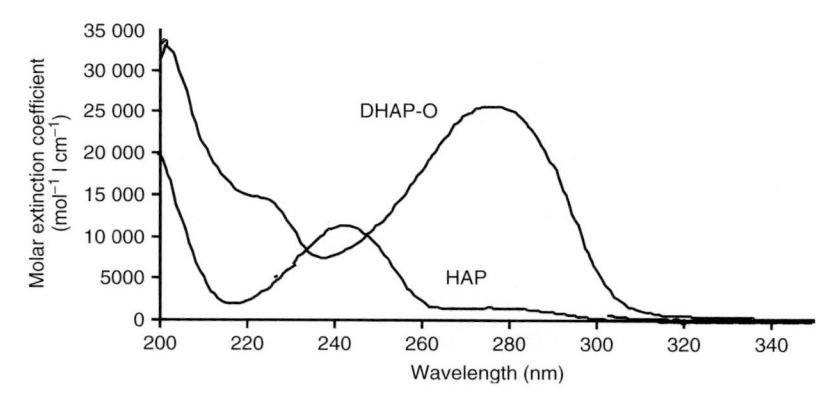

Figure 8.9 HAP-based difunctional photoinitiator: typical comparative absorption spectra in a mono- and a difunctional compound.

MO calculations show that the absorption properties must not be considered in any case as due to the sum of two HAP moieties. The n orbitals are localized only on one of the two benzoyl groups, whereas the π orbitals are partially delocalized over the entire molecule (Figure 8.10). This could explain why the $\pi\pi^*$ transitions are more affected than the $n\pi^*$ transitions when coupling the two HAP moieties. The delocalization of the π orbitals takes place through (i) the nonbinding n MO of the oxygen atom in DHAP-O (the partial charge transfer in the electronic transitions along the π system is already known in related compounds where the benzoyl group is substituted by a thioether, an amino, or an ether group) and (ii) a hyperconjugation in DHAP-C (the molecule is twisted, which allows the coupling of the two π systems).

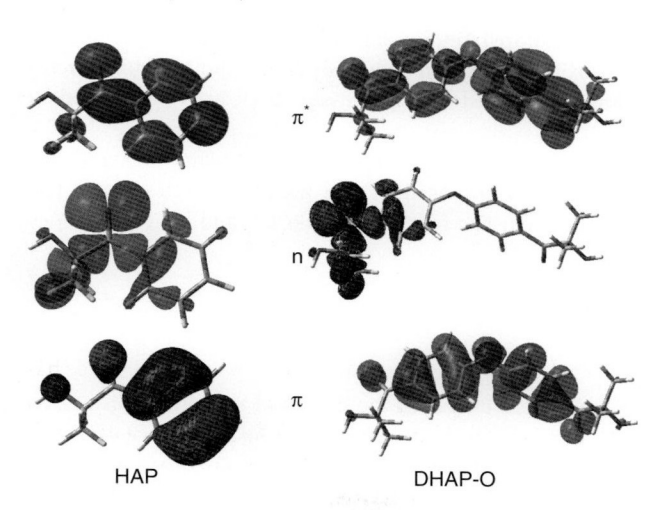

Figure 8.10 Molecular orbitals involved in a mono- and a HAP-based difunctional compound. (Source: From [105].)

The triplet-state lifetime is 7 ns for DHAP-O. The rate constant for the triplet-state quenching by MMA is 0.4×10^8 mol^{-1} 1s^{-1} compared to 2.5×10^8 mol^{-1} 1s^{-1} for HAP. The cleavage quantum yield is almost unity. The BDEs are 57.1 and 63.1 kcal mol^{-1} for DHAP-O and DHAP-C, respectively. No strong modification of the photochemical reactivity of these compounds compared to HAP was observed. This is in agreement with the almost unchanged practical efficiency of the photopolymerization reaction under a monochromatic light exposure and using the same absorbance. Under a polychromatic light excitation, higher rates of polymerization (compared to HAP) are obtained because of the higher amount of absorbed energy.

Other examples of dPIs or mPIs include bis [4-(2-hydroxy-isopropionyl)] ether [120] and oligomers containing an α-aminoalkylphenone chromophore [112].

Although one bond in a molecule is usually cleaved during a one-photon excitation process, the cleavage of two equivalent bonds can occur under a particular excitation through a concerted or a stepwise mechanism. This has been reported [121] in 1,8-bis [(4-benzoylphenoxy)methyl] naphthalene (where two benzophenone ethers are linked to naphthalene by a C–O bond) using a three-step excitation (Eq. 8.2).

$$S_0 \longrightarrow S_1 \text{ (308 nm) and } S_1 \longrightarrow T_1$$
$$T_1 \longrightarrow T_n \text{ (430 nm) and } T_n \longrightarrow R^\bullet : \text{first C–O cleavage}$$
$$R^\bullet \longrightarrow R'^\bullet \text{ (355 nm)} : \text{second C–O cleavage in the formed primary radical}$$

(8.2)

8.2.4
Modified Sulfonyl Ketones

The synthesis of substituted sulfonyl ketone derivatives and related compounds (bearing a thioether, a sulfoxide, a sulfonyl group, or a morpholino substituent) was largely carried out and the excited-state processes (β cleavage) described in several papers [122, 123].

The design of a bis-derivative linked by a sulfur atom R-S-R (where R stands for a sulfonyl ketone moiety) was also proposed. A C–S β cleavage still occurs. The absorption properties are enhanced. The reactivity (Table 8.5) and the PI efficiency are good.

8.2.5
Limit of the Substituent Effect

The investigation of the substituent effects in the design of new PIs is also fascinating. How can one understand the limit of such effects? As a matter of example, let us consider the series of compounds shown in **12**. The results reported in [108] appear as an interesting case to answer this question. Indeed, the introduction of the benzophenone thio moiety noticeably modifies the absorption of the sulfonyl ketone structure by a ∼30 nm red-shift (Section 8.5.4). The further introduction of an adequate X substituent on this new benzophenone-sulfonyl ketone structure leads to a shift as expected, the absorption toward longer wavelengths (up to 25 nm). The reactivity of these substituted compounds in their excited states is dependent on the substituents, which results in a change of the rate constants (cleavage, monomer, and amine interactions) and the ISC quantum yield.

X = Br, F, CN, CH$_3$, CH$_3$O, C$_6$H$_5$-S, Morpholine

12

The efficiency under monochromatic light decreases or is only slightly better than that obtained for X = H. The studied compounds, however, present a better absorption in the UV, near-UV/visible and, as a consequence, a better efficiency under a polychromatic light exposure than a reference system such as isopropylthioxanthone (ITX)/amine. True photosensitization processes are not very efficient as the energy transfer reactions exhibit a low efficiency.

MO calculations (Figure 8.11) confirm the evolution and the nature of the absorption bands and the absence of any striking effect on the triplet energy level. No change of the HOMO and LUMO is noted for X = Br, F, CH$_3$, or C$_6$H$_5$–S. For X = CN, a charge-transfer character occurs in the HOMO–LUMO transition. For X = morpholine, a HOMO$_{-1}$–LUMO transition is observed. The electron delocalization clearly reaches a limit.

These results outline the limit for a modification of a starting structure by introducing new substituents in order to increase the electron delocalization. Moreover, when the ground-state absorption shifts toward visible lights, the excited

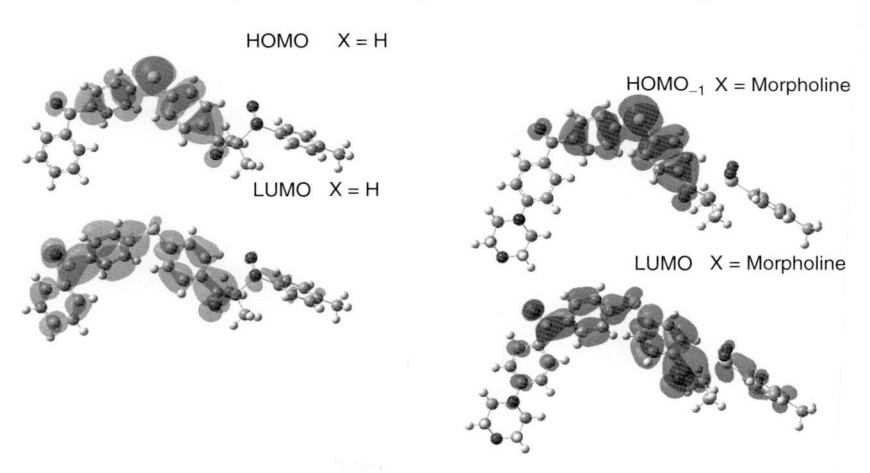

Figure 8.11 HOMO and LUMO of substituted benzophenone-sulfonyl ketones: X=H and X=morpholine. See **12**. (Source: From [108].)

states are often changed. The generally encountered problem is therefore to keep a high reactivity for the PI and/or to use a PS with a high energy triplet-state level and a PI with a low energy triplet-state level. When designing PSs with red-shifted absorption, the problem obviously relates to the concomitant lowering of the triplet levels.

8.2.6
Macrophotoinitiators

The design of copolymerizable, oligomeric-, polymeric-, or polymer-supported derivatives was largely exploited [9, 124, 125] in almost all families of PIs. The basic idea is to develop compounds (i) keeping a high reactivity, (ii) having a better solubility or compatibility in the monomer/oligomer matrix, a low volatility, (iii) exhibiting nonyellowing properties, and (iv) avoiding releasing odorous photolysis by-products [126–130]. Most of the new compounds involve a change of the hydroxyethoxy chain of compound (**10**), for example, (i) a polyethylene glycol moiety or an ionic group (better compatibility), (ii) a perfluorinated chain (better antisticking property), and (iii) an acrylated function or a chemical group that can link several PI moieties (better incorporation into the matrix and lower release of volatiles).

The particular reactivity of a variety of copolymerizable PIs based on the HAP skeleton (where an acrylate group is introduced) has been checked (Table 8.4) and discussed.

The reactivity of PIs carrying a long alkyl chain does not significantly differ from that of the unsubstituted parent compounds. Specific derivatives with a fluorinated substituent endcapped alkyl chain as well as a mixture of a lithium alkyl sulfonate and a suitable molecule bearing a sulfonic ester behave as surface-active PIs [4, 9] and help to reduce the detrimental effect of the air inhibition in the top layer of the coating.

Another commercially available PI relates to a base oligomeric molecule containing the hydroxyl alkylphenone moiety [131] (**13**), which exhibits interesting properties in UV-curing applications. The cleavage rate constant is slightly lower (Table 8.4) but the overall reactivity and the practical efficiency in industrial UV lines remain excellent. Moreover, the benzaldehyde loss after a 2000 mn heating time is reduced by a ~100-fold factor compared to that of the corresponding monomeric molecule.

13

Dendritic carbosilane-based macrophotoinitiators containing a lot of HAP-type moieties linked to the surface of the dendrimer [132], supramolecular-structured derivatives based on cyclodextrine, and DMPA moieties [133] or HAP moiety containing cleavable hyperbranched PI for water-borne coatings [134] might be promising PISs.

8.2.7
Supported Cleavable Photoinitiators

Supported PIs can be considered as PIs that are linked to a substrate (polymer and solid particle) in order to, for example, favor a grafting reaction or to incorporate a nanoparticle into the cured material. For example, cleavable PI (derived from HAP)-functionalized SiO_2 particles (14) allow starting the photopolymerization of a urethane diacrylate oligomer and forming a nanosilica/polyurethane nanocomposite [135]. Other miscellaneous systems are given in Sections 11.3 & 9.1.13.1.

8.3
Hydroxy Alkyl Heterocyclic Ketones

Few attempts have been made to change the benzoyl group of usual PIs for an electron-poor six-membered heterocyclic ring or an electron-rich five-membered ring. This was achieved in the hydroxyl alkyl ketone skeleton HAP. Compounds (15–17) containing a pyridinoyl, a furane, a thiophene, or a pyrrole chromophore instead of the benzoyl group as well as a new structure containing two furane moieties have been recently proposed [136–141].

The 3-pyridyl derivative 3NHAP is reported to have a better efficiency than the 2-pyridyl derivative 2NHAP. An electron-rich heterocycle (such as a thiazole) decreases the efficiency of the photoinitiation process [142]. The rate of polymerization is 10% higher with 3NHAP than for HAP itself. From the electronic structure analysis [143], the main difference between these compounds concerns the BDE: 71.2 kcal mol^{-1} for 2NHAP, which is about 8 kcal mol^{-1} higher than the relaxed triplet-state energy (63.1 kcal mol^{-1}), and 65.3 kcal mol^{-1} for 3NHAP, about 0.2 kcal mol^{-1} lower than the relaxed triplet state (65.5 kcal mol^{-1}). The cleavage is therefore less efficient in the case of 2NHAP. There is no difference, neither in the electronic transitions nor in the spin distribution of the radicals. At first sight, the reaction enthalpy (ΔH_r = BDE − E_T) seems to govern here the dissociation from the triplet state in contrast with the behavior of another family of compounds (Section 8.1.9).

8.4
Hydroxy Alkyl Conjugated Ketones

The interest of the introduction of an ynone or an enone as a chromophoric group was also checked [144, 145], for example, in **18–20** and 1,5-diphenyl-1,4-pentadiyn-3-one. This last compound is a very efficient PI [146]. The generation of radicals in the absence of co-initiators as well as the role of an added amine is complex. As revealed by time-resolved ESR and CIDNP, it likely involves a usual hydrogen abstraction, for example, with isopropanol, the propanol-2-ol-2-yl radical being able to add to an acrylate [147].

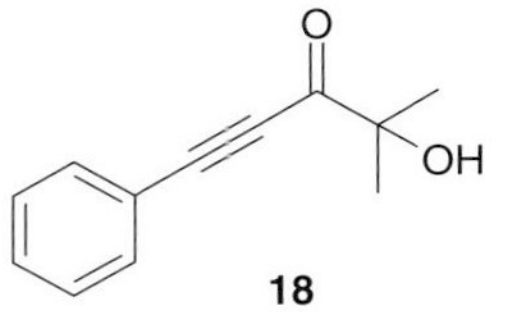

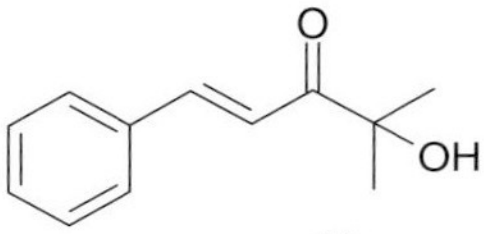

8.5
Benzophenone- and Thioxanthone-Moiety-Based Cleavable Systems

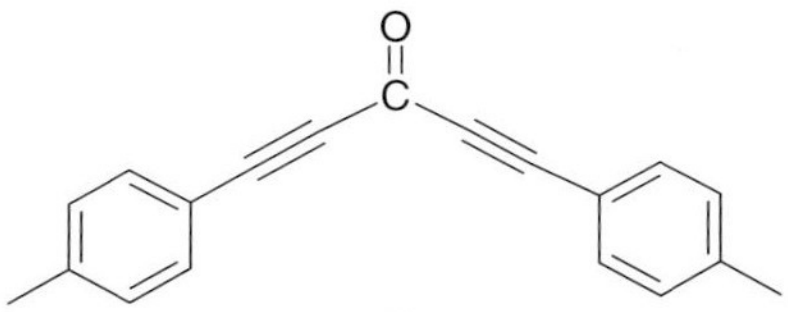

BP and thioxanthone (TX), largely employed as Type II PIs (Sections 9.1.3 and 9.1.4), present a near-UV absorption. These chromophores are also often encountered in a lot of compounds where the usual BP and TX photochemistry is totally changed.

8.5.1
Benzophenone Phenyl Sulfides

The bond cleavage of phenacyl phenyl sulfides has been recently reviewed [148]. The cleavage of the C–S bond in the benzoylphenyl phenyl sulfide [149] occurs in a rather long-lived triplet state (Table 8.5). An aryl and a thiyl radical are formed. This PI, however, works better in the presence of an amine (Section 9.1.12). The role of a sulfide group in a given structure has also been checked [150].

8.5.2
Ketosulfoxides

Ketosulfoxide derivatives where R is an alkyl or a phenyl group (Scheme 8.13) exhibit two cleavage processes [151]. The first one occurs in the triplet state (lifetime = 7.5 ns for R = methyl) and yields an aryl radical on the BP moiety and an aliphatic sulfinyl radical. The second one arises in the singlet state (lifetime ~3 ns) at the S-alkyl bond. The BP sulfinyl radical leads to a BP sulfonyl and a thiyl radical by disproportionation. The formed BP sulfonyl radical generates a BP radical through SO_2 release.

8.5.3
Benzophenone Thiobenzoates

Changing the benzoyl and the benzyl group (Section 8.1.8) for a BP in the thiobenzoate series Ar–C(=O)–S–R (Scheme 8.7) leads to a longer triplet-state lifetime (~500–700 ns). The cleavage process and the polymerization ability are less efficient [92, 93].

8.5.4
Benzophenone-Sulfonyl Ketones

A new and very efficient ketone, sulfonyl ketone bifunctional compound BPSK [152], exhibiting an excellent light absorption in the near-UV–visible wavelength range around 410 nm has been recently developed for the curing of printing inks (21). A rather slow β-cleavage occurs at the C–S bond [153]. The triplet state (which is located on the BP moiety) weakly interacts, however, with monomers (Table 8.5) so that a good initiation ability is found. Introducing various substituents at the para position of the BP moiety of BPSK has almost no effect as discussed in Section 8.2.5.

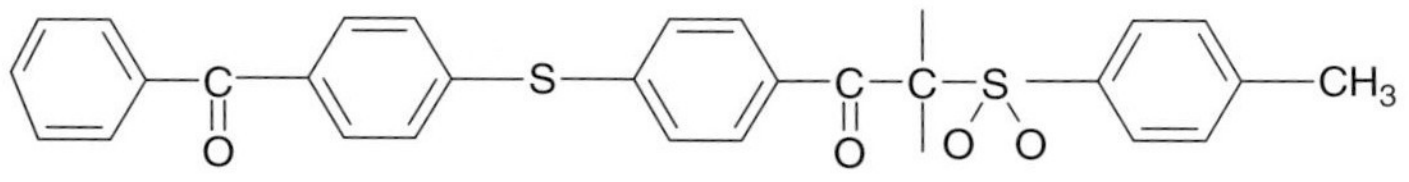

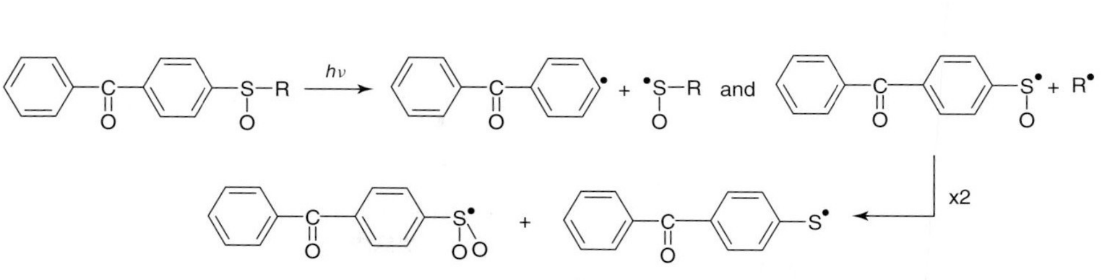

Scheme 8.13

8.5.5
Halogenated Derivatives

A photoinduced C-Br homolysis arises in 2-bromobenzophenone [153]. A cleavage of the C–Cl bond of 1,4 substituted TXs (Section 9.1.4) has been also reported as a parallel pathway of evolution of the excited states but the activity as a Type I PI is low.

8.5.6
Cleavable Benzophenone, Xanthone, and Thioxanthone Derivatives

Modification of the BP and TX structures (Sections 9.1.3, 9.1.4, 9.1.13) has been achieved for a long time through the introduction of adequate functional groups: in most cases, the modified BP or TX behave as the parent molecules (Type II PI) and do not generate new radicals.

The recent introduction of the silyl radical chemistry into the design of new PIs allowed the synthesis of cleavable compounds based on starting uncleavable structures such as in the BP [154] and TX [110, 155] series. This was realized through the introduction of a cleavable silyl moiety on the BP or TX skeletons such as in BP–O–Si, TX–O–Si, and TX–Si (22–24).

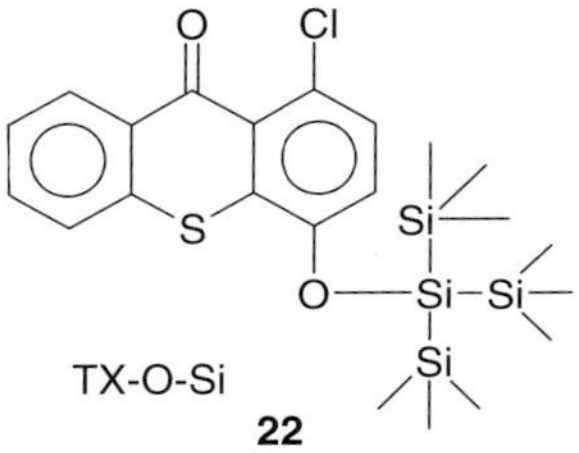

TX-O-Si
22

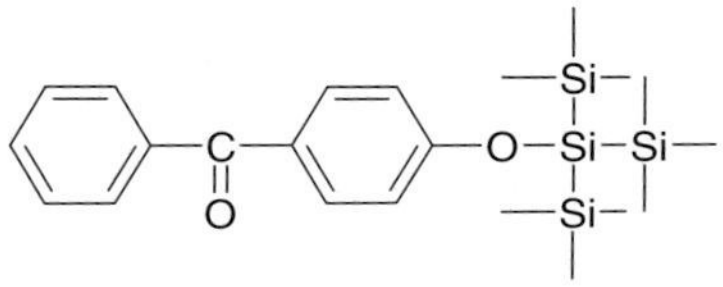

BP-O-Si
23

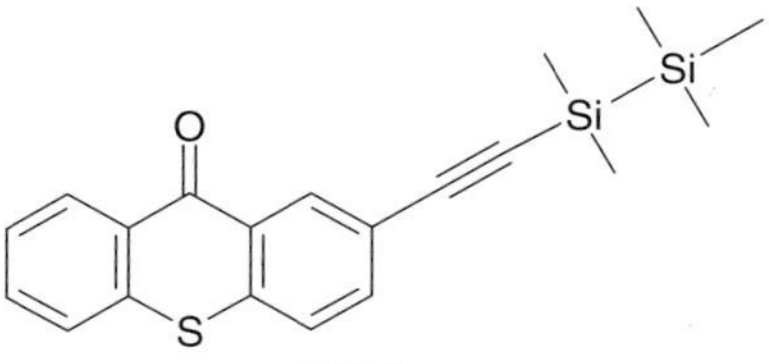

TX-Si
24

Compound BP–O–Si has a triplet lifetime of 500 ns, which is much shorter than the usual BP triplet-state lifetime. This result shows that this lifetime is probably governed by the Si–Si bond cleavage process (upper value for the cleavage rate constant: 2×10^6 mol^{-1} s^{-1}). For TX–O–Si, the lifetime (~5 µs) is much longer than that found for BP–O–Si supporting a lower dissociation rate constant for this compound ($k < 2 \times 10^5$ s^{-1}). This is ascribed to a reduced exothermicity for the cleavage from the triplet state [110].

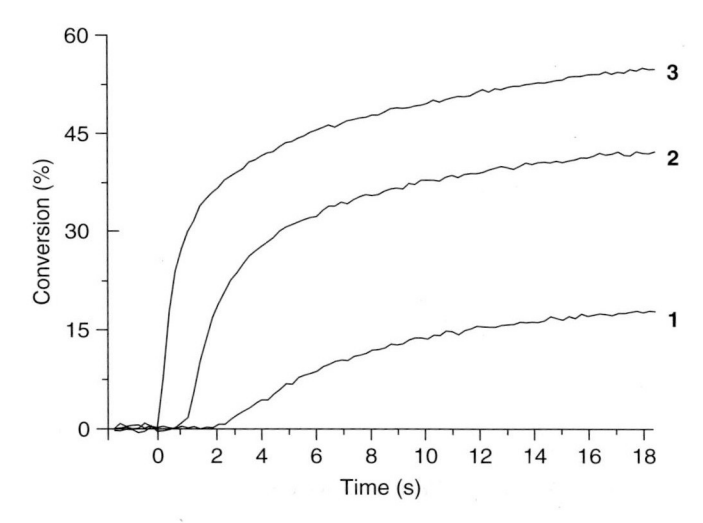

Figure 8.12 Radical photopolymerization ability of (1) 2-isopropylthioxanthone ITX (1% w/w), (2) TX–Si (1% w/w), (3) TX–Si/MDEA (1%/1% w/w) in an epoxy acrylate film under air. Hg–Xe lamp exposure. (Source: From [110].)

TX–Si exhibits a 10 nm red-shifted absorption (compared to TX), undergoes a triplet-state Si–Si cleavage, behaves as an efficient PI (Figure 8.12) and, compared to ITX, presents a reduced migration outside the cured film because of the incorporation of the TX fragments into the polymer chain. Addition of an amine still enhances the polymerization efficiency. Preliminary experiments suggested that a similar modification of the xanthone chromophore is promising.

8.6
Benzoyl Phosphine Oxide Derivatives

8.6.1
Compounds

Acylphosphine oxides and acylphosphonates that are efficient PIs were introduced in the early 1980s. The photochemistry of benzoyl phosphine oxides, benzoyl phosphonic acid esters, pivaloyl phosphonic acid esters, and pivaloyl diphenyl phosphine oxides has been studied in great detail [19, 156, 157]. Efficient radicals are produced in the triplet state through α-cleavage on a very short timescale. The 2,4,6-trimethyl benzoyl-diphenylphosphine oxide (TPO) compound was the interesting starting molecule of the series (Scheme 8.14). When a methyl substituent is present at the o-position as in TPO, a minor decomposition pathway involves the formation of a 1,4 biradical (Norrish II intramolecular hydrogen abstraction) and then of an enol. Reketonization regenerates the starting molecule.

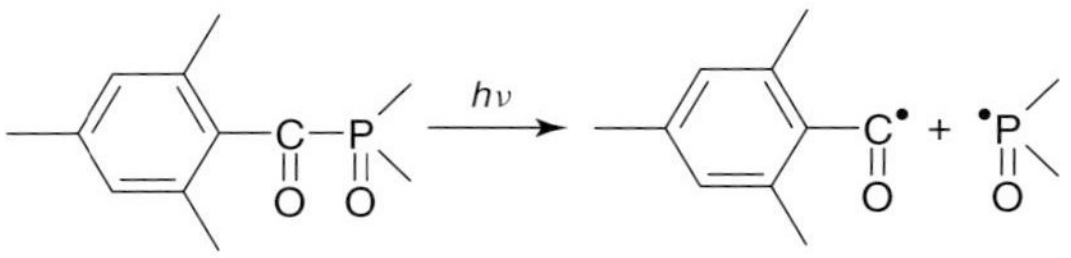

Scheme 8.14

Bis-acyl phosphine oxide derivatives BAPO, for example, **25** containing two substituted benzoyl moieties linked to the P atom were proposed later [9].

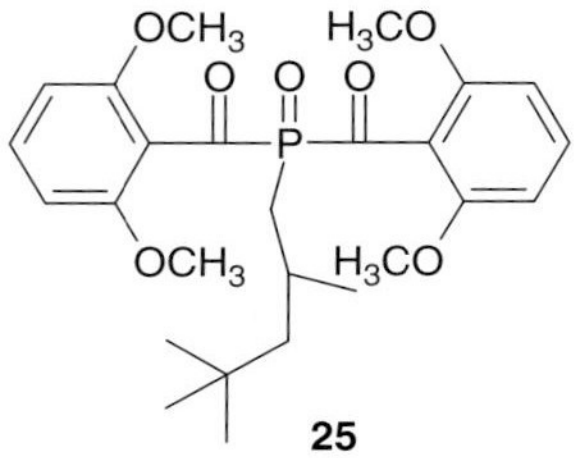

25

Polymeric [158–160] as well as hydrophilic phosphine oxide derivatives [161, 162], acyl-oxy and acyl-silyl phosphine oxides and structurally related compounds (phosphine sulfides, bis-phosphine sulfides [4, 22, 163], phosphonates [164]) have been reported. A co-initiator such as 2-mercaptothioxanthone usable with an acylphosphine oxide has also been mentioned [165]. The synthesis of tailor-made derivatives such as (i) camphorquinone-linked acylphosphine oxides [166] or (ii) substituted phosphine oxides, for example, bis 2,4,6-trimethylbenzoyl phenyl phosphine oxide derivatives that become water soluble, thanks to the introduction of two 3-{[2-(allyloxy)ethoxy] methyl} groups [167] was achieved for dental adhesive applications.

The photochemistry and photophysics of TPO and BAPO derivatives have been extensively investigated in the late 1990s by steady-state photolysis, fluorescence, phosphorescence, LFP, picosecond spectroscopy, and time-resolved ESR [168, 169]. When using TPO or BAPO derivatives, an oxygen inhibition of the radical polymerization is usually found at the surface. This is reinforced when using a low viscosity matrix under low light intensity (Figure 8.13).

8.6.2
Excited State Processes

A recent kinetic investigation [168] showed that the first excited singlet state of TPO decays in 120 ps and generates a triplet state that undergoes a fast cleavage within 100 ps (Table 8.5). This is an uncommon situation where the ISC rate limits the observed α-cleavage. The S_1–S_n transient absorption band has been recorded in pump-probe picosecond spectroscopy experiments [170].

The excited-state reactivity of BAPO derivatives is similar to that of TPO [168]. CIDEP [171] and CIDNP experiments [65] helped to clarify the cleavage mechanism.

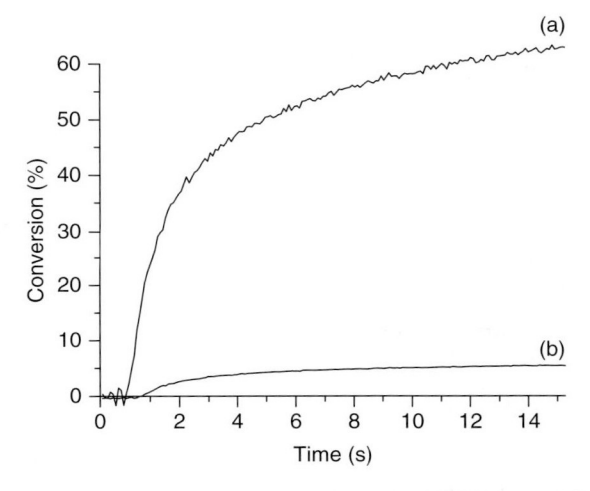

Figure 8.13 Typical photopolymerization profiles of trimethylolpropane triacrylate (TMPTA) in the presence of BAPO (**25**) in (a) laminate and (b) under air. Under the monochromatic light of a Hg lamp (366 nm ~5 mW cm^{-2}). (Source: From J. Lalevée, unpublished data.)

Both phosphine oxides and bis-phosphine oxides have an excellent reactivity because of the fast cleavage and the produced highly efficient phosphinoyl radical (Section 18.6).

A theoretical approach of the cleavage process of an acylphosphine oxide (phenyl-dimethyl phosphine oxide) through the calculation of the PESs was presented in [172]. The role of the conformational dynamics on the cleavage efficiency was well examined by the calculations of the rotation profiles along the P(O)–C(O) bond both in the S_1 and T_1 states. In fact, several conformations can be populated in the triplet state of such a molecule (as resulting from the excitation of several ground-state conformers or from the spread of conformers in S_1 or the conformational changes in T_1). The dihedral angle ϕ is equal to $0°$ when the P = O and the C=O are in an s–cis arrangement. The activation barrier for the cleavage is 8 kcal mol^{-1} for the relaxed conformation (that corresponds to an s-trans conformation of the O atoms; $\phi = 180°$) and only 4 kcal mol^{-1} for a higher energy conformation ($\phi = 60°$). This explains why the solvation that may influence the conformational change has an effect on the cleavage rate constant.

The phosphinoyl radicals add more efficiently to acrylate double bonds than the benzoyl radical [173, 174] (Section 18.6). The hydrogen abstraction ability of the benzoyl radical can be helpful for additional initiating reaction pathways in monomers containing labile hydrogen such as highly reactive acrylate monomers (HRAMs) (Section 4.2.1). It presumably explains the polymerization ability observed with monofunctional lactone containing acrylate monomers (**10–12**, Chapter 4) where the polymerization rate is six times higher than that obtained with the corresponding alkyl acrylate [175].

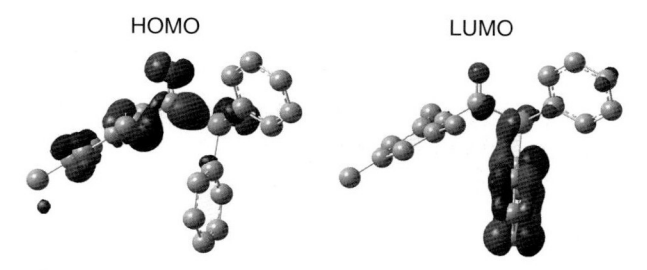

Figure 8.14 HOMO and LUMO in TPO at density functional theory (DFT) level. (Source: J. Lalevée, unpublished data.)

8.6.3
Absorption Properties and Photolysis

Bis-acyl phosphine oxides exhibit an excellent UV-near-visible absorption around 350–380 nm extending up to 420–440 nm together with remarkable molar extinction coefficients (300–800 $mol^{-1} l cm^{-1}$). The electronic transition corresponds to an $n\pi^*$ HOMO–LUMO transition (Figure 8.14) that is red-shifted and enhanced because of the conjugation with the phosphonyl group. The C=O and the C-P bonds in TPO lie in two different planes at 70° because of a favorable orbital overlap between the carbonyl and the phosphinoyl group.

This absorption property makes bis-phosphine oxides suitable in the curing of white-pigmented coatings [9, 176]. Moreover, contrary to usual PIs based on the benzoyl chromophore itself (Section 8.1), the decrease of the ground-state absorption and the absence of colored photolysis products when increasing the UV light exposure time enables the film to transmit more light so that thick samples can be easily cured (Figure 8.15). In the case of TPO, the mesityl–C(=O)–O–P(Ph)$_2$ photolysis product is observed [51].

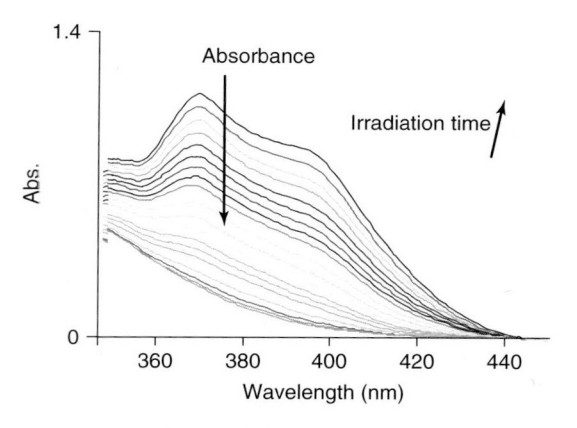

Figure 8.15 Evolution of the UV absorption of BAPO as a function of the irradiation time. From laboratory experiments.

8.6.4
Bis-acyl phosphine oxide/Phenolic Compound Interaction

As mentioned in Section 7.7, in wood coating UV curing, inhibition and retardation effects attributed to the release and migration of phenolic derivatives POH are observed. The kind of PI plays a decisive role as, for example, Type I PIs are less sensitive to the presence of POHs than Type II PIs. This problem has been studied in detail in the case of BAPO and could be considered as an example [177]. The different primary BAPO/POH interactions are summarized in (Eq. (8.3)). Reactions (1) and (2) are so fast that, despite the high rate constants of the S_1 or T_1 state/POH interaction ($10^9 - 10^{10}$ mol^{-1} 1s^{-1}), reaction (3) should not have a strong effect. Reaction (4) involving benzoyl type radicals exhibits a low rate constant ($\ll 10^5$ mol^{-1} 1s^{-1}) [178] contrary to reaction (5), which is more efficient ($\sim$ or $> 10^5$ mol^{-1} 1s^{-1}) and competes with the addition of the phosphinoyl radical to the monomer. This explains why, in cleavable PIs such as in the DMPA or HAP series, the POHs have no significant effect on the initiation step.

$$^1\text{BAPO} \longrightarrow {}^1\text{BAPO}^*(h\nu)$$
$$^1\text{BAPO}^* \longrightarrow {}^3\text{BAPO} \quad (1)$$
$$^3\text{BAPO} \longrightarrow \text{ArC}(=\text{O})^\bullet + \text{Ar}_1\text{Ar}_2\text{P}(=\text{O})^\bullet \quad (2)$$
$$^{1,3}\text{BAPO}^* + \text{POH} \longrightarrow \text{BAPO-H}^\bullet + \text{PO}^\bullet \quad (3)$$
$$\text{ArC}(=\text{O})^\bullet + \text{POH} \longrightarrow \text{ArC}(=\text{O})\text{-H} + \text{PO}^\bullet \quad (4)$$
$$\text{Ar}_1\text{Ar}_2\text{P}(=\text{O})^\bullet + \text{POH} \longrightarrow \text{Ar}_1\text{Ar}_2\text{P}(=\text{O})\text{-H} + \text{PO}^\bullet \quad (5) \quad (8.3)$$

8.7
Phosphine Oxide Derivatives

N-[(diphenyl phosphinyl) methyl]-N-methyl aniline $\text{Ph}_2\text{P}(=\text{O})\text{-CH}_2\text{-N}(\text{CH}_3)\text{-Ph}$ can be used as an efficient Type I PI. It presents an acceptable absorption (2530 mol^{-1} 1cm^{-1} at 296 nm and 11 270 at 253 nm). The light excitation leads to the two initiating radicals: $\text{Ph}_2\text{P}(=\text{O})^\bullet$ and $\text{Ph-N}(\text{CH}_3)\text{-CH}_2^\bullet$ (Sections 18.6 and 18.5) (J. Lalevée, unpublished data).

8.8
Trichloromethyl Triazines

The 2,4,6-tris(trichloromethyl)-1,3,5-triazine forms three molecules of trichloromethyl cyanurate and then leads to a chlorine and an NC–CCl$_2^\bullet$ radical through a thermal process. Substituted bis-(trichloromethyl)-1,3,5-triazines TZ (26) generate a chlorine radical and a dichloro carbon-centered radical through a C–Cl bond cleavage [179, 180]. Their absorption maxima are located around 400 nm. Variation of the chromophore allows tuning of the absorption wavelength

[181, 182].

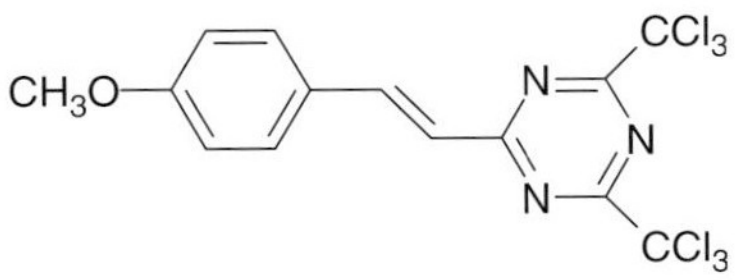

26

8.9
Biradical-Generating Ketones

Linear ketones possessing a γ-hydrogen (Scheme 8.15) can produce a biradical through an intramolecular hydrogen abstraction in the triplet state. This biradical can initiate the polymerization to some extent [183]. It also leads to a C_α–C_β bond cleavage that produces an alcene and an enol.

Cyclic ketones undergo a Norrish II reaction that creates a biradical. This has been reported in (i) cyclododecanones [184] where the C–C bond cleavage occurs at the C(=O) carbon and generates a biradical suitable for the initiation and (ii) dithiolactones (the formed biradical also leads to a cyclic dithione through a rearrangement and an intramolecular cyclization) [185].

The formation of biradicals has been exploited to increase the molecular weight of the obtained polymer. In fact, both radical centers on a biradical are able to initiate the polymerization. The dead polymer formed, once a growing chain is terminated, is still able to restart the initiation process. This has been demonstrated, for example, in the photopolymerization of styrene in the presence of cyclododecanones [184].

8.10
Peroxides

On light exposure, peroxides RO–OR, peroxycarbonates RO–C(=O)O–OR, peroxydicarbonates, or diacyl peroxides RC(=O)O–OC(=O)R undergo a O–O cleavage and finally lead to carbon-centered radicals. Three mechanisms have been proposed (Eq. (8.4)) as presented in [186]: (i) a single O–O bond cleavage followed by release of carbon dioxide, (ii) a stepwise two-bond cleavage (C–R and O–O) followed by decomposition of the carboxyl radical, and (iii) a concerted three-bond cleavage. The primary cleavage is suspected to occur in the singlet state. The O–O bond

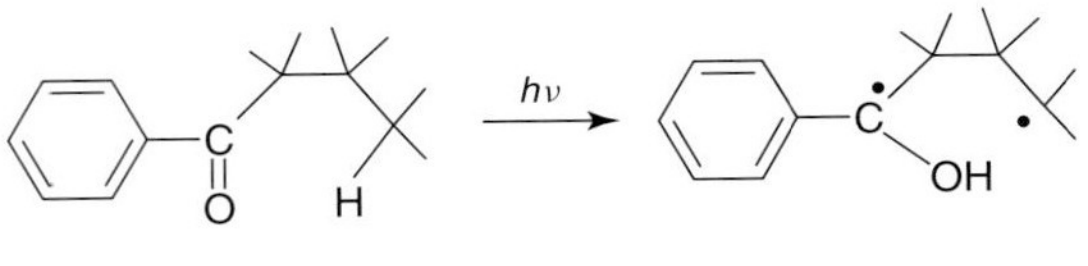

Scheme 8.15

energy is weak (~30 kcal mol^{-1}). The excited states and the photodissociation of hydroxymethyl hydroperoxide have been recently investigated [187].

$$RC(=O)O-OC(=O)R \longrightarrow RC(=O)O-OC(=O)R^*(h\nu)$$
$$RC(=O)O-OC(=O)R^* \longrightarrow RCOO^\bullet + RCOO^\bullet \text{ and } RCOO^\bullet \longrightarrow 2\,R^\bullet + 2CO_2$$
$$RC(=O)O-OC(=O)R^* \longrightarrow R^\bullet + CO_2 + RCO-O^\bullet \text{ and } RCO-O^\bullet \longrightarrow R^\bullet + CO_2$$
$$RC(=O)O-OC(=O)R^* \longrightarrow R^\bullet + CO_2 + CO_2 + R^\bullet \qquad (8.4)$$

A famous representative compound is the dibenzoyl peroxide Ph–C(=O)–O–O–C(=O)–Ph, which decomposes (rate constant: ~5 × 10^6 mol^{-1} l s^{-1} [188]) into (i) two phenyl radicals [189] after formation of the primary benzoyloxy radicals Ph–C(=O)–O$^\bullet$ and decarboxylation or (ii) a phenyl Ph$^\bullet$ radical and phenylbenzoate Ph–C(=O)–O–Ph [171] after cleavage, CO$_2$ release, and recombination. However, safety considerations limit the development of such peroxides in industrial applications.

Peroxides and hydroperoxides are thermally decomposed at ~60 and ~100 °C, respectively. This kind of reaction can occur in thick films (but not in thin films) where the exothermicity of the photopolymerization causes a temperature jump in the sample: it contributes to the presence of dark reactions in photocurable paints (Section 10.2.8).

8.11
Diketones

It is known that the cleavage of α-diketones such as benzil (Bz) ([190] and references therein) is very inefficient under a one-photon excitation (for Bz: triplet-state energy = 54 kcal mol^{-1}; BDE of the C–C bond = 66 kcal mol^{-1}). The cleavage of Bz arises, however, in an upper triplet state when using the consecutive absorption of a second photon by the lowest triplet state (Scheme 8.16).

Other nonconjugated diketones such as β- or γ-dibenzoyl derivatives do not cleave [191]. In 2,2-dibenzoyl methane, an enol form is preponderant. On irradiation, 2,2-dibenzoylpropane and dibenzoylethane are particularly stable. Norrish II cleavage arises in 1,3-dibenzoylpropane or 1,4-dibenzoylbutane.

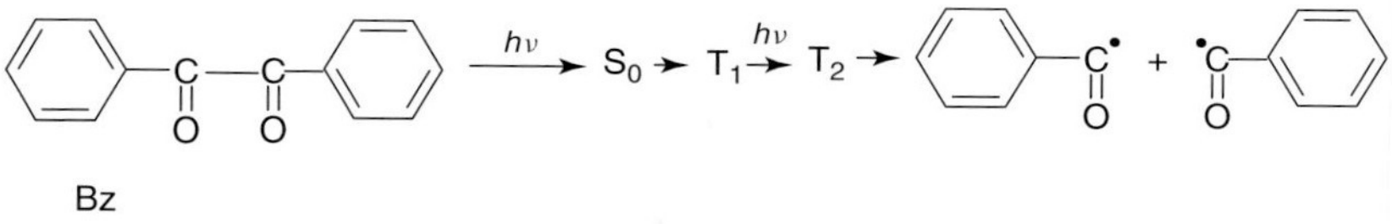

Bz

Scheme 8.16

8.12
Azides and Aromatic Bis-Azides

Azides R–N=N=N possess an N_3 group and decompose (mostly in their singlet state) into nitrenes [192–195]. They participate in cross-linking reactions through the addition of the nitrene to a double bond (Scheme 8.17). Their decomposition can be sensitized by TXs. A photochemical C–N cleavage occurs in CH_3N_3 [196]. An alkyl radical $R'^\bullet$ can also add to an azide: nitrogen is released and an $R'RN^\bullet$ radical is formed.

8.13
Azo Derivatives

Azo compounds $R_3CN=N-CR_3$ [171, 197, 198] are easily cleaved. They also release nitrogen (Eq. (8.5)). A concerted dissociation of the C–N bonds as well as a C–N bond cleavage followed by a fast fragmentation of the $R_3C-N=N^\bullet$ radical (and nitrogen elimination) was proposed. A well-known compound is the thermal AIBN (azo bis-isobutyro nitrile) initiator that can be also photochemically decomposed. Symmetrical azo compounds can initiate the microemulsion polymerization of MMA [199] or the photopolymerization of vinylpyrrolidone [200]. Solvent effects in homogeneous media have been explored [201].

$$R_3C-N=N-CR_3 \longrightarrow R_3C^\bullet + N_2 + {}^\bullet CR_3 \ (h\nu) \tag{8.5}$$

8.14
Disulfide Derivatives

Disulfides (such as dibenzoyl disulfide [202], dithiodiethanol [203], and tetraethylthiuram disulfide [204]) lead to two thiyl or dithiocarbamyl radicals (Eq. (8.6)), which exhibit a low sensitivity to oxygen (Sections 18.7 and 18.8) and allow the polymerization of an acrylate monomer [205]. Polysulfides RS–S–SR also leads to an S–S bond breaking. The formation of sulfur-centered radicals arises after a fast S–S

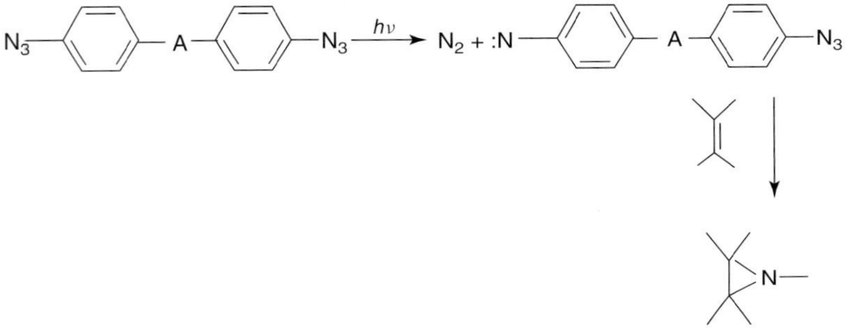

$$N_3-\bigcirc-A-\bigcirc-N_3 \xrightarrow{h\nu} N_2 + {:}N-\bigcirc-A-\bigcirc-N_3$$

Scheme 8.17

single-bond cleavage. In bis (p-aminophenyl) disulfide, the cleavage occurs within 40–100 fs [206].

$$R_1S-SR_2 \longrightarrow R_1S^\bullet + {}^\bullet SR_2 \ (h\nu) \tag{8.6}$$

8.15
Disilane Derivatives

The photodegradation of polysilane polymers or oligomers arising from a primary Si–Si cleavage has been reported as an efficient process. On the contrary, the photoinitiation ability of these silyl radicals is bad as exemplified by the low values determined for the initiation quantum yields in the presence of these polysilanes ($\Phi_i \sim 0.001$ compared to 0.25 for a benzoin ether). Oligosilanes [207] were mentioned a long time ago. In past years, polysilane structures exhibiting higher photoinitiation efficiencies were proposed ($\Phi_i \sim 0.1$) [208]. New disilane derivatives, for example, DSi_1–DSi_4 in **27–30** have been recently proposed as very efficient cleavable PIs [209].

DSi_1	DSi_2	DSi_3	DSi_4
27	**28**	**29**	**30**

The silyl radicals [210, 211] are generated through the homolytic cleavage of the Si–Si bond within $\ll 10$ ns for compound DSi_1; the dissociation quantum yield Φ_{diss} is ~ 1 and the quantum yield for the radical production ~ 2, thereby demonstrating the high efficiency of the cleavage process. The calculated (Si–Si) BDE, which is quite low compared to those of the C–C bonds (roughly 60 vs 85 kcal mol^{-1}), is in agreement with the fast cleavage observed. Contrary to the S_1 state, the triplet state T_1 is dissociative (as revealed by quantum mechanical calculations). The ISC quantum yield is ~ 1. As a consequence, the cleavage process probably mainly originates from T_1. The silyls efficiently add to MA (9×10^8 mol^{-1} 1s^{-1} for DSi_2). The particular reactivity of the silyl radicals especially under air is discussed in Section 18.9.

Poly(hydrosilane)s (PHS) or DSi_4 in **27–30** (that are also interesting co-initiators, as shown in Section 18.9) generate silyl and silylene radicals on UV light (Eq. (8.7)) [212].

$$[-(-H,R)Si-Si(-H,R)-Si(-H,R)-]_n \longrightarrow$$
$$[-(-H,R)Si^\bullet + {}^\bullet Si(-H,R)-Si(-H,R)-]_n \ (h\nu)$$

$$[-(-H,R)Si-Si(-H,R)-Si(-H,R)-]_n \longrightarrow$$
$$[-(-H,R)Si^{\bullet} + {}^{\bullet}Si^{\bullet}(-H,R) + {}^{\bullet}Si(-H,R)-]_n \ (h\nu) \qquad (8.7)$$

The Si–Si bond cleavage [213] was also used to bind disilanylene-oligothienylene polymers to a TiO$_2$ surface.

8.16
Diselenide and Diphenylditelluride Derivatives

Diphenyl, diaryl, or dialkyl diselenides (Eq. (8.8)) cleave according to a homolytic way [214]. The cleavage process has been investigated [215]. Diphenyl diselenides [216] as well as diphenyl ditelluride [217] can initiate the photopolymerization of acrylates. The formed radicals are rather persistent and can be used for controlled polymerization processes (Section 18.19).

$$R_1Se-SeR_2 \longrightarrow R_1Se^{\bullet} + {}^{\bullet}SeR_2 \ (h\nu) \qquad (8.8)$$

8.17
Digermane and Distannane Derivatives

Group 4B dimetals such as disilanes or polysilanes (see above), digermanes, and distannanes can be photochemically cleaved but their low absorption prevents an efficient yield in radicals [218]. The generated metal-centered radicals $R_3Ge^{\bullet}$ and $R_3Sn^{\bullet}$, however, efficiently add to acrylate (Section 18.13; $k_i > 3 \times 10^7$ mol^{-1} 1s^{-1}) and behave as excellent initiating species of free radical photopolymerization.

8.18
Carbon–Germanium Cleavable-Bond-Based Derivatives

Germanium-containing organometallic ketones were studied more than 15 years ago, for example, Ge2 in 31–34 [219]. This class of compounds was successfully reinvestigated [220] as PIs. Compared to TPO, compound Ge1 (32) exhibits a red-shifted $n\pi^*$ transition: 411 nm ($\varepsilon = 549$ mol^{-1} 1cm^{-1}) versus 380 nm (137 mol^{-1} 1cm^{-1}). The $\pi\pi^*$ transition is blue-shifted: 249 nm ($\varepsilon = 16120$ mol^{-1} 1cm^{-1}) versus 294 nm (3778 mol^{-1} 1cm^{-1}). A C–Ge cleavage occurs. The efficiency as a PI is excellent in dental composite applications.

Ge0	Ge1	Ge2	BisGe
31	32	33	34

Using the acylgermane Ge0 in **31–34** where the benzoyl was changed for an acetyl group and the three alkyl for three phenyl substituents, both higher excited-state energy levels and a higher stabilization of the germyl radical were obtained. As a consequence, the dissociation process was enhanced and a high photoinitiation ability was noted [221].

The germyl radicals are produced from the cleavage of the acylgermanes that occurs in the triplet state; a singlet-state contribution cannot be excluded. The overall dissociation quantum yield is close to 1. The triplet energy levels E_Ts and BDE(Ge-C)s were calculated as E_T = 62.9 and 58.2 kcal mol^{-1} and BDE(Ge–C) = 55.2 and 65.0 kcal mol^{-1} for Ge0 and Ge1, respectively. A singlet excited-state energy level E_{S1} ~72 kcal mol^{-1} is evaluated for Ge0. The BDE is clearly lower than E_{S1} and E_T, supporting an exothermic S_1 and T_1 dissociation in line with the efficient bond cleavage. For Ge1, the T_1 cleavage process appears as endothermic. On the other hand, owing to the more red-shifted absorption of Ge1, E_{S1} (Ge1) is < E_{S1} (Ge0). With a higher BDE(Ge–C), the S_1 dissociation of Ge1 is less exothermic than for Ge0. Both results probably also explain the lower decomposition observed for Ge1.

Bisgermyl derivatives BisGe (**34**) lead to a substantial red-shifted absorption from ~360 to ~510 nm (because of the participation of the Ge d-orbitals as shown in Figure 8.16) that extends up to the Ar$^+$ laser line [222]. A triplet cleavage of the Ge-C bond occurs within <100 ns. Two germyl radicals are generated as a fast decarbonylation takes place (Eq. (8.9)). The polymerization ability under a visible light irradiation is higher than that observed with a well-known reference titanocene derivative.

$$Ph_3GeCOGePh_3 \longrightarrow Ph_3Ge^\bullet + Ph_3GeCO^\bullet$$
$$Ph_3GeCO^\bullet \longrightarrow Ph_3Ge^\bullet + CO \quad (8.9)$$

The photoinduced decomposition of dibenzoyl diethyl germanes represents an interesting route (Scheme 8.18) to the formation of polygermanes [223] and the synthesis of block copolymers [224].

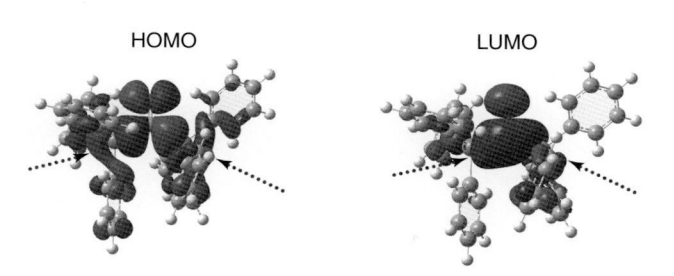

Figure 8.16 HOMO and LUMO for BisGe (**34**) at DFT level. (Source: From [222].)

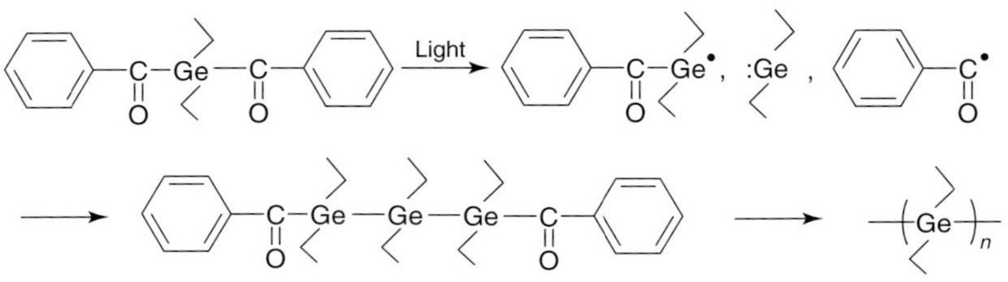

Scheme 8.18

8.19
Carbon–Silicon and Germanium–Silicon Cleavable–Bond-Based Derivatives

The cleavage of C–Si bonds is known [110, 226]. New compounds available in catalogs of fine chemicals have been also checked (J. Lalevée, unpublished data). They possess a Si-benzoyl, Si-phenyl, or a Si-Ge bond. Most of them could be considered as starting and promising structures that can be likely improved by further suitable modifications.

8.20
Silicon Chemistry and Conventional Cleavable Photoinitiators

The modification of usual cleavable PI structures by Si-containing substituents was encountered in (i) the HAP series (change of the phenyl ring or the chromophore, introduction of functional groups, for example, for improving the solubility in silicone media) [227], (ii) the benzoin derivatives (BZ) (such as benzoin trimethylsilyl ether BZeth having a O-Si(CH$_3$)$_3$ group instead of the usual OH) [228], and (iii) the BP and TX series (Section 8.5.7). In most cases, the modified HAP and BZ behave as the parent molecules.

A recent approach [110] consisted in the introduction of a silyl moiety into a usual structure of cleavable radical PI (HAP and BZ series) in order to generate silyl radicals (**35** and **36**).

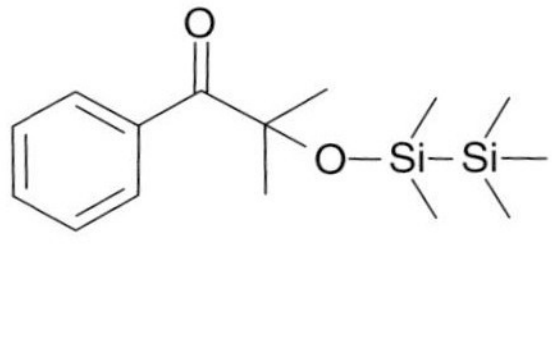

HAP-Si
35

BZ-Si
36

As observed by EST-ST, HAP-Si, and BZ-Si lead to (i) a benzoyl and a carbon-centered radical through α-cleavage and (ii) a carbon-centered radical and a silyloxy radical Si–Si–O• by β-cleavage, this last radical rearranging into an

Si–O–Si• radical. Compared to BZeth, the high efficiency of BZ-Si is ascribed to the presence of these silyl radicals: indeed, such a rearrangement in the absence of a Si–Si bond cannot obviously occur. This demonstrates that the introduction of an O-Si–Si fragment is absolutely required to get a high-reactive PI.

8.21
Sulfur–Carbon Cleavable-Bond-Based Derivatives

The search for new structures allowed proposing of sulfur–carbon-bond-containing compounds, for example, benzoyl- tetrazole or benzoyl-benzoxazole as PI (**37** and **38**) (J. Lalevée, unpublished data). The primary process in SC1 and SC2 is assigned to a C–S bond cleavage that generates a tetrazole and a mercaptobenzoxazole radical. Dissociation quantum yields ϕ_{diss} are 0.25 and 0.05. Their efficiency in the photopolymerization of an epoxyacrylate is relatively low (25% of that observed with DMPA).

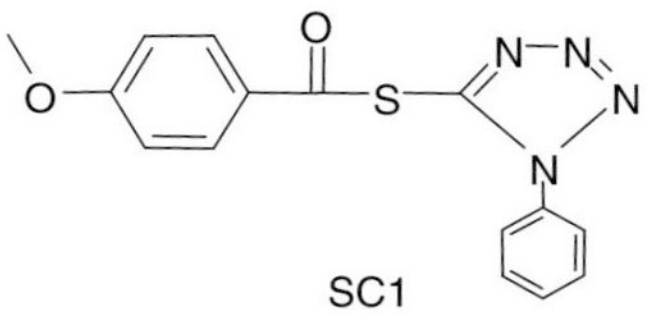

SC1
37

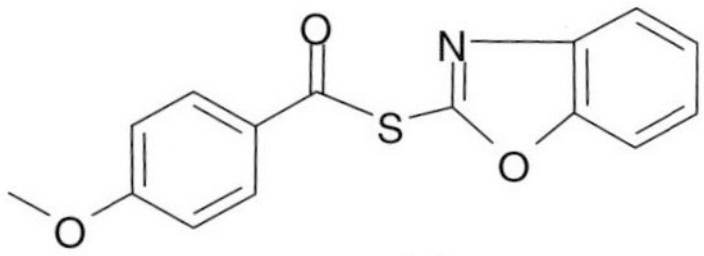

SC 2
38

8.22
Sulfur–Silicon Cleavable-Bond-Based Derivatives

Laser excitation of the compound (**39**) considered as an example of sulfur-silicon-bond-possessing compounds leads to a transient around 480 nm (typical of a sulfur-centered radical: Br–Ph–S•), which originates from an efficient S-Si bond cleavage [110]. A good-to-excellent polymerization initiating ability was found despite relatively low light absorption properties for $\lambda > 320$ nm.

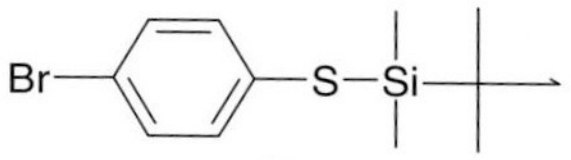

39

8.23
Peresters

In peresters Y–Ar–C(=O)O–OR [229–233], a very efficient cleavage of the O-O bond occurs. Then, a fast decarboxylation yields an aryl YAr• and an alkoxyl •OR radical as in BTTB (BP bearing four perester groups at the 2,2', 3,3' positions).

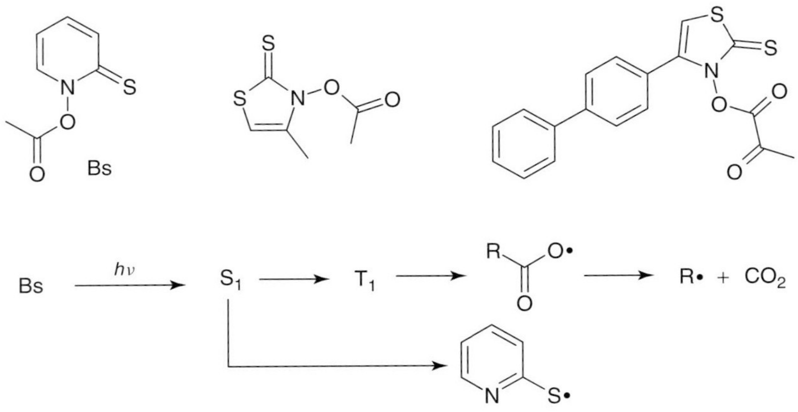

Scheme 8.19

8.24
Barton's Ester Derivatives

Barton's esters are based [234–237] on O-acyl derivatives of N-hydroxypyridine-2-thione and related compounds such as in Scheme 8.19. Working on the substitution of the five-membered ring or the nature of the alkyl radical derived from the O-acyl moiety might lead to more efficient and more stable compounds [238].

A theoretical approach of the cleavage process of Barton's esters has been developed in [238]. The cleavage of the N–O bond proceeds within the picosecond timescale. Both the first excited singlet state and the triplet state are dissociative (small activation energies about 5 and 7 kcal mol^{-1} for S_1 and T_1, respectively; computed BDE (N–O) = 28.2 kcal mol^{-1}; spectroscopic E_T = 53 kcal mol^{-1}). The direct excitation favors singlet-state dissociation and the photosensitization route triplet-state bond breaking [239]. The nitrogen-centered radical, resulting from the N-O cleavage, leads to a more stable pyridine thiyl radical. The acyl radical can decarboxylate and yields the R• radical (Scheme 8.19). A complex set of reactions is then involved, in particular a detrimental chain reaction that leads to a photolysis of the starting compound.

8.25
Hydroxamic and Thiohydroxamic Acids and Esters

The photodissociation of hydroxamic acid (Ha) and related compounds (Har) (40–48) was reported for a long time, although their photochemistry remained relatively unclear and strongly dependent on the considered chemical structures [236, 240, 241]. Thermal polymerization as well as photopolymerization of MMA and styrene, only reported by using related cyclic derivatives such as N-alkoxy phthalimides, naphthalimides, and succinimides, was shown to be effective. The

8.25 Hydroxamic and Thiohydroxamic Acids and Esters

decomposition of dibenzoyl hydroxylamines is considered as occurring through an N-O bond cleavage; photopolymerization of MMA was mentioned.

40–48

Hydroxamic acid derivatives [242] such as the ethers Heth (N-(alkyl-oxy) arylamides) and esters Hest (N-(acyl-oxy) arylamides) as well as thiohydroxamic derivatives [243] such the aryl (acyl-oxy) dithiocarbamate esters THest and the aryl (alkyl-oxy) dithiocarbamate ethers THeth shown in **40–48** were recently proposed and patented as a new family of PIs. Related compounds (**40–48**) such as S-aryl (acyl-oxy) and O-aryl (acyl-oxy) thiocarbamates, where the S–C = S group is changed for S–C=O (TCSest) and O–C=S (TCOest) and O-aryl (acyl-oxy) carbamates where the O–C=O group instead of S–C = S is introduced (Cest) were also recently considered as PI, for example, where R stands for naphthalene.

Few things are known about the photochemistry of (thio)hydroxamic compounds as they are mainly used in organic chemistry as radical generators in order to release suitable alkyl radicals after the N-O bond dissociation. In THest1 and THeth1 (Eq. (8.10)) [244], the singlet and triplet N-O cleavages are followed by a consecutive C–S cleavage (rather than two competitive N–O and C–S cleavages). Triplet sensitization is clearly feasible but the N–O cleavage mostly occurs. In Hest and Heth, a fast singlet N–O cleavage process leads to radicals (a triplet pathway contributes to some extent). In Hest, a O-C(=O) cleavage can occur. A decarboxylation is observed in the R-C(=O)–O• radicals. The cleavage of TcSest, TcOest, and Cest is rather unclear.

$$\text{THest} \longrightarrow \text{RS}^\bullet + \text{S=C=N-} + \text{R-C(=O)-O}^\bullet \quad (h\nu)$$
$$\text{THeth} \longrightarrow \text{RS}^\bullet + \text{S=C=N-} + \text{R-O}^\bullet \quad (h\nu)$$
$$\text{Hest} \longrightarrow \text{R-C(=O)-N}^\bullet + \text{R-C(=O)-O}^\bullet$$
$$\text{and } \text{R-C(=O)-NR-O}^\bullet + \text{R-(C=O)} \quad (h\nu)$$
$$\text{Heth} \longrightarrow \text{R-C(=O)-N}^\bullet + \text{R-O}^\bullet \quad (h\nu) \tag{8.10}$$

The polymerization was checked using a lot of derivatives as PIs (C. Dietlin et al., unpublished data) [244]. Ether derivatives are far less reactive than the corresponding esters. Rates of polymerization of epoxy acrylate/TPGDA (tetrapropyleneglycol diacrylate) in the presence of the best THest and Test derivatives are half that of DMPA. Addition of ITX leads to an efficiency close to DMPA. Structure/properties in photopolymerization reactions were also discussed.

8.26
Organoborates

Organoborates consist in a borate anion and a visible light-absorbing countercation (Scheme 8.20) so that the absorption can be elegantly tuned by changing this cation (e.g., cyanine and styrylpyridinium). After electron transfer and formation of a boranyl radical through an intramolecular process within the ion pair, an initiating R$^\bullet$ radical is generated [245–250]. When starting from a mixture of a cationic dye and a borate, the organoborate can also be considered as a Type II PI. Several works have been carried out on various hemicyanine dye-borate salt ion pairs [251, 252], styryl quinolinium borates [253], and styryl benzimidazolium phenyltributyl borates [254].

8.27
Organometallic Compounds

In the past, abundant literature was devoted to transition metal organometallic complexes [29] that have a specific interest owing to their absorption in the visible region (because of the presence of the metal d-orbitals) that can be tuned through a modification of the ligands. Few compounds work as Type I systems.

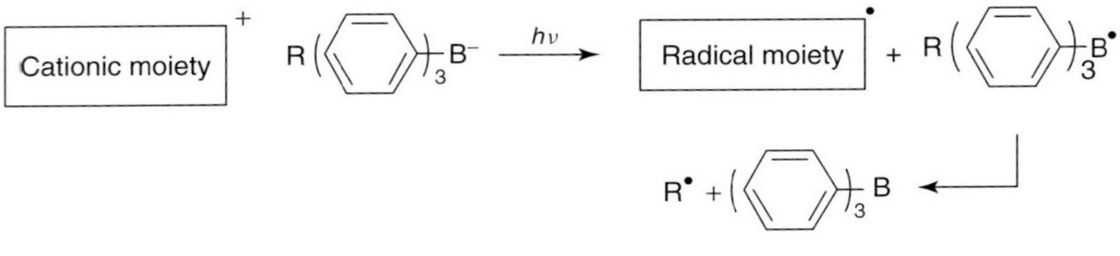

Scheme 8.20

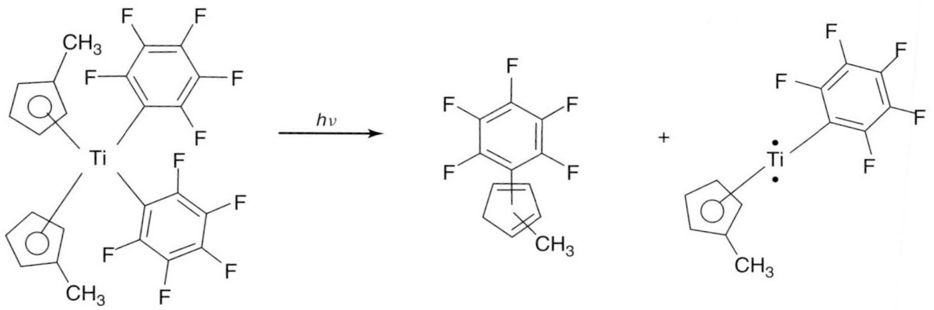

Scheme 8.21

8.27.1
Titanocenes

One of the main representative series is concerned with the titanocenes developed more that 25 years ago [29, 255]. The bis ($\eta - 5-2,4$–cyclopentadien-1-yl) bis[2,6-difluoro-3-(1H-pyrrole-1-yl)phenyl] titanium (TI) compound is the most widely used derivative. It exhibits an excellent absorption in the 400–500 nm range ($\varepsilon = 800 \text{ mol}^{-1} \text{ l cm}^{-1}$ in methanol at 450 nm) and is assumed to very efficiently eliminate a fluorinated aryl Ar$^\bullet$ and a cyclopentadienyl Cp$^\bullet$ radical yielding an unsaturated titanium species and an arylated cyclopentadiene. The unsaturated titanium structure adds to the CO group of one acrylate unit (maybe two) and generates the initiating radicals (Scheme 8.21). Recent ESR-ST experiments on TI [70] clearly evidence the formation of two carbon-centered radicals (Ar$^\bullet$ and Cp$^\bullet$) that recombine to form the main photolysis product. This supports possible additional initiation routes by these radicals. Photoinitiation properties [256] and excited-state processes of TI [257] have been reported. As in BAPO derivatives, the progressive disappearance of the long wavelength absorption on irradiation enables the curing of thick coatings [258].

8.27.2
Chromium Complexes

Tricarbonyl chromium complexes of phenalkyl trichloroacetates were recently shown to behave as Type I PI (**49**) [259]. The dissociation of a Cr-CO bond and the elimination of a carbonyl ligand allow a direct coordination of a chlorine atom of the trichloroacetate group to chromium. Subsequent oxidation and fragmentation form a 17e$^-$ Cr-arene complex bearing a carboxy dichloromethyl radical.

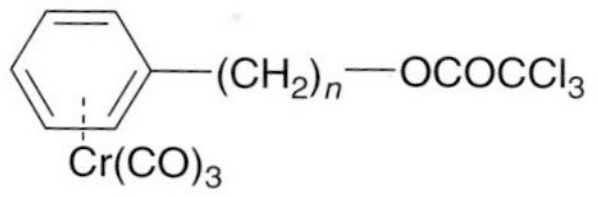

49

8.27.3
Aluminate Complexes

Stable dye–aluminate complexes have been developed [260] (**50**). They presumably work like the dye–borate systems through electron transfer from the dye to the aluminate and then release of an alkyl radical.

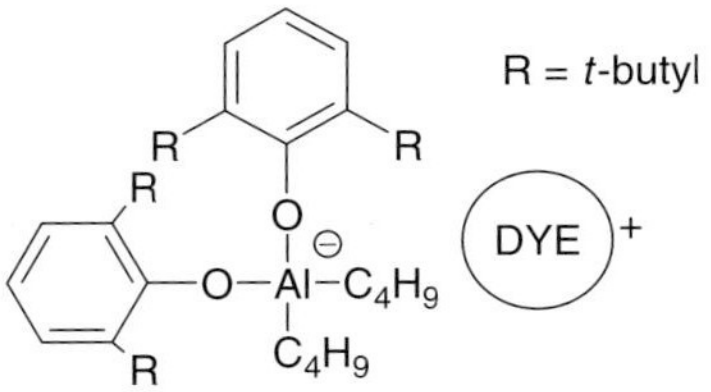

R = *t*-butyl

50

8.28
Metal Salts and Metallic Salt Complexes

Co, Fe, Ni, Cu, and Cr [261] based metal salts and metallic salt complexes are the most encountered systems; for example: transition metal ions Fe^{2+}, V^{2+}, V^{3+}, V^{4+}, UO_2^{2+} for photopolymerization in aqueous media or transition metal inorganic complexes such as $(acac)_2VOCl$, Cr(VI) salts, $Mn(acac)_3$, $Fe(oxalate)_3^{3-}$, $Ru(2,2'\ bipyridine)_3^{2+}$, $Co(acetonylacetnato)^3$, $NiCl_2$, $Ce^{4+}(OH^-)$, bis(aminoacid) Cu chelate, triarylalkyl Ag salt, and Hg bromide ion pair (see a review in [29]).

The decomposition of azidopentaamine cobalt (III) yields an initiating azide radical. In the same way, the reductive interaction between chromic acid and a suitable monomer such as acrylamide (AA) generates a radical on the monomer (Eq. (8.11)) [262]. Bimetallic systems such as chromium (VI)–copper (II) are suitable combinations to control the polymerization process [263]. Other systems are described in [264]. Chromium-based systems find applications in the hologram recording in panchromatic dichromatic gelatin and dichromated polyvinyl alcohol [265].

$$Co(NH_3)_5N_3 \longrightarrow N_3^{\bullet} + Co^{2+} + 5\ NH_3 \quad (h\nu)$$
$$HCrO^{4-} + AA \longrightarrow Cr(V)\ salt + AA^{\bullet} \quad (h\nu) \tag{8.11}$$

Nontransition metal complexes such as the $Pb^{2+}Cl^-$ or $Tl^{3+}SO_4^{2-}$ ion pairs for the photopolymerization of MMA in acidic aqueous solution have been mentioned [29]. Irradiation of the $Tl^{3+}\ SO_4^{2-}$ ion pair generates a sulfate radical.

8.29
Metal-Releasing Compound

A very recent work [266, 267] showed the intramolecular electron transfer occurring in a gold-functionalized 5-mercapto-2,2'-bithiophene derivative. A bithiophene

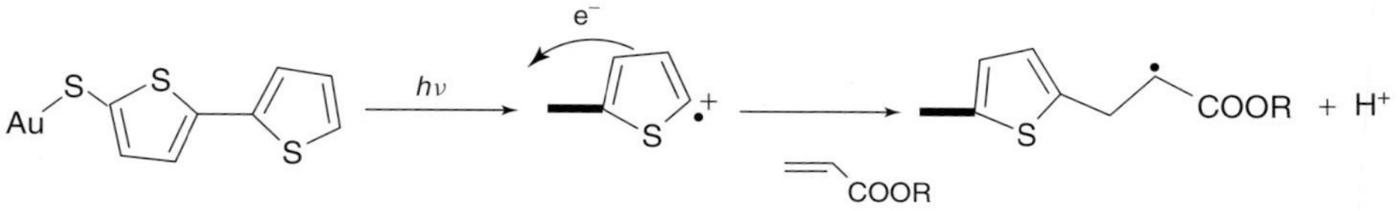

Scheme 8.22

radical cation is generated. The polymerization likely occurs by reaction of this species with the acrylate and then deprotonation (Scheme 8.22). This also allows the dispersion of the gold nanoparticles within the polymer matrix together with a good spatial distribution.

8.30
Cleavable Photoinitiators in Living Polymerization

8.30.1
Cleavable C–S- or S–S-Bond-Based Photoiniferters

The availability of photoiniferters usable in controlled radical photopolymerization reactions is rather limited [268, 269]. A photoiniferter behaves as a PI, a transfer agent, and a terminator. A renewal of interest was noted for the development of efficient compounds [270–275]. Examples of well-known cleavable photoiniferters In1 and In2 involving a C–S or S–S bond cleavage are shown in **51–56**.

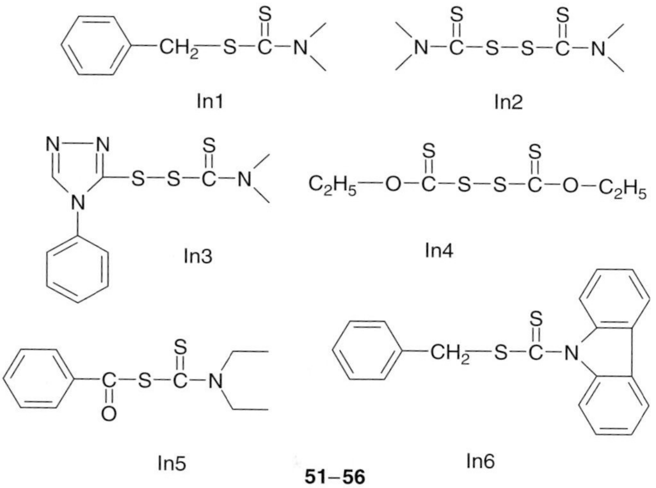

51–56

New compounds In3–In6 have been recently proposed (**51–56**). The particular tetrazole derivative In3 is noticeably attractive because of the generation of a tetrazole-thiyl radical, which presents a low selectivity and a high efficiency for the addition to electron-deficient or electron-rich monomers (values $> 10^7 \text{ mol}^{-1} \text{ l s}^{-1}$

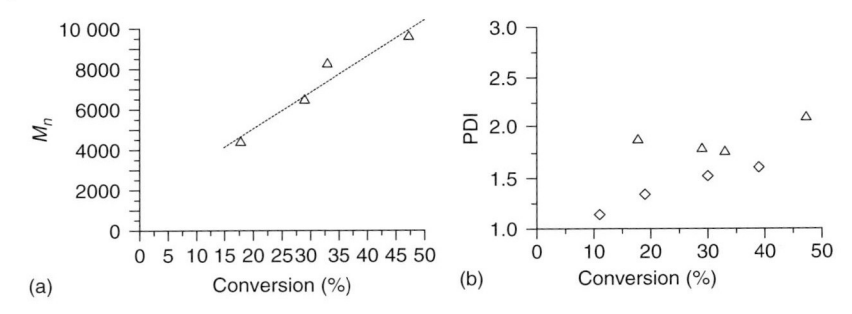

Figure 8.17 (a) Photopolymerization of MMA in bulk: number average molecular weight M_n versus conversion plot using In5 (1% w/w). (b) PDIs versus conversion plots; In5 (up triangle), In6 (diamond). (Source: From [225].)

for the addition to acrylonitrile, MA, MMA, AA, butyl allyl ether, vinyl ether, vinyl acetate, vinyl carbazole, and N-vinyl pyrrolidone) [276]. As a consequence, it leads to a great improvement for the control of the photopolymerization of a large set of monomers, that is, the polymerization is much faster than using In1 or In2 together without any loss of control.

The change of the benzyl for the benzoyl moiety in the In1-type structure (compound In5) leads to a drastic enhancement of R_p without any strong disadvantageous effects on the polymerization control [225] (Figure 8.17). A linear relationship Mn versus conversion is found in agreement with a controlled process. Reinitiation experiments with the formation of a block copolymer in the presence of dormant species and a controlled character for the reaction are also noted. Compound In6 leads to a very good control and a low polydispersity index (PDI) (PDI = 1.14 at 15% conversion and 1.5 at 50%; Mn value = 16 000 at 20% conversion). The significant improvement of the control mechanism with In6 (PDIs in the 1.1–1.5 vs 1.7–2.1 range for In6 vs In5, respectively) is ascribed in part to the increase in the light absorption by the dormant species. The higher persistent character of the dithiocarbamyl radical of In6 also plays a significant role.

The more stabilized sulfur-centered radical in In6 characterized by an important singly occupied molecular orbital (SOMO) delocalization (Figure 8.18) leads to a low reactivity toward oxygen ($k < 10^3 \text{ mol}^{-1} \text{ s}^{-1}$) or MMA ($k' < 10^3 \text{ mol}^{-1} \text{ s}^{-1}$) and a low ability to recombine with a 2,2,6,6, tetramethylpiperidine N-oxyl radical (TEMPO) radical ($k'' < 1.1 \times 10^5 \text{ mol}^{-1} \text{ s}^{-1}$). This behavior allows avoiding the competition with the side reactions generally encountered in other dithiocarbamyl radicals. Therefore, In6 is a better controlling agent. For comparison, in the case of the In2-derived radical, $k = 10^5 \text{ mol}^{-1} \text{ s}^{-1}$, $k' < 10^3 \text{ mol}^{-1} \text{ s}^{-1}$, $k'' = 5.5 \times 10^7 \text{ mol}^{-1} \text{ s}^{-1}$. The SOMO in In3 is more localized (Figure 8.18). The spin density on the sulfur atom is similar for In3 and In5 (~0.5).

All these compounds undergo a very fast C–S (in In1, In5, and In6) or S–S (in In2, In3, and In4) bond cleavage in the singlet state; the triplet state is dissociative. Photosensitized experiments support a triplet cleavage

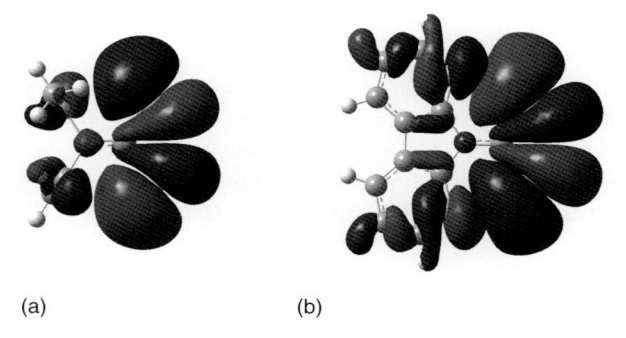

Figure 8.18 SOMOs of the dithiocarbamyl radicals derived from In1 (a) and In6 (b). (Source: From [225].)

(e.g., in In1, In5, and In6). The enhanced efficiency of compound In5 is ascribed to its better absorption properties combined with the higher reactivity of the benzoyl radical ($k = 10^5$ mol^{-1} l s^{-1}) compared to the benzyl radical ($k = 10^3$ mol^{-1} l s^{-1}) toward a methacrylate monomer unit.

A significant improvement of the control should be gained through a decrease in the polymerization rates that can be triggered by a decrease of the incident light intensity. This results in a lower instantaneous concentration in the polymeric radical and therefore allows a better control as the bimolecular recombination between two propagating radicals is reduced [225].

Using these photoiniferters, the formation of copolymers through a sequential approach can be achieved. For example, irradiation of In3 and MMA leads to the synthesis of the PMMA photoiniferter. If a second monomer such as styrene (St) is added, the photopolymerization restarts, thereby allowing the formation of a PMMA-polystyrene copolymer. Other copolymers can be probably formed using the low selectivity property of the tetrazole radical toward the monomers [277]. The insertion of a dormant species (R-M_n-T) in a polymer matrix is worthwhile.

PhSe$^\bullet$ and PhTe$^\bullet$ are also persistent radicals and can be used as controlling agents. Preliminary results (J. Lalevée, unpublished data) suggest that diphenyldiselenide and diphenylditelluride might be efficient photoiniferters. Interestingly, these structures are characterized by a correct absorption in the visible range.

Even if the use of dithiocarbamate compounds is very encouraging for surface modification, severe limitations have been pointed out: (i) dimerization of the produced radicals and/or initiation of extra chains leading to a dramatic loss of control and livingness and (ii) reproducibility concerns, poor quality in terms of topology, or composition of the surface pattern, difficulty to access covalent-bonded multicoat polymer layers. Hence, despite the potential of the dithiocarbamate technology, these drawbacks did not allow a more pronounced development. Therefore, in order to fulfill the growing need in the elaboration of photoinduced surface modification or micro/nanopatterned materials, the development of alternative compounds is still highly desired.

8.30.2
TEMPO-Based Alkoxyamines

O-alkoxyamines (**57–59**) incorporating a TEMPO moiety and based on a benzoyl or aDB chromophore Alk1 [278] or different chromophores covalently linked to the TEMPO-OH moiety Alk2 [279] are known. More recently, a new approach where the chromophore group is directly linked to the aminoxyl function Alk3 has been developed to facilitate the cleavage of the alkoxyamine [280].

The energy transfer from the chromophoric group to the C–O bond becomes efficient and the light absorption is better than in linear aryl nitroxide. Compounds Alk3 can be used as an efficient PI controller in nitroxide-mediated photopolymerization (NMP^2) (the dissociation process of Alk1 and Alk2 is not efficient enough). The dissociation leads to a carbon-centered radical that acts as the initiating species and a nitroxyl radical that plays the role of the controlling agent (Eq. (8.12)). The propagation step is described as the reversible dissociation between the macroalkoxyamine (dormant species) and the active species (carbon-centered propagating radical). Compound Alk3 is also an efficient PI (see below).

$$\begin{aligned} R'R''N-O-R &\longrightarrow R'R''N-O^\bullet + R^\bullet \ (h\nu) \\ R^\bullet + M &\longrightarrow \longrightarrow P_n^\bullet \\ R'R''N-O^\bullet + P_n^\bullet &\longrightarrow R'R''N-O-P_n \\ R'R''N-O-P_n &\longrightarrow P_n^\bullet + R'R''N-O^\bullet \ (h\nu) \end{aligned} \quad (8.12)$$

The dissociation of light-sensitive alkoxyamines was studied as a function of their chemical structures. The selectivity of the cleavable N-O versus C–O bond and the efficiency of the nitroxide formation are strongly related to the alkoxyamine structure (Section 18.19). The distance between the chromophore and the aminoxy group is a key parameter for an efficient pathway of radical generation as evidenced by the photopolymerization ability of these alkoxyamines. The control ability of TEMPO-based systems in photopolymerization is exemplified in Figure 8.19.

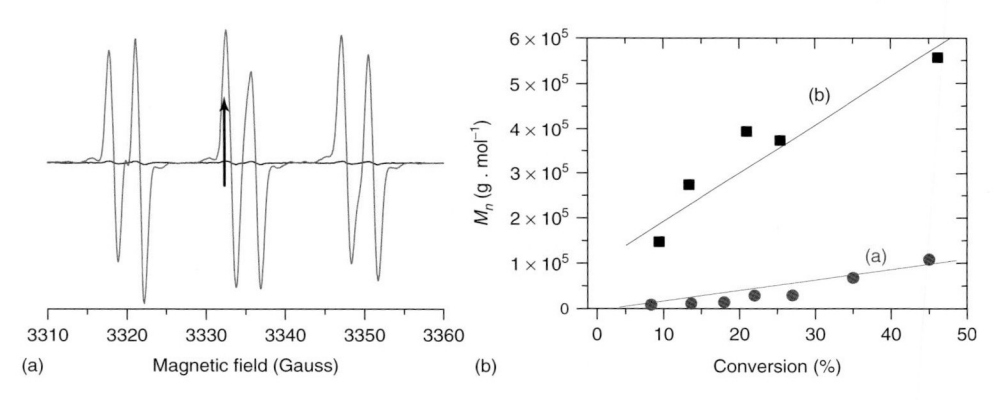

Figure 8.19 (A) ESR spectrum of Alk3 in *tert*-butyl benzene (before and after light irradiation: formation of the nitroxide). (B) M_n versus conversion (bulk photopolymerization of *n*-butyl acrylate at room temperature using 0.7 (a) or 0.07% (w/w) (b) of Alk3. Light intensity = 44 mW cm^{-2}.

In the photopolymerization of *n*-butyl acrylate, a linear increase of the average number molecular weight Mn with the conversion was also observed; 80% of conversion was obtained within 500 s. Moreover, a linear growth of the polymer chain combined with a partial living character has been evidenced: this is the first example of using efficient photosensitive nitroxides being able to fully achieve an NMP^2 process (Figure 8.19). Reinitiation experiments were achieved to highlight the formation of the dormant species (macroalkoxyamines). The ability of the macroalkoxyamines to reinitiate a second polymerization process is worthwhile for micropatterning applications as multilayered film photopolymerization was successfully carried out [280]. The macroalkoxyamine formed in the polymerization of the first layer acts as a macrophotoinitiator in the second layer, leading to covalently bonded multilayer devices. Spatially controlled 3D polymerization should be also feasible.

The development of new systems for the control of photopolymerization reactions is under way, for example, using substituted alkoxyamines for NMP^2 [281, 282], dithiocarbamates as RAFT agent [283] (Section 4.2.11).

8.31
Oxyamines

8.31.1
Alkoxyamines

Alkoxyamines [281] can also act as quite efficient PIs that cleave mostly in their singlet state on light exposure. Some derivatives exhibit polymerization ability close to that of the benzoin ether DMPA. The balance between a good PI behavior and a good control character is a function of the alkoxyamine structure.

8.31.2
Silyloxyamines

Although the overall excited-state mechanism is relatively complex, LFP and ESR experiments show that siloxyamines [284] undergo a Si–Si bond cleavage. The photoinitiating ability of silyloxy julolidine SiOa1 and silyloxy aniline SiOa2 (**60** and **61**) is promising but the absorption has to be improved. Addition of a silane (e.g., tris(trimethylsilyl)silane (TTMSS)) leads to a remarkable enhancement of the polymerization rate and monomer conversion.

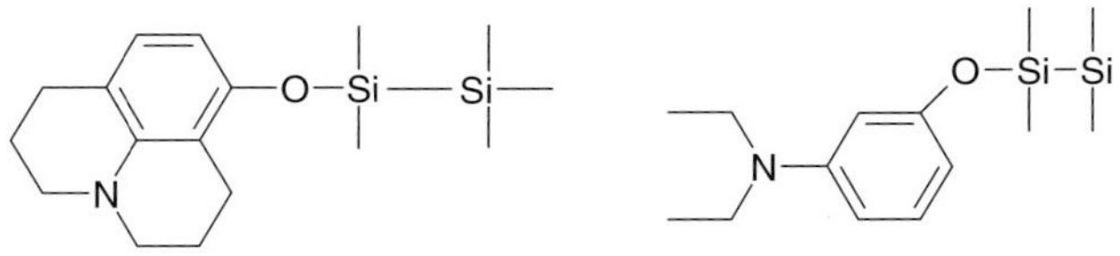

SiOa1
60

SiOa2
61

8.32
Cleavable Photoinitiators for Two-Photon Absorption

A two-photon absorption (TPA) corresponds to the simultaneous absorption of two photons, the absorbing material being characterized by a TPA cross section σ_2 (Section 5.2). Conventional one-photon UV-sensitive cleavable PIs such as bis-acyl phosphine oxides or benzoin ethers can be activated under an intense red light beam (e.g., around 900 or 700 nm) according to a TPA process. In that case, this excitation is considered as leading to the generation of the same initiating radicals as usually after one-photon absorption. Unfortunately, their σ_2s are rather low. Indeed, the main driving factors for getting high σ_2 are conjugation length, planarity of the chromophore, and strength of the donor and acceptor groups. New developments of suitable systems (e.g., (**62** and **63**) [285]) are under way [286–294].

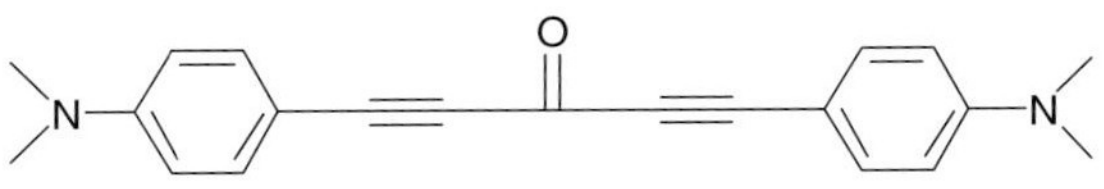

62

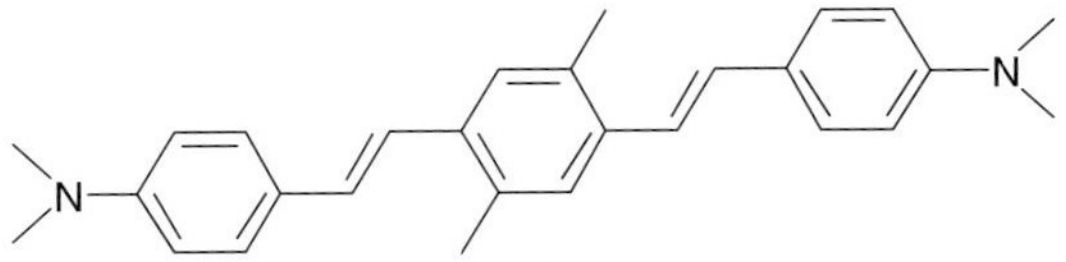

63

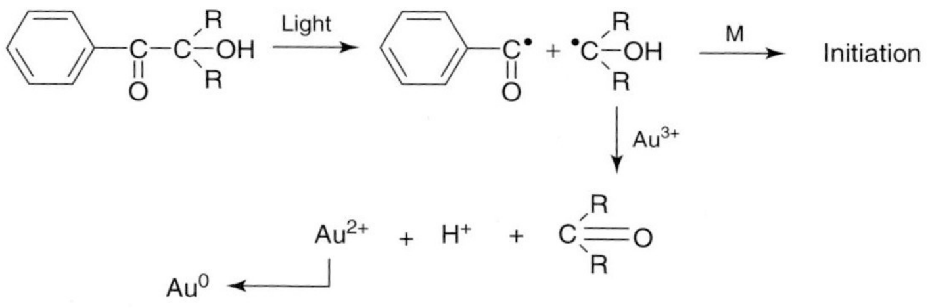

Scheme 8.23

8.33
Nanoparticle-Formation-Mediated Cleavable Photoinitiators

The search for the incorporation of nanoparticles in a polymer film is now well launched. One way might be to use a PI that cleaves under light and accordingly generates two radicals: one of them is used for the initiation, the other one engaged in a redox reaction with a metal salt. Both curing of the matrix and *in situ* formation of metal nanoparticles should be achieved. This could be realized (Scheme 8.23) using hydroxyalkyl phenyl ketone/Au salt [295], benzoin ether/Ag salt, PHS/Ag salt. The radical/metal salt reactivity and the particular features of the overall reactions are discussed in Chapter 19.

8.34
Miscellaneous Systems

Oxalic acid cleaves under light exposure. Oxalates undergo electron transfer to a metal center or the solvent. Phenacyl pyridinium oxalates were presented as interesting water-soluble PIs [296] that also cleave at the C-N bond according to a homolytic cleavage and/or an electron transfer between the pyridinium moiety and the oxalate. The pyridinyl cation radical is short lived. Pyridine and a bromine radical are formed.

The cleavage of dibenzyl ketones [297], ketoamides [298], alkylimides [299], aryloxy naphthalene [300], and ammonium tetraorganyl borate salts [301] has been reported.

Aliphatic ketones such as acetone (cleavage rate constant: $<10^6 \text{ s}^{-1}$ [302]; hydrogen abstraction rate constant: 10^6 and $4 \times 10^8 \text{ mol}^{-1} \text{ 1s}^{-1}$ by isopropanol and triethylamine, respectively [50]), have been proposed in photografting reactions [303].

8.35
Tentatively Explored UV-Light-Cleavable Bonds

Some of the following tentatively explored UV-light-cleavable bonds have been exploited in the photopolymerization area. Others might be considered as new possibilities.

Cyclic ketoimines (O=C–C=N– unit) undergo a Norrish I cleavage [304].

Boranes R_3B, in the presence of oxygen can cleave at the C-B bond and liberate a radical $R^\bullet$ ($R_3B + O_2 \rightarrow R_2BOO^\bullet + R^\bullet$). This very fast thermal reaction [305] can be useful in remote curing (Section 4.2.15).

Silicon boron compounds undergo a photochemically induced homolytic cleavage generating organosilyl radicals [306]. The Si-B-bond-based chromophore has been experimentally and theoretically investigated [307].

The cleavage of a carbone-indium bond has been mentioned [308].

Sulfenates RO–S–Ph and related compounds such as R = N–S–Ph lead to an O–S or N–S bond breaking [305].

Sulfur-containing azo derivatives such as phenylazo-4-biphenylsulfone (Ar–N=N–SO_2–Phi–Phi) undergo an N-S cleavage [202].

Thiosulfonates (–SO_2–SNa or –SO_2–S–) lead to an S-S bond cleavage. Azo thiosulfones (–N=N–S–SO_2–), azo sulfides (–N=N–S–), and sulfonyl sulfonamides (–SO_2–NH–SO_2–) exhibit an N-S bond breaking [309].

Xanthates RO–C (=S)–SR' generate R and R' radicals through the cleavage of the O-C and/or C–S bonds [305, 310]. Both the singlet and triplet states are reactive. Structurally similar derivatives RS–C(=S)–SR' and RS–C(=S)–NR'R" exhibit the same behavior [311]. In the same way, R–O–C(=S)–NR_2 presents an R-O cleavage, whereas the cleavage occurs at the N-N bond in R–O–C(=S)–N–NR_2 [305].

Oxothioates RC (=O)–C–C (=O)–SR' cleave at the C–C and C–S bonds [312].

Dithianes (e.g., $2 - \alpha$–hydroxy-benzyl-1,3-dithiane) can cleave but the process is easier in the presence of BP (in that case, cleavage and hydrogen abstraction occur as evidenced by the presence of benzaldehyde, the BP ketyl radical, and a radical on the dithiane cycle being formed from the C–C bond breaking [313]). A heterolytic cleavage of the C–S bond occurs in trithianes [314].

Hydrazines RR'N–NR"R''' and N-nitroso acylarylamines Ph–N (N=O)–C (=O)–R have low-energy N-N bonds that can be easily cleaved [315]. In nitroso thiols RS–N=O, the S-N bond is weak. Tetraphenyl hydrazine easily cleaves [316].

Triazenes R–N=N–NR'R" and pentaazadienes R–N=N–N–N=N–R' undergo a singlet and triplet N–N bond cleavage [317, 318].

Nitrosoamines RR'N–NO and nitrosoamides R–N (NO)–C (=O)–R' exhibit an N-N cleavage [305].

Generation of radicals can be also encountered in difluoroamines, S-nitrosothiols, [305] acyl-nitroindolines [319], dihydrobenzoxalones [320], thiophosphonates [321], phosphono di- or tri-thioates [322], phosphiranes [323], diphosphines [322], bis-phosphine sulfides [324], allyloxy tetrazole [325], chlorocarboranes [326], and bromoacrylates [327]. The cleavage of alkylditins [328] and Hg–C [329], O–Cl [330], and N–Cl [331] bonds has been also reported.

This chapter outlines the remarkable progress of the research during the past 20 years and the huge efforts to design novel and original cleavable PIs. Much has to be done. A recent general review on thermal and photochemical radical initiation was presented in [42].

References

1. Fouassier, J.P. and Rabek, J.F. (eds) (1990) *Lasers in Polymer Science and Technology: Applications*, CRC Press, Boca Raton, FL.
2. Pappas, S.P. (1986) *UV-Curing: Science and Technology*, Technology Marketing Corporation, Stamford, CT; Plenum Press, New York (1992).
3. Fouassier, J.P. and Rabek, J.F. (eds) (1993) *Radiation Curing in Polymer Science and Technology*, Chapman & Hall, London.
4. Fouassier, J.P. (1995) *Photoinitiation, Photopolymerization, and Photocuring*, Hanser, Münich.
5. Neckers, D.C. (1999) *UV and EB at the Millenium*, Sita Technology Ltd, London.
6. Davidson, S. (1999) *Exploring the Science, Technology and Application of UV and EB Curing*, Sita Technology Ltd, London.
7. Fouassier, J.P. (ed.) (1999) *Photosensitive Systems for Photopolymerization Reactions*, Trends in Photochemistry and Photobiology, Vol. 5, Research Trends, Trivandrum.
8. Fouassier, J.P. (ed.) (2001) *Light Induced Polymerization Reactions*, Trends in Photochemistry and Photobiology, Vol. 7, Research Trends, Trivandrum.
9. Dietliker, K. (2002) *A Compilation of Photoinitiators Commercially Available for UV Today*, Sita Technology Ltd, London.
10. Fouassier, J.P. (ed.) (2006) *Photochemistry and UV Curing*, Research Signpost, Trivandrum.
11. Mishra, M.K. and Yagci, Y. (eds) (2008) *Handbook of Vinyl Polymers*, CRC Press.
12. Allen, N.S. (ed.) (2010) *Photochemistry and Photophysics of Polymer Materials*, John Wiley & Sons, Inc., Hoboken.
13. Fouassier, J.P. and Allonas, X. (eds) (2010) *Basics of Photopolymerization Reactions*, Research Signpost, Trivandrum.
14. Green, W.A. (2010) *Industrial Photoinitiators*, CRC Press, Boca Raton, FL.
15. Fouassier, J.P. (2000) *Recent Research Development in Photochemistry and Photobiology*, vol. **4**, Research Trends, Trivandrum, pp. 51–74.
16. Fouassier, J.P. (1999) *Current Trends in Polymer Science*, vol. **4**, Research Trends, Trivandrum, pp. 131–145.
17. Fouassier, J.P. (2000) *Recent Research Development in Polymer Science*, vol. **4**, Research Trends, Trivandrum, pp. 131–145.
18. Fouassier, J.P. (1990) *Prog. Org. Coat.*, **18**, 227–250.
19. Schnabel, W. (1990) in *Lasers in Polymer Science and Technology*, vol. **II** (eds J.P. Fouassier and J.F. Rabek), CRC Press, Boca Raton, FL, pp. 95–143.
20. Urano, T. (2003) *J. Photopolym. Sci. Technol.*, **16**, 129–156.
21. Fouassier, J.P., Allonas, X., and Burget, D. (2003) *Prog. Org. Coat.*, **47**, 16–36.
22. Fouassier, J.P., Ruhlmann, D., Graff, B., Morlet-Savary, F., and Wieder, F. (1995) *Prog. Org. Coat.*, **25**, 235–271.
23. Fouassier, J.P., Ruhlmann, D., Graff, B., and Wieder, F. (1995) *Prog. Org. Coat.*, **25**, 169–202.
24. Fouassier, J.P. (1993) in *Radiation Curing in Polymer Science and Technology*, vol. 1 (eds J.P. Fouassier and J.F. Rabek), Elsevier, Barking, pp. 49–118.
25. Fouassier, J.P. (1993) in *Radiation Curing in Polymer Science and Technology*, vol. 2 (eds J.P. Fouassier and J.F. Rabek), Elsevier, Barking, pp. 1–62.
26. Lissi, E.A. and Encinas, M.V. (1991) in *Photochemistry and Photophysics*, vol. **IV** (ed. J.F. Rabek), CRC Press, Boca Raton, FL, pp. 221–293.
27. Fouassier, J.P. (1990) in *Photochemistry and Photophysics*, vol. **II** (ed. J.F. Rabek), CRC Press, Boca Raton, FL, pp. 1–26.
28. Timpe, H.J. (1993) in *Radiation Curing in Polymer Science and Technology*, vol. 2 (ed. J.P. Fouassier), Elsevier, Barking, pp. 529–554.
29. Cunningham, A.F. and Desobry, V. (1993) in *Radiation Curing in Polymer Science and Technology*, vol. 2 (eds J.P. Fouassier and J.F. Rabek), Elsevier, Barking, pp. 323–374.

30. Green, W.A. and Timms, A.W. (1993) in *Radiation Curing in Polymer Science and Technology*, vol. **2** (eds J.P. Fouassier and J.F. Rabek), Elsevier, Barking, pp. 375–434.
31. Carlini, C. and Angiolini, L. (1993) in *Radiation Curing in Polymer Science and Technology*, vol. **2** (eds J.P. Fouassier and J.F. Rabek), Elsevier, Barking, pp. 283–322.
32. Li Bassi, G. (1993) in *Radiation Curing in Polymer Science and Technology*, vol. **2** (eds J.P. Fouassier and J.F., Rabek), Elsevier, Barking, pp. 239–282.
33. Steiner, U.E. and Wolff, H.J. (1991) in *Photochemistry and Photophysics*, vol. **IV** (ed. J.F. Rabek), CRC Press, Boca Raton, FL, pp. 1–130.
34. Fouassier, J.P. and Lougnot, D.J. (1990) in *Lasers in Polymer Science and Technology*, vol. **II** (eds J.P. Fouassier and J.F. Rabek) CRC Press, Boca Raton, FL, pp. 145–168.
35. Timpe, H.J., Jockush, S., and Korner, K. (1993) in *Radiation Curing in Polymer Science and Technology*, vol. **2** (eds J.P. Fouassier and J.F. Rabek), Elsevier, Barking, pp. 575–602.
36. Fouassier, J.P., Allonas, X., and Lalevée, J. (2007) in *Macromolecular Engineering: From Precise Macromolecular Synthesis to Macroscopic Materials Properties and Applications*, vol. **1** (eds K. Matyjaszewski, Y. Gnanou, and L. Leibler), Wiley-VCH Verlag GmbH, Weinheim, pp. 643–672.
37. Allonas, X., Crouxte-Barghorn, C., Fouassier, J.P., Lalevée, J., Malval, J.P., and Morlet-Savary, F. (2008) in *Lasers in Chemistry*, vol. **2** (ed. M. Lackner), Wiley-VCH Verlag GmbH, pp. 1001–1027.
38. Arsu, N., Reetz, I., Yagci, Y., and Mishra, M.K. (2008) in *Handbook of Vinyl Polymers*, Part 2, Chapter 5 (eds M.K. Mishra and Y. Yagci), CRC Press, pp. 165–178.
39. Fouassier, J.P., Allonas, X., Lalevée, J., and Dietlin, C. (2010) in *Photochemistry and Photophysics of Polymer Materials* (ed. N.S. Allen), John Wiley & Sons, Inc., Hoboken, pp. 351–420.
40. Lalevée, J., El Roz, M., Allonas, X., and Fouassier, J.P. (2009) in *Organosilanes: Properties, Performance and Applications*, Chapter 6 (eds E. Wyman and M.C. Skief), Nova Science Publishers, Hauppauge, pp. 174–198.
41. Lalevée, J., Tehfe, M.A., Allonas, X., and Fouassier, J.P. (2010) in *Polymer Initiators*, Chapter 8 (ed. W.J. Ackrine), Nova Science Publishers, Hauppauge, pp. 201–224.
42. Lalevée, J. and Fouassier, J.P. in *Handbook of Radical Chemistry and Biology*, vol. **1**, Chapter 2 (eds A. Studer and C. Chatgilialoglou), Wiley-VCH Verlag GmbH, Weinheim, New York, pp. 34–57.
43. Guillaneuf, Y., Fouassier, J.P., Bertin, D., Lalevée, J., and Gigmes, D., in *Progress in Polymer Science* (ed. K. Matyjaszewski), to be published.
44. Fouassier, J.P., Lalevée, J., and Allonas, X., in *Polymer Handbook*, 5ème edn (ed. E. Grulke), John Wiley & Sons, Inc., in press.
45. Lalevée, J. and Fouassier, J.P. (2011) *Polym. Chem.*, **2**, 1107–1113.
46. Fouassier, J.P., Lalevée, J., Morlet-Savary, F., Allonas, X., and Ley, C. (2011) in *Materials, Advances in Dyes and Pigments* (ed. R. Becker), Hindawi Publishing Corporation, pp. 57–78.
47. Allen, N.S. (2007) *Photochemistry*, **36**, 232–297.
48. Dietliker, K., Hüsler, R., Birbaum, J.L., Ilg, S., Villeneuve, S., Studer, K., Jung, T., Benkhoff, J., Kura, H., Matsumoto, A., and Oka, H. (2007) *Prog. Org. Coat.*, **58**, 146–157.
49. Yagci, Y., Jockusch, S., and Turro, N.J. (2010) *Macromolecules*, **43**, 6245–6260.
50. Turro, N.J. (1978 and 1990) *Modern Molecular Photochemistry*, Benjamin, New York.
51. Colley, C.S., Grills, D.C., Besley, N.A., Jockusch, S., Matousek, P., Parker, A.W., Towrie, M., Turro, N.J., Gill, P.M.W., and George, M.W. (2002) *J. Am. Chem. Soc.*, **124**, 14952–14958.
52. Lewis, F.D., Hoyle, C.H., Magyar, J.G., Heine, H.-G., and Hartmann, W. (1975) *J. Org. Chem*, **40**, 488–492.
53. Woo, H.-G., Hong, L.-Y., Kim, S.-Y., Choi, Y.-K., Kook, S.-K., and

Ham, H.-S. (1995) *Bull. Korean Chem. Soc.*, **16**, 667–670.
54. Merlin, A. and Fouassier, J.P. (1981) *J. Chim. Phys.*, **78**, 267–275.
55. Lewis, F.D., Lautenbach, R.T., Heine, H.-G., Hartmann, W., and Rudolph, H. (1975) *J. Am. Chem. Soc.*, **97**, 1519–1525.
56. Lipson, M. and Turro, N.J. (1996) *J. Photochem. Photobiol., A: Chem.*, **99**, 93–96.
57. Dreeskamp, H. and Palm, W.U. (1990) in *Lasers in Polymer Science and Technology*, vol. II (eds J.P. Fouassier and J.F. Rabek), CRC Press, Boca Raton, FL, pp. 47–56.
58. Gunzler, F., Wong, E.H.H., Koo, S.P.S., Junkers, T., and Barner-Kowollik, C. (2009) *Macromolecules*, **42**, 1488–1493.
59. Voll, D., Junkers, T., and Barner-Kowollik, C. (2011) *Macromolecules*, **44**, 2542–2551.
60. Allonas, X., Lalevée, J., and Fouassier, J.P. (2003) *J. Photochem. Photobiol., A: Chem.*, **159** (2), 127–133.
61. Schnabel, W. and Sumiyoshi, T. (1985) in *New Trends in the Photochemistry of Polymers* (eds N.S. Allen and J.F. Rabek), Elsevier, pp. 69–85.
62. Lougnot, D.J., Fouassier, J.P., Merlin, A., Salvin, R., and Faure, J. (1981) *Nuovo Cimento*, **63**, 284–294.
63. Gour, H.A., Groenenboom, C.J., Hageman, H.J., Hakwoort, G.T.M., Osterholl, P., Overeem, T., Polman, R.I., and Van des Werf, S. (1984) *Makromol. Chem.*, **185**, 1795–1802.
64. Jaegermann, P., Lendzian, F., Rist, G., and Mobius, K. (1987) *Chem. Phys. Lett.*, **140**, 615–619.
65. Palm, W.U. and Dreeskamp, H. (1990) *J. Photochem. Photobiol., A: Chem.*, **52**, 439–450.
66. Heine, H.-G. (1972) *Tetrahedron Lett.*, **13**, 4755–4758.
67. Heine, H.-G., Hartmann, W., Kory, D.R., Magyar, J.G., Hoyle, C.E., McVey, J.K., and Lewis, F.D. (1974) *J. Org. Chem.*, **39**, 691–698.
68. Allonas, X., Lalevée, J., Morlet-Savary, F., and Fouassier, J.P. (2006) in *Photochemistry and UV Curing: New Trends* (ed. J.P. Fouassier), Research Signpost, pp. 9–16.
69. Sander, M.R. and Osborn, C.L. (1975) *Tetrahedron Lett.*, **5**, 415–418.
70. Criqui, A., Lalevée, J., Allonas, X., and Fouassier, J.P. (2008) *Macromol. Chem. Phys.*, **209**, 2223–2231.
71. Lalevée, J., Allonas, X., Jradi, S., and Fouassier, J.P. (2006) *Macromolecules*, **39**, 1872–1879.
72. Hageman, H.J. (1985) *Prog. Org. Coat.*, **13**, 123–150.
73. Fischer, H. and Radom, L. (2001) *Angew. Chem. Int. Ed.*, **40**, 1340–1349.
74. Tomioka, H., Takase, N., Maeyama, Y., Hida, K., Lemée, V., Fouassier, J.P., and Burget, D. (2001) *Res. Chem. Interm.*, **27**, 47–59.
75. Lemée, V., Burget, D., Fouassier, J.P., and Tomioka, H. (2000) *Eur. Polym. J.*, **36**, 1221–1230.
76. Fouassier, J.P., Ruhlmann, D., Zahouily, K., Angiolini, L., Carlini, C., and Lelli, N. (1992) *Polymer*, **33**, 3569–3573.
77. Ruhlman, D., Zahouily, K., and Fouassier, J.P. (1992) *Eur. Polym. J.*, **28**, 1063–1067.
78. Encinas, M.V., Rufs, A.M., Corrales, T., Catalina, F., Peinado, C., Schmith, K., Neumann, M.G., and Allen, N.S. (2002) *Polymer*, **43**, 3909–3913.
79. Christensen, J.E., Jacobine, A.F., and Scaiano, C.J.V. (1981) *Radiat. Curing*, **8**, 1–4.
80. Fouassier, J.P. and Lougnot, D.J. (1987) *J. Chem. Soc., Faraday Trans. 1*, **83**, 2935–2952.
81. Morlet-Savary, F., Allonas, X., Lalevée, J., Dietlin, C., Malval, J.P., and Fouassier, J.P. (2008) *J. Photochem. Photobiol., A: Chem.*, **197**, 342–350.
82. Leopold, D. and Fischer, H. (1992) *J. Chem. Soc., Perkin Trans.*, **2**, 513–517.
83. Desobry, V., Dietliker, K., Husler, R., Misev, L., Rembold, M., Rist, G., and Rutsch, W. (1989) *Radiation Curing of Polymeric Materials*, ACS Symposium Series, Vol. 247 (eds C.E. Hoyle and J.F. Kinstle), American Chemical Society, Washington, DC.
84. Eichler, J., Herz, C.P., Naito, I., and Schnabel, W. (1980) *J. Photochem. Photobiol., A: Chem.*, **12**, 225–234.

85. Jockusch, S., Landis, M.S., Freiermuth, B., and Turro, N.J. (2001) *Macromolecules*, **34**, 1619–1626.
86. Weber, M., Turro, N.J., and Beckert, D. (2002) *Phys. Chem. Chem. Phys.*, **4**, 168–172.
87. Allonas, X., Morlet-Savary, F., Lalevée, J., and Fouassier, J.P. (2006) *Photochem. Photobiol.*, **82**, 88–94.
88. Dauben, W.G., Salem, L., and Turro, N.J. (1975) *Acc. Chem. Res.*, **8**, 41–54.
89. Heine, H.-G., Hartmann, W., Lewis, F.D., and Lauterbach, R.T. (1976) *J. Org. Chem.*, **41**, 1907–1912.
90. Ruhlman, D., Fouassier, J.P., and Schnabel, W. (1992) *Eur. Polym. J.*, **28**, 287–292.
91. Hageman, H.J. (1998) *J. Photochem. Photobiol. A*, **117**, 235–238.
92. Tomioka, H., Takimoto, Y., Kawabata, M., Harada, M., Fouassier, J.P., and Ruhlmann, D. (1990) *J. Photochem. Photobiol., A: Chem.*, **53**, 359–372.
93. Morlet-Savary, F., Fouassier, J.P., and Tomioka, H. (1992) *Polymer*, **33**, 4202–4206.
94. Li Bassi, G., Cadona, L., and Broggi, F. (1986) *Proceedings of the Radcure '86*, Technical Paper 4-27, SME Ed., Dearborn, MI.
95. Fouassier, J.P., Lougnot, D.J., and Scaiano, J.C. (1989) *Chem. Phys. Lett.*, **160**, 335–340.
96. Fouassier, J.P., Lemée, V., Espanet, A., Burget, D., Morlet-Savary, F., Di Battista, P., and Li Bassi, G. (1997) *Eur. Polym. J.*, **33**, 881–896.
97. Ruhlman, D. and Fouassier, J.P. (1993) *Eur. Polym. J.*, **29**, 1079–1088.
98. Allonas, X., Lalevée, J., and Fouassier, J.P. (2004) *J. Photopolym. Sci. Technol.*, **17**, 29.
99. Hageman, H.J. and Oosterhoff, P. (2000) *Macromol. Chem. Phys.*, **201**, 1687–1690.
100. Fouassier, J.P. and Burr, D. (1990) *Macromolecules*, **23**, 3615–3619.
101. Berner, G., Rist, G., Rutsch, W., and Kirchmayer, R. (1985) *Proceedings of the Radcure, Basel*, Technical Paper FC85-446, SME Ed., Dearborn, MI.
102. Groenenboom, C.G., Hageman, H.J., Oosterhoff, P., Overeem, T., and Verbeek, J. (1997) *J. Photochem. Photobiol., A*, **107**, 261–269.
103. Hageman, H.J., Oosterhoff, P., and Verbeek, J. (1999) *J. Photochem. Photobiol. A*, **121**, 207–211.
104. Mallavia, R., Sastre, R., and Amat-Guerri, F. (1999) *J. Photochem. Photobiol. A*, **138**, 207–212.
105. Dietlin, C., Lalevée, J., Allonas, X., Fouassier, J.P., Visconti, M., Li Bassi, G., and Norcini, G. (2008) *J. Appl. Polym. Sci.*, **107**, 246–252.
106. Dietliker, K., Jung, T., Benkhoff, J., Kura, H., Geischeidt, G., and Rist, G. (2004) *Macromol. Symp.*, **217**, 77–97.
107. Ruhlman, D., Fouassier, J.P., and Wieder, F. (1992) *Eur. Polym. J.*, **28**, 1577–1582.
108. Lalevée, J., Allonas, X., Fouassier, J.P., Visconti, M., and Li Bassi, G. (2006) in *Photochemistry and UV Curing: New Trends* (ed. J.P. Fouassier), Research Signpost, Trivandrum, pp. 79–90.
109. Alberti, A., Benaglia, M., Macciantelli, D., Rossetti, S., and Scoponi, M. (2008) *Eur. Polym. J.*, **44**, 3022–3027.
110. Lalevée, J., Blanchard, N., Chany, A.C., Souane, R., El Roz, M., Graff, B., Allonas, X., and Fouassier, J.P. (2009) *Macromolecules*, **42**, 6031–6037.
111. Ruhlman, D. and Fouassier, J.P. (1992) *Eur. Polym. J.*, **28**, 591–599.
112. Ye, G., Yang, J., Zeng, Z., and Chen, Y. (2006) *J. Appl. Polym. Sci.*, **99**, 3417–3424.
113. Keskin, S. and Arsu, N. (2006) *Prog. Org. Coat.*, **57**, 348–351.
114. Kura, H., Oka, H., Ohwa, M., Matsumura, T., Kimura, A., Iwasaki, Y., Ohno, T., Matsumura, M., and Murai, H. (2005) *J. Polym. Sci., Part B: Polym. Phys.*, **43**, 1684–1695.
115. Visconti, M. and Cattaneo, M. (2000) *Prog. Org. Coat.*, **40**, 243–251.
116. Fouassier, J.P., Burr, D., and Wieder, F. (1991) *J. Polym. Sci., Part A: Polym. Chem.*, **29**, 1319–1327.
117. Vacek, K., Geimer, J., Beckert, D., and Mehnert, R. (1999) *J. Chem. Soc., Perkin Trans.*, **2**, 2469–2471.
118. Liska, R. (2002) *J. Polym. Sci., Part A: Polym. Chem.*, **40**, 1504–1518.
119. Visconti, M. (2006) in *Photochemistry and UV Curing: New Trends*

(ed. J.P. Fouassier), Research Signpost, Trivandrum, pp. 153–163.
120. Ye, G., Yang, J., Zhao, T., Rao, R., Zeng, Z., and Chen, Y. (2006) *J. Appl. Polym. Sci.*, **102**, 5297–5302.
121. Cai, X., Sakamoto, M., Yamaji, M., Fujitsuka, M., and Majima, T. (2007) *Chem.-A Eur. J.*, **13**, 3143–3149.
122. Fouassier, J.P., Lemée, V., Espanet, A., Burget, D., Morlet-Savary, F., Di Battista, P., and Li Bassi, G. (1997) *Eur. Polym. J.*, **33**, 881.
123. Lemée, V., Fouassier, J.P., Morlet-Savary, F., Burget, D., Di Battista, P., and Li Bassi, G. (1999) *Eur. Polym. J.*, **35**, 669–680.
124. Davies, W.D., Jones, F.D., Garrett, J., Hutchison, I., and Walton, G. (2000) *Surf. Coat. Int., Part B: Coat. Trans.*, **83**, 72–78.
125. Davies, W.D., Jones, F.D., Garrett, J., Hutchinson, I., and Walton, G. (2001) *Surf. Coat. Int., Part B: Coat. Trans.*, **84**, 213–222.
126. Jantas, R., Wodka, T., and Janowska, G. (2001) *Polimery*, **46**, 812–816.
127. Degirmenci, M., Cianga, I., and Yagci, Y. (2002) *Macromol. Chem. Phys.*, **203**, 1279–1284.
128. Tan, J., Wu, B., Yang, J., Zhu, Y., and Zeng, Z. (2010) *Polymer*, **51**, 3394–3401.
129. Degirmenci, M., Hizal, G., and Yagci, Y. (2002) *Macromolecules*, **35**, 8265–8270.
130. Corrales, T., Catalina, F., Peinado, C., and Allen, N.S. (2003) *J. Photochem. Photobiol., A: Chem.*, **159**, 103–114.
131. Fouassier, J.P., Lougnot, D.J., Li Bassi, G., and Nicora, C. (1989) *Polym. Commun.*, **30**, 245–248.
132. Wang, S.J., Fan, X.D., Liu, X., Kong, J., Liu, Y.Y., and Wang, X. (2007) *Polym. Int.*, **56**, 764–772.
133. Wang, Y., Jiang, X., and Yin, J. (2007) *J. Appl. Polym. Sci.*, **105**, 3817–3823.
134. Wang, J., Zhang, L., and Shi, G. (2010) *Tuliao Gongye*, **40**, 1–4.
135. Tan, H., Yang, D., Xiao, M., Han, J., and Nie, J. (2009) *J. Appl. Polym. Sci.*, **111**, 1936–1941.
136. Liska, R. (2001) *Heterocycles*, **55**, 1475–1486.
137. Liska, R. and Herzog, D. (2004) *J. Polym. Sci., Part A: Polym. Chem.*, **42**, 752–764.
138. Liska, R., Seidl, B., and Grabner, G. (2004) *Polym. Prepr.*, **45**, 75–76.
139. Liska, R. (2006) in *Photochemistry and UV Curing: New Trends* (ed. J.P. Fouassier), Research Signpost, Trivandrum, pp. 55–67.
140. Seidl, B., Liska, R., and Grabner, G. (2006) *J. Photochem. Photobiol. A: Chem.*, **180**, 109–117.
141. Liska, R., Knaus, S., and Wendrinsky, J. (1999) *Nucl. Instrum. Methods Phys. Res., Sec. B*, **151**, 290–292.
142. Liska, R. and Herzog, D. (2004) *J. Pol. Sci. Technol.*, **40**, 75–79.
143. Lalevée, J., Allonas, X., and Fouassier, J.P. (2005) Proceedings of the RadTech Europe, Barcelona, vol. 2, pp. 67–72.
144. Liska, R. and Seidl, B. (2005) *J. Polym. Sci., Part A: Polym. Chem.*, **43**, 101–111.
145. Seidl, B., Kalinyaprak-Icten, K., Fuss, N., Hoefer, M., and Liska, R. (2008) *J. Polym. Sci., Part A: Polym. Chem.*, **46**, 289–301.
146. Seidl, B. and Liskra, R. (2007) *Macromol. Chem. Phys.*, **208**, 44–54.
147. Rosspeintner, A., Griesser, M., Pucher, N., Iskra, K., Liska, R., and Gescheidt, G. (2009) *Macromolecules*, **42**, 8034–8038.
148. Yamaji, M., Wakabayashi, S., and Tobita, S. (2006) *Res. Chem. Interm.*, **32**, 749–758.
149. Fouassier, J.P. and Lougnot, D.J. (1990) *Polymer Commun.*, **31**, 418–421.
150. Andrzejewska, E. (2001) *Trends in Photochemistry and Photobiology*, vol. 7 (ed. J.P. Fouassier), Research Trends, Trivandrum, pp. 1–9.
151. Fouassier, J.P., Lougnot, D.J., and Avar, L. (1995) *Polymer*, **36**, 5005–5010.
152. Allonas, X., Grotzinger, C., Lalevée, J., Fouassier, J.P., and Visconti, M. (2001) *Eur. Polym. J.*, **37**, 897–906.
153. Moorthy, J.N. and Samanta, S. (2007) *J. Org. Chem.*, **72**, 9786–9789.
154. Lalevée, J., Blanchard, N., El Roz, M., Graff, B., Allonas, X., and Fouassier, J.P. (2008) *Macromolecules*, **41**, 4180–4186.

155. Lalevée, J., Blanchard, N., Fries, C., Tehfe, M.A., Morlet-Savary, F., and Fouassier, J.P. (2011) *Polym. Chem.*, **2**, 1077–1084.
156. Sumiyoshi, T., Schnabel, W., and Henne, A. (1986) *J. Photochem. Photobiol., A: Chem.*, **32**, 191–201.
157. Baxter, J.E., Davidson, R.S., and Hageman, H.J. (1988) *Polymer*, **29**, 1569–1574.
158. Castelvetro, V., Molesti, M., and Rolla, P. (2002) *Makromol. Chem. Phys.*, **203**, 1486–1496.
159. Wang, W., Wu, Q., Ding, L., Yang, Z., and Zhang, A. (2008) *J. Appl. Polym. Sci.*, **107**, 593–598.
160. Angiolini, L., Caretti, D., and Carlini, C. (1995) *J. Appl. Polym. Sci.*, **57**, 519–531.
161. Ullrich, G., Ganster, B., Salz, U., Moszner, N., and Liska, R. (2006) *J. Polym. Sci., Part A: Polym. Chem.*, **44**, 1686–1700.
162. de Groot, J.H., Dillingham, K., Deuring, H., Haitjema, H.J., van Beijma, F.J., Hodd, K., and Norrby, S. (2001) *Biomacromolecules*, **2**, 1271–1278.
163. Medsker, R.E., Sebenik, A., and Harwood, H.J. (2002) *Polym. Bull.*, **48**, 17–23.
164. Macarie, L., Manoviciu, I., Manoviciu, V., Dehelean, G., Ilia, G., Iliescu, S., Popa, A., and Plesu, N. (2002) *Rev. Chim.*, **53**, 568–571.
165. Keskin, S., Jockusch, S., Turro, N.J., and Arsu, N. (2008) *Macromolecules*, **41**, 4631–4634.
166. Ikemura, K., Ichizawa, K., Jogetsu, Y., and Endo, T. (2010) *Dent. Mater. J.*, **29**, 122–131.
167. Moszner, N., Lamparth, I., Angermann, J., Fischer, U.K., Zeuner, F., Bock, T., Liska, R., and Rheinberger, V. (2010) *Beilstein J. Org. Chem.*, **6** (26), 1–9.
168. Jockusch, S., Koptyug, I.V., McGarry, P.F., Sluggett, G.W., Turro, N.J., and Watkins, D.M. (1997) *J. Am. Chem. Soc.*, **119**, 11495–11501.
169. Sluggett, G.W., Turro, C., George, M.W., Koptyug, I.V., and Turro, N.J. (1995) *J. Am. Chem. Soc.*, **117**, 5148–5153.
170. Morlet Savary, F., Allonas, X., and Fouassier, J.P. (2006) in *Photochemistry and UV Curing: New Trends* (ed. J.P. Fouassier), Research Signpost, Trivandrum, pp. 69–78.
171. Murai, H. and Hayashi, H. (1993) in *Radiation Curing in Polymer Science and Technology*, vol. **2** (eds J.P. Fouassier and J.F. Rabek), Elsevier, Barking, pp. 63–154.
172. Spichty, M., Turro, N.J., Rist, G., Birbaum, J.L., Dietliker, K., Wolf, J.P., and Gescheidt, G. (2001) *J. Photochem. Photobiol., A: Chem.*, **142**, 209–213.
173. Gatlik, I., Rzadek, P., Gescheidt, G., Rist, G., Hellrung, B., Wirz, J., Dietliker, K., Hug, G., Kunz, M., and Wolf, J.P. (1999) *J. Am. Chem. Soc.*, **121**, 8332–8336.
174. Jockusch, S. and Turro, N.J. (1998) *J. Am. Chem. Soc.*, **120**, 11773–11777.
175. Lalevée, J., Allonas, X., and Fouassier, J.P. (2010) in *Basics and Applications of Photopolymerization Reactions*, vol. 2 (eds J.P. Fouassier and X. Allonas), Research Signpost, Trivandrum, pp. 89–100.
176. Berner, G., Kirchmayer, R., and Rist, G. (1978) *J. Oil Chem. Assoc.*, **61**, 105.
177. Dossot, M., Obeid, H., Allonas, X., Jacques, P., Fouassier, J.P., and Merlin, A. (2004) *J. Appl. Polym. Sci.*, **92**, 1154–1164.
178. Allonas, X., Dossot, M., Merlin, A., Sylla, M., Jacques, P., and Fouassier, J.P. (2000) *J. Appl. Polym. Sci.*, **78**, 2061–2069.
179. Kim, J.H. and Kim, H.L. (2001) *Chem. Phys. Lett.*, **333**, 45–50.
180. Kawamura, K. (2003) *Chem. Lett.*, **32**, 832–833.
181. Czech, Z., Butwin, A., and Kabatc, J. (2011) *J. Appl. Polym. Sci.*, **120**, 3621–3627.
182. Jiang, M., Wang, K., Ma, G., Jian, L., Juan, X., Yu, Q., and Nie, J. (2011) *J. Appl. Polym. Sci.*, **121**, 2013–2017.
183. Hamity, M. and Scaiano, J.C. (1975) *J. Photochem. Photobiol., A: Chem.*, **4**, 229–236.
184. Biondic, M., Giacopello, S., Bassells, R.E., Encinas, M.V., and Lissi, E.A. (1996) *J. Polym. Sci., Part A: Polym. Chem.*, **34**, 1941–1944.

185. Temkin, A.Y. (1999) *Lasers Eng.*, **9**, 239–259.
186. Gu, Z., Wang, Y., and Balbuena, P.B. (2006) *J. Phys. Chem.*, **110**, 2448–2454.
187. Eisfeld, W. and Francisco, J.S. (2008) *J. Chem. Phys.*, **128**, 174304-1–174304-7.
188. Buback, M., Kling, M., Schmatz, S., and Schroeder, J. (2004) *Phys. Chem. Chem. Phys.*, **6**, 5441–5455.
189. Chateauneuf, J., Lusztik, J., and Ingold, K.U. (1988) *J. Am. Chem. Soc.*, **110**, 2877–2882.
190. Mc Gimpsey, W.G. (1990) in *Lasers in Polymer Science and Technology*, vol. II (eds J.P. Fouassier and J.F. Rabek), CRC Press, Boca Raton, FL, pp. 77–94.
191. Wagner, P.J. and Frerking, H.W. (1995) *Can. J. Chem.*, **73**, 2047–2061.
192. Yasuda, N., Yamamoto, S., Wada, Y., and Yanagida, S. (2001) *J. Polym. Sci., Part A: Polym. Chem.*, **39**, 4196–4205.
193. Tsao, M.-L. and Platz, M.S. (2004) *J. Phys. Chem. A.*, **37**, 8984–8991.
194. Burdzinski, G., Gustafson, T.L., Hackett, J.C., Hadad, C.M., and Platz, M.S. (2005) *J. Am. Chem. Soc.*, **127**, 13764–13765.
195. Budyka, M.F. (2007) *High Energy Chem.*, **41**, 356–360.
196. Larson, C., Ji, Y., Samartzis, P.C., Quinto-Hernandez, A., Min Lin, J. Jr., Ching, T.T., Chaudhuri, C., Lee, S.-H., and Wodtke, A.M. (2008) *J. Phys. Chem. A*, **112**, 1105–1111.
197. Alvarez, J., Encinas, M.V., and Lissi, E.A. (1999) *Macromol. Chem. Phys.*, **200**, 2411–2415.
198. Sander, W., Strehl, A., and Winkler, M. (2001) *Eur. J. Org. Chem.*, **20**, 3771–3778.
199. Encinas, M.V., Lissi, E.A., Rufs, A.M., and Alvarez, J. (1998) *Langmuir*, **14**, 5691–5694.
200. Encinas, M.V., Lissi, E.A., and Quiroz, J. (1992) *Eur. Polym. J.*, **28**, 471–473.
201. Encinas, M.V., Lissi, E.A., and Martinez, C. (1996) *Eur. Polym. J.*, **32**, 1151–1154.
202. Allen, N.S. (1996) *J. Photochem. Photobiol. A*, **100**, 101–107.
203. Baruah, S.-R. and Kakati, D.K. (2006) *J. Appl. Polym. Sci.*, **100**, 1601–1606.
204. Rahane, S.B., Kilbey, S.M., and Metters, A.T. (2008) *Macromolecules*, **41**, 7440–7447.
205. Lalevée, J., Allonas, X., Zadoina, L., and Fouassier, J.P. (2007) *J. Polym. Sci., Polym. Chem.*, **45**, 2494–2502.
206. Bultmann, T. and Ernstig, N.P. (1996) *J. Phys. Chem.*, **100**, 19417–19420.
207. Semenov, V.V., Ladilina, E.Y., Cherepennikova, N.F., and Chesnokova, T.A. (2002) *Russ. J. Appl. Chem.*, **75**, 127–134.
208. Arsu, N., Hizai, G., and Yagci, Y. (1995) *J. Macromol. Chem., Macromol. Reports.*, **A32**, 1257–1263.
209. Lalevée, J., El Roz, M., Graff, B., Allonas, X., and Fouassier, J.P. (2007) *Macromolecules*, **40**, 8527–8530.
210. Lalevée, J., Allonas, X., and Fouassier, J.P. (2007) *J. Org. Chem.*, **72**, 6434–6439.
211. Lalevée, J., Dirani, A., El Roz, M., Allonas, X., and Fouassier, J.P. (2008) *Macromolecules*, **41**, 2003.
212. Lalevée, J., Shankar, R., Tehfe, M.A., Sahoo, U., and Fouassier, J.P. (2011) *Macromol. Chem. Phys., Macromol. Chem. Phys.*, **212**, 806–812.
213. Kajiwara, S., Ooyama, Y., Harima, Y., and Kakimoto, M. (2008) *Chem. Lett.*, **37**, 316–317.
214. Kwon, T.S., Suzuki, K., Takagi, K., Kunisada, H., and Yuki, Y. (2001) *J. Macromol. Sci., Pure Appl. Chem.*, **A38**, 591–604.
215. Ouchi, A., Liu, S., Li, Z., Kumar, S.A., Suzuki, T., Hyugano, T., and Kitahara, H. (2007) *J. Org. Chem.*, **72**, 8700–8706.
216. Jedrzejewska, B., Rafinski, Z., and Paczkowski, J. (2009) *Polimery*, **54**, 417–420.
217. Srivastava, A.K. and Tripathi, A. (2008) *Des. Monom. Polym.*, **11**, 83–95.
218. El Roz, M., Lalevée, J., Morlet-Savary, F., Allonas, X., and Fouassier, J.P. (2009) *Macromolecules*, **42**, 4464–4469.
219. Wakasa, M., Mochida, K., Sakaguchi, Y., Nakamura, J., and Hayashi, H. (1991) *J. Phys. Chem.*, **95**, 2241–2245.
220. Ganster, B., Fischer, U.K., Moszner, N., and Liska, R. (2008) *Macromol. Rapid Commun.*, **29**, 57–62.

221. Lalevée, J., Allonas, X., and Fouassier, J.P. (2009) *Chem. Phys. Lett.*, **469**, 293–303.
222. Tehfe, M., Blanchard, N., Fries, C., El Roz, M., Lalevée, J., Morlet-Savary, F., Allonas, X., and Fouassier, J.P. (2010) *Macromol Chem. Rapid Commun.*, **31**, 473–478.
223. Durmaz, Y.Y., Kukut, M., Monszner, N., and Yagci, Y. (2009) *Macromolecular*, **42**, 2899–2902.
224. Durmaz, Y.Y., Kukut, M., Moszner, R., and Yagci, Y. (2009) *J. Polym. Sci., Part A: Polym. Chem.*, **47**, 4793–4799.
225. Lalevée, J., Blanchard, N., El Roz, M., Allonas, X., and Fouassier, J.P. (2008) *Macromolecules*, **41**, 2347–2352.
226. Yamaji, M., Mikoshiba, T., and Masuda, S. (2007) *Chem. Phys. Lett.*, **438**, 229–233.
227. Kolar, A., Gruber, H.F., and Greber, G. (1994) *J. Macromol. Sci.*, **A31**, 305–318.
228. Woo, H.G., Hong, L.Y., Kim, S.H., Choi, Y.K., Kook, S.K., and Ham, H.S. (1995) *Bull. Korean Chem. Soc.*, **16**, 667–670.
229. Allen, N.S., Hardy, S.J., Jacobine, A.F., Glaser, D.M., Yang, B., Wolf, D., Catalina, F., Navaratnam, S., and Parsons, B.J. (1991) *J. Appl. Polym. Sci.*, **42**, 1169–1178.
230. Neckers, D.C., Abu Abdoun, I.I., and Thijs, L. (1984) *Macromolecules*, **17**, 282–288.
231. Shah, B.K. and Neckers, D.C. (2005) *J. Am. Chem. Soc.*, **170**, 195–201.
232. Shah, B.K., Gusev, A., Rodgers, M.A.J., and Neckers, D.C. (2004) *J. Phys. Chem. A*, **108**, 5926–5931.
233. Morlet-Savary, F., Wieder, F., and Fouassier, J.P. (1997) *J. Chem. Soc., Faraday Trans.*, **93**, 3931–3937.
234. Barton, D.H.R. and Zard, S.Z. (1986) *Pure Appl. Chem.*, **58**, 675–684.
235. Bales, B.C., Horner, J.H., Hueng, X., Newcomb, M., Crich, D., and Greenberg, M.M. (2001) *J. Am. Chem. Soc.*, **123**, 3623–3629.
236. Newcomb, M., Horner, J.H., Whitted, P.O., Crich, D., Huang, X., Yao, Q., and Zipse, H. (1999) *J. Am. Chem. Soc.*, **121**, 10685–10694.
237. Aveline, B.M., Kochevar, I.E., and Redmond, R.W. (1996) *J. Am. Chem. Soc.*, **118**, 10113–10123.
238. Dietlin, C., Allonas, X., Morlet-Savary, F., Fouassier, J.P., Visconti, M., Norcini, G., and Romagnano, S. (2008) *J. Appl. Polym. Sci.*, **109**, 825–833.
239. Allonas, X., Dietlin, C., Fouassier, J.P., Casiraghi, A., Visconti, M., Norcini, G., and Li Bassi, G. (2008) *J. Photopolym. Sci. Technol.*, **21**, 505–509.
240. Alam, M.M., Fujitsuka, M., and Ito, O. (2000) *Rec. Res. Dev. Phys. Chem.*, **4**, 369–381.
241. Bohne, C., Bosch, R., and Scaiano, J.C. (1990) *J. Org. Chem.*, **55**, 5414–5418.
242. Casiraghi, A., Visconti, M., Norcini, G., Dietlin, C., Allonas, X., Fouassier, J.P., and Li Bassi, G. (2010) US Patent 2010/0093883, Lamberti.
243. Romagnano, S., Visconti, M., Norcini, G., Li Bassi, G., Dietlin, C., Allonas, X., and Fouassier, J.P. (2010) US Patent 2010/0196826, Lamberti.
244. Allonas, X., Casiraghi, A., Dietlin, C., Fouassier, J.P., Lalevée, J., Li Bassi, G., Norcini, G., Romagnano, S., and Visconti, M. (2010) in *Basics of Photopolymerization Reactions*, vol. 1 (ed. J.P. Fouassier and X. Allonas), Research Signpost, Trivandrum, pp. 79–92.
245. Grinevich, O., Serguievski, P., Sarker, A.M., Zhang, W., Mejiritski, A., and Neckers, D.C. (1999) *Macromolecules*, **32**, 328–330.
246. Sarker, A.M., Sawabe, K., Strehmel, B., Kaneko, Y., and Neckers, D.C. (1999) *Macromolecules*, **32**, 5203–5209.
247. Kabatc, J., Jedrzejewska, B., and Paczkowski, J. (2003) *J. Polym. Sci., Part A: Polym. Chem.*, **41**, 3017–3026.
248. Kabatc, J., Pietrzak, M., and Paczkowski, J. (2002) *J. Chem. Soc., Perkin Trans.*, **2** (2), 287–295.
249. Jedrzejewska, B., Kabatc, J., Pietrzak, M., and Paczkowski, J. (2002) *J. Polym. Sci., Part A: Polym. Chem.*, **40**, 1433–1440.
250. Jedrzejewska, B., Pietrzak, M., and Rafinski, Z. (2011) *Polymer*, **52**, 2110–2119.
251. Kabatc, J., and Celmer, A. (2009) *Polymer*, **50**, 57–67.

252. Jedrzejewska, B. and Urbanski, S. (2010) *J. Appl. Polym. Sci.*, **118**, 1395–1405.
253. Jedrzejewska, B., Marcin, T., and Paczkowski, J. (2009) *Mater. Chem. Phys.*, **117**, 448–454.
254. Pietrzak, M., Jedrzejewska, B., and Paczkowski, J. (2009) *J. Polym. Sci., Part A: Polym. Chem.*, **47**, 4119–4129.
255. Roloff, A., Meier, K., and Riediker, M. (1986) *Pure Appl. Chem.*, **58**, 1267–1273.
256. Sabol, D., Gleeson, M.R., Liu, S., and Sheridan, J.T. (2010) *J. Appl. Phys.*, **107**, 53113 - 1–53113-8.
257. Klingert, B., Roloff, A., Urwyler, B., and Wirz, J. (1988) *Helv. Chim. Acta*, **71**, 1858–1864.
258. Finter, J., Riediker, M., Rohde, O., and Rotzinger, B. (1989) *Macromol. Chem. Makromol. Symp.*, **24**, 177–186.
259. Kundig, E.P., Xu, L.H., Kondratenko, M., Cunningham, A.F. Jr., and Kunz, M. (2007) *Eur. J. Inorg. Chem.*, **18**, 2934–2943.
260. Matsumoto, H., Yamaguchi, Y., Yanagihara, N., and Yamamoto, H. (1990) EP 90, 119, 767.
261. Neumann, M.G., Schmitt, C.C., and Rigoli, I.C. (2003) *J. Photochem. Photobiol., A: Chem.*, **159**, 145–150.
262. Billaud, C., Sarakha, M., and Bolte, M. (2000) *J. Polym. Sci., Part A: Polym. Chem.*, **38** (21), 3997–4005.
263. Maihlot, G. and Bolte, M. (1991) *J. Photochem. Photobiol., A: Chem.*, **56**, 387–396.
264. Jakubiak, J. and Rabek, J.F. (1999) *Polimery*, **44**, 447–461.
265. Djouani, F., Israel, Y., Frezet, L., Rivaton, A., Lessard, R.A., and Bolte, M. (2006) *J. Polym. Sci., Part A: Polym. Chem.*, **44**, 1317–1325.
266. Anyaogu, K.C., Cai, X., and Neckers, D.C. (2008) *Macromolecules*, **41**, 9000–9003.
267. Cai, X., Anyagu, K.C., and Neckers, D.C. (2007) *J. Am. Chem. Soc.*, **129**, 11324–11327.
268. Otsu, T. and Taraki, T. (1986) *Polym. Bull.*, **16**, 277–284.
269. Bertin, D., Boutevin, B., Gramain, P., Fabre, J.M., and Montginoul, C. (1998) *Eur. Polym. J.*, **34**, 85–90.
270. Ajayaghosh, A. and Francis, R. (1999) *J. Am. Chem. Soc.*, **121**, 6599–6606.
271. Qin, S.-H., Qin, D.-Q., and Qiu, K.-Y. (2001) *Chin. J. Polym. Sci.*, **19**, 441–445.
272. Rathore, K., Reddy, K.R., Tomer, N.S., Desai, S.M., and Singh, R.P. (2004) *J. Appl. Polym. Sci.*, **93**, 348–355.
273. Qin, S.H. and Qiu, K.Y. (2001) *Eur. Polym. J.*, **37**, 711–717.
274. Lalevée, J., El Roz, M., Allonas, X., and Fouassier, J.P. (2007) *J. Polym. Sci.*, **45**, 2436–2442.
275. Bhuyan, P.K. and Kakati, D.K. (2005) *J. Appl. Polym. Sci.*, **98**, 2320–2328.
276. Lalevée, J., Allonas, X., and Fouassier, J.P. (2006) *Macromolecules*, **39**, 8216–8218.
277. Lalevée, J., El Roz, M., Allonas, X., and Fouassier, J.P. (2010) in *Basics and Applications of Photopolymerization Reactions*, vol. 2 (eds J.P. Fouassier and X. Allonas), Research Signpost, Trivandrum, pp. 53–62.
278. Skene, W.G., Connolly, T.J., and Scaiano, J.C. (1999) *Tetrahedron Lett.*, **40**, 7297–7302.
279. Hu, S., Malpert, J.H., Yang, X., and Neckers, D.C. (2000) *Polymer*, **41**, 445–452.
280. Guillaneuf, Y., Bertin, D., Gigmes, D., Versace, D.L., Lalevée, J., and Fouassier, J.P. (2010) *Macromolecules*, **43**, 2204–2212.
281. Versace, D.L., Lalevée, J., Fouassier, J.P., Guillaneuf, Y., Bertin, D., and Gigmes, D. (2010) *J. Polym. Sci., Polym. Chem.*, **48**, 2910–2915.
282. Versace, D.L., Lalevée, J., Fouassier, J.P., Guillaneuf, Y., Bertin, D., and Gigmes, D. (2010) *Macromol. Chem. Rapid Commun.*, **31**, 1383–1388.
283. Tasdelen, M.A., Durmaz, Y.Y., Karagoz, B., Bicak, N., and Yagci, Y. (2008) *J. Polym. Sci., Part A: Polym. Chem.*, **46**, 3387–3395.
284. Versace, D.L., Tehfe, M.A., Lalevée, J., Fouassier, J.P., Casarotto, V., Blanchard, N., and Morlet-Savary, F. (2011) *J. Phys. Org. Chem.*, **24**, 342–350.
285. Pucher, N., Rosspeintner, A., Satzinger, V., Schmidt, V., Gescheidt, G.,

Stampfl, J., and Liska, R. (2009) *Macromolecules*, **42**, 6519–6528.

286. Tian, Y., Zhang, M., Yu, X., Xu, G., Ren, Y., Yang, J., Wu, J., Zhang, X., Tao, X., Zhang, S., and Jiang, M. (2004) *Chem. Phys. Lett.*, **388**, 325–329.

287. Yan, Y., Tao, X., Sun, Y., Xu, G., Wang, C., Yang, J., Zhao, X., Wu, Y., Ren, Y., Yu, X., and Jiang, M. (2004) *Mater. Sci. Eng., B: Solid-State Mater Adv. Technol*, **113**, 170–174.

288. Yan, Y., Tao, X., Sun, Y., Yu, W., Wang, C., Xu, G., Yang, J., Wu, Y., Zhao, X., and Jiang, M. (2004) *J. Mol. Struct.*, **733**, 83–87.

289. Boiko, Y. (2005) *J. Nonlinear Opt. Phys. Mater.*, **14**, 79–84.

290. Huang, Z.-L., Li, N., Sun, Y.-F., Wang, H.-Z., Song, H.-C., and Xu, Z.-L. (2003) *J. Mol. Struct.*, **657**, 343–350.

291. Zhou, G.Y., Wang, D., Tian, Y.P., Shao, Z.S., and Jiang, M.H. (2004) *Appl. Phys. B: Lasers Opt.*, **78**, 397–400.

292. Belfield, K.D., Bondar, M.V., Hernandez, F.E., Przhonska, O.V., and Yao, S. (2007) *J. Phys. Chem. B.*, **111**, 12723–12729.

293. Infuehr, R., Pucher, N., Heller, C., Lichtenegger, H., Liska, R., Schmidt, V., Kuna, L., Haase, A., and Stampfl, J. (2007) *Appl. Surf. Sci.*, **254**, 836–840.

294. Baldeck, P., Stephan, O., and Andraud, C. (2010) in *Basics of Photopolymerization Reactions*, vol. 3 (eds J.P. Fouassier and X. Allonas), Research Signpost, Trivandrum, pp. 200–220.

295. Marin, L., McGilvray, K.L., and Scaiano, J.C. (2008) *J. Am. Chem. Soc.*, **130**, 16572–16584.

296. Tasdelen, M.A., Karagoz, B., Bicak, N., and Yagci, Y. (2008) *Polym. Bull.*, **59**, 759–766.

297. Gibb, C.L.D., Sundaresan, A.K., Ramamurthy, V., and Gibb, B.C. (2008) *J. Am. Chem. Soc.*, **130**, 4069–4080.

298. Ma, C., Steinmetz, M.G., Kopatz, E.J., and Rathore, R. (2005) *J. Org. Chem.*, **70**, 4431–4442.

299. Sakayori, K., Shibasaki, Y., and Ueda, M. (2005) *J. Polym. Sci., Part A: Polym. Chem.*, **43**, 5571–5580.

300. Tanaka, K., Akimoto, R., Igarashi, T., and Sakurai, T. (2006) *J. Polym. Sci., Part A: Polym. Chem.*, **44**, 25–31.

301. YU, X., Chen, J., Yang, J., Zeng, Z., and Chen, Y. (2006) *J. Appl. Polym. Sci.*, **100**, 399–405.

302. Borkman, R.F. and Kearns, D.R. (1966) *J. Am. Chem. Soc.*, **88**, 3467–3472.

303. Linkun, Z., Irwan, G.S., Kondo, T., and Kubota, H. (1999) *Eur. Polym. J.*, **35**, 1557–1562.

304. Rodehorst, R.M. and Koch, T.H. (1975) *J. Am. Chem. Soc.*, **97**, 7298–7304.

305. Zard, S.Z. (2003) *Radical Reactions in Organic Synthesis*, Oxford University Press, New York.

306. Matsumoto, A. and Ito, Y. (2000) *J. Org. Chem.*, **65**, 5707–5711.

307. Araujo da Silva, J.C., Pillot, J.P., Birot, M., Desvergne, J.P., Liotard, D., Rayez, J.C., and Rayez, M.T. (2008) *J. Org. Chem.*, **693**, 2592–2596.

308. Hirashita, T., Hayashi, A., Tsuji, M., Tanaka, J., and Araki, S. (2008) *Tetrahedron*, **64**, 2642–2650.

309. Paczkowski, J., Kucybala, Z., Scigalski, F., and Wrzyszczynski, A. (2003) *J. Photochem. Photobiol., A: Chem.*, **159**, 115–125.

310. Ajayaghosh, A., Das, S., and George, M.V. (1993) *J. Polym. Sci., Part A: Polym. Chem.*, **31**, 653–659.

311. Assem, Y., Chaffey-Millar, H., Barner-Kowollik, C., Wegner, G., and Agarwal, S. (2007) *Macromolecules*, **40**, 3907–3913.

312. Hasegawa, T., Shimizu, T., Iwasaki, K., and Oguchi, S. (1983) *Bull. Chem. Soc. Jpn.*, **56**, 1869–1870.

313. Li, Z. and Kutateladze, A.G. (2003) *J. Org. Chem.*, **68**, 8236–8240.

314. Hug, G.L., Janeba-Bartoszewicz, E., Filipiak, P., Pedzinski, T., Kozubek, H., and Marciniak, B. (2008) *Pol. J. Chem.*, **82**, 883–892.

315. Denisov, E.T., Denisova, T.G., and Pokidova, T.S. (2003) *Handbook of Free Radical Initiators*, Wiley-Interscience.

316. Hirata, Y., Ohta, M., Okada, T., and Mataga, N. (1992) *J. Phys. Chem.*, **96**, 1517–1521.

317. Kyba, E.P. and Abramovitch, R.A. (1980) *J. Am. Chem. Soc.*, **102**, 735–740.

318. Budyka, M.F., Biktimirova, N.V., Gavrishova, T.N., and Laukhina, O.D.

(2005) *Russ. J. Phys. Chem.*, **79**, 1666–1671.
319. Papageorgiou, G., Lukeman, M., Wan, P., and Corrie, J.E.T. (2004) *Photochem. Photobiol. Sci.*, **3**, 366–373.
320. Ishida, S., Hashida, Y., Shizuka, H., and Matsui, K. (1979) *Bull. Chem. Soc. Jpn.*, **52**, 1135–1138.
321. Carta, P., Puljic, N., RIbert, C., Dhimane, A.L., Fensterbank, L., Lacôte, E., and Malacria, M. (2007) *Org. Lett.*, **9**, 1061–1063.
322. Leca, D., Fensterbank, L., Lacôte, E., and Malacria, M. (2005) *Chem. Soc. Rev.*, **34**, 858–865.
323. Hodgson, J.L., and Coote, M.L. (2005) *Macromolecules*, **38**, 8902–8910.
324. Medsker, R.E., Sebenik, A., and Hardwood, H.J. (2002) *Polym. Bull.*, **48**, 17–23.
325. Frija, L.M.T., Khmelinskii, I.V., Serpa, C., Reva, I.D., Fausto, R., and Cristiano, M.L.S. (2008) *Org. Biomol. Chem.*, **6**, 1046–1055.
326. Feng, D., Liu, J., Hitchcock, A.P., Kilcoyne, A.L.D., Tyliszczak, T., Riehs, N.F., Ruehl, E., Bozek, J.D., McIlroy, D., and Dowben, P.A. (2008) *J. Phys. Chem. A.*, **112**, 3311–3318.
327. Scherzer, T., Knolle, W., and Naumov, S. (2008) *Farbe Lack*, **114**, 40–44.
328. Togo, H. (2004) *Advanced Free Radical Reactions for Organic Synthesis*, Elsevier, Kindlington.
329. Parsons, A.F. (2000) *An Introduction to Free Radical Chemistry*, Blackwell Science, New York.
330. Nonhebel, D.C. and Walton, J.C. (1974) *Free Radical Chemistry*, Cambridge University Press, Cambridge.
331. Cessna, A.J., Sugamori, S.E., Yip, R.W., Lau, M.P., Snyder, R.S., and Chow, Y.L. (1977) *J. Am. Chem. Soc.*, **99**, 4044–4048.

9
Two-Component Photoinitiating Systems

This part is concerned with photoinitiating systems based on a combination of a photoinitiator (PI) with a coinitiator (which was originally and is still largely used in photopolymerization reactions) and a PI with a photosensitizer (PS) (which allows, in many cases, to efficiently excite the system at longer wavelengths). The performances are discussed in Chapter 17 and the reactivity in Chapter 18.

9.1
Ketone-/Hydrogen-Donor-Based Systems

9.1.1
Basic Mechanisms

The main classes of uncleavable PIs [1] are based on benzophenone (BP), thioxanthone (TX), camphorquinone (CQ), anthraquinone (AQ), ketocoumarin (KC), benzil (Bz), and substituted or structurally related compounds (**1–5**).

1–5

Basically [2], as largely encountered and used for many years, these systems usually work in their triplet state by a primary electron transfer in the presence of an amine AH (forming a charge transfer complex (CTC)) followed by a proton transfer, as shown in Scheme 9.1. The primary process is fast when using well-selected amines AH and efficiently competes with the monomer quenching. TXs and CQ can also react in their first excited singlet state.

Photoinitiators for Polymer Synthesis: Scope, Reactivity and Efficiency, First Edition.
Jean-Pierre Fouassier and Jacques Lalevée.
© 2012 Wiley-VCH Verlag GmbH & Co. KGaA. Published 2012 by Wiley-VCH Verlag GmbH & Co. KGaA.

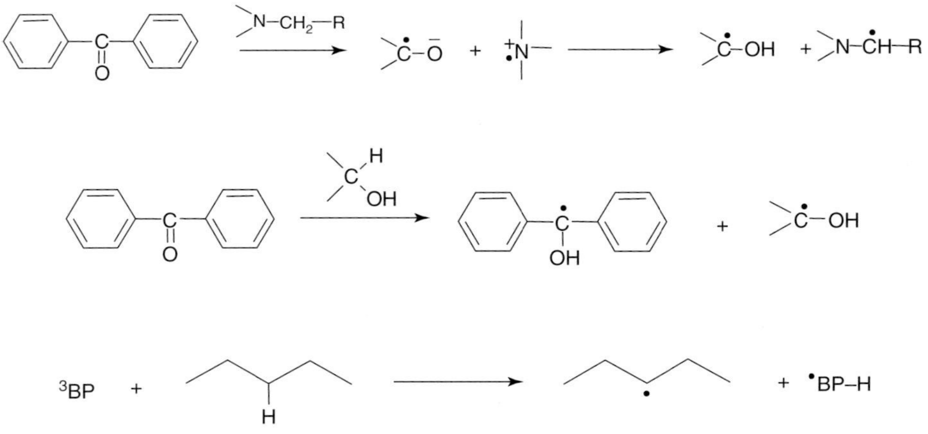

Scheme 9.1

The nπ* triplet states such as in BP (but not the ππ* triplet states, e.g., in fluorenone) can also work through a direct hydrogen abstraction reaction in the presence of a hydrogen donor DH such as an alcohol or tetrahydrofuran (THF) (Scheme 9.1). The hydrogen donor can also be a polymer chain possessing labile hydrogens. The hydrogen abstraction process creates a new radical. It allows a chain transfer reaction or/and a cross-linking reaction [3]. Many other hydrogen donors have been reported in the past years (see below and in Chapter 18).

The aminoalkyl A• and the D• radicals are the initiating radicals. The hydroxyl radical on the ketone (ketyl-type radical K•) is known as a *terminating agent* of the growing polymer chain as demonstrated in the BP [4] and TX [5] series (see Section 18.1 for the values of rate constants associated with K•). The properties of the excited diphenyl ketyl radical have been studied [6].

In PI/AH, the efficiency of the route leading to radicals is dependent on several factors: (i) the electron transfer rate constant, (ii) the ability of the CTC to undergo proton transfer (rather than a back electron transfer), (iii) the reactivity of the α-aminoalkyl radical (Section 18.5), (iv) the side reactions (especially those involving the ketyl- and the amine-derived radical) that consume both the excited triplet PI and the initiating radicals, and (v) the deactivation reactions of the precursor excited state. Presumably, ketone–amine systems are among the most complex photoinitiating combinations as no accurate and quantitative analysis of the involved processes can be generally done. The efficiency of the primary step of electron transfer is not the only driving factor when designing an efficient combination.

In PI/DH, the efficiency is governed by (i) the hydrogen abstraction rate constant and the regioselectivity of the process, (ii) the reactivity of the D• radical, (iii) the side reactions (of the ketyl and D• radicals), and (iv) the deactivation of the precursor excited state.

9.1.2
Hydrogen Donors

Many hydrogen donors have been proposed so far, and recent work leads to a revival of interest through the design of promising structures. The comparative data cited in Table 9.1 is discussed in the following paragraphs.

9.1.2.1 Amines

The most famous hydrogen donors remain the well-known amines (aliphatic or aromatic; secondary or tertiary). They present a quite satisfactorily reactivity (Section 18.5) and efficiency. Their role in photopolymerization reactions has been reviewed in [8]. Representative compounds very often used on a laboratory scale are methyl diethanol amine (MDEA), triethyl amine (TEA), and so on. Some examples of industrially widely used low-molecular-weight amines (e.g., ethyl dimethylaminobenzoate (EDB), 4-dimethylamino benzoic acid 2-ethyl hexyl ester (EHA) ...) are displayed in **6–12**.

[Chemical structures of EDB, SA$_1$, SA$_2$, EHA, NPG, AmAcry, and AHp labeled **6–12**]

Substituted amines (SA$_1$ or SA$_2$ derived from EDB, N-phenyl glycine (NPG)), amine functional acrylates, oligomeric and polymeric amines AHp, low-viscosity-amine-modified polyether acrylates (AmAcry) or copolymerizable amines such as a polyethyleneoxide derivative AHpe [9] are also encountered (**6–12**) [10–12]. Compared to the parent compounds, they accelerate the curing reaction presumably because of better compatibility.

Table 9.1 Typical hydrogen donors and rate constants for H-transfer to the ketone triplet state. The ketyl radical (BPH$^\bullet$) quantum yields for the ^{3}BP/hydrogen donor reaction are given in brackets (column 2).

Hydrogen donors	$k(^3BP)$ in 10^7 mol^{-1} l s^{-1}	$k(^3ITX)$ in 10^7 mol^{-1} l s^{-1}	$k(^3CQ)$ in 10^7 mol^{-1} l s^{-1}
Ethyldimethylaminobenzoate	390	–	–
Triethylamine	310 (0.94)	410	–
Mercaptobenzoxazole	150 (0.5)	610	6.5
Isopropanol	0.16	–	–
Tris(trimethylsilyl)silane (TMS)$_3$Si-H	10.2 (0.95)	4.1 (0.7)	0.56
Tris(trimethylsilyl)germane (TMS)$_3$GeH	105 (0.81)	4.2 (0.8)	0.71
Ph$_3$Sn-H	46 (0.74)	–	–
Et$_3$N → BH$_3$	9.1 (0.73)	–	–

Source: From Ref. [7].

Other interesting amine type coinitiators are phenylglycine derivatives such as NPG (**10**). Glycine derivatives are amino acids, biologically less toxic and nonallergic compounds suitable for specific applications, for example, in dental restorative formulations [13]. A carbon-centered radical is formed after electron transfer, proton transfer from the methylene group, and decarboxylation. Because of this fast decarboxylation, an interesting feature of NPG is the absence of a back electron transfer as unfortunately encountered in most ketone/amine couples (Section 18.5).

The design of covalently bound amine/ketone systems in order to maintain the amine in the close vicinity of the PI has received particular attention (Section 9.1.13); this results in a better efficiency, although the excited state reactivity is almost unchanged.

Hindered amine light stabilizer (HALS) used as light stabilizers in photocurable formulations are also able to react with ketone triplet states through electron transfer and yield an initiating radical. The ability of suitable HALSs to act as coinitiators such as in BP (or CQ)/HALS systems was checked [14]. On the other hand, when using a Type II PI (e.g., chlorothioxanthone (CTX)/amine) or a Type I PI (e.g., DMPA or PMK) in bulk aerated media (e.g., epoxy acrylate (EP)/hexanediol acrylate 60 : 40 w/w), the PI/HALS interaction does not compete with the PI/amine or the PI cleavage. The yield in initiating radicals is therefore not affected. This explains the absence of any effect on R_p when incorporating a HALS (2%) into a formulation [15]. A review on the photochemistry of HALS was done in [16].

Amine terminated poly (propylene imine) dendrimers of third through fifth generations were proposed [17]. In methylmethacrylate (MMA) photopolymerization, branched and star-shaped polymer PMMA structures are therefore expected.

Compared to the utilization of a usual amine, larger molecular weights and higher T_g's were obtained.

The investigation of a polymeric amine bearing an alkyl chain (that can mimic a chain length corresponding to 8–10 acrylate monomer units) and the parent unsubstituted compound [18] shows that no striking difference in reactivity occurs. Its diffusion in a viscous medium may affect, however, the rate constant of addition to a monomer. The formed macroradical could also be used, in a first approach, to test the chain length effect in a polymerization (Section 18.5).

In addition to their role as an electron/proton donor, amines are able to reduce the oxygen inhibition. This effect is usually explained on the basis of the reaction already described in Scheme 7.2, but from a kinetic point of view, this proposed mechanism is now questioned (Section 18.5).

Type II PIs are more sensitive to the presence of phenolic compounds (POHs) than Type I PI's. In BP/MDEA, the POHs efficiently react with the BP triplet state ($0.3–12 \times 10^9$ $mol^{-1} 1 s^{-1}$). This photoreduction process competes with the MDEA interaction (thereby decreasing the production of the initiating aminoalkyl radicals) and leads to phenoxy radicals, partly responsible for the inhibition and retardation effects [19]. No significant interaction occurs between the phenols and the ketyl radical ($\sim 10^4$ $mol^{-1} 1 s^{-1}$) or the aminoalkyl radicals (rate constants $<10^5$ $mol^{-1} 1 s^{-1}$; see a discussion of the mechanism in Section 18.5).

9.1.2.2 Thio Derivatives

Mercaptans (thiols) represent an important class of hydrogen donors (see a review in [20]). They yield a thiyl radical. For a long time, however, they have suffered from decisive drawbacks such as bad smell. New developments overcame this problem, and compounds such as mercaptobenzoxazole (MBO), mercaptobenzimidazole (MBI), and mercaptobenzothiazole (MBT) (**13–15**) are industrially used. The role of the mercaptans [21] and the design of newly reactive compounds based on iniferter structures [22] described in Section 8.30 have been outlined.

MBO
13

MBT
14

MBI
15

Phenyl thioacetic acid tetraalkyl ammonium salt Ph–S–CH$_2$ COO^{-+}NR$_4$ is a good coinitiator that can be used with BP [23]. The primary electron transfer from the sulfur atom is followed by a fast decarboxylation leading to a Ph–S–CH$_2^{\bullet}$ radical.

Trithianes such as 2,4,6-trimethyl-1,3,5 trithiane (**16**) are very efficient coinitiators [24]. After electron transfer with BP, the CTC undergoes either the usual proton transfer leading to a ketyl radical and an α-(alkylthio) alkyl radical or an in-cage escape process that generates a ketyl radical anion and a sulfur-centered radical cation.

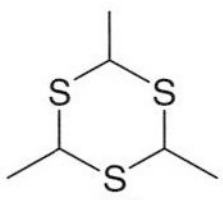

9.1.2.3 Benzoxazines

Benzoxazines (17) were shown to be good hydrogen donors [25]. They have the great advantage of incorporating a benzoxazine unit into the polymer through the attack of the aminoalkyl radical to the monomer double bond. The further thermal ring-opening process that is known in benzoxazines can be operative here to achieve thermal curing. Therefore, PI/benzoxazine is an interesting system for a dual-cure procedure. Thick samples can be photopolymerized.

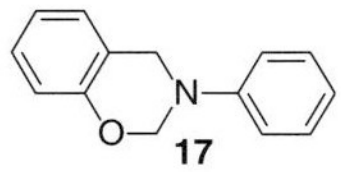

9.1.2.4 Aldehydes

Aldehydes (ALDs) can act as hydrogen donors [26, 27] (BP + RCHO → BP − H$^\bullet$ + RCO$^\bullet$). Interestingly, the triplet state of ALDs leads to hydrogen abstraction reactions, that is, ALDs can also behave as interesting one-component Type II photoinitiating systems (Scheme 9.2). The acyl radicals are expected to be the major initiating structures compared to the ketyl radicals (Sections 18.3 and 18.1).

9.1.2.5 Acetals

Cyclic acetals have been proposed as substitutes of amines (that can be responsible for toxicity and mutagenicity) in biocompatible applications. Linear acetals such as benzaldehyde dimethylacetal and acetaldehyde diethylacetal or cyclic acetals such as 1,3-benzodioxole or 6-benzoyl-1,3-benzodioxolane (18–21) can be used in the presence of BP or CQ [28–31]. The mechanism is not totally understood [32]. A carbon-centered radical is presumably formed on the acetal through hydrogen abstraction. Sesamin, which is a natural product (present in sesame seed), is an interesting bifunctional acetal [33].

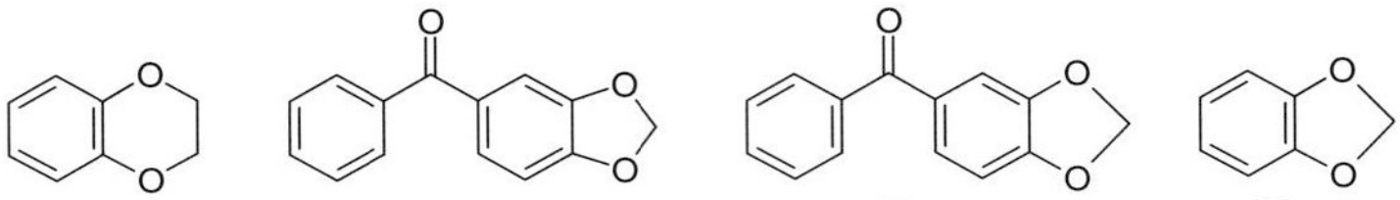

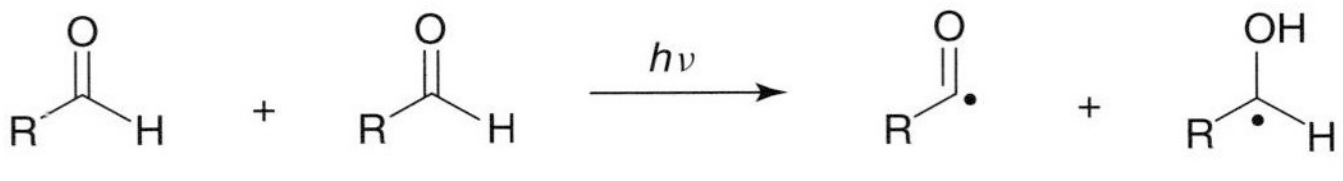

Scheme 9.2

9.1.2.6 Hydroperoxides

Hydroperoxides (ROOH) can be used as coinitiators (such as in the presence of CQ [34] or other selected diketones [35], Section 9.3.10). The primary processes involve a hydrogen abstraction. Hydroxyl and peroxyl radicals are formed. Monomer M initiation occurs according to Eq. (9.1).

$$\begin{aligned}
CQ &\rightarrow CQ^* \quad (h\nu) \\
{}^3CQ^* + ROOH &\rightarrow CQ\text{-}H^\bullet + ROO^\bullet \text{ or/and } CQ + RO^\bullet + OH^\bullet \\
RO^\bullet(+OH^\bullet) + M &\rightarrow M_{-H}^\bullet \rightarrow \rightarrow \rightarrow \text{polymer} \\
{}^3CQ^* + M &\rightarrow M_{-H}^\bullet(+CQ\text{-}H^\bullet) \rightarrow \rightarrow \rightarrow \text{polymer} \\
ROO^\bullet + M &\rightarrow M_{-H}^\bullet(+ROOH) \rightarrow \rightarrow \rightarrow \text{polymer}
\end{aligned} \quad (9.1)$$

9.1.2.7 Silanes

Silanes are known as *good hydrogen donors* in photochemistry. The silyl radical chemistry, which is largely encountered in organic synthesis or hydrosilylation reactions, has not been successfully used in photopolymerization until the recent years.

9.1.2.7.1 Silanes as Coinitiators

Silane derivatives R_3SiH, for example, tris(trimethylsilyl)silane (TTMSS) (22) were recently proposed as very efficient coinitiators for radical photopolymerization [36, 37]. This silyl radical chemistry introduced in the photopolymerization area has opened up new doors as already shown in other examples (Section 8.20).

TTMSS
22

Silanes in the presence of a PI such as BP, isopropylthioxanthone (ITX), or CQ or a dye such as eosin are highly reactive and, in many cases, even better than a reference amine coinitiator such as EDB (**6–12**). The ability of the PI/silane to initiate the polymerization under air is excellent (better than with PI/amine), and the inhibiting effect of oxygen is reduced or eliminated. An enhancement of the polymerization rates Rp under air is even sometimes noted (Figure 9.1). This may be one of the first real cases where a light-induced radical photopolymerization reaction is accelerated under air rather than inhibited. For example, using ITX/TTMSS, Rp increases by a twofold factor when going from a laminate to an aerated medium.

Compared to polysilanes $-(R_1SiR_2)n-$ (with R = alkyl or aryl), poly(hydrosilane)s $[-(R_1Si\text{-}H)n-]$ possess many labile hydrogens leading to very high H abstraction rate constants, for example, in the presence of BP (typically $10^9 \text{ mol}^{-1}\text{l s}^{-1}$). In addition to their Type I PI behavior (Section 8.15), they can also work as interesting coinitiators [38].

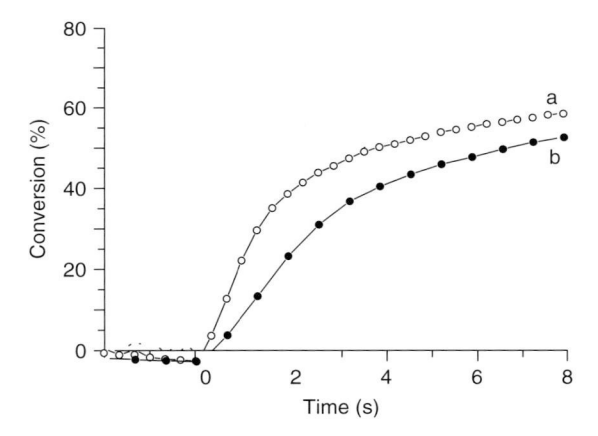

Figure 9.1 Conversion versus time curves for the photopolymerization of an epoxy acrylate. Photoinitiating system: ITX/TTMSS 1%/1% w/w. In air (a) and laminated conditions (b). (Source: From J. Lalevée, unpublished data.)

9.1.2.7.2 Silanes and Oxygen Inhibition

The role of oxygen and the reaction mechanisms in silyl-based photoinitiating systems (see references in [39]) that is discussed in detail in Section 18.9 lies in several key reactions (Scheme 9.3): (i) active silyls $R_3Si^\bullet$ are produced in the PI/silane system, (ii) the $R_3Si^\bullet$ radicals and the other radicals $R^\bullet$ overcome the oxygen inhibition through the oxygen consumption and the scavenging of the peroxyls, and (iii) new silyls are thus formed in the medium and the total amount of interesting $R_3Si^\bullet$ increases so that oxygen becomes a mediator in the silyl production. This explains the decrease of the oxygen inhibition effect in the polymerization.

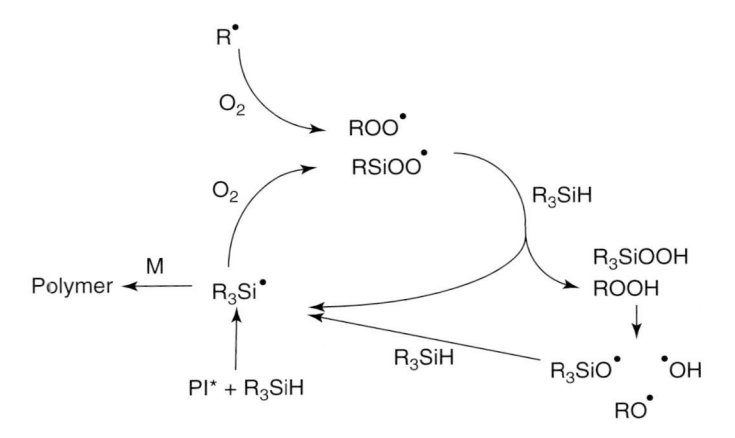

Scheme 9.3

9.1.2.7.3 Silanes as Leading to an *In Situ* Hydrophobic Coating Property

Interestingly, using ketones together with a hydride-terminated polysiloxane (such as polymethyl hydrosiloxane with a Si–H functionality = 24), strong *in situ* hydrophobic surface properties were achieved for the EP coating [40]. The surface hardness was very similar to that obtained with an amine, but the surface character properties were drastically changed, as revealed by the contact angles of the top surface (85°/101° vs. 50° for EDB). This behavior was ascribed to the presence of the siloxane units, well known for their hydrophobic properties. Indeed, the migration of the silane groups toward the air-exposed surface leads to oxygen consumption and an increase of the siloxane concentration and therefore, to a hydrophobic surface. The contact angle for the bottom of the film is close to that found with EDB. As a small amount of siloxane (1% w/w) into an acrylate matrix already leads to a hydrophobic surface, this could be of interest compared to the design, synthesis, and use of acrylate-modified silicones.

9.1.2.7.4 Silanes and Silane-Ene or Silane-Acrylate Chemistry

The change of thiols for silanes in a thiol-ene- or a thiol-acrylate-type process gives access to silane-ene or silane-acrylate chemistry [41]. For example, conversions of 100% and about 25% for different multifunctional acrylates and diphenylsilane, respectively, were reached and excellent final tack-free polymers obtained. The polymerization rates were not affected by oxygen. The same step-growth addition mechanism, as observed in thiol-ene polymerization, presumably holds true in silane-ene photopolymerization (Eq. (9.2)). In the first step, the initiating species (silyl radicals) are generated by reaction of the silane with a PI through hydrogen abstraction. The silyl radical adds to the double bond and leads to a carbon-centered radical, which then regenerates a silyl radical by a hydrogen abstraction reaction with a silane. This silane-ene process can probably be extended to nonacrylate monomers such as allyl ethers or vinyl ethers. The design of materials with new physical, chemical, and mechanical properties is expected using this new chemistry.

$$\begin{aligned}
&PI + R_3Si\text{-}H \rightarrow PI^{\bullet}\text{-}H + R_3Si^{\bullet} \quad (h\nu) \\
&R_3Si^{\bullet} + M \rightarrow R_3Si\text{-}M^{\bullet} \\
&R_3Si\text{-}M^{\bullet} + R_3Si\text{-}H \rightarrow R_3Si\text{-}M\text{-}H + R_3Si^{\bullet}
\end{aligned} \quad (9.2)$$

9.1.2.7.5 Silanes and the Future

The increased knowledge in the silyl photochemistry and chemistry allows the design of new silyl-radical-generating structures (Section 8.20) for wider and more convenient applications. New developments in surface modification, hydrosilylation, or living photopolymerization are under way.

9.1.2.8 Silylamines

Combining the two amine and silane functional groups in a single molecule denoted as a silylamine has been achieved (see, e.g., (23–27)). This structure can benefit from (i) the highly efficient reaction of the radical formation through an electron/proton transfer typical of the amine moiety as well as (ii) the higher propensity of the silane moiety to convert peroxyl radicals into new initiating structures.

SiA$_1$	SiA$_2$	SiA$_3$	SiA$_4$	SiA$_5$
23	24	25	26	27

An excellent efficiency (Figure 9.2) has been found using the BP (or ITX, CQ, thiopyrrylium salt TP$^+$ derivative)/silylamine systems under air (J. Lalevée, to be published). These coinitiators are better than the EDB reference for photopolymerization both under air and in laminate.

The mechanism can be depicted as in Eq. (9.3) where the silylamines are denoted HA-SiH (amine AH and silane SiH functions). It confirms the role of the amine/silane moieties. The regioselectivity of the C–H versus Si–H hydrogen abstraction in silylamines (J. Lalevée, to be published) is discussed in Section 18.5. The AH function is very efficient for the interaction with the PI excited state (*PI) leading to an aminoalkyl radical. The SiH function is useful to convert peroxyls into new initiating structures in photopolymerization under air. For SiA$_1$, the process for the peroxyl conversion is a S$_H$2 reaction (Section 9.1.2.9).

$$*PI + HA\text{-}SiH \rightarrow PIH^\bullet + {}^\bullet A\text{-}SiH$$
$$^\bullet A\text{-}SiH + M \rightarrow HSi\text{-}A\text{-}M^\bullet + M \rightarrow \rightarrow \rightarrow \text{polymer}$$
$$^\bullet A\text{-}SiH \text{ (or -}M^\bullet) + O_2 \rightarrow HSi\text{-}A\text{-}O_2^\bullet \text{ (or -}MO_2^\bullet)$$
$$HSi\text{-}A\text{-}O_2^\bullet \text{ (or-}MO_2^\bullet) + HA\text{-}SiH$$
$$\rightarrow HSi\text{-}A\text{-}O_2H \text{ (or -}MO_2H) + HA\text{-}Si^\bullet$$
$$HA\text{-}Si^\bullet + M \rightarrow \rightarrow \rightarrow \text{polymer} \quad (9.3)$$

9.1.2.9 Metal-(IV) and Amine-Containing Structures

Multivalent-atom-containing structures, (**28–31**) such as tetrakis (dimethylamido)silane, tetrakis(dimethylamido) tin(IV), tetrakis(diethylamido) titanium(IV), and tetrakis(dimethylamido) zirconium(IV) were proposed as coinitiators in BP-based Type II systems [42].

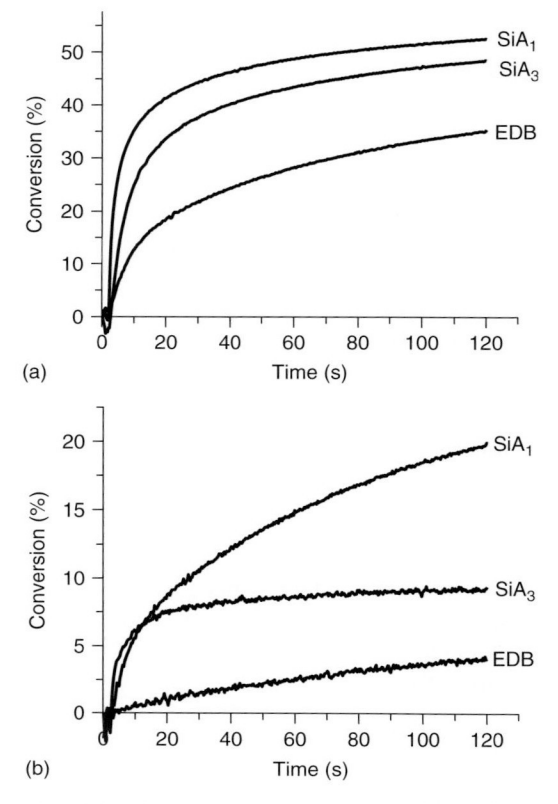

Figure 9.2 Conversion versus time curves for the photopolymerization of TMPTA in laminate (a) and under air (b). Photoinitiating system: CQ/col (3%/1% w/w). col = EDB, SiA$_1$, or SiA$_3$. Xe lamp exposure (I_0 = 60 mW × cm^{-2}; λ > 400 nm. Sample thickness = 20 μm). (Source: From [42].)

SiA$_1$ 28   TiA$_1$ 29   TiA$_2$ 30   ZrA$_1$ 31

Remarkable performances were achieved under air. The BP/EDB initiating system usually does not work, but the BP/ZrA$_1$ is very efficient: this system increases the final conversion by an ~fivefold factor (from 10.7 to 49.3%) and the polymerization rates by an ~100-fold factor, leading to a tack-free coating. Such a behavior is also found for SiA$_1$, TiA$_1$, and TiA$_2$ (Figure 9.3).

These results are accounted for by the presence of (i) the usual noticeable efficient addition reaction of the aminoalkyl radical to oxygen close to the diffusion limit and (ii) a bimolecular homolytic substitution S$_H$2 reaction that allows to

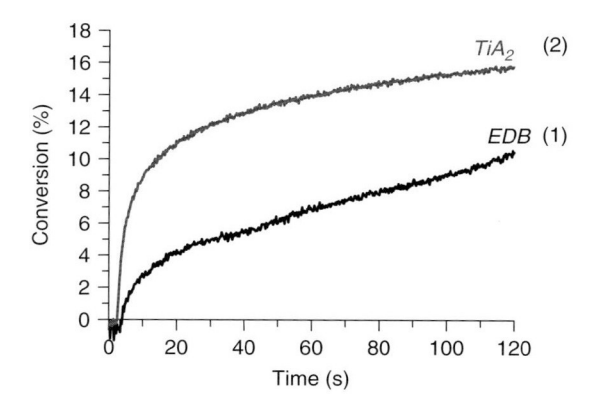

Figure 9.3 Photopolymerization of TMPTA under air. Photoinitiating system: (1) BP/EDB (1/1.7% w/w) and (2) BP/tetrakis(diethylamido)titanium(IV) (1/0.8% w/w). Hg-Xe lamp; $I_0 = 44$ mW cm^{-2}. (Source: From Ref. [42].)

efficiently convert peroxyls into new initiating radicals (Eq. (9.4)). The S_H2 reaction corresponds to the addition of a peroxyl to a metal–nitrogen bond, which results in an aminyl radical and a peroxide. The aminyl (which is not a good initiating structure, see Section 18.12), fortunately, further reacts with the starting compounds to form an aminoalkyl, which is a highly efficient initiating radical. The S_H2 reaction is presented in detail in Sections 18.10 and 18.11.

$$ROO^{\bullet} + X(N(CH_3)_2)_4 \rightarrow ROO\text{-}X(N(CH_3)_2)_3$$
$$+ {}^{\bullet}N(CH_3)_2 \quad (X = Si, Ti, Zr\ldots)$$
$$^{\bullet}N(CH_3)_2 + X(N(CH_3)_2)_4 \rightarrow HN(CH_3)_2$$
$$+ ((CH_3)_2N)_3 XN(CH_3)CH_2^{\bullet} \tag{9.4}$$

9.1.2.10 Silyloxyamines

Suitable silyloxyamines bearing an amine group [42] as those described in Section 8.31 also behave as efficient H donors (better than the reference amine EDB).

9.1.2.11 Germanes and Stannanes

In the same way, the possibility of using stannanes and new germanes (**32–34**) has been explored [43, 44]. Interaction rate constants with ^{3}BP leading to R$_3$Ge$^{\bullet}$ (or R$_3$Sn$^{\bullet}$) and BPH$^{\bullet}$ are high (6.2 × 10^7 and 48 × 10^7 mol^{-1} l s^{-1} for GeH and SnH, respectively) together with excellent quantum yields of radical formation (0.69–0.74). The bond dissociation energies (BDEs) are relatively low (79.9 to 73.8 kcal mol^{-1}). Germyl and stannyl radicals (Section 18.13) are excellent initiating species. The proposed structures are characterized by efficiencies similar to or better than those noted with EDB. For SiGeH, better hydrogen-properties than GeH are found.

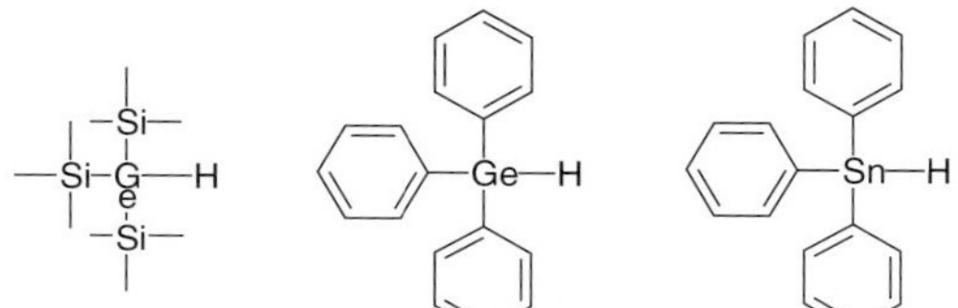

9.1 Ketone-/Hydrogen-Donor-Based Systems

SiGeH
32

GeH
33

SnH
34

The reaction in the presence of a ketone (e.g., using the CQ/germane system) is slightly accelerated under air rather than being inhibited. The behavior is similar to that of silanes (Section 9.1.2.7).

9.1.2.12 Phosphorus-Containing Compounds

New highly efficient coinitiators based on phosphorus-containing compounds (**35** and **36**) for acrylate photopolymerization were recently proposed (J. Lalevée, to be published). For example, in the presence of BP, they are highly reactive and lead, in many cases, to similar or even better polymerization rates than those obtained with the reference amine coinitiator EDB.

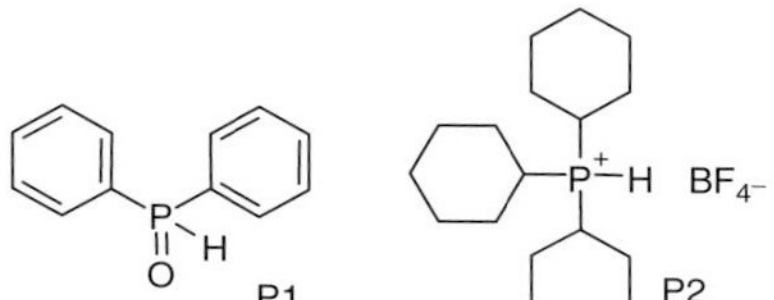

P1
35

P2
36

The primary process (Eq. (9.5)) corresponds to a hydrogen abstraction reaction from relatively labile P–H bonds in phosphine oxides and phosphonates that generate a phosphinoyl radical (low P–H BDEs have been calculated for P1 : 83.3 kcal mol^{-1}). For the trialkylphosphine salt P2, the BDE is higher (95.5 kcal mol^{-1}). An electron transfer occurs. The generated phosphoniumyl cation radicals ($R'_3P^{\bullet+}$) are considered (based on other literature data) as the radical-initiating species. The interaction with peroxyl radicals $t-\text{Bu}-\text{OO}^{\bullet}$ (e.g., 224 mol^{-1} l s^{-1} for P1) as determined by K-ESR (kinetic electron spin resonance) converts inefficient peroxyl radicals (formed during the photopolymerization under air) to a hydroperoxide t–Bu–OOH and a phosphinoyl radical $R_2P(O)^{\bullet}$.

$$PI + R_2P(=O)\text{-}H \rightarrow R_2P(=O)^{\bullet} + PI\text{-}H^{\bullet}(h\nu)$$
$$PI + R'_3P^+\text{-}H \rightarrow R'_3P^{+\bullet} + PI\text{-}H^{\bullet}(h\nu) \tag{9.5}$$

9.1.2.13 Borane Complexes

Various borane complexes (BC) (**37–40**) were proposed for the first time as new highly efficient coinitiators for acrylate photopolymerization. The reactivity of such compounds is driven by the dissociation energy of the B–H bond. The radical stability is related to the degree of spin delocalization from the boron to the Lewis base (the more the delocalization, the weaker is the B–H bond). Amine and phosphine ligands containing BCs L → BH_3 exhibit a relatively high BDE (92–105 kcal mol^{-1}) [45]. The planarity, the substituent sizes, and the spectroscopic nature of the radical play a substantial role. New structures derived from N-heteroaryl boranes [46] and N-heterocyclic carbene boranes NHC-BH_3 [47] allow to decrease the BDE (B–H) down to 70–85 kcal mol^{-1}. The effect of the Lewis base coordination on the reactivity of the amine borane derivatives (where the ligand is TEA, 2-picoline, N-heteroaryl (4-dimethylaminopyridine), quinoline, and diphenyl phosphine) [48] and the structure/reactivity relationships in NHC-BH_3 [49] has also been checked. A new water-soluble borane has been introduced [50].

BC_1 37 BC_2 38 BC_3 39 BC_4 40

In the presence of a ketone such as BP, efficient hydrogen abstraction occurs and generates a ketyl and an initiating boryl radical (Eq. (9.6)): high rate constant (0.2–70 × 10^8 M^{-1} s^{-1}) and quite high radical formation quantum yields (up to ∼1) are noted. Boranes (like silanes) are able to scavenge peroxyl radicals.

$$^3BP + L \rightarrow BH_3 \rightarrow BP\text{-}H^\bullet + L \rightarrow BH_2^\bullet$$
$$^3BP + NHC\text{-}BH_3 \rightarrow BP\text{-}H^\bullet + NHC\text{-}BH_2^\bullet$$
$$ROO^\bullet + NHC\text{-}BH_3 \text{ (or } L \rightarrow BH_3\text{)}$$
$$\rightarrow NHC\text{-}BH_2^\bullet \text{ (or } L \rightarrow BH_2^\bullet\text{)} + ROO\text{-}H \quad (9.6)$$

The ketone/BC Type II PI leads to better polymerization rates than those obtained with a usual reference amine (Figure 9.4), particularly under air. Moreover, the reaction in aerated conditions is quite efficient. The behavior of the boryl radicals under air is similar to that described above for the silyls. The BC photochemistry and the boryl radical reactivity are discussed in detail in Section 18.14.

9.1.2.14 Alkoxyamines

Alkoxyamines (ALKs) have been already encountered in free radical photopolymerization as Type I PI and in nitroxide mediated photopolymerization (NMP^2) (Sections 8.30 and 8.31). They can also behave as Type II PI. For example, compound **41** presumably works with BP or ITX through a direct hydrogen abstraction [51]. A sensitized C–O cleavage is not excluded. By-side reactions can also occur.

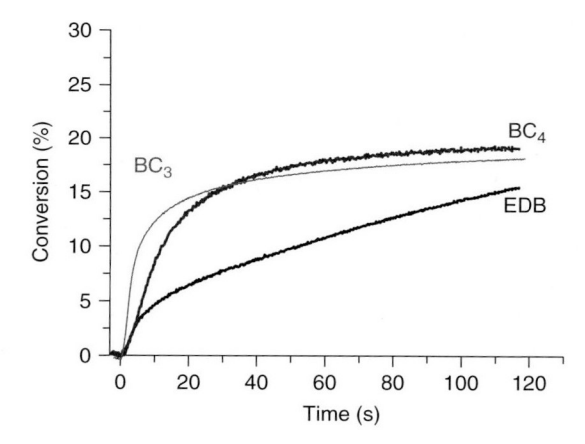

Figure 9.4 Radical photopolymerization ability of various BP/coinitiator couples (1/1% w/w; in ethoxylated pentaerythritol tetraacrylate (EPT), under air); the coinitiators are EDB, BC$_3$, or BC$_4$. (Source: From Ref. [49].)

41

Thus, the efficiency of the PI/ALK systems is mostly driven by the absorption of PI, the competition between reactions (1), (2), and (3) in Eq. (9.7), and the addition properties of the radicals to the monomer double bond. A potentially interesting way should be the finding of adequate hydrogen acceptors that do not belong to the ketone family to avoid the detrimental ketyl radical formation.

$$^3\text{ITX (or }^3\text{BP)} + \text{ALK} \rightarrow \text{ITX-H}^\bullet \text{ (or BP-H}^\bullet\text{)} + \text{ALK-H}^\bullet \quad (1)$$
$$^3\text{ITX (or }^3\text{BP)} + \text{ALK} \rightarrow \text{ITX (or BP)} + ^3\text{ALK}$$
$$\text{and } ^3\text{ALK} \rightarrow \text{C-O bond cleavage} \quad (2)$$
$$^3\text{ITX (or }^3\text{BP)} + \text{R'R''N-O}^\bullet \rightarrow \text{ITX-H}^{\bullet-} \text{ (or BP-H}^{\bullet-}\text{)}$$
$$+ \text{R'R''N}^+(=O) \quad (3) \quad (9.7)$$

9.1.2.15 Monomers

A monomer carrying a labile hydrogen can act as a hydrogen donor. For example, CQ (see above) or BP alone can initiate an acrylate photopolymerization reaction, the alkyltype radical formed on the monomer acting as the initiating radical (Eq. (9.8)). This initiation process exhibits a relatively low efficiency. Other examples include light absorbing amines (Section 9.3.17) or alkylphenylglyoxylates (APGs)

(Section 9.1.8).

$$^3BP + R-CH_2-CH=CH-COOCH_3 \rightarrow R-CH^\bullet-CH=CH-COOCH_3 + BP-H^\bullet$$
$$R-CH^\bullet-CH=CH-COOCH_3 + R-CH_2-CH=CH-COOCH_3$$
$$\rightarrow \text{radical addition to double bond} \quad (9.8)$$

9.1.2.16 Photoinitiator Itself

An excited PI containing a labile hydrogen can lead to a bimolecular reaction with a ground state PI molecule. For example, this is encountered in some TX derivatives such as in the thiol-group-containing TX-SH (Eq. (9.9)) where a thiyl and a ketyl radical are formed; this is also true for ALDs (Section 9.1.13.6).

$$TX-SH \rightarrow TX-SH^* \quad (h\nu)$$
$$^3TX-SH + TX-SH \rightarrow TX(-H)^\bullet \text{-SH} + TX-S^\bullet \quad (9.9)$$

9.1.2.17 Alcohols and THF

The reaction of the ketone/alcohol and ketone/THF systems (Scheme 9.1) is rather slow ($k \sim 10^5 - 10^6$ mol^{-1} l s^{-1}). Combinations between ketones and these H donors do not find widespread applications owing to the weak interaction between the two molecules, which does not prevent the O_2 and the monomer quenching of the ketone triplet state.

9.1.2.18 Polymer Substrate

Polymer substrates when possessing a labile hydrogen such as polyethylene, polynorbornene, ethylene-propylene-diene monomers (EPDMs), or poly dicylopentenyl acrylates (poly DCPA) can act as a coinitiator (Scheme 9.1). The hydrogen abstraction process creates a new radical on the substrate, which allows the grafting of a polymer chain [52]. The main drawback here in such a simultaneous grafting method is the presence of an important homopolymerization due to the hydrogen abstraction between BP and the monomer.

To circumvent this limitation, a sequential method for photoinduced living graft polymerization was proposed [53]. In the first step in the absence of monomer, under near-UV light, BP abstracts a hydrogen atom on the substrate Sub–H (e.g., a polypropylene microfiltration membrane) and forms Sub$^\bullet$ and BPH$^\bullet$. Then, BPH$^\bullet$ recombines with Sub$^\bullet$ and leads to the surface-active PI (Sub-BPH). In the second step in the presence of monomer and under short-wavelength UV light, the newly formed C–C bond in Sub-BPH is cleaved: the Sub$^\bullet$ radical initiates the acrylate polymerization.

A review paper recently covered the photografting of polymeric materials [54].

9.1.2.19 Silicon-Hydride-Terminated Surface

The preparation of ideal silicon-hydride-terminated surfaces H–Si(111) has attracted much interest in recent years as they constitute an excellent potential substrate for nanotechnology. It is known that, under UV light irradiation (typically $\sim$300 nm), a Si–H bond cleavage occurs at the H–Si(111) surface [55]. The

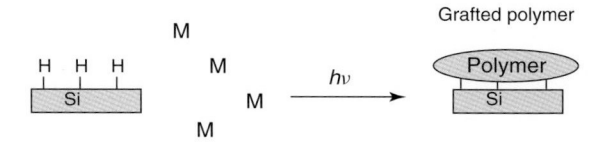

Scheme 9.4

mechanism is, however, unclear. True free radical photopolymerization of multifunctional acrylates using a H–Si(111) as PI (Scheme 9.4) has been very recently reported ([56], see references therein). The same grafting reaction can occur on H–Si powder surface that plays the role of a co-PI.

9.1.3
Benzophenone Derivatives

9.1.3.1 Benzophenone

9.1.3.1.1 Absorption
The BP series has certainly been the most largely studied family of PI. BP [57] exhibits the same kind of molecular orbitals (n, π, π^*) and energy levels (S_0, S_1, S_2, T_1, T_2) as the benzoyl chromophore. It presents an n $\rightarrow$ π^* (maximum absorption $\sim$350 nm, $\varepsilon = 100$ mol^{-1} l cm^{-1} in cyclohexane) and a $\pi \rightarrow \pi^*$ (maximum absorption = 250 nm, $\varepsilon = 20\,000$ mol^{-1} l cm^{-1}) electronic transition, which have been extensively described.

9.1.3.1.2 Excited States
The BP photochemistry is well known, and many interaction rate constants with a lot of additives are accessible [57]. The fluorescence quantum yield Φ_f is $< 10^{-4}$, S_1 lifetime is <10 ps, and S_1 energy $= 75$ kcal mol^{-1}. The lowest lying nπ^* triplet state T_1 ($E_T = 69$ kcal mol^{-1}) is formed from T_2 after the S_1 to T_2 intersystem crossing (ISC) (quantum yield $\Phi_{isc} = 1$): this has been reconsidered, as a recent paper suggests an ISC from S_1 to an upper vibrational level of T_1 coupled with the T_2 state [58]. Self-quenching (bimolecular reaction between a ground BP and an excited ^{3}BP, $k = 8 \times 10^5$ mol^{-1} l s^{-1} in acetonitrile) and triplet–triplet annihilation (bimolecular reaction between two excited ^{3}BP, $k = 2 \times 10^{10}$ mol^{-1} l s^{-1} in freon) occur in addition to the ISC from T_1 to S_0 (major process) and phosphorescence (which is almost inexistent at room temperature, $\Phi_p = 0.9$ at 77 K); nπ^* triplet states are less prone to self-quenching than $\pi\pi^*$ states. This explains why the T_1 state decays in the 1–100 μs range as a function of the excitation light intensity and the BP concentration. The interaction of ^{3}BP with neat isopropanol, THF, TEA, and methyl acrylate (MA) in benzene occurs with rate constants of 10^6, 4×10^6, 2×10^9, and 5.5×10^6 mol^{-1} l s^{-1}, respectively. Other quenching rate constants by monomers and amines are reported in Table 9.2. When the lowest lying triplet state in the BP series remains nπ^*, the rate constants increase with the solvent polarity.

Table 9.2 Examples of available data concerning usual noncleavable photoinitiators based on benzophenones: triplet state energy levels and rate constants of interaction k_q with oxygen, monomer, and amine (eventually isopropanol or THF).

Compound	E_T (kcal mol^{-1})	$10^{-6} k_q$ (mol^{-1} l s^{-1}) oxygen	$10^{-6} k_q$ (mol^{-1} l s^{-1}) monomer	$10^{-6} k_q$ (mol^{-1} l s^{-1}) H-donor
benzophenone	69	5500	66 (MMA) 34 (AN) 5.4 (VA) 78 (VE) 5.5 (MA) 400 (VP) 3330 (STY)	1300 (MDEA) 250 (benzylamine) 2300 (dibenzylamine) 1100 (tribenzylamine) 4.2 (THF) 1.6 (isopropanol)
4-OCH$_3$ benzophenone	—	—	150 (MMA)	2000 (MDEA)
4-SCH$_3$ benzophenone	—	—	2.5 (MMA)	1200 (MDEA)
4-S-phenyl benzophenone	—	—	2 (MMA)	1200 (MDEA)
4-N(CH$_3$)$_2$ benzophenone	—	—	1 (MMA)	200 (MDEA)

Structure	Col2	Col3
4-phenylbenzophenone	61	270 (MDEA) 0.05 (MMA) 0.01 (neat IP)
4,4'-diphenoxybenzophenone	—	1500 (MDEA) 180 (MMA) 1500 (NPG)
4-(diethylamino)-4'-(diethylamino)benzophenone type (with N(C₂H₅)₂ and N(C₂H₅)₂)	—	0.3 (BA) 13 (MMA) 0.4 (BMA) —
4-[(CH₂N(CH₃)₃)⁺Cl⁻]benzophenone	—	1500 (MDEA) 490 (AA)
4-(CH₂SO₃⁻Na⁺)benzophenone	—	1550 (MDEA) 880 (AA)

MDEA, methyldiethanolamine; NPG, N-phenylglycine; BEA, alkyl p-dimethylamino benzoate; MA, methylacrylate; MMA, methylmethacrylate; AN, acrylonitrile; IP, isopropanol; VA, vinylacetate; VE, vinylether. See text.
Source: From Refs. [1, 57, 59].

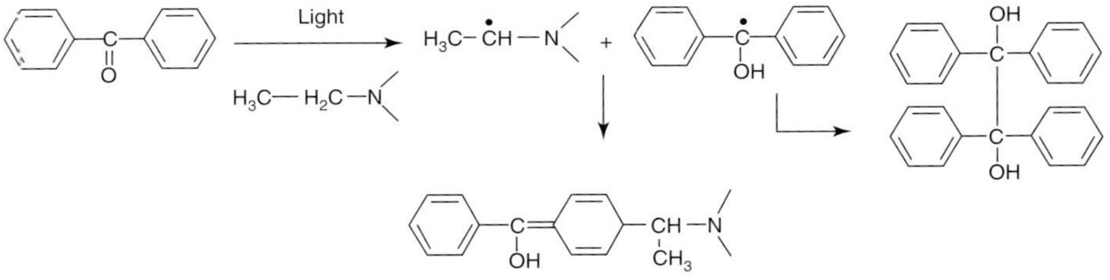

Scheme 9.5

9.1.3.1.3 Photolysis

The photolysis of BP in the presence of hydrogen donors (e.g., alcohol or amine) mostly leads to the formation of pinacol (dimerization of two ketyl radicals) and colored products (absorption maximum around 330 nm) referred to as light absorbing transients (LATs). A yellowing of the medium is generally observed. The LATs are formed from an attack of the ketyl radical by the aminoalkyl at the para position of the phenyl ring (Scheme 9.5). The mechanism is quite complex (see e.g., [60] and references therein). Oxygen plays a role in the further evolution of the formed LATs. The role of the medium (solvent, micelle, cyclodextrin (CD), and poly(ethylene-vinyl alcohol) film) as well as the thermal stability have been studied.

9.1.3.2 Aminobenzophenones

Several aminobenzophenones [2] such as 4,4′-dimethylamino benzophenone Mischler's ketone (MK) or 4,4′-diethylamino benzophenone (EAB) were proposed. For MK [57] and EAB [61], an enhanced and red-shifted absorption is noted (e.g., $\varepsilon = 44\,000\ \text{mol}^{-1}\,\text{l}\,\text{cm}^{-1}$ in cyclohexane at 365 nm for EAB). The S_1 state is short (18 ps for MK in benzene). The triplet state becomes $\pi\pi^*$, although a charge transfer contribution as well as a thermally populated $n\pi^*$ state can be expected. The Φ_{isc}s are [∼0.95] in benzene for MK and EAB as measured by LIPAC (0.3–0.5 for EAB as determined by laser flash photolysis (LFP) in acetonitrile). For EAB, $E_T = 70\ \text{kcal}\,\text{mol}^{-1}$, and the self-quenching rate constant is $1 \times 10^{10}\ \text{mol}^{-1}\,\text{l}\,\text{s}^{-1}$. The ^{3}MK/TEA interaction in benzene occurs with a rate constant of $7 \times 10^7\ \text{mol}^{-1}\,\text{l}\,\text{s}^{-1}$ (3×10^6 and $1.5 \times 10^9\ \text{mol}^{-1}\,\text{l}\,\text{s}^{-1}$ for ^{3}EAB/MDEA and ^{3}EAB/NPG in acetonitrile). The quenching by acrylate monomers is low ($k = 0.3 \times 10^6\ \text{mol}^{-1}\,\text{l}\,\text{s}^{-1}$ for EAB/butyl acrylate). The ^{3}EAB/oxygen interaction is highly efficient: $k = 1.25 \times 10^{10}\ \text{mol}^{-1}\,\text{l}\,\text{s}^{-1}$ in acetonitrile. A fast electron transfer in the ^{3}EAB/iodonium salt couple occurs ($6 \times 10^9\ \text{mol}^{-1}\,\text{l}\,\text{s}^{-1}$ in acetonitrile for bis-(4-tert-butylphenyl) iodonium triflate (BIP-T); $\Delta G = -0.93$ eV).

9.1.3.3 Other Benzophenones

In 4-phenyl benzophenone [57], the molar extinction coefficients are 22 200 and 215 $\text{mol}^{-1}\,\text{l}\,\text{cm}^{-1}$ in cyclohexane at 284 and 365 nm, respectively. The triplet state is $\pi\pi^*$ due to an intramolecular energy transfer to the biphenyl moiety ($E_T = 61\ \text{kcal}\,\text{mol}^{-1}$, $\Phi_{isc} = 1$, quenching by neat cyclohexane: $2 \times 10^3\ \text{mol}^{-1}\,\text{l}\,\text{s}^{-1}$). The quenching rate constant by MMA in benzene is $2 \times 10^6\ \text{mol}^{-1}\,\text{l}\,\text{s}^{-1}$ (vs. $6.6 \times 10^7\ \text{mol}^{-1}\,\text{l}\,\text{s}^{-1}$ for BP).

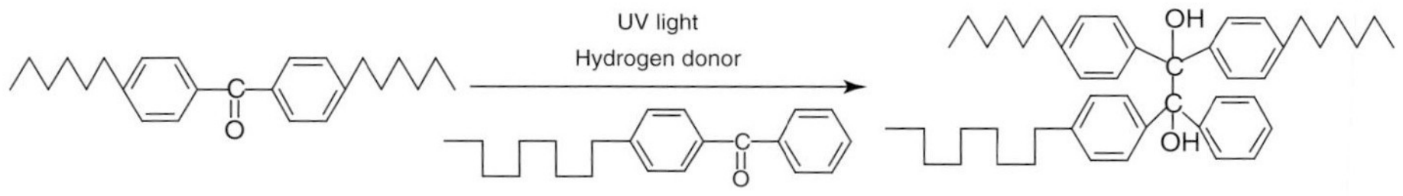

Scheme 9.6

The change in the excited state properties and in the polymerization ability observed in BPs modified by a thioether, an ether, or an amino substituent has been discussed in [62] and the role of an acrylate moiety linked to BP in [63]. Interaction rate constants with amines and monomers are presented in Table 9.2.

Trimethylsilylbenzophenones [64], alkylthiobenzophenones [65], aryloxy benzophenones [66], maleimidophenoxyl benzophenones [67], benzoyl benzophenones, terephtalophenones [64], and thiobenzophenones [68] were mentioned to exhibit some interesting initiating properties.

The photochemical coupling of polymer chains containing a BP unit allows to form a block copolymer (Scheme 9.6) [69].

9.1.3.4 Photopolymerization Activity

The photopolymerization activity of many substituted BP derivatives has been checked and extensively discussed as a function of hydrogen donors such as THF [2], amines [70–74], thiols [75–78], aminoacids [79], phenylglycine [80], and so on. The role of the BP structure [81], the search for new amines [82], the synthesis of BP piperazine [83] or thiol functionalized BPs [84], the effect of water [85], the use of a biphotonic excitation [86], the participation of higher excited states [87, 88], and the excited state dynamics [89] have also been investigated.

9.1.4
Thioxanthone Derivatives

9.1.4.1 Thioxanthone

TX, which has also been extensively studied in photochemistry (see, e.g., references in [90–94]), presents an interesting absorption at 377 nm in cyclohexane (with a high extinction coefficient), which extends up to 430 nm (see Figure 9.5). The S_1 state is $\pi\pi^*$ and the S_2 $n\pi^*$. When the medium polarity increases, the following dramatic changes are observed: (i) the S_2 is destabilized compared to S_1, (ii) the ISC quantum yield decreases (0.85 in acetonitrile and 0.56 in methanol), and (iii) the fluorescence quantum yield increases (4.1×10^{-3} in acetonitrile and 0.12 in methanol). This was attributed to a proximity effect through a decrease of the internal conversion because of a vibronic perturbation of S_1 by S_2. The two low-lying electronic absorption bands exhibit large modifications when the TX conformation (planar/non planar) changes. The stability difference being less than 1 kcal mol^{-1} suggests a dynamical motion of TX. In the frame of a dynamical model, the absorption spectrum is quite well explained as well as the dipole moment. The experimental S_0 dipole moment is 2.65 D. The S_1/S_0 dipole moment change is calculated as 4.2 D (planar TX) and 0.6 D (nonplanar TX): the experimental value

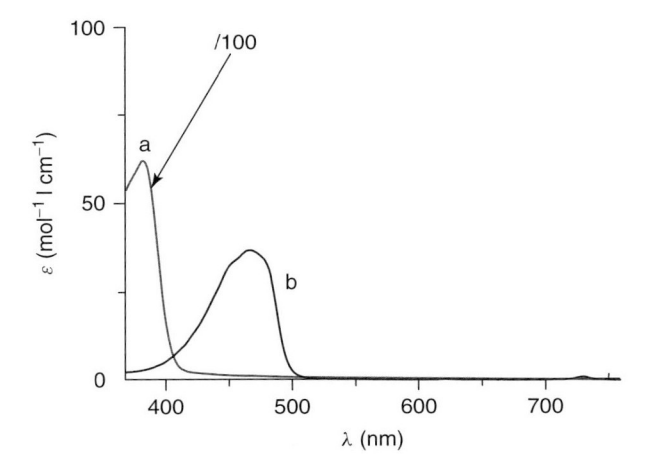

Figure 9.5 UV–visible absorption spectra of (a) 2-isopropylthioxanthone ITX and (b) camphorquinone CQ in acetonitrile. (Source: From J. Lalevée, unpublished data.)

is close to the average value. Self-quenching is more efficient in TX than in BP (2×10^9 mol^{-1} l s^{-1}) [57].

9.1.4.2 Substituted Thioxanthones

9.1.4.2.1 Compounds
Usual derivatives on industrial grounds are substituted at the 2 position as in 2-CTX or 2-isopropyl thioxanthone ITX. The red-shifted absorption (molar extinction coefficients, 3950 and 5180 mol^{-1} cm^{-1} for CTX and ITX, respectively, at 360 nm and 6700 mol^{-1} l cm^{-1} for CTX at 388 nm) makes these compounds very attractive for the UV curing of heavily pigmented white coatings where the pigment screens out a large part of the incident photons. The effect of other usual small substituents (methyl and alkyl chain) has been checked.

Compared to CTX or ITX, the substitution at the 1- and 4-position of the TX skeleton (**42**) has a larger effect [95]. Dehalogenation has been reported [64]. Photolysis under steady state conditions as well as nanosecond laser spectroscopy suggests a possible mechanism for a C–Cl bond breaking. Excited state interactions with amines on the picosecond time scale are rather complex [96].

Incorporation of a hydrocarbon moiety allows to extend the photosensitivity toward longer wavelengths as in TX-anthracene, TX-carbazole, and TX-fluorene (see in Section 9.1.13). These compounds can work as a Type II PI in the presence

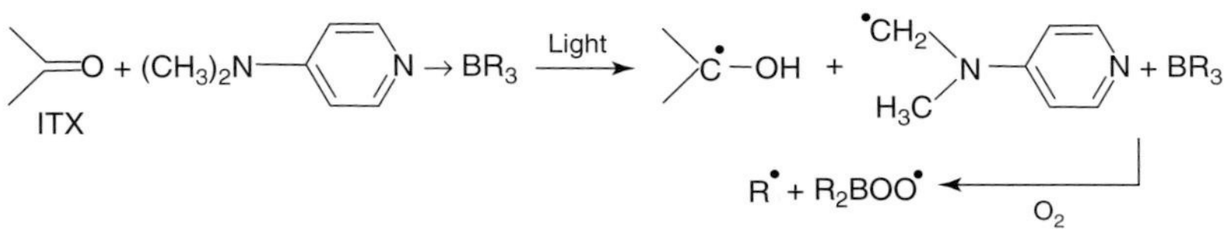

$$\text{ITX} \ \rangle\!=\!\!O + (CH_3)_2N\!-\!\!\langle\rangle\!-\!N \to BR_3 \xrightarrow{\text{Light}} \ \rangle\!\overset{\bullet}{C}\!-\!OH + \ \overset{\bullet CH_2}{\underset{H_3C}{}}\!\!N\!-\!\!\langle\rangle\!-\!N + BR_3$$

$$R^{\bullet} + R_2BOO^{\bullet} \xleftarrow{O_2}$$

Scheme 9.7

of a hydrogen donor (usual electron/proton transfer) or Type I PI. A TX containing a UV absorber moiety (based on a triaza cyclopentyl cycle) has been developed [97]. Silyl containing BP or TX also allows to design efficient Type I and Type II systems (Sections 8.20 and 8.5.6).

Works on the tentative elaboration of reactivity/efficiency relationships have been carried out in a lot of TXs, most of them being substituted at the 2-position [98]. A recent paper [99] reviews the available literature.

Using ITX (or other electron acceptor ketones such as BP) in the presence of a borane–amine complex (Scheme 9.7) allows to not only produce an aminoalkyl radical as usual but also liberate *in situ* a borane BR_3 [100]. This latter process is favored in the presence of oxygen and mostly occurs at the surface when working in an aerated medium. The oxygen inhibition is reduced as the borane efficiently scavenges an oxygen molecule to form a borylperoxyl and an alkyl radical that can contribute to the initiation. In the absence of oxygen, the addition of the borane to the excited ketone $R'_2\,C{=}O$ forms a $R^{\bullet}$ and a $R'_2\,C^{\bullet} - OBR_2$ radical. The surface hardness under nitrogen is higher than that obtained with a classical ketone/amine system.

9.1.4.2.2 Photolysis

As in BP (see above), the photolysis of the TX derivatives in the presence of hydrogen donors such as in the largely used ITX/amine system leads to yellowing of the solution. Based on earlier works of the photoreduction of TX by amines [101], this can be attributed to secondary reactions involving the coupling of two TX ketyl radicals and the generation of photoproducts derived from the amine radical partner. LATs are also formed through the same complex mechanism (Scheme 9.5) involving here the ITX ketyl and the aminoalkyl radicals. It was recently shown that the monomer can eventually participate in the formation of the photoproducts: this has explained the reduced polymerization ability of maleic anhydride in contrast with that of N-substituted maleimide [102].

9.1.4.2.3 Interaction Rate Constants: Solvent and Viscosity Effect

Interaction rate constants k_q with monomers and hydrogen donors are reported in Table 9.3. In compounds exhibiting a pure $n\pi^*$ (BP) or $\pi\pi^*$ (e.g., naphthalene for comparison) triplet state, the values of k_q increase with the solvent polarity (e.g., 0.9 and $5.5 \times 10^8 \text{ mol}^{-1}\,\text{l}\,\text{s}^{-1}$ for BP/MMA in benzene and methanol). In xanthones (XTs) and TXs, however, the $n\pi^*$ and $\pi\pi^*$ triplet states are very close [103]. A change of the spectroscopic character of the lowest lying triplet state (in XT) or a mixing of the two states (in TX) occurs when changing the solvent, and the rate constants usually decrease with a solvent polarity increase. This has

Table 9.3 Examples of available data concerning thioxanthone derivatives: triplet state energy levels and rate constants of interaction k_q with oxygen, monomer, and amine.

Compound	E_T (kcal mol^{-1})	$10^{-6} k_q$ (mol^{-1} l s^{-1}) oxygen	$10^{-6} k_q$ (mol^{-1} l s^{-1}) monomer	$10^{-6} k_q$ (mol^{-1} l s^{-1}) H-donor
(chlorothioxanthone)	62	3000	2 (MMA) 0.4 (AN) 0.02 (VA) 1.6 (VE)	5000 (BEA)
(isopropylthioxanthone)	61	3300	3 (MMA)	2600 (BEA)
ETX	58.5	—	0.2 (MMA)	3 (BEA)
(dodecylthioxanthone)	—	—	—	1400 (BEA) 440 (dibenzylamine) 40 (tribenzylamine)

BEA, butoxyethyl p-dimethylamino benzoate; MMA, methylmethacrylate; AN, acrylonitrile; VA, vinylacetate; VE, vinylether. See text.
In toluene.
Source: From Refs. [1, 2, 59].

Table 9.4 Triplet state absorption maxima λ_m and interaction rate constants k_{tr} of a 1-methylester thioxanthone with TPMK in various solvents. TPMK is defined in Scheme 8.11.

Monomer media	λ_m (nm)	$10^{-6} k_{tr}$ (mol^{-1} l s^{-1})
Toluene	~670	240
Methanol	~620	630
Acetonitrile	~640	410
Methanol/ethylene glycol 20 : 80	–	575
Methanol/ethylene glycol 60 : 40	–	420
Ethylene glycol	–	180

Source: From Refs. [1, 105].

been studied in detail for TX/MMA and TX/N-vinyl-2-pyrrolydone [104]: the three solvatochromic parameters (π^*, α, and β) contribute to nearly the same extent to the observed rate constant decrease (e.g., 80×10^6 mol^{-1} l s^{-1} in cyclohexane to 0.9×10^6 mol^{-1} l s^{-1} in methanol).

A blue shift of the triplet state absorption maximum is observed on change of the polarity and the hydrogen bonding ability of the medium as well as a noticeable evolution of the interaction rate constants with an additive (Table 9.4). The results obtained in a methanol–ethylene glycol mixture mostly reflect the viscosity effect as the polarity is almost kept constant [1, 105].

The same holds true in bulk. An indication of the viscosity effect on the monomer quenching has been illustrated by measuring the triplet state lifetime τ_T of a TX in various monomer media (Table 9.5 from values reported in [1, 105]). For example, $\tau_T = 50$ ns in toluene/MMA 7M and 650 ns in bulk trimethylolpropane triacrylate (TMPTA). As revealed by the shift of the triplet absorption (e.g., from 640 nm in Tol/HDDA (hexanediol diacrylate)/EP (25/50/25) to 620 nm in HDDA/EP (50/50)), a contribution of the medium polarity has to be taken into account and therefore explains why the trend $\tau_T = f$ (viscosity) is not straightforward. In such media, the interaction rate constants of the TX with an additive are also affected (e.g., in PS/PI systems); this is well observed in Tables 9.4 and 9.5.

9.1.5
Diketones

9.1.5.1 Aromatic Diketones

Biacetyl is a well-known model of α-diketones in photochemistry [57]: $\Phi_f = 0.025$, $\Phi_p = 0.05$, S_1 lifetime = 14 ns, $\Phi_{isc} = 1$, $E_T = 56$ kcal mol^{-1}, $n\pi^*$ triplet state, T_1 lifetime ~500 µs. Bz is another model [106] that can also cleave in an upper triplet state (Section 8.11). Among various other α-diketones [107], suberone (SU) exhibits interesting properties (**43–45**). Its photochemistry is well known [108]. The efficiency of the 1-phenyl-1,2-propanedione (PPD)/amine system (**43–45**) has been recently revisited [109].

Table 9.5 Triplet state lifetimes τ_T of 1-methylester thioxanthone in different monomer media. Interaction rate constants k_{tr} with TPMK.

Monomer media	Viscosity (cp)	τ_T (ns)	$10^{-6} k_{tr}$ (mol^{-1} l s^{-1})
TMPTA/EP (66/33)	185	2200	31
HDDA/EP (50/50)	58	2500	22
TMPTA	40	650	80
Tol/EP (25/75)	9.5	1500	35
Tol/TMPTA (25/75)	5.8	270	135
Tol/HDDA/EP (25/50/25)	5.8	850	68
HDDA	5.2	650	95
Tol/pentaerythritol tetraacrylate (PETA) (50/50)	3.1	220	40
Tol/HDDA (50/50)	1.2	1000	140
Tol/MMA 7M	0.53	50	240

EP, epoxy acrylate; HDDA, hexanediol diacrylate; tol, toluene.
Source: From Ref. [105].

PPD   SU   Bz
               43–45

Bz exhibits a UV absorption spectrum consisting of a main band localized at ~250 nm ($\pi \rightarrow \pi^*$ transition) and a shoulder at ~380 nm that extents up to 450 nm (n $\rightarrow \pi^*$ transition). The increase of the dihedral angle in SU induces a blue shift of the absorption spectrum. Bz and SU exhibit a transoïdal and cisoidal conformation, respectively. The triplet states of Bz and SU keep an $n\pi^*$ character; SU has a triplet energy of 56.5 kcal mol^{-1}. The Bz triplet state is quite insensitive to 2-propanol, the upper limit value of the quenching rate constant ($<1.7 \times 10^4$ mol^{-1} l s^{-1}) being similar to that of biacetyl. On the opposite, SU exhibits a high rate constant (2.7×10^7 mol^{-1} l s^{-1}).

The efficiency of Bz [110] and other α-diketones in acrylate photopolymerization has been evaluated [2, 111]. Carefully selected α-diketones such as SU (**44**) allow an excellent efficiency in the near-UV-visible region for the photopolymerization of clear, thick molded objects [112].

Other types of nonconjugated diketones such as the β (e.g., diethyl malonate), γ, or δ-dibenzoyl derivatives) have been seldom investigated. The 1,2-dibenzoylethane has a triplet lifetime about 50 ns [113]. The efficiency of dibromo derivatives of dibenzoylmethane has been checked [114].

9.1.5.2 Camphorquinone
Another particular α-diketone is CQ (Table 9.6). This compound [115] absorbs the light in the UV ($\pi \rightarrow \pi^*$ transition) and exhibits a slight absorption at 470 nm

with a low extinction coefficient indicating an n → π* transition (see Figure 9.5). The S_1 lifetime is 18 ns. The ISC quantum yield Φ_{isc} is close to 1 ($\Phi_f = 3.3 \times 10^{-3}$ in benzene), $E_T = 51.6$ kcal mol^{-1}; in degassed solution and under a low light excitation, the triplet lifetime is long (20 μs). The oxygen quenching is rather low (k = 1.9×10^8 mol^{-1} l s^{-1}). On addition of an amine such as MDEA, both the S_1 and T_1 are quenched (3.7×10^9 and 6.1×10^8 mol^{-1} l s^{-1} respectively) and leads to an aminoalkyl and a ketyl radical. The role of the amine in the CQ/AH system used in dental restorative resins was largely studied. The behavior and the efficiency/reactivity of CQ as a PI have been investigated in detail [116, 117], specially in CQ/N,N-dimethyl-p-toluidine/triethyleneglycol dimethacrylate [118].

9.1.6
Ketocoumarins

Ketocoumarins (KCs) such as 3–3'-carbonyl-bis-7-diethylaminocoumarin KC [119] (2) present an interesting near-visible absorption. A triplet state is formed from a weakly fluorescent singlet state ($\Phi_f = 0.08$; $\Phi_{isc} = 0.92$). It reacts with electron donors (see Table 9.6) such as thiol, amine, phenoxy acetic acid, alkoxy pyridinium salts.

9.1.7
Coumarins

Coumarins are an interesting class of laser dyes. Various structures are able (i) to behave as Type II PIs (in that case, they work in their first excited singlet state [120]) or (ii) to photosensitize the perester or bisaryl imidazole decomposition (Section 10.2.5). Examples are 7-diethylamino-4-methyl coumarin C1 or 3-(2'-benzothiazoryl)-7-diethylaminocoumarin C6 (46 and 47). They have high Φ_f(0.9 for C6). The ground state absorption of C1 has a maximum at 367 nm ($\varepsilon = 26\,880$ mol^{-1} l cm^{-1}), a singlet state lifetime of 3.4 ns in acetonitrile, a singlet state energy $E_{S1} = 2.8$ eV, and redox potentials $E_{ox} = +1$ eV and $E_{red} = -2.1$ eV. The interaction rate constants [121] are 1.6×10^{10} mol^{-1} l s^{-1} and 1.2×10^9 mol^{-1} l s^{-1} for C1/NPG and C6/MBO, respectively.

C1
46

C6
47

9.1.8
Alkylphenylglyoxylates

In their triplet state, alkylphenylglyoxylates (APGs) react with a monomer bearing an abstractable hydrogen (Scheme 9.8) and form a radical on the monomer. The

226 | *9 Two-Component Photoinitiating Systems*

Table 9.6 Examples of available data concerning other usual noncleavable photoinitiators: triplet state energy levels and interaction rate constants with oxygen, monomer, amine, and thiol.

Compound	E_T (kcal mol^{-1})	$10^{-6} k_q$ (mol^{-1} l s^{-1}) oxygen	$10^{-6} k_q$ (mol^{-1} l s^{-1}) monomer	$10^{-6} k_q$ (mol^{-1} l s^{-1}) H-donor
(camphorquinone)	51.6	190	—	600 (MDEA)
(benzil)	55	—	2 (MMA)	2000 (MDEA)
(3-benzoylchromone)	—	—	840 (styrene) 110 (MMA) 470 (AN)	—
KC	58	—	11 (MMA)	950 (MBO) 2500 (MDEA) 2000 (NPG)

9.1 Ketone-/Hydrogen-Donor-Based Systems

Structure					
fluorenone	55	—	—	0.35 (MMA)	850 (MDEA) / 10 (TEA)
thioxanthone	65	—	—	—	—
xanthone	75	—	—	—	—
acridine	56	—	—	—	—

MDEA, methyldiethanolamine; NPG, N-phenylglycine; MBO, mercaptobenzoxazole; MMA, methylmethacrylate; AN, acrylonitrile. See text.
Source: From Refs. [1, 2, 59].

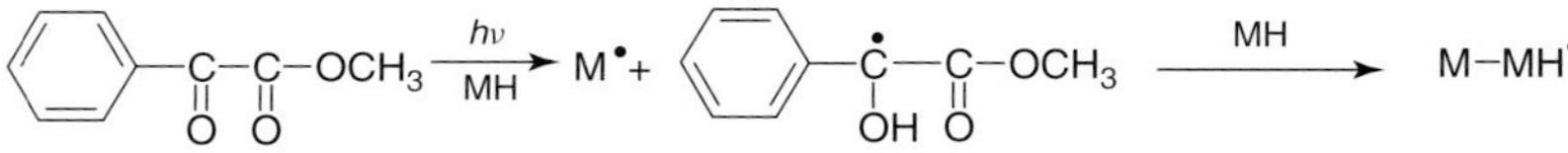

Scheme 9.8

APG–monomer interaction occurs with $k = 5 \times 10^5 \text{ mol}^{-1} \text{ l s}^{-1}$ in the case of TMPTA. This bimolecular hydrogen abstraction is likely the primary step in the initiation process of the polymerization. A ketyl radical (which subsequently leads to a benzoyl radical and a ketone) is also formed. The same hydrogen abstraction reaction can occur between two APG molecules (self-quenching) yielding still the ketyl on one APG and a carbon-centered radical on the other APG [122]. A competitive reaction is the formation of a short-lived (~16 ns) triplet 1,4-biradical through a Norrish II reaction. According to a multistep process, the decomposition of this biradical results in the generation of an ALD and α-hydroxyphenyl ketene (which yields benzaldehyde after CO release). The triplet state is rather long lived and, at low APG concentration, the formation rate constant of the biradical is $\sim 10^6 \text{ s}^{-1}$. On increasing the APG concentration, the lifetime dramatically decreases (APG–APG interaction rate constant $k \sim 10^6 - 10^8 \text{ mol}^{-1} \text{ l s}^{-1}$). Recent works ([123] and references therein) using TR-FTIR in the 1350–2350 cm^{-1} range (30 ns time resolution) shows that (i) two triplet state conformers are present, only one of them being able to undergo the Norrish II process, (ii) the biradical is too short lived to be able to participate in the radical formation, and (iii) a minor direct Norrish I cleavage also occurs. The photochemistry of fluorenone oxime phenylglyoxylate has also been studied through TR-FTIR, ESR, and *ab initio* calculations [124].

9.1.9
Other Type II Ketone Skeletons

A lot of other ketones have been checked, and almost all of the available ketone structures have been considered such as flavone, anthrone [103], quinone, benzoquinone [125], naphtoquinone and quinoline [2], benzodioxinone (BD) [126], azidoxanthone and azidofluorene [127], polycyclic aromatic ketones (such as tetralone derivatives) [2], XTs [128], decahydroacrydine-1,8-diones [129], flavins [130], as well as phenanthrene quinines [131].

The reactivity of ketones (e.g., XT) in ionic liquids, the addition of radicals to MMA in liquid, and supercritical carbon dioxide [132] are being studied.

9.1.9.1 Anthraquinones

Studies of the reactivity/efficiency relationships have also been carried out using AQs [133–136]. The 2-vinylcarbonyloxymethyl AQ was proposed as an unsaturated PI for the curing of PSA [137].

DFT calculations explain the photophysical and photochemical properties of AQ [138]. The photochemistry of substituted AQs can be found in [139].

9.1.9.2 Fluorenones

Fluorenone [57] has a low-lying triplet state T_1 ($E_T = 55$ kcal mol^{-1}). The T_1 quenching rate constant by TEA is 10^7 mol^{-1} l s^{-1} in cyclohexane. The photopolymerization in the presence of TEA [140] as well as the curing of PSA in the presence of fluorenone/amines [141] was studied.

9.1.10
Aldehydes

ALDs such as formaldehyde and acetaldehyde abstract a hydrogen atom on water or alcohol in photografting reactions in water-borne media [142]. Other various substituted aromatic ALDs were more recently proposed [26, 143] in conjunction with tertiary amines (Ar–CHO + $R_3N \rightarrow$ ArCHOH$^\bullet$ + $R_2NR_{-H}^\bullet$) or a quaternary ammonium salt of a tertiary amine. They can also work as one-component Type II PIs (Section 9.1.13.6).

9.1.11
Aliphatic Ketones

Some aliphatic ketone (butanone, pentanone)/water (or alcohol) systems also work as Type II PIs [144].

9.1.12
Cleavable Ketones as Type II Photoinitiators

Some ketones having rather long-lived triplet states and already mentioned as Type I cleavable systems (Sections 8.1.9, 8.1.10, and 8.2.4) such as benzophenone phenyl sulfide (BMS), sulfonyl ketones, oxysulfonyl ketones, and benzophenone ketosulfone (BPSK) (21, Chapter 8) exhibit an increased efficiency in the presence of a coinitiator. Acetone also behaves as a Type II system.

9.1.13
Tailor-Made Type II Ketones

9.1.13.1 Low- and High-Molecular-Weight Compounds and Macrophotoinitiators

The design of Type II compounds having specific tailor-made properties has received considerable attention as in Type I systems. New functionalities are achieved, for example (i) the incorporation of a chromophore enhances the visible light absorption, (ii) a low extractability and a reduction of the migration risk when the cured material is in contact with food or the human body can be achieved, and (iii) a close vicinity of the PI and the H-donor or the PI and an acrylate double bond allows both a better compatibility of the photoinitiating system in the film and a better interaction (which is less dependent on the diffusion) between PI and the coinitiator or the initiating radical and the monomer.

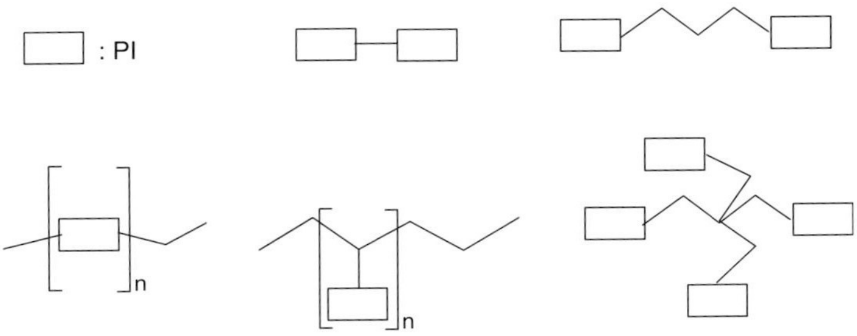

Scheme 9.9

9.1.13.1.1 Examples

Points (i)–(iii) have been realized using the following different strategies through the design of Type II macrophotoinitiators (Scheme 9.9) that correspond to a molecular or macromolecular assembly containing a more or less important number of PI units (difunctional, oligomeric, polymeric, or dendritic PIs).

Incorporation of a Photoinitiator and Eventually a Coinitiator in the Same (Macro) Molecule This is illustrated by the amine containing ketone-based PIs in the BP or TX derivatives that possess a grafted coinitiator group such as a thiol, an amine, or an acetic acid (**48–51**) [145–150]. They can also be considered as one-pot Type II systems (Section 9.1.13.6). The synthesis of main-chain oligomeric or polymeric PIs allows to enhance the solubility/miscibility and to decrease the volatility [151]; these systems operate as usual two-component Type II systems.

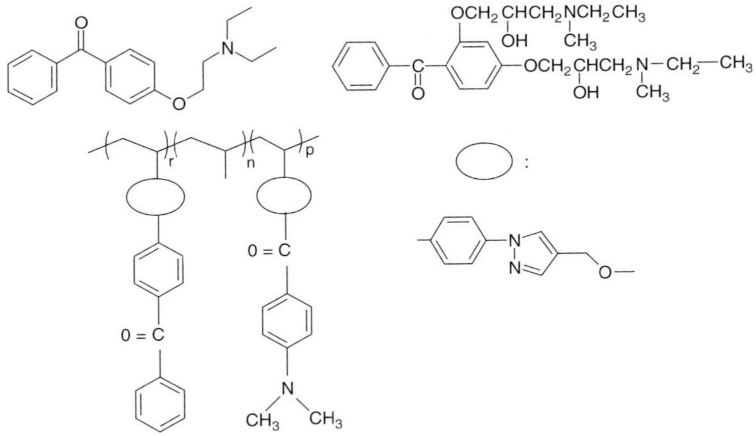

48–51

Many representative systems have been synthesized, for example, thiol-containing BP derivatives [152, 153], covalently bonded CQ/amine [154], BP/phenylglycine [155], BP/piperazine [83], amino alkyl ketone/thiol [84, 156], low-viscosity hyperbranched polymeric PI with built-in amine coinitiators

PIhyp (**52** and **53**) [157, 158], sulfur-containing polymeric PI bearing a side chain BP and an amine [159], macrophotoinitiator/coinitiator systems and maleimide-group-containing polymerizable BP derivatives [160], polymeric PI with a pendant amine moiety and a pendant PI moiety (BP, TX, fluorenone, CQ) [2, 161–169], covalently bonded amine and side chain ketone moieties [170–172], amphipathic polymeric TX TXam [173] (**52** & **53**), polymeric 4-maleimido-4′-thio-phenylbenzophenone [174], as well as silsesquioxanes functionalized with MMA and an amine (usable as a coinitiator in the presence of a dye) [175]

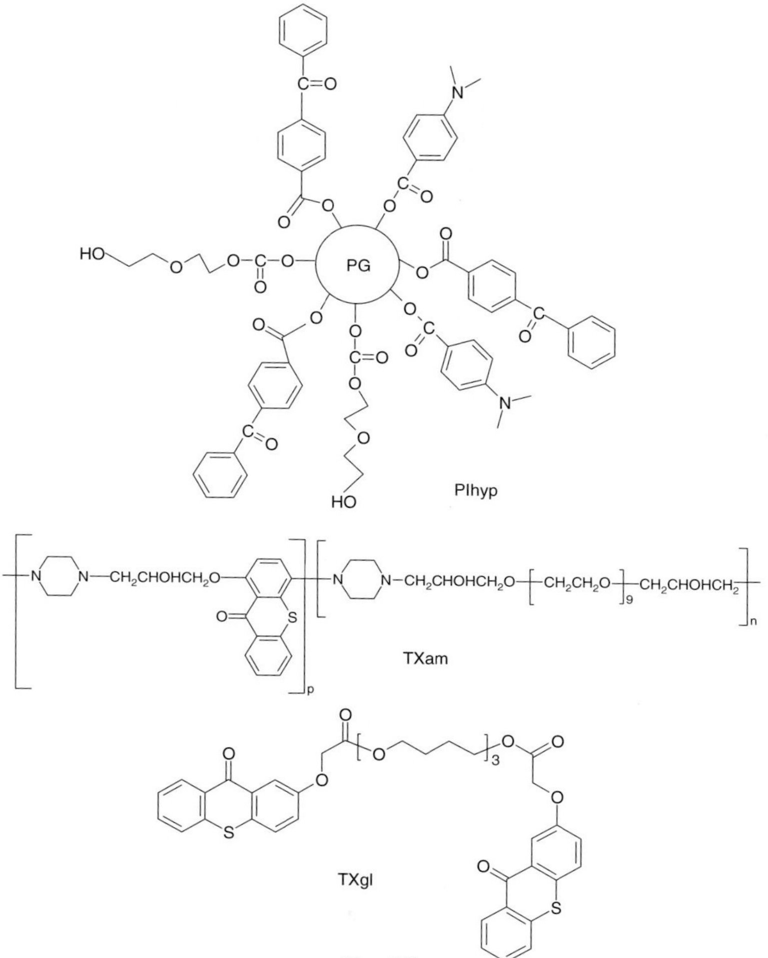

PIhyp

TXam

TXgl

52 and **53**

Examples of complex architectures of PIs include oligomeric phenyl (or BP, TX) glyoxylates [176] such as TXgl in **52** and **53**, photosensitive-group-modified cubic spherosilicates [177], or silsesquioxane PI (comprising a photoactive moiety such as a BP and an amine group bonded to a polyhedral oligomeric silsesquioxane) [178] (which allows the synthesis of inorganic-organic starlike polymers), hyperbranched polymeric TXs [179], polymeric Michler's ketones [180], as well as immobilized BPs on cellulose [181]. Polymeric compounds are continuously proposed ([182–189] and references therein).

Introduction of a Chromophoric Group into a Ketone Structure For example, a nitrostilbene linked to an alkoxybenzophenone [190] increases the conjugation and allows a red-shifted absorption. Similarly, BP moiety containing bifunctional PIs in which the absorption is totally changed as in Type I cleavable BPSK (Section 8.5.4) also works as Type II PI.

Design of Copolymerizable PI For example, acrylated BPs (**54**) allow to anchor the PI into the cured film [191, 192] and to increase the migration stability. Copolymerizable one-component Type II PIs have been designed [193]. From the view point of their excited state processes and polymerization ability, both the reactivity and the efficiency of copolymerizable PIs is fairly close to that of the parent related compounds.

54

9.1.13.1.2 Reactivity

When PI/AH Type II systems are introduced on the same polymer chain (Poly) as PI-Poly-AH, the polymerization efficiency is usually lower than that of PI-Poly/AH, presumably owing to a more favorable and detrimental recombination of the two polymeric PI − H• and A• radicals, which decreases the amount of available radicals for the monomer addition reaction (Scheme 9.10). The situation is different in PS-Poly-PI structures where (i) the close vicinity of PS with a cleavable PI increases the energy transfer efficiency and (ii) a mobile low-molecular-weight radical is formed after the cleavage of the PI moiety, which reduces the possible recombination with the counterpolymeric radical.

Excited state processes have been investigated in various systems such as in polysiloxanes with pendent TX and amine moieties [169], polymer backbones bearing side chain TX and morpholinoacetophenone moieties [194], or CQ and amine moieties linked to an acrylated polymer chain [195]. In this last example, when CQ is grafted on a polymeric structure as a pendent group pCQ and reacts with a polymeric amine pAH or a monomeric amine mAH, the reactivity is almost not affected (e.g., 2×10^8 and 1.3×10^8 mol^{-1} l s^{-1} for the CQ/mAH and pCQ/mAH interaction, respectively, [196]). In the polymeric structure p[CQ-co-AH], the singlet

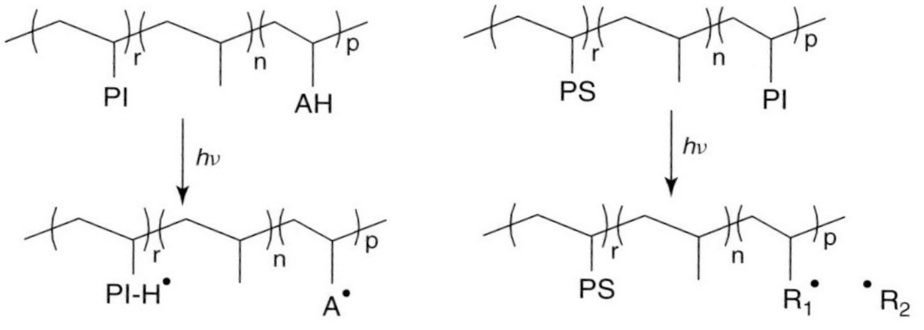

9.1 Ketone-/Hydrogen-Donor-Based Systems

Scheme 9.10

and triplet state lifetimes are 10 ns (instead of 16 ns for CQ) and 60 ns, respectively, thereby evidencing a fast quenching of the CQ units. This was mostly ascribed to the fact that in such a system, the interaction is less affected by the viscosity as less diffusion is required compared to the case where low-molecular-weight compounds are used. On the contrary, the rate of polymerization of a HDDA/butylacrylate mixture remains higher (by approximately a twofold factor; the induction period is noticeably lower) in the presence of CQ/mAH, pCQ/mAH, CQ/pAH compared to pCQ/pAH, or p[CQ-co-AH]; the final conversions, however, are almost the same (~60%). The hindering effect of the polymer chain in p[CQ-co-AH] likely reduces the radical mobility and favors the recombination of the ketyl radical with an aminoalkyl or a polymer radical. These results illustrate the limitation in the design of too complicated PI architectures.

9.1.13.2 Water-Soluble Compounds

Various structures derived from water-soluble neutral or ionic BPs, Bzs, and TXs (55–58) have been extensively proposed in the past [197]. More recently, a lot of studies have (i) described the excited state properties in the presence of water-soluble monomers, monomers in direct and reverse micelles or microemulsions, aqueous dispersions of monomers, and water-based coatings [198–204]; (ii) proposed new water-soluble TX derivatives [145, 205], polymeric TXs with glucamine as a coinitiator unit [206], carbohydrate group containing BPs [207]; and (iii) reported the photopolymerization of CD/acrylate host/guest complex using TX-catechol diacetic acid [208].

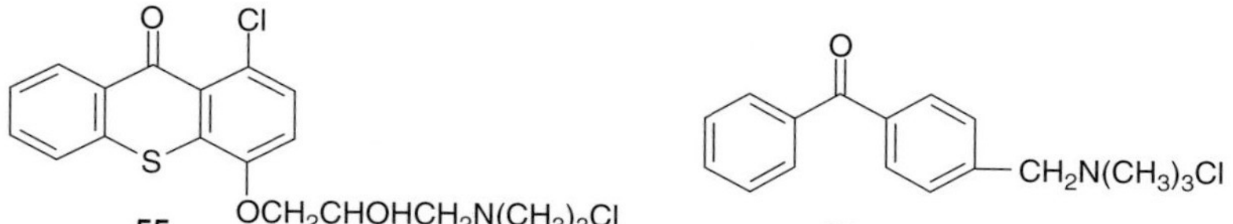

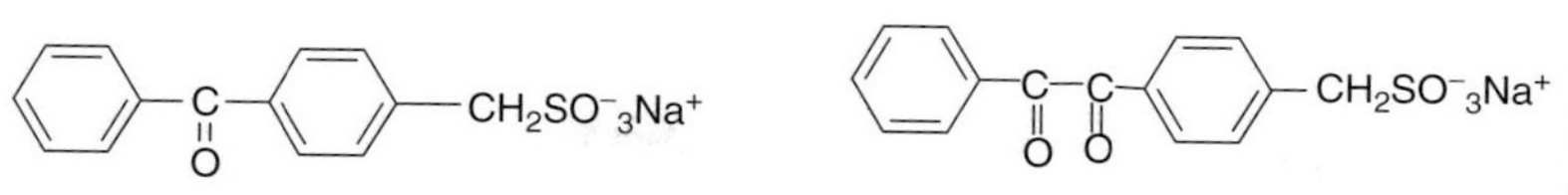

Table 9.7 Examples of available data for water-soluble photoinitiators: triplet state energy levels, interaction rate constants k_q, k_e with acrylamide AA (monomer), and methyldiethanolamine (MDEA as a H-donor). See text.

Compound	$10^{-6} k_q$ in $mol^{-1}\,l\,s^{-1}$ monomer	$10^{-6} k_e$ in $mol^{-1}\,l\,s^{-1}$ H-donor
Ph-CO-CO-C$_6$H$_4$-CH$_2$SO$_3^-$Na$^+$	0.06	450
Ph-CO-CO-C$_6$H$_4$-CH$_2$N(CH$_3$)$_3^+$Cl$^-$	0.05	380
Thioxanthone-OCH$_2$CH$_2$CH$_2$N(CH$_3$)$_3^+$	<0.1	37
Thioxanthone-OCH$_2$COOH	<0.1	65
Ph-CO-C$_6$H$_4$-CH$_2$N(CH$_3$)$_3^+$Cl$^-$	490	1500
Ph-CO-C$_6$H$_4$-CH$_2$SO$_3^-$Na$^+$	880	1550

Source: From Ref. [198].

The excited state reactivity is often similar to that of the parent compound [198]. On light exposure, the abstraction of an electron from the amine by the lowest lying triplet state of the PI is followed by a hydrogen transfer to give the ketyl radical and the amine-derived radical (Table 9.7). The efficiency of the acrylamide photopolymerization is high. An important influence of the pH on the rate of polymerization and a loss of activity, observed in acidic media for TXs, is certainly due to the protonation of the amine, which prevents the charge transfer from taking place. This effect is overcome by using cleavable water-soluble or hydrophilic PIs.

TX-fluorene carboxylic acid and carboxylate (Section 9.1.13.6) allows a long wavelength excitation for the formation of water-borne coatings.

9.1.13.3 Two-Photon Absorption Photoinitiators

Usual one-photon UV-sensitive noncleavable PIs can be activated in the red part of the spectrum and are expected to lead to the generation of the same usual initiating radicals. This was achieved with the BP/amine or fluorenone/amine systems [209]. Some particular ketones usable as multiphoton PIs have been synthesized (**59–61**) [210, 211]. For example, the compound TPA1 containing two electron-donating groups linked by a conjugated chain is sensitive in the 800–1000 nm range. The initiation mechanism is not fully understood. An electron transfer with the monomer might be the primary process. Newly developed cross-conjugated PIs with bathochromic shifts exhibit one- and two-photon activity [212].

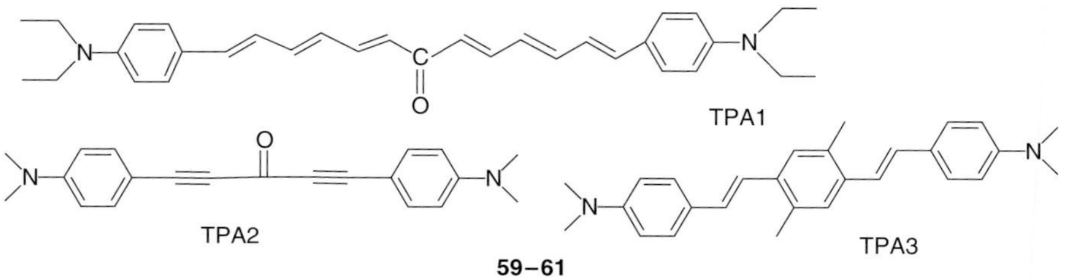

59–61

Linking one or two CQ units (R1 = CQ or R1 = R2 = CQ) as an energy acceptor to a 2,7-bis(diphenylamino)fluorene (FL) moiety as an energy donor allows to generate an excited CQ through an intramolecular singlet–singlet energy transfer process from FL. Then, in the presence of an amine, the usual CQ/amine interaction yields an aminoalkyl radical and photopolymerization occurs [213]. As fluorene derivatives exhibit relatively large two-photon absorption cross sections, the system can efficiently absorb the NIR light. Other supramolecular structures have been proposed (see in [214] and references therein).

9.1.13.4 Photomasked Photoinitiator

The design of a photochemically released masked PI (BP when using BD as the starting compound in Scheme 9.11) was proposed to increase the shelf life [126]. In BD/amine, BP can only be liberated on irradiation (then, BP reacts with the amine). This idea can also avoid the use of a conventional PI and cross-linker two-pack system as the concomitantly formed ketene is an efficient cross-linking agent of suitable monomers [215]. A review on the ability of BDs in various

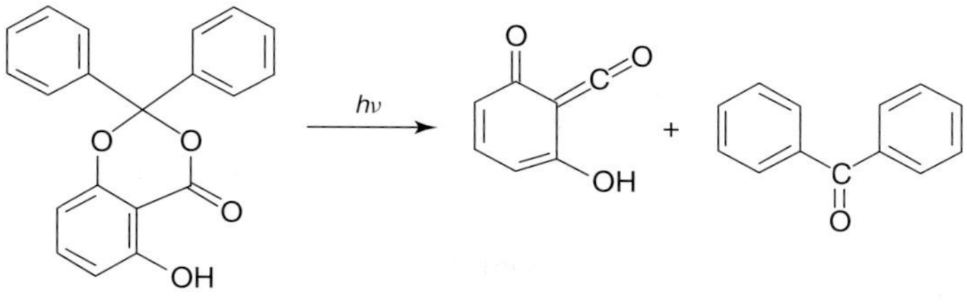

Scheme 9.11

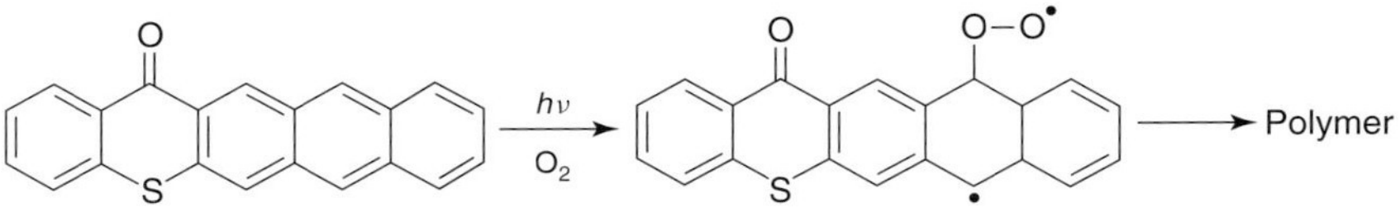

Scheme 9.12

applications has been recently reported [216] (step-growth photopolymerization, photoinduced synthesis of polyesters, free radical and cationic photopolymerization, and photoinduced curing).

9.1.13.5 Oxygen Self-Consuming Thioxanthone Derivatives

A TX-anthracene (where an anthracene moiety replaces one of the two phenyl rings in the TX structure) gives an interesting PI [217]. It consumes oxygen during the photopolymerization through the formation of an endoperoxide on the anthracene moiety (Scheme 9.12), which further thermally and photochemically decomposes into free radicals [218]. Photopolymerization does not proceed under nitrogen. This system is described as even better than a TX/amine reference system.

9.1.13.6 Low-Molecular-Weight One-Component Systems

In the absence of any hydrogen donor, Type II one-component mono- and bimolecular PIs work through an intramolecular and an intermolecular hydrogen abstraction, respectively. Many polymeric Type II PIs where the PI and coinitiator moieties are grafted onto the same backbone (see above the macrophotoinitiators in Section 9.1.13.1) behave as Type II one-component monomolecular systems.

The following sections enumerate the low-molecular-weight Type II one-component mono- and bimolecular PIs

9.1.13.6.1 Thioxanthone Derivatives

For example, a ketyl and an initiating thiyl radical are generated in TX-SH (**62**) through a ^{3}TX-SH/TX-SH bimolecular reaction [219]. In addition, this allows to convert the peroxyls (formed in a polymerization carried out under air) into a hydroperoxide through a ROO$^\bullet$/TX $-$ SH hydrogen abstraction and to generate a new thiyl radical TX $-$ S$^\bullet$ (and ROOH), thereby resulting in higher monomer conversions in aerated media.

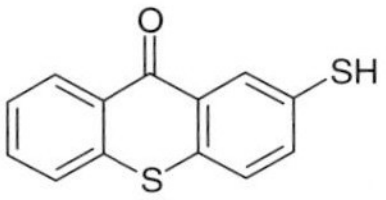

62

This concept was then extended in recently designed long-wavelength-absorbing TXs. In the absence of hydrogen donors, TX-fluorene carboxylic acid derivatives [220] work as Type I PIs. This occurs through an intramolecular hydrogen abstraction followed by a CO_2 elimination that generates a phenyl radical on the fluorene moiety (Scheme 9.13). At high concentration, an intermolecular hydrogen

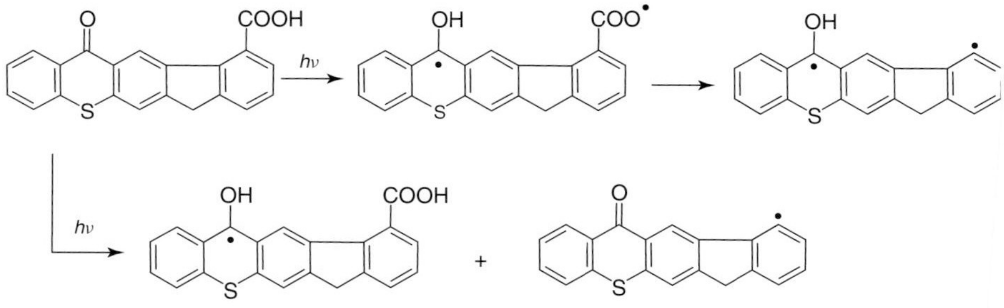

Scheme 9.13

abstraction between two molecules that generates both a ketyl- and a phenyl-type radical can be operative.

The same kind of mechanism has been reported in bifunctional TX acetic acid [221] and TX-carbazole derivatives (**63** and **64**) [222].

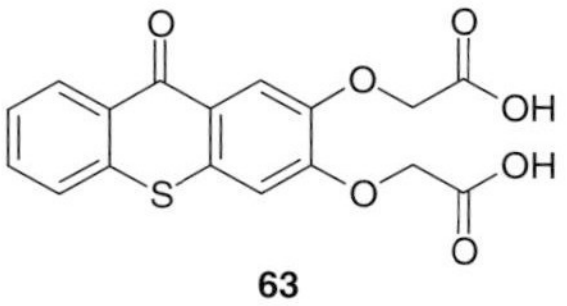

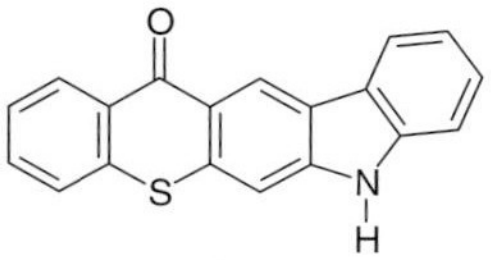

9.1.13.6.2 Benzophenone Derivatives

Benzoyl benzodioxolane (BBD) (**65** and **66**) was mentioned as a one-component system for the photopolymerization of styrene in bulk, but the initiation mechanism was not discussed [223]. Another derivative bisbenzo-[1,3]dioxol-5-yl methanone (BBDOM) exhibits a better absorption ($\varepsilon = 1540$ vs. $490 \, \text{mol}^{-1} \, \text{l cm}^{-1}$ at 365 nm), generates radicals through a hydrogen transfer between two molecules, and leads to an efficiency higher than that of BP/EDB (it can also work as a Type II system) [224].

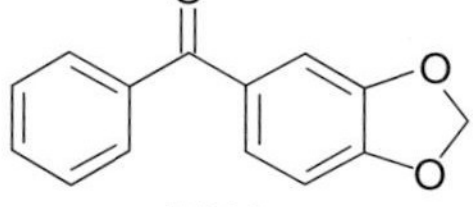

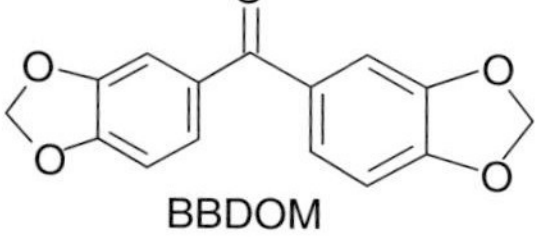

9.1.13.6.3 Aldehyde Derivatives

Aldehydes (ALDs) belong to a unique class of one-component systems that naturally work through a bimolecular primary process (in addition to their coinitiator ability, see also Section 9.1.10). Benzaldehyde has been mentioned as a PI [143]. It absorbs at 325 nm. In its triplet state, a bimolecular reaction with ground state benzaldehyde yields a benzoyl radical and a ketyl radical (rate constant $\sim 10^8 \, \text{mol}^{-1} \, \text{l s}^{-1}$). In the

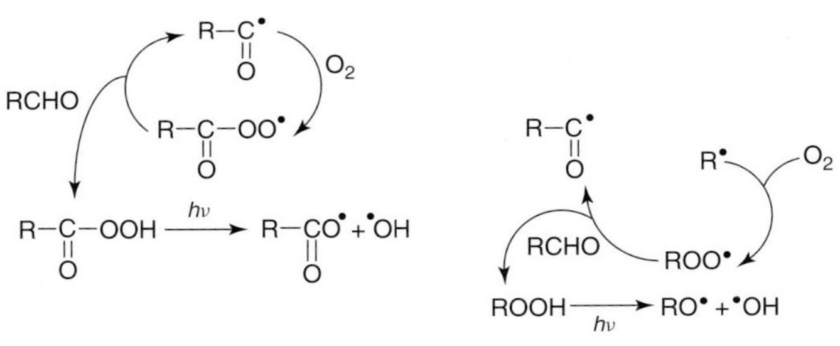

Scheme 9.14

presence of air, an interesting reaction (Scheme 9.14) consumes oxygen and regenerates a benzoyl radical [225], that is, the formed acylperoxyls or peroxyls abstract a hydrogen atom from the ALD. The acylperoxyls/ALD interaction rate constants are high ($\sim 10^4$ mol^{-1} l s^{-1}) compared to those of alkylperoxyls/ALD (~ 1–10^2 mol^{-1} l s^{-1}).

9.2
Dye-Based Systems

9.2.1
Dye/Amine Systems

Many compounds are used in dye-sensitized photopolymerization reactions [2, 226–229]. Strictly speaking, a dye is a substance used to color materials. In the photopolymerization area, the term *dye* is, however, very often used in a broader sense and could refer to any visible light absorbing molecule. Since the early compounds proposed in the pioneering work described in [230], classical examples are found in the numerous available dyes and colored molecules [2, 231–233] such as xanthenic dyes (Rose Bengal (RB) and eosin [234–236]), azines, thiazines (methylene blue), thionine, acridines, N-methylacridone, phenosafranines, thiopyronines, polymethines (cyanines, merocyanines, carbocyanines), pyrylium and thiopyrylium salts [237–242], pyrromethenes (PYRs) [243], fluorones [244], squarylium salts [245], julolidine dyes [246], quinoxaline dyes [247], styrylquinolium iodides [248, 249], and bisimidazoles (see below).

Other miscellaneous systems [2, 250] include phenoxazones, quinolinones, phtalocyanines, benzopyranones, rhodanines, RB peroxybenzoate, crystal violet, benzofuranone derivatives, dimethyl aminostyryl benzothiazolinium iodides, porphyrins, tetrathiafulvalenes, aminobenzylidene carbonyl derivatives, dibenzoylbenzene derivatives, quinoline imidazopyridinium salts [251], safranine [252], riboflavin [237], phenoxazines [253], styryl dyes [254], and naphthalene benzimidazoles [255].

Only few reducible dyes such as xanthene and acridine dyes can directly react with electron-deficient monomers, but the efficiency remains low. As a consequence, most of common dyes behave as Type II PIs in the presence of an amine.

For example (**67** and **68**; Eq. (9.10)), in the case of RB or eosin as a dye D, a photoreduction of the dye D occurs through an electron transfer with an electron donor (such as a tertiary amine), which yields a semireduced form of the dye $D^{\bullet-}$ and then a protonated semireduced form $DH^{\bullet}$ (leuco form). This process is accompanied by the generation of an initiating aminoalkyl radical. The $DH^{\bullet}$ radical both disproportionates into D and DH_2 and scavenges the growing polymer radical. The D excited states can also directly generate $D^{\bullet-}$ and the semioxidized form $D^{\bullet+}$ through self-quenching or triplet annihilation. The acidity of the medium plays a significant role in the ground state properties (coexistence of various protonated forms) and in the excited state processes (the $DH^{\bullet}$ radical can also be generated through an acid–base equilibrium between a proton and $D^{\bullet-}$).

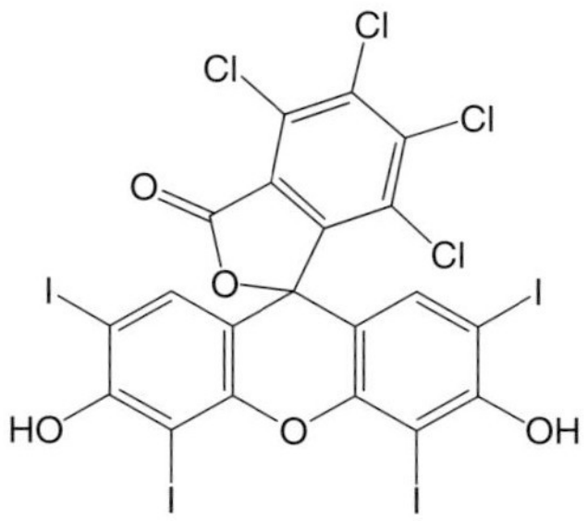

RB
67

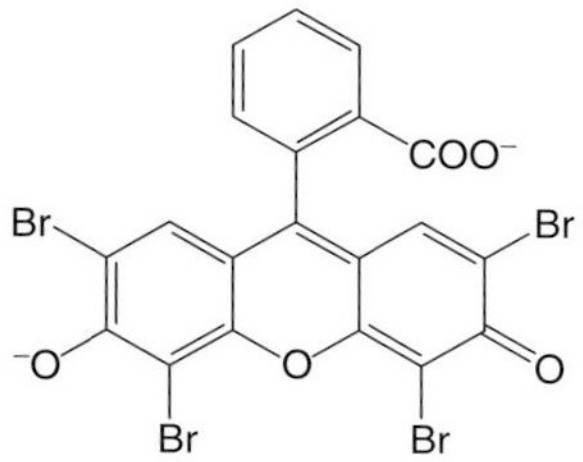

Eo
68

$$^{1,3}D + AH \rightarrow D^{\bullet-} + AH^{\bullet+} \rightarrow DH^{\bullet} + A^{\bullet}$$
$$DH^{\bullet} + O_2 \rightarrow D + HO_2^{\bullet}$$
$$DH_2 + O_2 \rightarrow D + H_2O_2 \qquad (9.10)$$

In aerated media, the quenching of the D triplet state yields singlet oxygen and the superoxide anion. Moreover, oxygen reacts with DH (and forms the hydroperoxyl radical) or DH_2 (and generates hydrogen peroxide); in both reactions, the starting dye D is regenerated (**67** and **68**; Eq. (9.10)) [230, 256]. This mechanism has been recently re-examined and confirmed [257]. The regeneration is more or less important as a function of the possible competitive reactions with other additives. This interesting property (almost no dye consumption) is counterbalanced by the lack of bleaching of the formulation and a final coloration of the coating (due to the absorption of the dye and the photolysis products involving the aminoalkyl radical). The mechanisms encountered in various dye-based systems have been reported, for example, in [258].

9.2.2
Dye/Coinitiator Systems

In addition to amines, other electron/hydrogen donors include [232]: NPGs, benzyltrimethylstannanes, thiols, arylsulfinates, sulfur compounds [259], 1,3-diketone enolates, phosphines, carboxylates, borates, organotins, and amino acids (in the presence of phenoxazines [239, 260]).

The reaction of a dye with electron acceptors is generally less usual, but it has been reported for a xanthenic dye in the presence of iodonium, sulfonium, or phosphonium salts; polyhalogenated hydrocarbons; p-nitrobenzoyl halide; or organic peroxides. A phenyl radical is generated in the dye/iodonium salt system. The ferrocenium salts are also an example of electron acceptors (see below).

Organotin compounds such as benzyltin cleave to produce a benzyl radical and a tin-centered radical after electron transfer with a dye [232].

A singlet oxygen sensitizer (such as zinc tetraphenylporphyrins) in the presence of a metal reducing agent (such as vanadylnaphtenate or vandylacetonate) and a singlet oxygen acceptor leads to a high efficiency for a radical photopolymerization in very thin films under the Ar^+ laser line [261].

The dyes PYR (**69**) exhibit an absorption around 500 nm and are usually considered as photoacid generators (Chapter 14). They also behave, however, as radical PIs when used with peresters (e.g., BTTB described in Section 8.23) [262]. The mechanisms have been fully investigated. Decomposition of BTTB occurs after an efficient singlet state quenching of PYR in a way similar to that encountered with thiopyrylium salts (Section 9.3.4).

69

The photophysical behavior of aminochalcone dyes based on the julolidine moiety (**70**) in the presence of a radical generating reagent has been described in [228].

70

9.2.3
Improvement of Dye/Amine Systems

The search for efficient systems under visible light exposure usable in various applications [263] has induced a lot of works aiming at improving the dye/additive interaction. The exchange of the cation of the dye (e.g., Na^+ in the case of RB) for a suitable cation being able to induce an intramolecular electron transfer leads to stable complexes. This was exemplified by RB $(Fc(+))_2$, which corresponds to a complex between RB and two ferrocenium salt moieties Fc(+) [264]. The polymerization efficiency is better than that observed when using the free species (see the mechanism in Section 9.3.7). In the same way, an iodonium [265] or a bipyridinium cation was also shown to form a complex with eosin and RB. The association constants between a dye and various electron donors have been determined [266].

The synthesis of Eosin and amine covalently bound statistical copolymers was recently reported to improve the migration stability in cured coatings [267].

9.2.4
Dye/Amine Water-Soluble Systems

Water-soluble systems based on dye/amine (e.g., xanthene dye/amine [268] or $Ru(bpy)_3^{2+}$/amine [269]) have been proposed for the photopolymerization of acrylamide. This last system involves an electron transfer from the amine to the metal-to-ligand charge transfer state of $Ru(bpy)_3^{+2}$ followed by a proton transfer yielding an aminoalkyl radical (Eq. (9.11)).

$$Ru(bpy)_3^{2+} + R\text{-}CH_2\text{-}N(R'_2) \longrightarrow \longrightarrow \longrightarrow Ru(bpy)_3^{+} + R\text{-}CH^{\bullet}\text{-}N(R'_2) \text{ (light)}$$
(9.11)

Another approach that could be more versatile to get water-soluble PIs is based on the encapsulation of the hydrophobic moiety of an oil-soluble coinitiator (an iodonium salt IO here) into a cyclodextrin cavity (CD) (CD is a cyclic oligosaccharide containing 1,4-glucopyranose units that presents a torus-shaped structure with a hydrophobic cavity and a hydrophilic exterior surface) [270]. A 2 : 1 host-to-guest complex $IO\text{-}(CD)_2$ is formed, which associates with a water-soluble xanthenic dye D (e.g., eosin or RB) to generate a stable $D/IO\text{-}(CD)_2$ binary complex that favors the D/IO interaction.

9.2.5
Kinetic Data

Kinetic data are available in a lot of papers. Examples of redox, excited states properties, and interaction rate constants of some selected and representative dyes with typical hydrogen donors, oxygen, and monomers are gathered in Table 9.8.

9.3
Other Type II Photoinitiating Systems

9.3.1
Maleimide/Amine and Photoinitiator/Maleimide

N-substituted maleimides absorb near 308 nm ($\varepsilon = 500\text{–}700 \text{ mol}^{-1} \text{ l cm}^{-1}$) and have a singlet lifetime $\sim$10 ns, an ISC quantum yield $\sim$0.2, and triplet lifetimes lying in the 100–800 ns range. The interaction between an excited maleimide and an amine (Scheme 9.15) occurs at a high rate constant close to the diffusion [274]

Table 9.8 Redox and excited states properties (singlet and triplet energy level). Interaction rate constants of selected dyes with representative hydrogen donors. See text.

Compound	E_{ox} (V)	E_{red} (V)	E_{S1} (eV)	E_{T1} (eV)	k mol^{-1} l s^{-1} MDEA	k mol^{-1} l s^{-1} O$_2$	k mol^{-1} l s^{-1} TZ	k mol^{-1} l s^{-1} MMA
Eosin-Y	0.6	−1.05	2.33	1.84	$1.5 \times 10^9 (S_1)$; $8 \times 10^8 (S_1)$; $4.3 \times 10^6 (T_1)$	—	—	—
Rose Bengal	0.65	−1.0	2.2	1.8	$9.4 \times 10^8 (S_1)$; $4 \times 10^5 (T_1)$	—	$6 \times 10^9 (S_1)$; $1.7 \times 10^7 (T_1)$	—
Thiopyrylium salt	>2	−0.39	2.3	—	—	$1.9 \times 10^9 (T_1)$	—	$<1.9 \times 10^8 (T_1)$
Phenosafranin	0.9	−0.65	2.34	1.77	—	—	—	—
Methylene blue	0.05	−1.2	1.89	1.50	—	—	—	—
Pyrromethene dye	1.21	−1.18	2.48	—	$1.3 \times 10^{10} (S_1)^a$; $6.6 \times 10^9 (T_1)^a$	—	$5.4 \times 10^9 (S_1)$	—

[a]The amine is ethyl dimethylaminobenzoate. TZ is 2,4,6-tris(trichloromethyl)-1,3,5 triazine.
Source: From papers [271–273] cited below in Chapter 10.

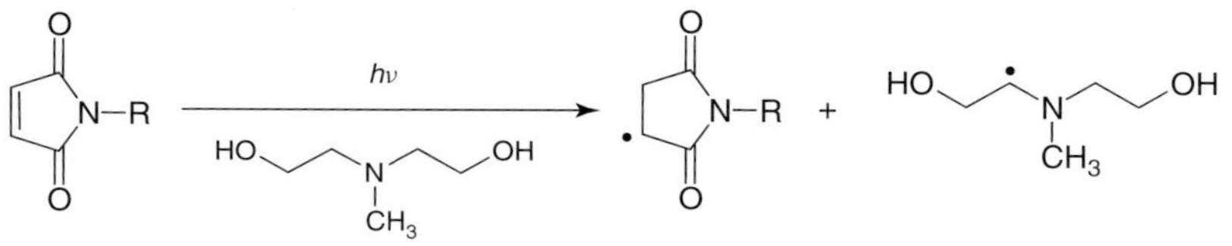

Scheme 9.15

and leads to two initiating radicals through an electron/proton transfer from the amine to the C=C bond of the maleimide [275, 276].

The maleimide can also act as a coinitiator. For example, a strong interaction (almost diffusion controlled) is observed between BP and the =N–R moiety of the maleimide [274, 277]: two radicals are thus generated through hydrogen transfer. Other data concerned with the ketone/maleimide behavior are reported in [278–282]. The LFP of bismaleimides is described in [283]. Isopropylthioxanthone in the presence of maleic anhydride (but not with N-substituted maleimides) leads to a photolysis product that decreases the overall efficiency [102].

Maleimides can also be immobilized on a polyethylene surface (**71**) [284]: this is another example of supported PI (Section 8.2.7). On light exposure, the initiating radical formed on the maleimide in the presence of acrylic monomers allows to obtain a covalently bonded surface polymer layer.

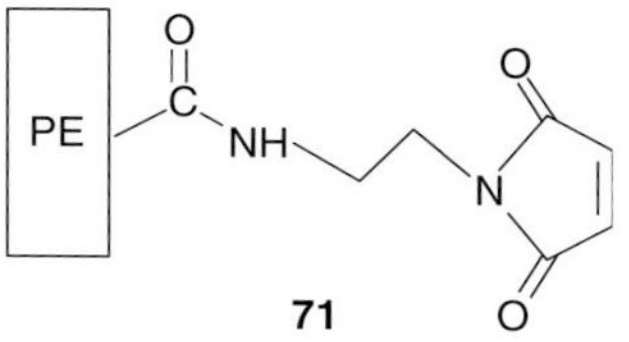

9.3.2
Donor/Acceptor Systems

Several systems have been mentioned. For example, under irradiation, morpholine and two bromine molecules yields a complex [285]. The further interaction with an acrylate M generates a radical (Scheme 9.16). In the same way, the CTC between morpholine and sulfur dioxide can initiate the MMA photopolymerization [286].

The p-ethoxy-p-cyanopyridinium salt in the presence of 1,2,4-trimethoxybenzene leads to a complex where an intramolecular electron transfer creates the trimethoxybenzene radical cation and the substituted pyridinyl radical, which then liberates an ethoxyl initiating radical. Borate salts in the presence of onium salts also yield a complex whose decomposition forms radicals [287].

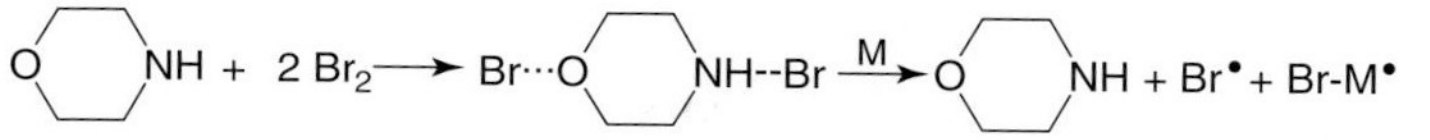

Scheme 9.16

9.3.3
Bisarylimidazole Derivative/Additive

Bisimidazole derivatives hexaarylbiimidazole (HABI) (the bis 2,4,5-triphenylimidazole is named lophine and sometimes represented by L_2) are largely used in the laser imaging area (**72**). They exhibit a very fast cleavage in the S_1 state, leading to two colored lophyl radicals $L^•$.

72

The lophyl does not efficiently add to acrylate double bonds. The presence of a hydrogen donor is necessary. For example, the unreactive lophyl reacts with electron/hydrogen donors such as mercaptans RSH (an initiating thiyl radical is formed as shown in Eq. (9.12) or NPG. It is a unique example of a highly cleavable compound that cannot work as a one-component system. Big efforts are still oriented to the design and the study of more efficient derivatives [288–294]. The lophyls have been studied in dry films by ESR [295].

$$\text{Cl-HABI (L2)} \rightarrow 2\,L^• \text{ and } L^• + RSH \rightarrow LH + RS^• \qquad (9.12)$$

The absorption band of Cl-HABI presents interesting molar extinction coefficients (~400 mol^{-1} l cm^{-1} at 366 nm); the absorption maximum is located at 265 nm ($\varepsilon = 27\,900$ mol^{-1} l cm^{-1}). The ISC is very low. The S_1 and T_1 state lie at 3 and 2.31 eV, respectively. The low bond energy between the two imidazoyl moieties accounts for the very fast cleavage. As deduced from picosecond spectroscopy, the cleavage occurs in 80 fs. A conformational change of $L^•$ is observed in the picosecond time range [296]. The homolysis quantum yield is 0.6 in dichloromethane and 0.95 in methanol. The dissociation quantum yield was evaluated as $\Phi_{rad} \sim 2$ in benzene: two lophyls are formed for each absorbed photon. The triplet state that could be formed in photosensitized experiments through energy transfer (see below) is also dissociative. The $L^•$ radical is very long lived and exhibits an absorption around 540 nm in benzene. Its reactivity is low (e.g., 1.1×10^5 mol^{-1} l s^{-1} with M3O in ethylacetate). Typical rate constants characterizing the $L^•$ are given in Section 18.15.

9.3.4
Pyrylium and Thiopyrylium Salts/Additive

The pyrylium P^+ and thiopyrylium TP^+ salts (**73** and **74**) match quite perfectly the excitation by a 488 nm laser light and have been often proposed in the past (see references, e.g., in [297, 298]). They exhibit a broad absorption band around 420 and 450 nm for P^+ and TP^+, respectively. The thiopyrilium salt TP^+ has a short-lived excited singlet state ($\tau_S = 1.0$ ns; $\Phi_{fluo} = 0.10$) and a quite high ISC quantum yield ($\Phi_{isc} = 0.30$ for the dimethoxy triphenylpyrylium structure). Its triplet state decays in the microsecond time range and is quenched by oxygen ($k = 1.9 \times 10^9$ M^{-1} s^{-1}). The pyrylium salt P^+ is a strongly fluorescent compound ($\tau_S = 5.5$ ns; $\Phi_{fluo} = 0.59$; $\Phi_{isc} < 0.1$).

TP⁺
73

P⁺
74

Amines decompose the salts through a nucleophilic addition to the S^+ or O^+ site, and therefore, the TP^+ or P^+/amine systems, despite promising capabilities in the field of visible-light-induced polymerization, cannot be really used in applications.

Addition of a diphenyl iodonium salt or a bromo compound such as CBr_4 to a thiopyrylium salt leads to an electron transfer process resulting in the generation of radicals [297].

When combined with peresters, TP^+s lead to highly sensitive and efficient formulations for the initiation of radical photopolymerization reactions. Their photophysical properties with regard to their use as PSs for the decomposition of peroxides have been explored and discussed. The mechanism is complicated (see [299] and references therein). It involves a $^3TP^+$/BTTB (benzophenone tetraperester BP-(COOOR)$_4$) interaction (rate constant $\sim 6 \times 10^6$ mol^{-1} l s^{-1}), a fast decarboxylation, and the release of an alkoxyl radical $RO^\bullet$. The low yield in thiopyranyl radicals suggests an energy transfer through electron exchange rather an electron transfer. In TP^+/benzoyl peroxide (BPO), an energy transfer and an O–O bond cleavage have been substantiated. Instead of peresters, borates appear as useful additives for printing plates on argon laser light exposure [300].

Recently, new coinitiators (silanes, thiols, disulfides) were proposed [301] in conjunction with (thio)pyrylium salts. The TP^+/5,5′-dithiobis-1-phenyl-1H-tetrazole system is particularly more efficient than the reference eosin/amine combination.

These coinitiators can overcome the thermal instability drawbacks mentioned above.

Electron transfer as confirmed by ESR occurs between TP^+ and, for example, disulfides ($\Delta G > -0.2$) according to Eq. (9.13) where $TP^\bullet$ is the 2,4,6-trimethoxyphenyl thiabenzene radical. The generation of a disulfide radical anion is thermodynamically unfavorable. A triplet–triplet sensitization process, however, cannot be ruled out, and a further cleavage process of ^{3}RS-SR is expected.

$$^3TP^+ + RS\text{-}SR \rightarrow TP^\bullet + RS\text{-}SR^{\bullet +} \rightarrow TP^\bullet + RS^\bullet + RS^+$$
$$^3TP^+ + RS\text{-}SR \rightarrow TP^+ + {}^3RS\text{-}SR \rightarrow TP^+ + 2\,RS^\bullet \quad (9.13)$$

9.3.5
Ketone/Ketone-Based Systems

In ketone/ketone-based systems, energy transfer is rather seldom encountered because of the obvious difficulty to have the PS triplet state level higher than that of PI. As revealed by CIDNP [302], the most famous energy transfer example (Eq. (9.14)) was encountered more than 20 years ago between a TX derivative such as ITX and the morpholinoketone derivative TPMK (defined in Scheme 8.3).

$$ITX \rightarrow {}^1ITX\,(h\nu) \quad \text{and} \quad {}^1ITX \rightarrow {}^3ITX$$
$$^3ITX + TPMK \rightarrow {}^3TPMK + ITX \quad \text{and} \quad {}^3TPMK \rightarrow \text{radicals} \quad (9.14)$$

An electron transfer (experimentally supported) also occurs. It generates an amino radical cation on the morpholino moiety $TPMK^{\bullet +}$ and a ketyl radical anion $ITX^{\bullet -}$ and leads to an aminoalkyl radical on the morpholino moiety and a ketyl radical. Another process involving an electron-transfer-assisted cleavage process might be (i) cleavage of the radical cation $TPMK^{\bullet +}$ into the methyl thiobenzoyl radical and an immonium cation and (ii) back electron transfer between this last cation and $ITX^{\bullet -}$, finally leading to the morpholino radical. This process seems to be ruled out. The energy/electron transfer balance is a function of the TX derivatives and the medium [302]. LFP experiments support the dual process.

Other examples of various ketone/ketone couples can be found in [303, 304]. Electron transfer in 2-ethoxythioxanthone/MPPK (see the formula in Scheme 8.12) was recently supported by electron spin resonance spin trapping (ESR-ST) [305].

The ITX/TPMK system is particularly efficient in a white pigmented lacquer [302]. The attained cure speeds are: 50 vs. 20 m s^{-1} (when TPMK is used alone). The complete through-cure of a blue silk screen ink (7 μm thick) is achieved at a cure speed $\sim$30 m mn^{-1} (5 m mn^{-1} in the absence of ITX).

A close vicinity between the two partners is interestingly achieved when both PI and PS are linked together. The sensitized cleavage in covalently linked TX derivative/cleavable PI was recently investigated and discussed [306]. Polymeric PIs bearing pendent TX and morpholinoacetophenone moieties [168, 307] lead to an efficient energy transfer process that appears as much higher than that observed for the corresponding low-molecular-weight model mixtures (this was ascribed to the close vicinity of the donor–acceptor couple). An improvement of the photoinitiating activity in the acrylate polymerization is thus achieved.

In the formerly proposed mixture of BP and Michler's ketone (bis (4,4'-dimethylamino) benzophenone), an electron transfer obviously arises between the two compounds as the Michler's ketone can behave as an amine and yields an aminoalkyl radical.

In BP/trimethyl benzophenone TMBP/amine AH, energy transfer occurs from BP to TMBP. The TMBP triplet state is more reactive, thereby explaining the higher efficiency of BP/TMBP/AH compared to BP/AH or TMBP/AH [2].

In the CQ/aminobenzophenone (ABP) two-component system, both CQ and ABP absorb the light. An aminoalkyl radical on ABP and a ketyl radical on CQ are formed [250].

Another ketone/ketone two-component system refers to the TX (or ITX)/phosphineoxide derivative system [308]. The direct and sensitized photolysis of phosphine oxides has been nicely studied by time-resolved ESR [309]. LFP gives a quenching rate constant of 4.5×10^9 mol^{-1} l s^{-1} for the ^{3}ITX/BAPO interaction. The free energy change calculated when the ITX triplet state acts as an electron acceptor ($\Delta G_{et} > +0.88$ eV) or an electron donor (ΔG_{et} is around $+0.3$ eV) suggests that the photoinduced electron transfer is not favorable and explains why no ITX ketyl radical is observed. By contrast, the energy transfer, which generates the ^{3}BAPO state and then the benzoyl and the phosphinoyl radicals, is slightly exothermic as E_T(BAPO) $= 61$–63 kcal mol^{-1} and E_T(ITX) $= 63$–64 kcal mol^{-1}). The polymerization efficiency increases when using ITX/BAPO due to both the higher quantity of photons absorbed by the formulation and this energy transfer process.

In the more recent BAPO/2-mercaptothioxanthone TX-SH system [310], the light is absorbed by TX-SH. Energy transfer between the TX-SH triplet state and BAPO leads to a cleavage of BAPO.

9.3.6
Organometallic Compound/Ketone-Based Systems

An example of a bond cleavage via an electron transfer reaction was demonstrated in the ruthenium trisbipyridine/morpholino ketone system [311]. On excitation of the organometallic compound (Eq. (9.15)), electron transfer occurs and the readily formed radical cation on the morpholino moiety NR$_2$ either undergoes a proton transfer or cleaves into an imino cation and an initiating benzoyl radical.

$$RuL_3^{2+} + \text{Benzoyl-CH (-phenyl)} \text{-} NR_2 \rightarrow RuL_3^+$$
$$+ \text{Benzoyl-CH (-phenyl)-}NR_2^{\bullet+} \text{ (light)}$$
$$\text{Benzoyl-CH (-phenyl)-}NR_2^{\bullet+} \rightarrow \text{Benzoyl-C}^\bullet \text{ (-phenyl)-}NR_2 + H^+$$
$$\text{Benzoyl-CH (-phenyl)-}NR_2^{\bullet+} \rightarrow \text{Benzoyl}^\bullet + {}^+CH \text{ (-phenyl)-}NR_2$$
$$^+CH \text{ (-phenyl)-}NR_2 + RuL_3^+ \rightarrow RuL_3^{2+} + CH^\bullet\text{(-phenyl)-}NR_2$$

(9.15)

9.3.7
Organometallic Complex/Additive

9.3.7.1 Organometallic Derivatives

Except few cases where organometallic compounds operate as Type I PI (Section 8.27), organometallic complexes work as Type II systems (see a review in [312]), for example, see in [313–319]. Few things have been published in recent years [313]. Examples include (i) metal carbonyls ($Mn_2(CO)_{10}$, $Re_2(CO)_{10}$, $Cr(CO)_6$, $W(CO)_6$, $Ru_3(CO)_{13}$, $ArCr(CO)_3$, $Os_3(CO)_{12}$) and half-sandwich metal carbonyls ($[CpFe(CO)_2]_2$, $[CpMo(CO)_3]_2$) or (ii) metallocenes (ferrocene, cobaltcene) in conjunction with organic halides (or bromides). The $CCl_3^{\bullet}$ radical is responsible for the initiation, for example, as in Eq. (9.16). In the absence of a coinitiator, the photopolymerization is almost not observed.

$$Mn_2(CO)_{10} \rightarrow\,^{\bullet}Mn(CO)_5 + \,^{\bullet}Mn(CO)_5 \,(h\nu) \quad (1)$$
$$^{\bullet}Mn(CO)_5 + CCl_4 \rightarrow Mn(CO)_5Cl + CCl_3^{\bullet}$$
$$Cr(\eta^6\text{-arene})(CO)_3 + CCl_4 \rightarrow \ldots + CCl_3^{\bullet} + \ldots \quad (2)$$
$$Cp_2Fe + CCl_4 \rightarrow Cp_2Fe^+\,Cl^- + CCl_3^{\bullet} \quad (3) \quad (9.16)$$

In the $Cr(\eta^6\text{-arene})(CO)_3$ complex (Eq. (9.17))/organic chloro compound systems, a good correlation between the polymerization ability and the metal–arene and metal–carbonyl bond strength was observed. The highest efficiency was found, however, when the organic chloro compound is tethered to the complex. The photoinitiation mechanism is complicated [315]. It involves (through a Cr-C cleavage and a CO loss) the formation of an exciplex between (benzene) tricarbonylchromium and an acrylate monomer (M), leading to a $Cr(\eta^2\text{-acrylate})(\eta^6\text{-benzene})(CO)_2$ complex that further interacts with carbon tetrachloride to yield benzene, 2 CO, CrCl, and the initiating trichloromethyl radical.

$$ArCr(CO)_3 \rightarrow *ArCr(CO)_3 \,(h\nu)$$
$$*ArCr(CO)_3 + M \rightarrow ArCr(CO)_2M + CO$$
$$ArCr(CO)_2M + CCl_4 \rightarrow\rightarrow\rightarrow\,^{\bullet}CCl_3 + Ar + 2\,CO + CrCl + M \quad (9.17)$$

Instead of chloro compounds, other coinitiators such as olefins, fumarates, maleic anhydride, and iodonium salt have been proposed (Eq. (9.18)).

$$R_4Pt + CF_2 = CF_2 \rightarrow R_4Pt\,CF_2\text{-}CF_2^{\bullet}\,(h\nu)$$
$$[CpFe(CO)_2]_2 + PhI_2{}^+\,X^- \rightarrow [CpFe(CO)_2]_2{}^{+\bullet}X^- + Ph^{\bullet} + PhI\,(h\nu)$$
$$(9.18)$$

9.3.7.2 Ferrocenium Salts

Examples of transition metal complexes without carbonyl-ligand-based systems concern, for example, the dimethyl (2,2′-bipyridyl) platinum salt (usable with tetrafluoroethylene) or the ferrocenium salt Fc(+). Ferrocenium salts Fc(+) based on η^5-cyclopentadienyl η^6-arene Fe(II) hexafluorophosphate, where the arene is cumene (**75**), dichlorobenzene, fluorene, xanthene, or thioxanthene [320], are usually considered as cationic PIs (Section 12.3).

75

They can, however, initiate a radical photopolymerization [312]. Indeed, a beneficial interaction with oxygen enhances the final conversion under air compared to a reaction in laminate (Eq. (9.19)). This was ascribed either to (i) an oxidation of Fc(+) into Fc(2+) and then a complexation of Fc(2+) with an acrylate M or (ii) an induced decomposition of the peroxidic compounds (inherently present in the medium) by Fc(+) [320].

$$Fc(+) + O_2 \rightarrow Fc(2+) + O_2^{\bullet-} \quad (h\nu)$$
$$Fc(2+) + M \rightarrow\rightarrow Fc(+) + M^\bullet$$
$$Fc(+) + ROOH \rightarrow Fc(2+) + RO^\bullet + OH^- \quad (h\nu) \quad (9.19)$$

Based on this scheme, an added hydroperoxide (such as cumene hydroperoxide CumOOH) can obviously act as a coinitiator (Eq. (9.20)).

$$Fc(+) + CumOOH \rightarrow Fc(2+) + CumO^\bullet + OH^- \quad (9.20)$$

Suitable dyes (such as RB) and ferrocenium Fc(+) salts (Section 9.2) can efficiently interact. Isolation of a RB − (Fc(+))$_2$ complex where two ferrocenium cations replace the cations of a dianionic dye such as RB has been achieved [264]. A simplified mechanism is shown in Eq. (9.21). The excited complex undergoes a fast intra-ion-pair electron transfer between RB and Fc(+), which prevents the oxygen quenching of ^{1,3}RB. The generated reduced form Fc(0) is a 19 e$^-$ complex highly reactive toward an amine AH, CumOOH, and oxygen.

$$RB\text{-}(Fc(+))_2 \rightarrow RB\text{-}(Fc(+))_2^* \quad (h\nu)$$
$$RB\text{-}(Fc(+))_2^* \rightarrow RB\text{-}Fc(+)^{\bullet+} + Fc(0)$$
$$Fc(0) + AH \,(CumOOH, O_2) \rightarrow\rightarrow A^\bullet \,(CumO^\bullet, O_2^{\bullet-}) \quad (9.21)$$

Non-transition metal complexes such as tetraethyl lead (for the photopolymerization of acrylonitrile) or triethyl aluminum have been mentioned [312]. For example, triethyl aluminum and MMA form an adduct. The ground state adduct/excited adduct interaction yields radicals.

Other systems involve ferrocene/carbon tetrachloride (Eq. (9.22)) or alkyl chloride [R826] (a trichloromethyl radical is formed with CCl_4) and ferrocene/carbon tetrabromide.

$$(C_5H_5)_2Fe + CCl_4 \rightarrow (C_5H_5)_2Fe^+ \; CCl_4^- \; (h\nu)$$
$$(C_5H_5)_2Fe^+ \; CCl_4^- \rightarrow (C_5H_5)_2Fe^+ \; Cl^- + CCl_3^\bullet \qquad (9.22)$$

9.3.8
Metal Carbonyl/Silane

The irradiation of metal carbonyl/onium salt systems (e.g., the [cyclopentadienyl Fe $(CO)_2]_2$/diaryliodonium hexafluorophosphate combination) generates a phenyl radical.

An interesting progress has been recently achieved using transition metal carbonyls (e.g., $Mn_2(CO)_{10}$, $Re_2(CO)_{10}$, $[CpFe(CO)_2]_2$) and silanes (e.g., TTMSS (**22**)) [321], thereby avoiding the presence of the highly volatile and carcinogenic carbon tetrachloride. Significant absorptions arise at 342 and 390 nm ($Mn_2 (CO)_{10}$), 313 nm ($Re_2 (CO)_{10}$), and 346, 410, and 514 nm ($[CpFe(CO)_2]_2$). As studied by LFP and ESR for $Mn_2(CO)_{10}$, the mechanism involves the formation of the $Mn(CO)_5^\bullet$ radical, the peroxidation of this radical, and the hydrogen abstraction of $Mn(CO)_5^\bullet$ and $Mn(CO)_5^- OO^\bullet$ by the silane (Eq. (9.23)).

$$Mn_2(CO)_{10} \rightarrow 2\, Mn(CO)_5^\bullet \; (h\nu)$$
$$Mn(CO)_5^\bullet + O_2 \rightarrow Mn(CO)_5\text{-}OO^\bullet$$
$$Mn(CO)_5^\bullet + R_3Si\text{-}H \rightarrow Mn\text{-}H(CO)_5 + R_3Si^\bullet \qquad (9.23)$$

The polymerization initiating ability of these systems is good [321] but can be strongly enhanced by addition of hydroperoxides (Section 10.2.6).

9.3.9
Photosensitizer-Linked Photoinitiator or Coinitiator-Based Systems

Attempts have been made to incorporate the PI and PS (the energy or electron donor) in the same molecule (Section 9.1.13). An old PI/PS example (but representative of many proposed systems) is shown in **76–78** [194]. A more recent example corresponds to a dye linked to a triazine derivative D-TZ (**76–78**). An intramolecular electron transfer yields a radical anion on the triazine moiety which results in the production of a radical [322, 323]. Another example relates to an Eosin linked to an O-acyloxime Eo-Ox (**76–78**) where radicals are generated according to a complex mechanism [324].

Other examples can be found in xanthene dye–thiol [325], carbazole–triazine [326], various dye–ketone systems [327]. Very often, the practical efficiency of the dye–donor linked PI is higher than that of a physical mixture of the dye and donor moieties.

9.3.10
Photoinitiator/Peroxide- or Hydroperoxide-Based Systems

In the PI/peroxide (or hydroperoxide) systems where useful effects can occur [2], oxygen-centered radicals are generated. The true mechanism is not always clear (Sections 18.10 and 18.11). Many investigations show that the mechanism of the sensitized cleavage of O–O bonds can involve various pathways as shown Eq. (9.24) in the case of peroxides such as energy transfer, electron transfer (with t-Bu-OOt-Bu; $E_{ox} = 2.07$ V and $E_{red} = -2.1$ V [328]), and production of vibrationally excited ground state peroxides, H atom transfer (according to the nature of the PI, the type of peroxide, and the medium). One or two oxyl radicals are finally formed. In ^{3}TX/di-$tert$-buylperoxide, the rate constant is relatively low (2×10^6 mol^{-1} l s^{-1}), and the process is ascribed to a triplet–triplet energy transfer [328]. In the case of BPO, the benzoyloxyl decarboxylates and forms a phenyl radical. Acylperoxides are more easily reduced than alkylperoxides, but the energy transfer pathway remains important except when using quite easily oxidized PSs.

$$PI^* + ROOR \rightarrow PI + ROOR^* \rightarrow PI + RO^\bullet + RO^\bullet$$
$$PI^* + ROOR \rightarrow PI^{\bullet+} + ROOR^{\bullet-} \rightarrow PI^{\bullet+} + RO^\bullet + RO^- \qquad (9.24)$$

For example (see references in [329–331]), the quenching of aromatic hydrocarbon singlet states by hydroperoxides and peroxides is explained on the basis of an energy transfer to a ground state peroxide with a O–O bond length greater than that of the equilibrium value. N-vinyl carbazole in its excited singlet state was shown to react with BPO through an energy transfer in benzene and an electron transfer in dichloromethane. The quenching of triplet PSs (such as anthracene, phenanthrene, acetonaphtanone, methoxy- and cyanobenzophenones, and propiophenone) by BPO involves the contribution of an electronic energy transfer to a O–O bond dissociative state as well as a charge transfer interaction via an exciplex. The results obtained for the singlet state quenching of polycyclic hydrocarbons, substituted naphtalenes, and anthracenes by BPO are consistent with a mechanism involving the formation of exciplexes with a charge transfer character.

The quenching of triplet excited molecules (e.g., BP) by various diaroyl peroxides results in an energy transfer process (thanks to the high donor triplet energy), whereas the CQ/peroxide ROOR interaction rather leads to an electron transfer with formation of a peroxide radical anion ROOR$^{\bullet-}$ that further cleaves into an oxyl radical RO$^\bullet$ and an anion RO$^-$ (Eq. (9.24)).

Very often, hydroperoxides ROOH undergo a hydrogen transfer in the presence of ketone triplet states (or even with radicals in suitable cases), which generates a peroxyl radical ROO$^\bullet$ (Section 18.11). The same holds true with other diketones such as in the SU (43–45)/ROOR (or ROOH) systems mentioned as efficient visible photoinitiating systems [112].

In the same way, in the 1-benzoyl-cyclohexanol HAP/BP system under air, it was proposed [332] that the radicals formed after the α-cleavage of HAP consume oxygen and generate a peroxyl radical. The decomposition of the further generated hydroperoxide ROOH can be sensitized by BP (Eq. (9.25)) through energy transfer (electron transfer contributes to some extent) as discussed in Section 18.11. Another process based on the presence of acetone might involve a rearrangement of the isopropylketyl peroxyl radical that forms acetone and a HO$_2^\bullet$ radical. The oxygen consumption and the oxyl/hydroxyl radical generation account for a decrease of the oxygen inhibition observed in photopolymerization reactions under air.

$$\text{HAP*} \rightarrow \text{radicals } R_i^\bullet$$
$$R_i^\bullet + O_2 \rightarrow R_iOO^\bullet \quad \text{and} \quad R_iOO^\bullet + \text{H-donor} \rightarrow R_iOOH + \text{donor}^\bullet$$
$$\text{BP*} + R_iOOH \rightarrow \text{BP} + RO^\bullet + OH^\bullet \tag{9.25}$$

9.3.11
Photoinitiator/Disulfide or Group 4B Dimetal Derivatives

Efficient BP (or ITX, CQ, eosin, thiopyrylium salt TP$^+$)/disulfides RS-SR (or group 4B dimetal derivatives R$_3$M-MR$_3$) systems have been recently reported [37]. Thiyl (Section 9.3.4) and metal-centered radicals are formed. On excitation of disilanes, polysilanes, digermanes, distannanes in the presence of BP or TP$^+$, a metal–metal bond cleavage is observed. The generated R$_3$Si$^\bullet$, R$_3$Ge$^\bullet$, and R$_3$Sn$^\bullet$ radicals efficiently add to acrylate (between $2 \times 10^7 - 9 \times 10^8$ mol^{-1} l s^{-1}) and behave as excellent initiating species of free radical photopolymerization (Section 18.13).

Depending on PI and the dimetal, electron transfer or energy transfer can occur (Eq. (9.26)). The formed triplet states are dissociative and lead to radicals. The R_3M-MR_3 BDEs are quite low (36.4–61.3 kcal mol^{-1}) in agreement with this efficient cleavage process.

$$^3BP + R_3M\text{-}MR_3 \rightarrow BP^{\bullet-} + (R_3M\text{-}MR_3)^{\bullet+} \rightarrow BP^{\bullet-} + R_3M^{\bullet} + R_3M^+$$
$$\text{and } BP^{\bullet-} + R_3M^+ \rightarrow BP + R_3M^{\bullet}$$
$$^3BP + R_3M\text{-}MR_3 \rightarrow BP + {}^3(R_3M\text{-}MR_3) \rightarrow BP + 2\,R_3M^{\bullet}$$
$$^3TP^+ + R_3M\text{-}MR_3 \rightarrow TP^{\bullet} + (R_3M\text{-}MR_3)^{\bullet+} \rightarrow TP^{\bullet} + R_3M^{\bullet} + R_3M^+$$

(9.26)

The very good photopolymerization efficiency in aerated conditions has been already discussed when using silanes, germanes, and stannanes in Type II systems (Section 9.1.2.7). The same behavior is obviously observed when the $R_3Si^{\bullet}$, $R_3Ge^{\bullet}$, and $R_3Sn^{\bullet}$ radicals are generated from dimetals. In addition, a particular feature of the R_3M-MR_3 compounds is their involvement in a homolytic substitution reaction SH$_2$ (Eq. (9.27)) where the fast conversion of a peroxyl into an efficient metal-centered radical is achieved. The high rate constant ($\sim$800 mol^{-1} l s^{-1} for a Si-Si unit), compared to the hydrogen abstraction rate constant (6 mol^{-1} l s^{-1}) in the corresponding ROO$^{\bullet}$/EDB system, demonstrates the interest of this reaction to overcome the oxygen inhibition.

$$R'OO^{\bullet} + R_3M\text{-}MR_3 \rightarrow R'OO\text{-}MR_3 + R_3M^{\bullet} \quad (9.27)$$

9.3.12
Photoinitiator/Phosphorus-Containing Compounds

As revealed by ESR-ST and LFP, the fragmentation of a phosphoniumyl cation radical $R_3''P^{+\bullet}$ from phosphinites (Pho1, Pho2) and phosphites (Pho3) in **79–81** (J. Lalevée, to be published) in the presence of BP generates a phosphinoyl radical.

Pho1
79

Pho2
80

Pho3
81

The interaction rate constants of the different excited states with Pho1, Pho2, and Pho3 are high in agreement with an electron transfer reaction (Eq. (9.28)). This reaction is more efficient in ^{3}BP/Pho1 (or Pho2) than in ^{3}BP/Pho3 as phosphites exhibit a higher oxidation potential than phosphinites (1.28, 1.33, and >1.8 V vs. saturated calomel electrode (SCE) for Pho1, Pho2, and Pho3, respectively). Using $^3TP^+$, the electron transfer is highly exothermic with Pho1-Pho3 but noticeably less favorable when using isopropyl thioxanthone or CQ.

$$PI + R'_2P\text{-}OR \rightarrow PI^{\bullet-} + R'_2P^{\bullet+}\text{-}OR \quad (h\nu)$$
$$R'_2P^{\bullet+}\text{-}OR \rightarrow R'_2P(=O)^{\bullet} + R^+ \quad (9.28)$$

As evidenced by the investigation of the t-BuO$^{\bullet}$/Pho1 (or Pho2, Pho3) interaction, a phosphoranyl radical t-Bu-O $-$ P$^{\bullet}$R$'_2$(OR) is formed. In a photopolymerizable medium under air, this interesting reaction converts oxyls into phosphoranyls (Eq. (9.29)) that are inefficient initiating radicals. However, they further consume oxygen and the formed peroxyls are then destroyed by the starting compound.

$$R_a\text{-}O^{\bullet} + R'_2POR \rightarrow R_a\text{-}O\text{-}P^{\bullet}R'_2(OR)$$
$$R_a\text{-}O\text{-}P^{\bullet}R'_2(OR) + O_2 \rightarrow R_a\text{-}OO^{\bullet} + R'_2P(O)(OR)$$
$$R_a\text{-}OO^{\bullet} + R'_2POR \rightarrow R_a\text{-}O^{\bullet} + R'_2P(O)(OR) \quad (9.29)$$

9.3.13
Photoinitiator/Onium Salts

Iodonium salts are good electron acceptors. In the presence of an excited donor, a diphenyl iodinium radical cation is formed. Then, it easily cleaves into phenyl iodide and a phenyl radical that behaves as a good initiating radical (Section 18.1). This was, for example, achieved with BP [333] (Eq. (9.30)), decahydroacridine-1,8-diones [334, 335] (sulfonium salts can also be decomposed in that case), and other ketones, dyes (e.g., xanthenic dyes), and organometallic compounds (Section 10.2.9). The same holds true in the thiophene/onium salt couple [336] and should be observed in the ALD/iodonium salt system (^{3}aldehyde $+ Ph_2I^+ \rightarrow$ aldehyde$^{\bullet+} + PhI + Ph^{\bullet}$) due to the electron donor capability of suitable ALDs [26].

$$^3BP + Ph_2I^+ \rightarrow BP^{\bullet+} + Ph_2I^{\bullet} \quad \text{and} \quad Ph_2I^{\bullet} \rightarrow PhI + Ph^{\bullet} \quad (9.30)$$

Electron transfer also occurs on irradiation of gold nanoparticles functionalized with 5-mercapto-2,2′−bithiophene (BT) in the presence of an iodonium salt [337] (the formed BT-Au radical cation subsequently aggregates via charge recombination).

Phenyl radicals can also be generated in BP/phosphonium salt [338].

9.3.14
Photosensitizer/Titanocenes

The decomposition of titanocenes TI (see formula in Scheme 8.21) in the presence of coumarins has been reported [121]. Generation of the titanocene radical anion is favored through electron transfer ($\Delta G = -1.11$ eV for coumarin C1 (**46, 47**); $E_{ox} = +1.5$ eV and $E_{red} = -1.01$ eV for TI). Nothing is known on the evolution of this species. The formation of radicals is expected [339] as the photopolymerization is more efficient with C1/TI than with TI alone.

9.3.15
Type I Photoinitiator/Additive: Search for New Properties

In a general way, the search and the use of an additive to a given cleavable PI was very often encountered for the improvement of, for example, the top layer photopolymerization under air, the bottom layer photopolymerization, the surface/body cure extent, and the surface hardness properties. This was achieved by mixing two or more PI, adding hydroperoxides, amines, and surface-active compounds. The incorporation of an additive into a Type I PI remains, however, a fascinating challenge to get a significant improvement of the rate of polymerization and the conversion.

A very efficient and recent procedure [44] is the use of R_3X-H additives where X = Si, Ge, Sn. Polymerization profiles are shown in Figure 9.6 in the case of DMPA, HAP, BAPO, or TI (see abbreviations in Tables 8.1 and 8.5 and Scheme 8.21) as PIs and TTMSS (22) as additive for the photopolymerization of a low-viscosity monomer (TMPTA) under air and on low light irradiation.

The striking feature is the enhancement of the efficiency under air that is ascribed to fast oxygen consumption, quenching of the peroxyls by R_3X-H, and regeneration of an X radical, for example, Eq. (9.31) in the case of BAPO. Interestingly, this improvement is strongly related to the rate constants of the peroxyl conversion, that is, higher rate constants leading to better final conversions. The observed high conversion of the Si-H function is also a strong evidence for the peroxyl/silane interaction. The dramatic improvement with DMPA, HAP, and TI is still explained on the basis of a similar mechanism.

$$\text{BAPO} \rightarrow \text{phosphinoyl radical P}^\bullet + \text{ benzoyl radical C}^\bullet \ (h\nu)$$
$$\text{P}^\bullet \text{ (or C}^\bullet\text{)} + \text{O}_2 \rightarrow \text{POO}^\bullet \text{ (or COO}^\bullet\text{)}$$
$$\text{P}^\bullet \text{ (or C}^\bullet\text{)} + n\text{M} \rightarrow \text{P-Mn}^\bullet \text{ (or C-Mn}^\bullet\text{)}$$
$$\text{P-Mn}^\bullet \text{ (or C-Mn}^\bullet\text{)} + \text{O}_2 \rightarrow \text{P-MnOO}^\bullet \text{ (or C-MnOO}^\bullet\text{)}$$
$$\text{POO}^\bullet \text{ (or COO}^\bullet\text{)} + \text{R}_3\text{Si-H} \rightarrow \text{P-OOH (or COOH)} + \text{R}_3\text{Si}^\bullet$$
$$\text{P-MnOO}^\bullet \text{ (or C-MnOO}^\bullet\text{)} + \text{R}_3\text{Si-H} \rightarrow \text{P-MnOOH (or C-MnOOH)} + \text{R}_3\text{Si}^\bullet$$

(9.31)

All these results outline the interest of R_3XH in Type I PI where (i) the usually formed radicals after the cleavage of PI are involved in the initiation process and (ii) R_3XH behaves as a high performance additive to convert inefficient peroxyls into new additional efficient initiating radicals (e.g., silyls) and thereby to reduce the detrimental oxygen inhibition effect. Some of these systems allow a sunlight curing of acrylates.

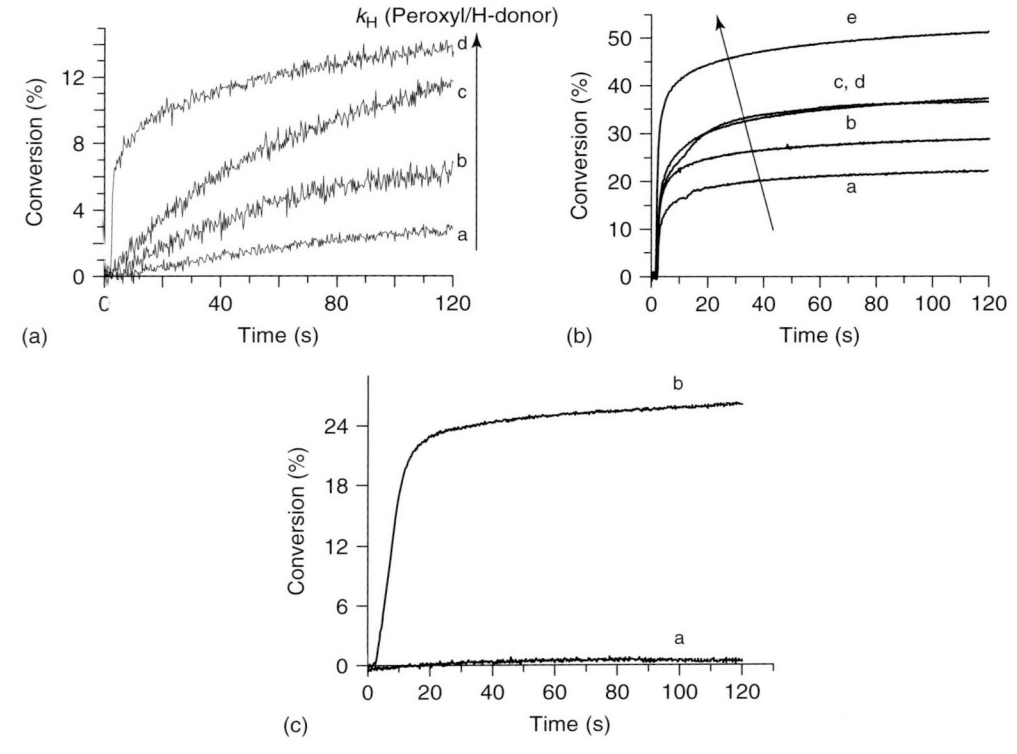

Figure 9.6 Conversion versus time curves for the photopolymerization of: (A) an epoxy acrylate in the presence of (a) BAPO 1% w/w alone, (b) BAPO/TTMSS, (c) BAPO/tris-(trimethylsilyl)germane, and (d) BAPO/triphenyltin hydride, (1%/3% w/w); (B) TMPTA using (a) DMPA, (b) DMPA/diphenylsilane, (c) DMPA/triphenylgermane, (d) DMPA/TTMSS, (e) DMPA/triphenyltin hydride, (1%/3% w/w); $I_0 = 22$ mW × cm^{-2}; sample thickness $= 20$ µm; and (C) TMPTA in the presence of (a) Ti, (b) Ti/tris-(trimethylsilyl)silane (0.1%/3% w/w); $I_0 = 60$ mW·cm^{-2}; xenon lamp; 300 nm < λ < 800 nm; sample thickness $= 20$ µm. Under air. See text. (Source: From Ref. [44].)

9.3.16
Nanoparticle-Formation-Mediated Type II Photoinitiators

The incorporation of nanoparticles in a polymer film has been achieved using the eosin/amine/Ag salt combination [340]. This should also be done using other Type II PI/metal salt systems, for example, BP/amine/gold salt as suggested by the mechanism proposed in Eq. (9.32) [341]. The presence of the metal salt allows to (i) scavenge the detrimental ketyl radical (from this point of view, BP/amine/gold salt might also be considered as a three-component system) and (ii) generate *in situ* metal nanoparticles. As in Type I PI, the radical/metal salt reactivity, the UV–visible absorption, the interactions between the three partners, and the stability of the formulation are crucial (Chapter 19).

$$BP + AH \rightarrow BP\text{-}H^\bullet + A^\bullet \ (h\nu)$$
$$BP\text{-}H^\bullet + Au^{3+} \rightarrow BP + H^+ + Au^{2+}$$
$$Au^{2+} + R^\bullet \rightarrow \rightarrow \rightarrow Au^0 \tag{9.32}$$

9.3.17
Miscellaneous Two-Component Systems

9.3.17.1 Light Absorbing Amine/Monomer

On light exposure at rather short wavelengths (254 or 313 nm), some amines (designed as light absorbing amines) such as dimethylamino ethyl benzoate (DMAEB) can be directly excited; then they react with an acrylate monomer M [250]. The monomer acts as an electron acceptor. From the resulting [DMAEB$^{\bullet+}$ M$^{\bullet-}$] radical ion pair and proton transfer, a radical is created on the monomer and an aminoalkyl radical DMAEB $-$ H$^\bullet$ is obviously formed.

9.3.17.2 Nitro Compound/Amine

Nitro aromatic compounds (e.g., nitroanilines [342] and nitronaphtylamines [342–344] and nitroacetanilide [345]) exhibit a good absorption in the UV range. In p-nitroaniline/AH (TEA) and nitroacetanilide/AH, an initiating A$^\bullet$ radical is formed as in Eq. (9.33).

$$H_2N\text{-}Ph\text{-}NO_2 + AH \rightarrow H_2N\text{-}Ph\text{-}NO_2H^\bullet + A^\bullet \tag{9.33}$$

9.3.17.3 Hydrocarbon/Amine

Electron transfer occurs in the pyrene Py/AH (such as TEA) system [346] and generates the pyrene radical anion Py$^{\bullet-}$ and the amine radical cation AH$^{\bullet+}$. The initiating aminoalkyl A$^\bullet$ radical is assumed to be generated according to Eq. (9.34). Bromoacetylpyrene was also proposed [347].

$$Py + AH \rightarrow Py^{\bullet-} + AH^{\bullet+} \ (h\nu)$$
$$Py^{\bullet-} + MMA \rightarrow Py + MMA^{\bullet-}$$
$$MMA^{\bullet-} + AH^{\bullet+} \rightarrow MMA\text{-}H^\bullet + A^\bullet \tag{9.34}$$

9.3.17.4 Photosensitizer/Triazine Derivative

Several kinds of compounds (**82** and **83**) [229] can photosensitize the decomposition of triazine derivatives through electron transfer. PS-bound triazines have also been prepared.

The mechanism is likely based on an electron transfer followed by a cleavage of the triazine radical anion (Eq. (9.35)).

$$PS + Tz \rightarrow PS^{\bullet +} + Tz^{\bullet -}$$
$$Tz^{\bullet -} \rightarrow Tz^{\bullet}_{(-Cl)} + Cl^- \qquad (9.35)$$

9.3.17.5 Other Systems

Other systems have been also proposed, for example, phosphine oxides/fluorescent optical brighteners [348]. The mechanism has not been studied.

References

1. Fouassier, J.P., Ruhlmann, D., Graff, B., Morlet-Savary, F., and Wieder, F. (1995) *Prog. Org. Coat.*, **25**, 235–271.
2. Fouassier, J.P. (1995) *Photoinitiation, Photopolymerization, and Photocuring*, Hanser, Münich.
3. Eiselé, G., Fouassier, J.P., and Reeb, R. (1999) *Angew. Macromol. Chem.*, **264**, 10–19.
4. Bloch, H., Ledwith, A., and Taylor, A.R. (1971) *Polymer*, **12**, 271–275.
5. Anderson, D.G., Davidson, R.S., and Elvery, J.J. (1996) *Polymer*, **37**, 2477–2484.
6. Johnston, L.J., Lougnot, D.J., and Scaiano, J.C. (1986) *Chem. Phys. Lett.*, **129**, 205–210.
7. Lalevée, J. and Fouassier, J.P. (2012) in *Handbook of Radical Chemistry and Biology*, vol. 1, Chapter 2 (eds A. Studer and C. Chatgilialoglou), Wiley-VCH Verlag GmbH, Weinheim, New York, pp. 34–57.
8. Davidson, R.S. (1993) in *Radiation Curing in Polymer Science and Technology*, vol. 3 (eds J.P. Fouassier and J.F. Rabek), Elsevier, Barking, pp. 153–176.
9. Tasdelen, M.A., Moszner, N., and Yagci, Y. (2009) *Polym. Bull.*, **63**, 173–183.
10. Anderson, D. (2005) *Proceedings of the Radtech Europe Conference*, Vincentz Network, Hannover, pp. 435–444.
11. Allen, N.S., Marin, M.C., Edge, M., Davies, D.W., Garrett, J., and Jones, F. (2001) *Polym. Degrad. Stab.*, **73**, 119–139.
12. Xu, H., Wu, G., and Nie, J. (2008) *J. Photochem. Photobiol. A: Chem.*, **193**, 254–259.
13. Kucybala, Z., Pietrzak, M., Paczkowski, J., Linden, L.A., and Rabek, J.F. (1996) *Polymer*, **37**, 4585–4591.
14. Fouassier, J.P., Ruhlmann, D., and Erddalane, A. (1993) *Macromolecules*, **26**, 721–728.
15. Rabek, J.F., Linden, L.A., Fouassier, J.F., Nie, J., Andrzejewska, E., Paczkowski, J., Jakubiak, J., Wrzyszczynski, A., and Sionkowska, A. (1999) in *Trends in Photochemistry and Photobiology*, vol. 5 (ed. J.P. Fouassier), Research Trends, Trivandrum, pp. 51–62.
16. Bortolus, P. (1993) in *Radiation Curing in Polymer Science and Technology*, vol. 2 (eds J.P. Fouassier and J.F. Rabek), Elsevier, Barking, pp. 603–636.
17. Tasdelen, M.A., Demirel, A.L., and Yagci, Y. (2007) *Eur. Polym. J.*, **43**, 4423–4430.
18. Lalevée, J., Allonas, X., and Fouassier, J.P. (2008) *Chem. Phys. Lett.*, **466**, 227–230.
19. Allonas, X., Dossot, M., Merlin, A., Sylla, M., Jacques, P., and Fouassier, J.P. (2000) *J. Appl. Polym. Sci.*, **78**, 2061–2069.
20. Andrzejwska, E. (2006) in *Photochemistry and UV Curing: New Trends* (ed. J.P. Fouassier), Research Signpost, Trivandrum, pp. 127–141.
21. Lalevée, J., El Roz, M., Morlet-Savary, F., Allonas, X., and Fouassier, J.P. (2009) *Macromol. Chem. Phys.*, **210**, 311–319.

22. Lalevée, J., El Roz, M., Allonas, X., and Fouassier, J.P. (2007) *J. Polym. Sci.*, **45**, 2436–2442.
23. Bartoszewicz, J., Hug, G.L., Pietrzak, M., Kozubek, H., Paczkowski, J., and Marciniak, B. (2007) *Macromolecules*, **40**, 8642–8648.
24. Marciniak, B., Andrzejewska, E., and Hug, G.L. (1998) *J. Photochem. Photobiol., A: Chem.*, **112**, 21–28.
25. Tasdelen, M.A., Kiskan, B., and Yagci, Y. (2006) *Macromol. Rapid Commun.*, **27**, 1539–1544.
26. Birbaum, J.L., Dietliker, K., Freihermuth, B., Kuntz, M., Oka, H., and Wolf, J.P. (2004) PCT Patent 104051, Ciba.
27. Aydin, M. and Arsu, N. (2006) *Prog. Org. Coat.*, **56**, 338–342.
28. Shi, S., Gao, H., Wu, G., and Nie, J. (2007) *Polymer*, **48**, 2860–2865.
29. Ouchi, T., So, N., and Komatsu, Y. (1980) *J. Macromol. Sci.*, **A14**, 277–285.
30. Wang, K. and Nie, J. (2009) *J. Photochem. Photobiol., Part A: Chem.*, **204**, 7–12.
31. Wang, K., Ma, G., Qin, X.-H., Xiao, M., and Nie, J. (2010) *Polym. J.*, **42**, 450–455.
32. Xiao, P., Lalevée, J., Allonas, X., Ley, C., El Roz, M., Fouassier, J.P., Shi, S.Q., and Nie, J. (2010) *J. Polym. Sci., Polym. Chem.*, **48**, 5758–5766.
33. Wang, K., Yang, D., Xiao, M., Chen, X., Lu, F., and Nie, J. (2009) *Acta Biomater.*, **5**, 2508–2517.
34. Nie, J., Andrzejewska, E., Rabek, J.F., Linden, L.A., Fouassier, J.P., Paczkowski, J., Scigalski, F., and Wrzyszczynski, A. (1999) *Macromol. Chem. Phys.*, **200**, 1692–1701.
35. Grotzinger, C., Burget, D., Fouassier, J.P., Richard, G., Primel, O., and Yean, S. (2006) Proceedings of the RadTech USA International Conference.
36. Lalevée, J., Allonas, X., and Fouassier, J.P. (2007) *J. Org. Chem.*, **72**, 6434–6439.
37. Lalevée, J., Dirani, A., El Roz, M., Allonas, X., and Fouassier, J.P. (2008) *Macromolecules*, **41**, 2003.
38. Lalevée, J., Shankar, R., Tehfe, M.A., Sahoo, U., and Fouassier, J.P. (2011) *Macromol. Chem. Phys., Macromol. Chem. Phys.*, **212**, 806–812.
39. Lalevée, J., El Roz, M., Allonas, X., and Fouassier, J.P. (2009) in *Organosilanes: Properties, Performance and Applications*, Chapter 6 (eds E. Wyman and M.C. Skief), Nova Science Publishers, Hauppauge, pp. 174–198.
40. Lalevée, J., El Roz, M., Allonas, X., and Fouassier, J.P. (2009) *Prog. Org. Coat.*, **65**, 457–461.
41. El Roz, M., Lalevée, J., Allonas, X., and Fouassier, J.P. (2008) *Macromol. Chem. Rapid Commun.*, **29**, 804–808.
42. El Roz, M., Lalevée, J., Allonas, X., and Fouassier, J.P. (2010) *Macromolecules*, **43**, 2219–2227.
43. Lalevée, J., Dirani, A., Graff, B., El Roz, M., Allonas, X., and Fouassier, J.P. (2008) *J. Polym. Sci., Polym. Chem.*, **46**, 3042–3047.
44. El Roz, M., Lalevée, J., Allonas, X., and Fouassier, J.P. (2009) *Macromolecules*, **42**, 8725–8732.
45. Lalevée, J., Allonas, X., and Fouassier, J.P. (2009) *Macromolecules*, **41**, 9057–9062.
46. Lalevée, J., Blanchard, N., Tehfe, M.A., Chany, A., and Fouassier, J.P. (2010) *Chem. Eur. J.*, **16**, 12920–12927.
47. Tehfe, M.A., Lalevée, J., Fouassier, J.P., Fensterbank, L., Lacote, E., Malacria, M., and Curran, D.P. (2010) *Macromolecules*, **43**, 2261–2267.
48. Lalevée, J., Blanchard, N., Chany, A., Tehfe, M.A., Allonas, X., and Fouassier, J.P. (2009) *J. Org. Chem. Phys.*, **22**, 986–993.
49. Tehfe, M.A., Monot, J., Brahmi, M.M., Curran, D.P., Malacria, M., Fensterbank, L., Lacote, E., Lalevée, J., and Fouassier, J.P. (2011) *Polym. Chem.*, **2**, 625–631.
50. Tehfe, M.A., Monot, J., Brahmi, M.M., Curran, D.P., Malacria, M., Fensterbank, L., Lacote, E., Lalevée, J., and Fouassier, J.P. (2012) *ACS Macro Lett.*, **1**(1), 92–95.
51. Versace, D.L., Lalevée, J., Fouassier, J.P., Guillaneuf, Y., Bertin, D., and Gigmes, D. (2010) *J. Polym. Sci., Polym. Chem.*, **48**, 2910–2915.
52. Wu, Q. and Qu, B. (2001) *Polym. Eng. Sci.*, **41**, 1220–1226.

53. Ma, H., Davis, R.H., and Bowman, C.N. (2000) *Macromolecules*, **33**, 331–335.
54. Muftuogli, A.E., Tasdelen, M.A., and Yagci, Y. (2010) in *Photochemistry and Photophysics of Polymer Materials* (ed. N.S. Allen), John Wiley & Sons, Inc., Hoboken, pp. 509–540.
55. Buriak, J.M. (2002) *Chem. Rev.*, **102**, 1271–1282.
56. Souane, R., Lalevée, J., Allonas, X., and Fouassier, J.P. (2010) *Macromol. Mater. Eng.*, **295**, 351–354.
57. Turro, N.J. (1978 and 1990) *Modern Molecular Photochemistry*, Benjamin, New York.
58. Aloise, S., Ruckebusch, C., Blanchet, L., Réhaut, J., Buntix, G., and Huvenne, J.P. (2008) *J. Phys. Chem.*, **112**, 224–231.
59. Schnabel, W. (1990) in *Lasers in Polymer Science and Technology*, vol. II (eds J.P. Fouassier and J.F. Rabek), CRC Press, Boca Raton, FL, pp. 95–143.
60. Scully, A.D., Horsham, M.A., Aguas, P., and Murphy, J.K.G. (2008) *J. Photochem. Photobiol. A: Chem.*, **197**, 132–140.
61. Allonas, X., Fouassier, J.P., Kaji, M., and Murakami, S. (2003) *Photochem. Photobiol. Sci.*, **2**, 224.
62. Rulhman, D. and Fouassier, J.P. (1992) *Eur. Polym. J.*, **28**, 591.
63. Rulhman, D., Zahouily, K., and Fouassier, J.P. (1992) *Eur. Polym. J.*, **28**, 1063.
64. Allen, N.S. (1996) *J. Photochem. Photobiol. A*, **100**, 101–107.
65. Wang, H., Wei, J., Jiang, X., and Yin, J. (2006) *Macromol. Chem. Phys.*, **207**, 1080–1086.
66. Nagarajan, R., Bowers, J.S. Jr., Wu, Z., Cui, H., Bao, R., Jonsson, S., McCartney, R., Pittman, C.U. Jr., Cao, L., and Ran, R. (1999) *Surf. Coat. Int.*, **82**, 344–347.
67. Wang, H., Shi, Y., Wei, J., Jiang, X., and Yin, J. (2006) *J. Appl. Polym. Sci.*, **101**, 2347–2354.
68. Tanaka, M., Yago, T., and Wasaka, M. (2009) *Chem. Lett.*, **38**, 1086–1087.
69. Temel, G., Aydogan, B., Arsu, N., and Yagci, Y. (2009) *J. Polym. Sci., Part A: Polym. Chem.*, **47**, 2938–2947.
70. Viswanathan, K., Hoyle, C.E., Joensson, E.S., Nason, C., and Lindgren, K. (2002) *Macromolecules*, **35**, 7963–7967.
71. Valderas, C., Bertolotti, S., Previtali, C.M., and Encinas, M.V. (2002) *J. Polym. Sci., Part A: Polym. Chem.*, **40**, 2888–2893.
72. Paczkowski, J. and Neckers, D.C. (2001) *Electron. Transfer. Chem.*, **5**, 516–585.
73. Kabatc, J., Jedrzejewska, B., and Paczkowski, J. (2006) *Macromol. Mater. Eng.*, **291**, 646–654.
74. Wu, G.Q., Shi, S., Xiao, P., and Nie, J. (2007) *J. Photochem. Photobiol., A: Chem.*, **188**, 260–266.
75. Lalevée, J., Allonas, X., Zadoina, L., and Fouassier, J.P. (2007) *J. Polym. Sci., Polym. Chem.*, **45**, 2494–2502.
76. Andrzejewska, E., Zych-Tomkowiak, D., Bogacki, M.B., and Andrzejewski, M. (2004) *Macromolecules*, **37**, 6346–6354.
77. Andrzejewska, E., Zych-Tomkowiak, D., Andrzejewski, M., Hug, G.L., and Marciniak, B. (2006) *Macromolecules*, **93**, 3777–3785.
78. Andrjewska, E. (2006) in *Photochemistry and UV Curing: New Trends* (ed. J.P. Fouassier), Research Signpost, Trivandrum, p. 127.
79. Scigalski, F. and Paczkowski, J. (2005) *J. Appl. Polym. Sci.*, **97**, 358–365.
80. Ullrich, G., Burtscher, P., Salz, U., Moszner, N., and Liska, R. (2006) *J. Polym. Sci., Part A: Polym. Chem.*, **44**, 115–125.
81. Shi, S., Gao, H., Xiao, P., and Nie, J. (2006) *Polym. Prepr.*, **47**, 333.
82. Wu, G. and Nie, J. (2006) *J. Photochem. and Photobiol., A: Chem.*, **183**, 154–158.
83. Xiao, P., Wang, Y., Dai, M., Wu, G., Shi, S., and Nie, J. (2008) *Polym. Adv. Tech.*, **19**, 409–413.
84. Hai, S., Zhou, H., Matsushima, H., and Hoyle, C.E. (2008) Proceedings of the RadTech, Chicago.
85. Encinas, M.V., Lissi, E.A., Rufs, A.M., Altamirano, M., and Cosa, J.J. (1998) *Photochem. Photobiol.*, **68**, 447–452.
86. Woodward, J.R., Lin, T.S., Sakaguchi, Y., and Hayashi, H. (2002) *Mol. Phys.*, **100**, 1235–1244.

87. Cai, X., Sakamoto, M., Hara, M., Sugimoto, A., Tojo, S., Kawai, K., Endo, M., Fujitsuka, M., and Majima, T. (2003) *Photochem. Photobiol. Sci.*, **2** (11), 1209–1214.
88. Cai, X., Sakamoto, M., Fujitsuka, M., and Majima, T. (2005) *Chem.- A Eur. J.*, **11**, 6471–6477.
89. Palit, D.K. (2005) *Res. Chem. Interm.*, **31** (1–3), 205–225.
90. Rubio-Pons, O., Serrano-Andres, L., Burget, D., and Jacques, P. (2006) *J. Photochem. Photobiol., A: Chem.*, **179**, 298–304.
91. Corrales, T., Peinado, C., Catalina, F., Neumann, M.G., Allen, N.S., Rufs, A.M., and Encinas, M.V. (2000) *Polymer*, **41**, 9103–9109.
92. Ley, C., Morlet-Savary, F., Jacques, P., and Fouassier, J.P. (2000) *Chem. Phys.*, **255**, 335–346.
93. Allonas, X., Ley, C., Bibaut, C., Jacques, P., and Fouassier, J.P. (2000) *Chem. Phys. Lett.*, **322**, 483–490.
94. Angulo, G., Grij, J., Vauthey, E., Serrano-Andres, L., Rubio-Pons, O., and Jacques, P. (2010) *Chem. Phys. Chem.*, **11**, 480–488.
95. Allen, N.S., Salleh, N.G., Edge, M., Corrales, T., Shah, M., Catalina, F., and Green, A. (1997) *J. Photochem. Photobiol., A: Chem.*, **103**, 185–189.
96. Allen, N.S., Salleh, N.G., Edge, M., Shah, M., Ley, C., Morlet-Savary, F., Fouassier, J.P., Catalina, F., Green, A., Navaratnam, S., and Parsons, B.J. (1999) *Polymer*, **40**, 4181–4193.
97. Esen, D.S., Karasu, F., and Arsu, N. (2011) *Prog. Org. Coat.*, **70**, 102–107.
98. Encinas, M.V., Rufs, A.M., Corrales, T., Catalina, F., Peinado, C., Schmith, K., Neumann, M.G., and Allen, N.S. (2002) *Polymer*, **43**, 3909–3913.
99. Catalina, F., Peinado, C., Sastre, R., Mateo, J.L., and Allen, N.S. (1989) *J. Photochem. Photobiol., A: Chem.*, **47**, 365–372.
100. Fedorov, A.V., Ermoshkin, A.A., Mejiritski, A., and Neckers, D.C. (2007) *Macromolecules*, **40**, 3554–3560.
101. Yates, S.F. and Schuster, G.B. (1984) *J. Org. Chem.*, **49**, 3349–3356.
102. Mullings, M.E., Walder, C.D., McDonough, A., and Cavitt, T.B. (2007) *J. Coat. Technol. Res.*, **4**, 255–261.
103. Scaiano, J.C. (1980) *J. Am. Chem. Soc.*, **102**, 7747–7751.
104. Fouassier, J.P., Jacques, P., and Encinas, M.V. (1988) *Chem. Phys. Lett.*, **148**, 309–312.
105. Fouassier, J.P., Ruhlmann, D., Graff, B., and Wieder, F. (1995) *Prog. Org. Coat.*, **25**, 169–202.
106. Wintgens, V., Netto-Ferreira, J.C., and Scaiano, J.C. (2002) *Photochem. Photobiol. Sci.*, **1**, 184–189.
107. Malval, J.P., Dietlin, C., Allonas, X., and Fouassier, J.P. (2007) *J. Photochem. Photobiol., A: Chem.*, **192**, 66–73.
108. Netto-Ferreira, J.C. and Scaiano, J.C. (1989) *J. Photochem. Photobiol., A: Chem.*, **48**, 345–352.
109. Asmussen, S. and Vallo, C. (2009) *J. Photochem. Photobiol., A: Chem.*, **202**, 228–234.
110. Encinas, M.V. and Lissi, E. (1984) *J. Polym. Sci., Part A: Polym. Chem.*, **22**, 2469–2477.
111. Okutsu, T., Ooyama, M., Hiratsuka, H., Tsuchiya, J., and Obi, K. (1999) *J. Phys. Chem. A.*, **104** (2), 288–292.
112. Grotzinger, C., Burget, D., Fouassier, J.P., Rirchard, G., Primel, O., and Yean, S. (2006) Proceedings of the RadTech, USA International Conference.
113. Kochevar, I.E. and Wagner, P.J. (1972) *J. Am. Chem. Soc.*, **94**, 3859–3865.
114. Bosch, P., Del Monte, F., Mateo, J.L., and Davidson, R.S. (1993) *J. Photochem. Photobiol., A: Chem.*, **73**, 197–204.
115. Jakubiak, I., Allonas, X., Fouassier, J.P., Sionkowka, A., Andrzejewska, E., Rabek, J., and Linden, L. (2003) *Polymer*, **44**, 5219–5226.
116. Pyszka, I., Kucybala, Z., and Paczkowski, J. (2004) *Makromol. Chem. Phys*, **205**, 2371–2375.
117. Allonas, X., Fouassier, J.P., Angiolini, L., and Caretti, D. (2001) *Helv. Chim. Acta*, **84**, 2577–2588.
118. Nie, J., Linden, L.A., Rabek, J.F., Fouassier, J.P., Morlet-Savary, F., Scigalski, F., Wrzyszczynski, A., and Andrzejewska, E. (1998) *Acta Polym.*, **49**, 145–161.

119. Fouassier, J.P. and Wu, S.K. (1992) *J. Appl. Polym. Sci.*, **44**, 1779–1786.
120. Satpati, A.K., Nath, S., Kumbhakar, M., Maity, D.K., Senthilkumar, S., and Pal, H. (2008) *J. Mol. Struct.*, **878**, 84–94.
121. Allonas, X., Fouassier, J.P., Kaji, M., and Miasaka, M. (2000) *J. Photopolym. Sci. Tech.*, **13**, 237.
122. Fedorov, A.V., Danilov, E.O., Rodgers, M.A.J., and Neckers, D.C. (2001) *J. Am. Chem. Soc.*, **123**, 5136–5137.
123. Merzlikine, A.G., Voskresensky, S.V., Danilov, E.O., Neckers, D.C., and Fedorov, A.V. (2007) *Photochem. Photobiol. Sci.*, **6**, 608–613.
124. Kolano, C., Bucher, G., Wenk, H.H., Jaeger, M., Schade, O., and Sander, W. (2004) *J. Phys. Org. Chem.*, **17**, 207–214.
125. Gan, D., Jia, M., Vaughan, P.P., Falvey, D.E., and Blough, N.V. (2008) *J. Phys. Chem. A.*, **112**, 2803–2809.
126. Tasdelen, M.A., Kumbaraci, V., Talinli, N., and Yagci, Y. (2006) *Polymer*, **47**, 7611–7614.
127. Novikova, E., Kolendo, A., Syromyatnikov, V., Avramenko, L., Prot, T., and Golec, K. (2001) *Polimery*, **46**, 406–413.
128. Ley, C., Morlet-Savary, F., Fouassier, J.P., and Jacques, P. (2000) *J. Photochem. Photobiol.*, **137**, 87–92.
129. Ulrich, S., Timpe, H.J., Morlet-Savary, F., and Fouassier, J.P. (1993) *J. Photochem.*, **74**, 165.
130. Encinas, M.V. and Previtali, C.M. (2006) *Compr. Ser. Photochem. Photobiol. Sci.*, **6**, 41–59.
131. Shurygina, M.P., Kurskii, Y.A., Chesnokov, S.A., and Abakumov, G.A. (2008) *Tetrahedron*, **64**, 1459–1466.
132. Forbes, M.D.E. and Yashiro, H. (2007) *Macromolecules*, **40**, 1460–1465.
133. Shah, M., Allen, N.S., Edge, M., Navaratnam, S., and Catalina, F. (1996) *J. Appl. Polym. Sci.*, **62**, 319–340.
134. Allen, N.S., Pullen, G., Shah, M., Edge, M., Weddell, I., Swart, R., and Catalina, F. (1995) *Polymer*, **36**, 4665–4674.
135. Encinas, M.V., Majmud, C., and Lissi, E.A. (1990) *J. Polym. Sci., Part A: Polym. Chem.*, **28**, 2465–2474.
136. Tozuka, M., Igarashi, T., and Sakurai, T. (2009) *Polym. J.*, **41** (9), 709–714.
137. Czech, Z. (2006) *Polimery*, **51**, 754–757.
138. Shen, L., Ji, H.F., and Zhang, H.Y. (2008) *Theochem*, **851**, 220–224.
139. Sarma, S.J. and Jones, P.B. (2010) *J. Org. Chem.*, **75**, 3806–3813.
140. Encinas, M.V., Lissi, E.A., Rufs, A.M., and Previtali, C.M. (1994) *J. Polym. Sci., Part A: Polym. Chem.*, **32**, 1649–1655.
141. Zbigniev, C., Butwin, A., and Kabatc, J. (2011) *Eur. Polym. J.*, **47**, 225–229.
142. Song, A., Zhao, D., Rong, R., Zhang, L., and Wang, H. (2011) *J. Appl. Polym. Sci.*, **119**, 629–635.
143. Le Gern, J., Farge, H., Allonas, X., Lalevée, J., and Fouassier, J.P. (2009) WO/2009/071844, Ciba Spec.
144. Wang, H.L., Brown, H.R., and Li, Z.R. (2007) *Polymer*, **48**, 939–946.
145. Corrales, T., Catalina, F., Allen, N.S., and Peinado, C. (2006) in *Photochemistry and UV Curing: New Trends* (ed. J.P. Fouassier), Research Signpost, Trivandrum, pp. 31–45.
146. Wang, Y., Jiang, X., and Yin, J. (2009) *Eur. Polym. J.*, **45** (2), 437–447.
147. Wei, J., Liu, F., Lu, Z., Song, L., and Cai, D. (2008) *Polym. Adv. Technol.*, **19**, 1763–1770.
148. Aydin, M., Arsu, N., and Yagci, Y. (2003) *Macromol. Rapid Commun.*, **24**, 718–723.
149. Aydin, M., Arsu, N., Yagci, Y., Jockusch, S., and Turro, N.J. (2005) *Macromolecules*, **38**, 4133–4138.
150. Temel, G., Aydogan, B., Arsu, N., and Yagci, Y. (2009) *Macromolecules*, **42**, 6098–6106.
151. Temel, G., Karaca, N., and Arsu, N. (2010) *J. Polym. Sci., Part A: Polym. Chem.*, **48**, 5306–5312.
152. Wei, J. and Liu, F. (2009) *Macromolecules*, **42**, 5486–5491.
153. Matsushima, H., Hait, S., Li, Q., Zhou, H., Shirai, M., and Hoyle, C.E. (2010) *Eur. Polym. J.*, **46**, 1278–1287.
154. Ullrich, G., Herzog, D., Liskra, R., Burtscher, P., and Moszner, N. (2004) *J. Polym. Sci., Part A: Polym. Chem.*, **42**, 4948–4963.
155. Jauk, S. and Liska, R. (2005) *Macromol. Rapid Commun.*, **26**, 1687–1692.
156. Kura, H., Oka, H., Ohwa, M., Matsumura, T., Kimura, A.,

Iwasaki, Y., Ohno, T., Matsumura, M., and Murai, H. (2005) *J. Polym. Sci., Part B: Polym. Phys.*, **43**, 1684–1695.

157. Chen, Y., Loccufier, J., Vanmaele, L., and Frey, H. (2007) *J. Mater. Chem.*, **17**, 3389–3392.

158. Wei, J., Jiang, X.S., and Wang, H.Y. (2007) *Macromol. Chem. Phys.*, **208**, 2303–2311.

159. Wang, H., Wei, J., Jiang, X., and Yin, J. (2007) *J. Photochem. Photobiol. A: Chem.*, **186**, 106–114.

160. Sandholzer, M., Schuster, M., Liska, R., Turecek, C., Varga, F., Stelzer, F., and Slugovc, C. (2007) *Polym. Prepr.*, **48**, 925–926.

161. Carlini, C. and Angiolini, L. (1993) in *Radiation Curing in Polymer Science and Technology*, vol. **2** (eds J.P. Fouassier and J.F. Rabek), Elsevier, Barking, pp. 283–322.

162. Tan, J., Wu, B., Yang, J., Zhu, Y., and Zeng, Z. (2010) *Polymer*, **51**, 3394–3401.

163. Corrales, T., Catalina, F., Peinado, C., Allen, N.S., Rufs, A.M., Bueno, C., and Encinas, M.V. (2002) *Polymer*, **43**, 4591–4597.

164. Jiang, X., Xu, H., and Yin, J. (2003) *Polymer*, **45**, 133–140.

165. Jiang, X. and Yie, J. (2004) *Macromol. Rapid Commun.*, **25**, 748–752.

166. Jiang, X. and Yin, J. (2004) *J. Appl. Polym. Sci.*, **94**, 2395–2400.

167. Wei, J., Wang, H., Jiang, X., and Yin, J. (2006) *Makromol. Chem. Phys.*, **207**, 1752–1763.

168. Fouassier, J.P., Ruhlmann, D., Zahouily, K., Angiolini, L., Carlini, C., and Lelli, N. (1992) *Polymer*, **33**, 3569–3573.

169. Pouliquen, L., Coqueret, X., Morlet-Savary, F., and Fouassier, J.P. (1995) *Macromolecules*, **28**, 8028–8034.

170. Wei, J., Wang, H., Jiang, X., and Yin, J. (2007) *Makromol. Chem. Phys.*, **208**, 287–294.

171. Wei, J., Wang, H., and Yin, J. (2007) *J. Polym. Sci., Part A: Polym. Chem.*, **45**, 576–587.

172. Wang, H., Wei, J., Jiang, X., and Yin, J. (2007) *Polym. Int.*, **56**, 200–207.

173. Jiang, X., Luo, J., and Yin, J. (2009) *Polymer*, **50**, 37–41.

174. Wei, J. and Wang, B. (2011) *Macromol. Chem. Phys.*, **212**, 88–95.

175. Gomez, M.L., Fasce, D.P., Williams, R.J.J., Erra-Balsells, R., Kaniz Fatema, M., and Nonami, H. (2008) *Polymer*, **49**, 3648–3656.

176. Bertens, F., Qingjin, Y., and Gu, D. (2005) Proceedings of the RadTech Europe, Vol. 1, pp. 471–478.

177. Holtzinger, D. and Kickelbick, G. (2002) *J. Polym. Sci., Part A: Polym. Chem.*, **40**, 3858–3872.

178. Frey, M., Frossard, C., Studer, K., Powell, K., Birbaum, J.L., and Muller, M. (2010) WO 2010/0636120A1, BASF.

179. Wen, Y., Jiang, X., Liu, R., and Yin, J. (2009) *Polymer*, **50**, 3917–3923.

180. Wen, Y., Jiang, X., and Yin, J. (2009) *Prog. Org. Coat.*, **66**, 65–72.

181. Hong, K.H., Liu, N., and Sun, G. (2009) *Eur. Polym. J.*, **45**, 2443–2449.

182. Xiao, P., Wang, Y., Dai, M., Shi, S., Wu, G., and Nie, J. (2008) *Polym. Eng. Sci.*, **48**, 884–888.

183. Gacal, B., Akat, H., Balta, D.K., Arsu, N., and Yagci, Y. (2008) *Macromolecules*, **41**, 2401–2405.

184. Hou, G., Shi, S., Liu, S., and Nie, J. (2008) *Polym. J.*, **40**, 228–232.

185. Rufs, A.M., Valdebenito, A., Rezende, M.C., Bertolotti, S., Previtali, C., and Encinas, M.V. (2008) *Polymer*, **49**, 3671–3676.

186. Wu, G., Zeng, S., Ou, E., Yu, P., Lu, Y., and Xu, W. (2010) *Mater. Sci. Eng., C: Mat. Biol. Appl.*, **30**, 1030–1037.

187. Ye, G., Yang, J., Zeng, Z., and Chen, Y. (2010) in *Basics of Photopolymerization Reactions*, vol. **1** (eds J.P. Fouassier and X. Allonas), Research Signpost, Trivandrum, pp. 49–60.

188. Wen, Y., Jiang, X., Liu, R., and Yin, J. (2011) *Polym. Adv. Technol.*, **22**, 598–604.

189. Temel, G., Enginol, B., Aydin, M., Karaca Balta, D., and Arsu, N. (2011) *J. Photochem. Photobiol. A: Chem.*, **219**, 26–31.

190. Gao, F., Peng, H., Yang, L., Liu, X., Wang, J., Hu, N., Xie, T., Li, H., and Zhang, S. (2009) *Polym. Technol.*, **20**, 1010–1016.

191. Xiao, P. Dai, M., and Nie, J. (2008) *J. Appl. Polym. Sci.*, **108**, 665–670.
192. Xiao, P., Zhang, H., Dai, M., and Nie, J. (2009) *Prog. Org. Coat.*, **64**, 510–514.
193. Xiao, P., Shi, S., and Nie, J. (2008) *Polym. Adv. Technol.*, **19**, 1305–1310.
194. Angiolini, L., Caretti, D., Carlini, C., Corelli, E., Fouassier, J.P., and Morlet-Savary, F. (1995) *Polymer*, **36**, 4055–4060.
195. Angiolini, L., Caretti, D., and Salatelli, E. (2000) *Macromol. Chem. Phys.*, **201**, 2646–2653.
196. Allonas, X., Fouassier, J.P., Angiolini, L., and Caretti, D. (2001) *Helv. Chim. Acta*, **84**, 2577.
197. Green, W.A. and Timms, A.W. (1993) in *Radiation Curing in Polymer Science and Technology*, vol. 2 (eds J.P. Fouassier and J.F. Rabek), Elsevier, Barking, pp. 375–434.
198. Fouassier, J.P. and Lougnot, D.J. (1990) in *Lasers in Polymer Science and Technology*, vol. II (eds J.P. Fouassier and J.F. Rabek) CRC Press, Boca Raton, FL, pp. 145–168.
199. Capek, I. (2001) *Trends Photochem. Photobiol.*, **7**, 147–157.
200. Fouassier, J.P., Lougnot, D.J., Zuchowicz, I., Green, P.N., Timpe, H.J., Kronfeld, K.P., and Muller, U. (1987) *J. Photochem.*, **36**, 347–363.
201. Lougnot, D.J., Turck, C., and Fouassier, J.P. (1989) *Macromolecules*, **22**, 108–116.
202. Corales, T., Catalina, F., Allen, N.S., and Peinado, C. (2005) *J. Photochem. Photobiol., A: Chem.*, **169**, 95–100.
203. Lougnot, D.J. and Fouassier, J.P. (1988) *J. Polym. Sci., Part A: Polym. Chem.*, **26**, 1021–1028.
204. Lougnot, D.J., Jacques, P., Fouassier, J.P., Casal, H.L., Kim-Thuan, N., and Scaiano, J.C. (1985) *Can. J. Chem.*, **63**, 3001–3006.
205. Balta, D.K., Temel, G., Aydin, M., and Arsu, N. (2010) *Eur. Polym. J.*, **46**, 1374–1379.
206. Jiang, X. and Yin, J. (2008) *Makromol. Chem. Phys.*, **209**, 1593–1600.
207. Liska, R., Knaus, S., Gruber, H., and Wendrinsky, J. (2000) *Surf. Coat. Int.*, **83**, 297–303.
208. Karasu, F., Balta, D.K., Liska, R., and Arsu, N. (2010) *J. Inclusion Phenom. Macrocyclic Chem.*, **68**, 147–153.
209. Belfield, K.D., Ren, X., Van Stryland, E.W., Hagan, D.J., Dubiskovsky, V., and Miesak, E.J. (2000) *J. Am. Chem. Soc.*, **122**, 1217–1218.
210. Belfield, K. (2005) *Proceedings of the Radtech Europe Conference*, vol. 2, Vincentz Network, Hannover, pp. 75–82.
211. Pucher, N., Rosspeintner, A., Satzinger, V., Schmidt, V., Gescheidt, G., Stampfl, J., and Liska, R. (2009) *Macromolecules*, **42**, 6519–6528.
212. Heller, C., Pucher, N., Seidl, B., Kalinyaprak, K.R., Ullrich, G., Kuna, L., Satzinger, V., Schmidt, V., Lichtenegger, L., Stampfl, J., and Liska, R. (2007) *J. Polym. Sci., Part A: Polym. Chem.*, **45**, 3280–3291.
213. Jin, M., MalvalL, J.P., Versace, D.L., Morlet-Savary, F., Chaumeil, H., Defoin, A., Allonas, X., and Fouassier, J.-P. (2008) *Chem. Commun.*, 6540–6542.
214. Baldeck, P., Stephan, O., and Andraud, C. (2010) in *Basics of Photopolymerization Reactions*, vol. 3 (eds J.P. Fouassier and X. Allonas), Research Signpost, Trivandrum, pp. 200–220.
215. Tasdelen, M.A., Kumbaraci, V., Talinli, N., and Yagci, Y. (2007) *Macromolecules*, **40**, 4406–4408.
216. Tasdelen, M.A., Kumbaraci, V., Talinli, N., and Yagci, Y. (2010) in *Basics of Photopolymerization Reactions*, vol. 1 (eds J.P. Fouassier and X. Allonas), Research Signpost, Trivandrum, pp. 75–88.
217. Balta, D.K., Arsu, N., Yagci, Y., Jockusch, S., and Turro, N.J. (2007) *Macromolecules*, **40**, 4138–4141.
218. Balta, D.K., Arsu, N., Yagci, Y., Sundare, A.K., Jockusch, S., and Turro, N.J. (2011) *Macromolecules*, **44**, 2531–2535.
219. Cokbaglan, L., Arsu, N., Yagci, Y., Jockusch, S., and Turro, N.J. (2003) *Macromolecules*, **36**, 2649–2653.
220. Yilmaz, G., Aydogan, B., Temel, G., Arsu, N., Moszner, N., and Yagci, Y. (2010) *Macromolecules*, **43**, 4520–4526.

221. Karasu, F., Arsu, N., Jockusch, S., and Turro, N.J. (2009) *Macromolecules*, **42**, 7318–7323.
222. Yilmaz, G., Tuzun, A., and Yagci, Y. (2010) *J. Polym. Sci., Part A: Polym. Chem.*, **48**, 5120–5125.
223. Wang, K., Ma, G., Yin, R., Nie, J., and Yu, Q. (2010) *Mater. Chem. Phys.*, **124**, 453–457.
224. Cui, R., Wang, K., Ma, G., Qian, B., Yang, J., Ju, Q., and Nie, J. (2011) *J. Appl. Polym. Sci.*, **120**, 2754–2759.
225. Zaikov, G.E., Howard, J.A., and Ingold, K.U. (1969) *Can. J. Chem.*, **47**, 3017–3021.
226. Fouassier, J.P. (2000) *Recent Research Development in Photochemistry and Photobiology*, vol. **4**, Research Trends, Trivandrum, pp. 51–74.
227. Fouassier, J.P., Allonas, X., and Burget, D. (2003) *Prog. Org. Coat.*, **47**, 16–36.
228. Krongauz, V. and Trifunac, A. (eds) (1994) *Photoresponsive Polymers*, Chapman & Hall, New York.
229. Monroe, B.M. and Weed, G.C. (1993) *Chem. Rev.*, **93**, 435–448.
230. Oster, G. (1954) *Nature*, **173**, 300–304.
231. Paczkowski, J. and Neckers, D.C. (2001) in *Electron Transfer in Chemistry*, vol. **5** (ed. I.R. Gould), Wiley-VCH Verlag GmbH, pp. 516–545.
232. Eaton, D.F. (1986) in *Advances in Photochemistry*, vol. **13** (eds D.H. Hammond and G.S. Gollnick), John Wiley & Sons, Inc., New York, pp. 427–487.
233. Neckers, D.C. and Valdes-Aguilera, O.M. (1993) in *Advances in Photochemistry*, vol. **18** (eds D.H. Hammond and G.S. Gollnick), John Wiley & Sons, Inc., New York, p. 315.
234. Bi, Y. and Neckers, D.C. (1993) *J. Photochem. Photobiol. A: Chem.*, **74**, 221–228.
235. Mallavia, R., Amat-Guerri, F., Fimia, A., and Sastre, R. (1994) *Macromolecules*, **27**, 2643–2646.
236. Popielarz, R. and Vogt, O. (2008) *J. Polym. Sci., Part A: Polym. Chem.*, **46**, 3519–3532.
237. Bertolotti, S.G., Previtali, C.M., Rufs, A.M., and Encinas, M.V. (1999) *Macromolecules*, **32**, 2920–2924.
238. Encinas, M.V., Rufs, A.M., Bertolotti, S., and Previtali, C.M. (2001) *Macromolecules*, **34**, 2845–2847.
239. Villegas, L., Encinas, M.V., Rufs, A.M., Bueno, C., Bertolotti, S., and Previtali, C.M. (2001) *J. Polym. Sci., Part A: Polym. Chem.*, **39**, 4074–4082.
240. Kabatc, J. and Paczkowski, J. (2009) *J. Polym. Sci., Part A: Polym. Chem.*, **47**, 4636–4654.
241. Kabatc, J. and Paczkowski, J. (2010) *Dyes Pigm.*, **86**, 133–142.
242. Soppera, O., Turck, C., and Lougnot, D.J. (2009) *Opt. Lett.*, **34**, 461–464.
243. Garcia, O., Costela, A., Garcia-Moreno, I., and Sastre, R. (2003) *Makromol. Chem. Phys.*, **204**, 2233–2239.
244. Klimtchuk, E., Rodgers, M.A.J., and Neckers, D.C. (1992) *J. Phys. Chem.*, **96**, 9817–9820.
245. Yong, H., Zhou, W., Liu, G., Zhen, L.M., and Wang, E. (2000) *J. Photopolym. Sci. Technol.*, **13**, 253–258.
246. Liu, A.D., Trifunac, A.D., and Krongrauz, V. (1992) *Chem. Phys. Lett.*, **198**, 200–205.
247. Kucybala, Z. and Paczkowski, J. (1999) *J. Photochem. Photobiol., A: Chem.*, **128**, 135–138.
248. Kabatc, J. and Paczkowski, J. (2010) *J. Appl. Polym. Sci.*, **117**, 2669–2675.
249. Kabatc, J., Krzyzanowska, E., Jedrzejewska, B., Pietrzak, M., and Paczkowski, J. (2010) *J. Appl. Polym. Sci.*, **118**, 165–172.
250. Fouassier, J.P. (2000) *Recent Research Development in Polymer Science*, vol. **4**, Research Trends, Trivandrum, pp. 131–145.
251. Pyszka, I. and Kucybala, Z. (2007) *Polymer*, **48**, 959–965.
252. Encinas, M.V., Rufs, A.M., Neumann, M.G., and Previtali, C.M. (1996) *Polymer*, **37**, 1395–1398.
253. Gomez, M.L., Previtali, C.M., Montejano, H.A., and Bertolotti, S.G. (2007) *J. Photochem. Photobiol., A: Chem.*, **188**, 83–89.
254. Sokolowska, J., Podsiadly, R., and Stoczkiewicz, J. (2008) *Dyes Pigm.*, **77**, 510–514.
255. Podsiadly, R., Kolinska, J., and Sokolowska, J. (2008) *Color. Technol.*, **124**, 79–85.

256. Delzenne, G. (1960) *J. Polym. Sci., Part A: Polym. Chem.*, **48**, 347–352.
257. Avens, H.J. and Bowman, C.N. (2009) *J. Polym. Sci., Part A: Polym. Chem.*, **47**, 6083–6094.
258. Neumann, M.G., Schmitt, C.C., and Goi, B.E. (2005) *J. Photochem. Photobiol. A: Chem.*, **174**, 239–245.
259. Pietrzak, M., Gluszak, A., and Wrzyszczynski, A. (2007) *Polimery*, **52**, 768–771.
260. Villegas, M.L., Bertolotti, S.G., Previtali, C.M., and Encinas, M.V. (2005) *Photochem. Photobiol.*, **81**, 884–890.
261. Breslow, D.S., Simpson, D.A., Kramer, B.D., Schwartz, R.J., and Newburg, N.R. (1987) *Ind. Eng. Chem. Res.*, **26**, 2144–2148.
262. Urano, T., Ohno, E., Sakamoto, K., Wada, M., Ito, S., Ono, N., Suzuki, S., and Yamaoka, T. (2000) *Imaging Sci. J.*, **48**, 147–152.
263. Mauguière-Guyonnet, F., Burget, D., and Fouassier, J.P. (2006) *Prog. Org. Coat.*, **57**, 23–32.
264. Grotzinger, C., Burget, D., Jacques, P., and Fouassier, J.P. (2001) *J. Appl. Polym. Sci.*, **81**, 2368–2376.
265. Linden, S.M. and Neckers, D.C. (1998) *J. Am. Chem. Soc.*, **110**, 1257–1261.
266. Kim, D., Scranton, A.B., and Stansbury, J.W. (2009) *J. Polym. Sci., Part A: Polym. Chem.*, **47**, 1429–1439.
267. Sandholzer, M., Schuster, M., Varga, F., Liska, R., and Slugovc, C. (2008) *J. Polym. Sci., Part A: Polym. Chem.*, **46**, 3648–3661.
268. Encinas, M.V., Rufs, A.M., Bertolotti, S.G., and Previtali, C.M. (2009) *Polymer*, **50**, 2762–2767.
269. Rivarola, C.R., Biasutti, M.A., and Barbero, C.A. (2009) *Polymer*, **50**, 3145–3152.
270. Li, S., Wu, F., and Wang, E. (2009) *Polymer*, **50**, 3932–3937.
271. Fouassier, J.P. and Chesneau, E. (1991) *Makromol. Chem.*, **192**, 1307–1315.
272. Grotzinger, C., Burget, D., Jacques, P., and Fouassier, J.P. (2001) *Macromol. Chem. Phys.*, **202**, 3513–3522.
273. Grotzinger, C., Burget, D., Jacques, P., and Fouassier, J.P. (2003) *Polymer*, **44**, 3671–3677.
274. Hoyle, C.E., Viswanathan, K., Clark, S.C., Miller, C.W., Nguyen, C., Jonsson, S., and Shao, L. (1999) *Macromolecules*, **32**, 2793–2795.
275. Nguyen, C.K., Smith, R.S., Cavitt, T.B., Hoyle, C.E., Jonsson, S., Miller, C.W., and Pappas, S.P. (2001) *Polym. Prepr.*, **42**, 707–708.
276. Wang, H., Wei, J., Jiang, X., and Yin, J. (2006) *Polym. Int.*, **55**, 930–937.
277. Burget, D., Mallein, C., and Fouassier, J.P. (2004) *Polymer*, **45**, 6561–6567.
278. Hoyle, C.E., Clark, C., Jonsson, S., and Shimose, M. (1997) *Polymer*, **38**, 5695–5698.
279. Clark, C., Jonsson, S., and Hoyle, C.E. (1997) *Polym. Prepr.*, **38**, 363–364.
280. Sonntag, J.V. and Knolle, W. (2000) *J. Photochem. Photobiol., A*, **136**, 133–139.
281. Miller, C.W., Joensson, E.S., Hoyle, C.E., Viswanathan, K., and Valente, E.J. (2001) *J. Phys. Chem. B*, **105**, 2707–2717.
282. Nguyen, C.K., Cavitt, T.B., Hoyle, C.E., Kalyanaraman, V., and Jonsson, S. (2003) *Photoinitiated Polymerization*, ACS Symposium Series, Vol. **847**, American Chemical Society, pp. 27–40.
283. Hoyle, C.E., Clark, S.C., Viswanathan, K., and Jonsson, S. (2003) *Photochem. Photobiol. Sci.*, **2**, 1074–1079.
284. Silva, R., Muniz, E.C., and Rubira, A.F. (2009) *Langmuir*, **25**, 873–880.
285. Ghosh, P. and Pal, G. (1998) *Eur. Polym. J.*, **34**, 677–681.
286. Gosh, P. and Pal, G. (1998) *J. Polym. Sci., Part A: Polym. Chem.*, **36**, 1973–1979.
287. Hizal, G., Emigloru, S.E., and Yagci, Y. (1998) *Polym. Int.*, **47**, 391–396.
288. Catalina, F., Peinado, C., Blanco, M., Allen, N.S., Corrales, T., and Lukac, I. (1998) *Polymer*, **39**, 4399–4408.
289. Bendyk, M., Jedrzejewska, B., Paczkowski, J., and Linden, L.-A. (2002) *Polimery*, **47**, 654–656.
290. Shi, Y.-T., Yin, J., Kaji, M., and Yori, H. (2006) *Polym. Int.*, **55**, 330–339.
291. Shi, Y., Yin, J., Kaji, M., and Yori, H. (2006) *Polym. Eng. Sci.*, **46**, 474–479.
292. Allonas, X., Obeid, H., Fouassier, J.P., Kaji, M., Ichihashi, Y., and

Murakami, Y. (2003) *J. Photopolym. Sci. Technol.*, **16**, 123–128.

293. Liu, A.D., Trifunac, A.D., and Krongrauz, V. (1992) *J. Phys. Chem.*, **96**, 207–211.
294. Shi, Y., Wang, Y., Jiang, X., Yin, J., Kaji, M., and Hanako, H. (2007) *J. Appl. Polym. Sci.*, **105**, 2027–2035.
295. Caspar, J.V., Khudyakov, I.V., Turro, N.J., and Weed, G.C. (1995) *Macromolecules*, **28**, 636–641.
296. Satoh, Y., Ishibashi, Y., Ito, S., Nagasawa, Y., Miyasaka, H., Chosrowjan, H., Taniguchi, S., Mataga, N., Kato, D., Kikuchi, A., and Abe, J. (2007) *Chem. Phys. Lett.*, **448**, 228–231.
297. Morlet-Savary, F., Fouassier, J.P., Matsumoto, T., and Inomata, K. (1994) *Polym. Adv. Technol.*, **5**, 56–62.
298. Parret, S., Morlet-Savary, F., Fouassier, J.P., Inomata, K., and Matsumoto, T. (1998) *J. Chem. Soc., Faraday Trans.*, **94**, 745.
299. Morlet-Savary, F., Wieder, F., and Fouassier, J.P. (1997) *J. Chem. Soc., Faraday Trans.*, **93**, 3931.
300. Komarova, E.Y., Ren, K., and Neckers, D.C. (2002) *Langmuir*, **18**, 4195–4197.
301. El Roz, M., Morlet-Savary, F., Lalevée, J., Allonas, X., and Fouassier, J.P. (2008) *J. Polym. Sci., Polym. Chem.*, **46** (22), 7369–7375.
302. Rist, G., Borer, A., Dietliker, K., Desobry, V., Fouassier, J.P., and Ruhlmann, D. (1992) *Macromolecules*, **25**, 4182–4193.
303. Ruhlman, D. and Fouassier, J.P. (1993) *Eur. Polym. J.*, **29**, 27–34.
304. Ruhlman, D. and Fouassier, J.P. (1993) *Eur. Polym. J.*, **29**, 505–512.
305. Alberti, A., Benaglia, M., Macciantelli, D., Rossetti, S., and Scoponi, M. (2008) *Eur. Polym. J.*, **44**, 3022–3027.
306. Dietliker, K., Broillet, S., Hellrung, B., Rzadek, P., Rist, G., Wirz, J., Neshchadin, D., and Gescheidt, G. (2006) *Helv. Chem. Acta*, **89**, 2211–2225.
307. Angiolini, L., Caretti, D., Corelli, E., and Carlini, C. (1995) *J. Appl. Polym. Sci.*, **55**, 1477–1488.
308. Dossot, M., Obeid, H., Allonas, X., Jacques, P., Fouassier, J.P., and Merlin, A. (2004) *J. Appl. Polym. Sci.*, **92**, 1154–1164.
309. Williams, R.M., Khudyakov, I.V., Purvis, M.B., Overton, B.J., and Turro, N.J. (2000) *J. Phys. Chem. B*, **104**, 10437–10443.
310. Keskin, S., Jockusch, S., Turro, N.J., and Arsu, N. (2008) *Macromolecules*, **41**, 4631–4634.
311. Lee, L.Y., Xi, C., Giannotti, C., and Whitten, D.G. (1986) *J. Am. Chem. Soc.*, **113**, 2304.
312. Cunningham, A.F. and Desobry, V. (1993) in *Radiation Curing in Polymer Science and Technology*, vol. **2** (eds J.P. Fouassier and J.F. Rabek), Elsevier, Barking, pp. 323–374.
313. Kundig, E.P., Xu, L.H., Kondratenko, M., Cunningham, A.F. Jr., and Kunz, M. (2007) *Eur. J. Inorg. Chem.*, **18**, 2934–2943.
314. Yagci, Y. and Hepuzer, Y. (1999) *Macromolecules*, **32**, 6367–6370.
315. Bamford, C.H. and Mullik, S.U. (1977) *J. Chem. Soc., Faraday Trans. 1*, **73**, 1260–1270.
316. Sato, T., Katayose, T., and Seno, M. (2000) *J. Appl. Polym. Sci.*, **79**, 166–175.
317. Rivarola, C.R., Bertolotti, S.G., and Previtali, C.M. (2001) *J. Polym. Sci., Part A: Polym. Chem.*, **39**, 4265–4273.
318. Gosh, P. and Pal, G. (1997) *Eur. Polym. J.*, **33**, 1695–1700.
319. Alvarez, J., Lissi, E.A., and Encinas, M.V. (1998) *J. Polym. Sci., Part A: Polym. Chem.*, **36**, 207–208.
320. Browser, R. and Davidson, S. (1994) *J. Photochem. Photobiol., A: Chem.*, **77**, 269–276.
321. Tehfe, M.A., Lalevée, J., Gigmes, D., and Fouassier, J.P. (2010) *J. Polym. Sci., Polym. Chem.*, **48**, 1830–1837.
322. Kawamura, K. and Kato, K. (2004) *Polym. Adv. Technol.*, **15**, 324–328.
323. Kawamura, K. (2004) *J. Photochem. Photobiol., Part A: Chem.*, **162**, 329–338.
324. Burget, D., Fouassier, J.P., Amat-Guerri, F., Mallavia, R., and Sastre, R. (1999) *Acta Polym.*, **50**, 337–346.
325. Scigalski, F. and Paczkowski, J. (2001) *Polimery*, **46**, 613–621.

326. Kawamura, K., Aotani, Y., and Tomioka, H. (2003) *J. Phys. Chem. Part B*, **107**, 4579–4586.
327. Liska, R. (2004) *J. Polym. Sci., Part A: Polym. Chem.*, **42**, 2285–2301.
328. Engel, P.S., Woods, T.L., and Page, M.A. (1983) *J. Phys. Chem.*, **87**, 10–13.
329. Abuin, E.B., Lissi, E.A., and Encinas, M.V. (1988) *J. Polym. Sci., Part A: Polym. Lett.*, **26**, 501–504.
330. Ingold, K.U., Johnston, L.J., Lusztyk, J., and Scaiano, J.C. (1984) *Chem. Phys. Lett.*, **110**, 433–438.
331. Takahara, S., Sakuragi, H., Tokumaru, K., Kitamura, A., and Shiga, A. (1985) *J. Photochem. Photobiol. A: Chem.*, **31**, 239–244.
332. Pappas, S.P. (1986) *UV-Curing: Science and Technology*, Technology Marketing Corporation, Stamford, CT; Plenum Press, New York (1992).
333. Timpe, H.J., Kronfeld, K.P., Lammel, U., and Fouassier, J.P. (1990) *J. Photochem. Photobiol. A: Chem.*, **52**, 111–118.
334. Timpe, H.J., Ulrich, S., and Fouassier, J.P. (1993) *J. Photochem. Photobiol. A: Chem.*, **73**, 139–150.
335. Timpe, H.J., Ulrich, S., Decker, C., and Fouassier, J.P. (1993) *Macromolecules*, **26**, 4560–4566.
336. Beyazit, S., Aydogan, B., Osken, I., Ozturk, T., and Yagci, Y. (2011) *Polym. Chem.*, **2**, 1185–1189.
337. Anyaogu, K.C., Cai, X., and Neckers, D.C. (2008) *Photochem. Photobiol. Sci.*, **7**, 1469–1472.
338. Bartoszewicz, J., Hug, G.L., Pietrzak, M., Kozubek, H., Paczkowski, J., and Marciniak, B. (2008) *J. Polym. Sci., Part A: Polym. Chem.*, **24**, 8013–8022.
339. Criqui, A., Lalevée, J., Allonas, X., and Fouassier, J.P. (2008) *Macromol. Chem. Phys.*, **209**, 2223–2231.
340. Balan, L., Malval, J.P., Schneider, R., Le Nouen, D., and Lougnot, D.J. (2010) *Polymer*, **51**, 1363–1369.
341. Marin, L., McGilvray, K.L., and Scaiano, J.C. (2008) *J. Am. Chem. Soc.*, **130**, 16572–16584.
342. Encinas, M.V., Rufs, A.M., Norambuena, E., and Giannotti, C. (2000) *J. Polym Sci., Part A: Polym. Chem.*, **38**, 2269–2273.
343. Costela, A., Garcia-Moreno, I., Dabrio, J., and Sastre, R. (1997) *J. Polym. Sci., Part A: Polym. Chem.*, **35**, 3801–3812.
344. Costela, A., Garcia-Moreno, I., Garcia, O., and Sastre, R. (2001) *Chem. Phys. Lett.*, **347**, 115–120.
345. Encinas, M.V., Rufs, A.M., Norambuena, E., and Giannotti, C. (1997) *J. Polym. Sci., Part A: Polym. Chem.*, **35**, 3095–3100.
346. Encinas, M.V., Majmud, C., Lissi, E.A., and Scaiano, J.C. (1991) *Macromolecules*, **24**, 2111–2112.
347. Mishra, A. and Daswal, S. (2007) *Int. J. Chem. Kinet.*, **39**, 261–267.
348. Wilczak, W.A. (1998) US Patent 5, 707, 781, Bayer Corp., PA.

10
Multicomponent Photoinitiating Systems

In this Part, the design of high-performance PIs has been largely achieved using cleavable PI (Type I) and two-component PI (Type II). The addition of a third or a fourth compound to Type II systems leads to even more efficient three- or four-component PIs. Many multicomponent photoinitiating systems have been proposed so far in order to improve the efficiency of the radical production (see a recent review in [1]). They are very often used in the design of high-speed photopolymers [2, 3].

10.1
Generally Encountered Mechanism

Many multicomponent photoinitiating systems behave according to the following mechanism proposed a long time ago (Scheme 10.1). The C1/C2 combination (where C1 is the absorbing species) is used to produce the initiating radical R1; the concomitantly formed radical R2 is usually a scavenger of the growing polymer chain. The idea consists in eliminating this R2 radical by using an appropriate quencher C3. An increase in the polymerization-initiating ability is expected.

The following systems obey such a mechanism.

10.1.1
Ketone/Amine/Onium Salt or Bromo Compound

The simplified mechanism involved in a system [4], where C1 is isopropylthioxanthone (ITX), C2 an amine AH, and C3 carbon tetrabromide, is shown in Eq. (10.1). Here, a ketyl-type radical on ITX is formed and then scavenged by the bromo compound. A tribromomethyl radical is generated in addition to the aminoalkyl radical. Finally, the ketone is recovered. When using an onium salt instead of CBr_4, a reactive phenyl radical is created.

$$ITX + AH \rightarrow ITX-H^{\bullet} + A^{\bullet} \quad (h\nu)$$
$$ITX-H^{\bullet} + CBr_4 \rightarrow ITX + H^+ + CBr_3 + Br^- \quad (10.1)$$

The same mechanism as (Eq. (10.1)) holds true in the ketocoumarin (KC) (1–5, Chapter 9)/amine (N-phenyl glycine NPG)/onium salt (diphenyl iodonium Ph_2I^+)

Photoinitiators for Polymer Synthesis: Scope, Reactivity and Efficiency, First Edition.
Jean-Pierre Fouassier and Jacques Lalevée.
© 2012 Wiley-VCH Verlag GmbH & Co. KGaA. Published 2012 by Wiley-VCH Verlag GmbH & Co. KGaA.

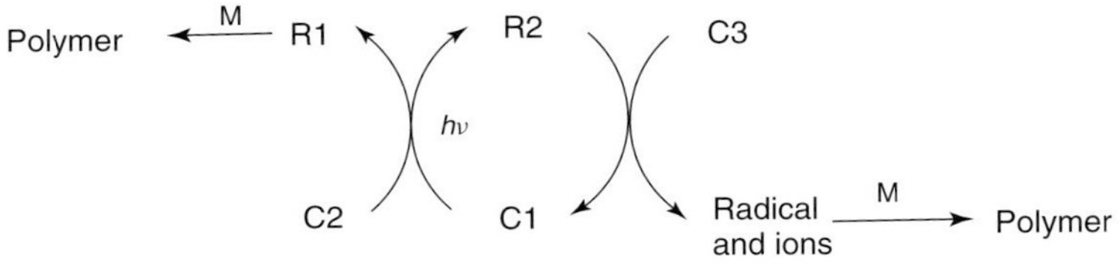

```
Polymer  ←—M—  R1  ╳  R2  ╳  C3
                    hν
                C2      C1      Radical —M→ Polymer
                                and ions
```

Scheme 10.1

system [5]. Photosensitivities of 0.2, 0.9, and 1.2 mJ cm^{-2} were obtained using an acrylate matrix on a Ar$^+$ laser exposure at 488 nm in the presence of the KC/NPG/Ph$_2$I$^+$, KC/Ph$_2$I$^+$, and KC/NPG combinations, respectively.

10.1.2
Dye/Amine/Onium Salt or Bromo Compound

An example is displayed in Eq. (10.2) for the Eosin (Eo)/amine/diphenyliodonium salt system [6]. In addition to the aminoalkyl radical, a phenyl radical is generated. Carbon tetrabromide, tetrahydrofuran (THF) or a keto-oxime derivative (PDO) has also been used instead of the onium salt.

$$Eo + AH \rightarrow Eo - H^{\bullet} + A^{\bullet} \; (h\nu)$$
$$Eo - H^{\bullet} + Ph_2I^+ \rightarrow Eo + H^+ + PhI + Ph^{\bullet} \quad (10.2)$$

A more or less similar behavior was described in methylene blue/amine/onium salt [7, 8], dye (e.g., merocyanine, safranine)/amine/bromo compound (or onium salt) [9, 10], thioxanthene dye/amine/onium salt (or bromo compound) [11], safranine/amine/onium salt [12], and fluorone dye (**1**)/amine/onium salt [13].

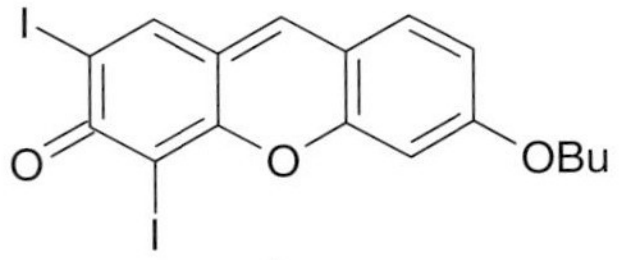

1

In the acridinium cation Acr$^+$/dihydropyridine pyH$_2$/onium salt Ph$_2$I$^+$ [14], the mechanism is more complicated.

10.1.3
Ketone/Amine/Imide Derivatives

Maleimide, maleic anhydride, or arylnaphtalimides have also been proposed as additives [15]. For example, this was basically described in benzophenone/amine/maleimide as shown in Scheme 10.2 [16]. An initiating radical is formed on the amine and the maleimide [17]. An alternative mechanism lies on an energy transfer leading to the maleimide triplet state, which then reacts with the amine through an electron/proton transfer. In that case, no radical is formed on the maleimide. The ITX/maleic anhydride interaction is discussed in [18].

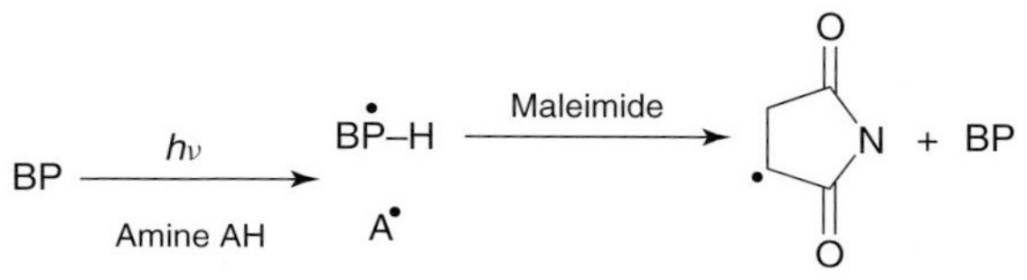

BP $\xrightarrow{h\nu}$ BP-H $\xrightarrow{\text{Maleimide}}$ [succinimidyl radical structure] + BP

Amine AH → A•

Scheme 10.2

10.1.4
Ketone/Ketone/Amine

As observed in ITX/benzophenone sulfonylketone BPSK (**21**, Chapter 8)/amine, such a system combines an electron-transfer reaction and a ketyl radical quenching (Eq. (10.3)) as above. In addition, an energy transfer between the two ketones (e.g., ITX and BPSK) can contribute to some extent [19].

$$^{1,3}\text{ITX} + \text{AH} \rightarrow\rightarrow \text{A}^\bullet + \text{ITX} - \text{H}^\bullet$$
$$\text{ITX} - \text{H}^\bullet + \text{BPSK} \rightarrow \text{ITX} - \text{H}^+ + \text{BPSK}^{\bullet-} \rightarrow \text{ITX} + \text{H}^+ + \text{BPSK}^{\bullet-}$$
$$^3\text{ITX} + \text{BPSK} \rightarrow \text{ITX} + {}^3\text{BPSK}$$
$$^3\text{BPSK} \rightarrow \text{radicals}$$
$$^3\text{ITX} + \text{BPSK} \rightarrow \text{ITX}^{\bullet-} + \text{BPSK}^{\bullet+} \rightarrow \text{radicals, by - side reactions} \quad (10.3)$$

10.1.5
Dye/Amine/Triazine Derivative

The mechanism is still roughly similar when using Rose Bengal (RB). The protonated semireduced form of the dye RBH• formed by the dye/amine AH electron transfer is deactivated by the triazine derivative (Tz) (Eq. (10.4)) [20, 21]. Moreover, a dye/Tz electron transfer competes with the usual dye/AH interaction, thereby rendering the mechanism more complex.

$$^{1,3}\text{RB} + \text{AH} \rightarrow \text{A}^\bullet + \text{RB} - \text{H}^\bullet$$
$$\text{RB} - \text{H}^\bullet + \text{Tz} \rightarrow \text{RB} - \text{H}^+ + \text{Tz}^{\bullet-} \rightarrow \text{RB} + \text{H}^+ + \text{Tz}^{\bullet-} \quad (10.4)$$

Other various dyes have been checked, such as eosin, erythrosine, acriflavine, acridine orange, safranin, phenosafranin, and fluorone derivatives.

10.2
Other Mechanisms

Other three- or four-component systems involve more or less different mechanisms (some of them have been already described in earlier works [22] and references therein).

10.2.1
Triazine-Derivative-Containing Three-Component Systems

Other triazine-based systems work according to a mechanism different from that shown in Section 10.1.5. For example, excellent results were achieved using the pyrromethene dye (PYR) (**69**, Chapter 9)/amine AH (e.g., EDB)/1,3,5-triazine derivative (Tz) system whose mechanism is described in Eq. (10.5) [23]. An aminoalkyl $A^\bullet$ radical and a triazine-derived radical $Tz^\bullet_{(-Cl)}$ are formed. The beneficial effect of Tz in this system results from the secondary reactions that regenerate the dye and lead to new $A^\bullet$ and $Tz^\bullet_{(-Cl)}$ radicals.

$$^{1,3}PYR + AH \rightarrow PYR^{\bullet-} + AH^{\bullet+}$$
$$^{1,3}PYR + Tz \rightarrow PYR^{\bullet+} + Tz^{\bullet-}$$
$$PYR^{\bullet-} + AH^{\bullet+} \rightarrow A^\bullet + H^+ + PYR$$
$$PYR^{\bullet-} + Tz \rightarrow PYR + Tz^{\bullet-}$$
$$Tz^{\bullet-} \rightarrow Tz^\bullet_{(-Cl)} + Cl^-$$
$$PYR^{\bullet+} + AH \rightarrow AH^{\bullet+} + PYR \rightarrow A^\bullet + H^+ + PYR \quad (10.5)$$

The mechanism involved in cyanine dye (Cy^+)/butyltriphenylborate anion (Bo^-)/1,3,5-triazine halogenated derivative (Tz) has also been described in [24, 25]. Electron transfer occurs between the cyanine dye and the borate (Eq. (10.6)). Then, a reaction between the cyanine radical and Tz regenerates the dye and forms a $Tz^{\bullet-}$ radical anion that rapidly fragments into an initiating triazinyl radical and a halogen anion.

$$Cy^+ + Bo^- \rightarrow Cy^\bullet + Bo^\bullet$$
$$Cy^\bullet + Tz \rightarrow Cy^+ + Tz^{\bullet-} \rightarrow Cy^+ + Tz^\bullet_{(-Cl)} + Cl^- \quad (10.6)$$

10.2.2
Dye/Ketone/Amine

The dye (crystal violet, phenosafranine, methylene blue, thiopyronine)/amine/ketone (acetophenone, benzophenone, thioxanthone, 4,4′-bis(dimethylamino) benzophenone) combination was reported many years ago. The dye–ketone interaction is complex and depends on the considered couple [26].

10.2.3
Dye/Amine/Metal Salt

The processes involved in the dye (methylene blue (MB))/amine AH/cobalt salt (Co acetylacetonate, hexammine cobalt) systems have also been described [27]. They involve a redox cycle reaction between the reduced leuco dye and the cobalt salt

(Eq. (10.7)). The dye and the cobalt salt are regenerated, and oxygen is consumed.

$$MB^* + AH \rightarrow MBH^\bullet + A^\bullet$$
$$MBH^\bullet + Co\ (III)complex \rightarrow MB + Co\ (II)\ complex + H^+$$
$$Co\ (II)\ complex + O_2 \rightarrow \rightarrow Co\ (III)\ complex + \ldots \quad (10.7)$$

10.2.4
Dye/Borate/Additive

The additive can be a triazine derivative (see above), borate, or thiol. In the dye (carbocyanine)/butyltriphenylborate anion/methyl picolinium salt system (Scheme 10.3) [28], electron transfer between the dye (D^+) and the borate (as electron donor) takes place. The formed boranyl radical cleaves into a butyl radical and borane. The C–B bond in such boranyl radicals are known to dissociate within ~250 fs [29]. The dye radical undergoes an electron transfer with the methyl picolinium cation (as electron acceptor). A radical is thus generated. Then, a fast C–O cleavage yields the alkoxyl radical.

The same kind of mechanism occurs in dye (styrylbenzothiazolinium derivatives)/butyltriphenylborate anion/alkoxypyridinium salt systems [30]. After the primary electron transfer, the dye radical/alkoxypyridinium cation generates an alkoxy pyridinyl radical where a subsequent fast N–O cleavage yields an alkoxyl radical.

The cyanine dye/borate salt/heteroaromatic thiol system allows the design of high-speed photopolymers [31].

10.2.5
Photosensitizer/Cl-HABI/Additive

The decomposition of Cl-HABI derivatives can be produced by a photoinduced electron-transfer process. Efficient photosensitizers under exposure at 488 or 514 nm (Ar^+ laser), 568 nm (Kr^+ laser), or 632.8 nm (HeNe laser line) are based

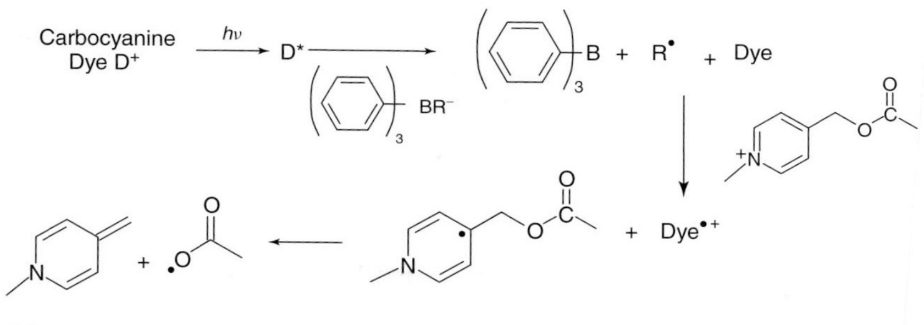

Scheme 10.3

on julolidine derivatives [32], for example, JAW1 and JAW2 julolidine derivative (JAW) in 2.

[Chemical structures of JAW1, JAW2, and NAS]

2

The photosensitization of the Cl-HABI decomposition in the presence of an aminostyryl dye such as a diethylamino styryl naphto thiazole derivative NAS (2) was largely studied and discussed [33, 34]. The mechanism is based on a primary electron transfer allowing the further generation of lophyl radicals (Eq. (10.8)); by-side reactions occur. The NAS/Cl-HABI/thiol, as well as the JAW/Cl-HABI/thiol system, is very efficient for visible laser-induced photopolymerization at 488 nm.

$$JAW + Cl-HABI \rightarrow JAW^{\bullet+} + L_2^{\bullet-} \ (h\nu) \quad (1)$$
$$NAS + Cl-HABI \rightarrow NAS^{\bullet+} + L_2^{\bullet-} \ (h\nu) \quad (2)$$
$$\text{and } L_2^{\bullet-} \rightarrow L^{\bullet} + L^- \quad (10.8)$$

The excitation of diethyl amino benzophenone (EAB) in the presence of Cl-HABI also leads to L$^{\bullet}$ (through the fast cleavage of the Cl-HABI radical anion (Eq. (10.9)). The ΔG is calculated as -0.38 eV; a small contribution of energy transfer cannot be totally excluded. The reaction between the lophyl anion ($E_{ox} = 0.29$ V) and the EAB radical cation generates a L$^{\bullet}$ radical ($\Delta G = -0.66$ eV). The dissociation quantum yield is close to that observed under the direct excitation of Cl-HABI: $\Phi_{Rad} \sim 2$ [35].

$$^3EAB + Cl-HABI \rightarrow EAB^{\bullet+} + L_2^{\bullet-} \rightarrow EAB^{\bullet+} + L^{\bullet} + L^-$$
$$EAB^{\bullet+} + L^- \rightarrow EAB + L^{\bullet} \quad (10.9)$$

The reaction mechanisms have been also explored in the coumarin or KC/Cl-HABI/mercaptobenzoxazole system [36, 37]. The KC triplet state undergoes a fast electron transfer with Cl-HABI ($k = 1.9\ 10^9\ \text{mol}^{-1}\text{s}^{-1}$. For example (Eq. (10.10)), when the sensitizer is coumarin C1 (**46** and **47**, Chapter 9), an electron transfer occurs in the C1-excited singlet state ($k = 2.3\ 10^{10}\ \text{mol}^{-1}\text{s}^{-1}$ in acetonitrile; $\Delta G = -0.73$ eV). The formed Cl-HABI radical anion leads to a lophyl radical, which then reacts through hydrogen transfer with a thiol RSH. The C1/RSH interaction can also participate in the initiation process. The role of new thiols and disulfides has been recently studied [38]. The C1/Cl-HABI/tetrazolethiol derivative system exhibits a higher reactivity than that of the photosensitizer/Cl-HABI/N-phenylglycine commonly used systems.

$$C1 + Cl-HABI\ (L_2) \rightarrow C1^{\bullet+} + L^{\bullet} + L^{-}\ (h\nu) \quad (1)$$
$$C1 + RSH \rightarrow C1^{\bullet-} + RSH^{\bullet+}\ (h\nu)$$
$$\text{and } RSH + L^{\bullet} \rightarrow RS^{\bullet} + L-H \quad (2) \quad (10.10)$$

10.2.6
Metal Carbonyl Compound/Silane/Hydroperoxide

The addition of a hydroperoxide such as cumene hydroperoxide CumOOH to a suitable metal carbonyl (e.g., (3))/silane (tris(trimethyl)silylsilane (TTMSS) (22, Chapter 9)) combination [39] leads to a significant improvement in the polymerization profiles. The absorption spectra of the metal carbonyls have maxima located in the UV (315–388 nm) and spread over the visible range; for $Cp_2Mo_2(CO)_6$, the molar extinction coefficient is 750 mol^{-1} cm^{-1} at 532 nm. Tungsten derivatives can also decompose hydroperoxides thanks to the formation of, for example, the $CpW(CO)_3^{\bullet}$ radical [40].

3

The metal carbonyl/TTMSS systems basically work as those described in Section 9.3.8: (i) cleavage of the metal–metal bond in the Mo and Ru derivatives or CO loss in the Fe and Cr derivatives and (ii) hydrogen abstraction with the silane. The ability of the formed Mo- or Ru-centered radicals and the transient $ArCr(CO)_2$ or $Fe_2(CO)_8$ species to decompose both the hydroperoxides (that coexist in the irradiated aerated medium) and the introduced CumOOH allows generation of the additional radicals (Eq. (10.11)).

$$Cp_2Mo_2\ (CO)_6 \rightarrow 2\ CpMo\ (CO)_3^{\bullet}\ (h\nu)$$
$$CpMo\ (CO)_3^{\bullet} + CumOOH\ (and\ ROOH) \rightarrow CumO^{\bullet}\ (and\ RO^{\bullet})$$
$$+ CpMoOH\ (CO)_3$$
$$CumO^{\bullet}\ (and\ RO^{\bullet}) + R_1R_2R_3Si-H \rightarrow CumOH\ (and\ ROH) + R_3Si^{\bullet}$$
$$(10.11)$$

As shown in Figure 10.1, it can be noted that the $Cp_2Mo_2(CO)_6$/TTMSS/ROOH system exhibits a good initiating ability under air contrary to $Cp_2Mo_2(CO)_6$/TTMSS [39].

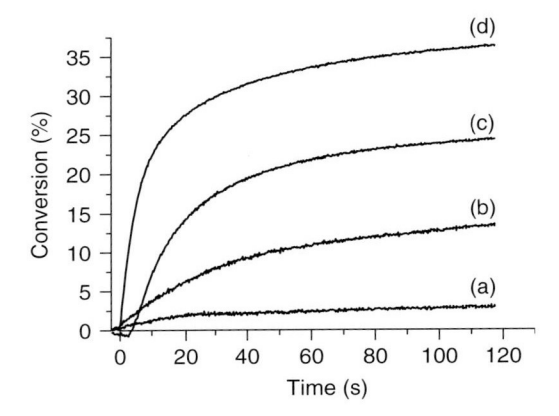

Figure 10.1 Photopolymerization profiles of TMPTA on a diode laser irradiation at 473 nm in the presence of (a) $Cp_2Mo_2(CO)_6$ (1% w/w), (b) $Cp_2Mo_2(CO)_6$/TTMSS (1%/3% w/w), (c) $Cp_2Mo_2(CO)_6$/CumOOH (1%/3% w/w), and (d) $Cp_2Mo_2(CO)_6$/CumOOH/TTMSS (1%/3%/3% w/w). In laminates. (Source: From [39].)

10.2.7
Photosensitizer/Amine/HABI Derivatives/Onium Salt

Another example concerns the EAB/amine/Cl-HABI/onium salt Ph_2I^+ combination. The rather complex mechanism has been recently discussed [35, 36, 41, 42]. Two kinds of initiation processes can arise: (i) the usual direct photoinitiation process based on Cl-HABI and (ii) the photosensitized initiation process in which EAB acts as a photosensitizer and Cl-HABI as a PI. For example, under a monochromatic light exposure at 366 nm, EAB absorbs 10 times more light than Cl-HABI (for an overall absorption of 90%).

The observed processes are described in Eq. (10.12). In Cl-HABI/NPG, one could expect an electron transfer between Cl-HABI and NPG, which results in the generation of a radical ion pair and then in the formation of an amine-derived radical and a lophyl radical. This bimolecular process, however, should not compete efficiently with the fast generation of $L^{\bullet}$ ($\sim 5 \times 10^8$ s^{-1}) followed by the formation of $A^{\bullet}$ through the $L^{\bullet}$/NPG interaction. In EAB/NPG, the mechanism involves a well-known electron transfer followed by a proton transfer (2): a $NPG_{-H}^{\bullet}$ radical is formed. The ^{3}EAB/Ph_2I^+ interaction leads to a decomposition of the iodonium salt (1). The ^{3}EAB/Cl-HABI interaction is very strong and is mostly ascribed to an electron-transfer process (3). Moreover, the $NPG^{\bullet}$/Cl-HABI and $L^{\bullet -}$/$EAB^{\bullet +}$ (or Ph_2I^+) interactions lead to an additional production of $L^{\bullet}$ radical. New $NPG_{-H}^{\bullet}$ radicals are also generated from the $EAB^{\bullet +}$/NPG interaction. Lastly, the ketyl radical of EAB (EAB-H$^{\bullet}$) are scavenged in the reactions with Cl-HABI and Ph_2I^+. The efficiency of the polymerization photoinitiation step is strongly dependent on the true rate constants in the monomer medium (and not necessarily on those obtained in solution; Section 5.7.3 and Chapter 17) and the irradiation conditions (wavelengths and intensity).

$$^3EAB + Ph_2I^+ \rightarrow EAB^{\bullet+} + Ph^\bullet + PhI \quad (1)$$
$$^3EAB + NPG \rightarrow EAB - H^\bullet + NPG_{-H}{}^\bullet \quad (2)$$
$$^3EAB + Cl\text{-}HABI \rightarrow EAB^{\bullet+} + L_2^{\bullet-} \quad (3a)$$
$$^3EAB + Cl - HABI \rightarrow EAB + 2L^\bullet \quad (3b)$$
$$EAB^{\bullet+} + NPG \rightarrow EAB + H^+ + NPG^\bullet_{-H}$$
$$L_2^{\bullet-} \rightarrow L^\bullet + L^-$$
$$L_2^{\bullet-} + Ph_2I^+ \rightarrow L_2 + Ph^\bullet + PhI$$
$$L_2^{\bullet-} + EAB^{\bullet+} \rightarrow L_2 + EAB$$
$$L^\bullet + NPG \rightarrow LH + NPG^\bullet_{-H}$$
$$EAB - H^\bullet + L_2 \rightarrow EAB + L_2^{\bullet-} + H^+ \quad (10.12)$$

10.2.8
Dye/Ferrocenium Salt/Hydroperoxide

The irradiation of the dye (D)/ferrocenium salt Fc(+) (75, Chapter 9)/hydroperoxide ROOH system yields a neutral radical structure Fc(0) through electron transfer with the dye (Eq. (10.13)) [43]. Fc(0) has a low ionization potential (4–5 eV) and easily reacts with oxygen to form the oxygen super anion as discussed above. Hydroperoxyl and peroxyl radicals are then generated.

$$D + Fc(+) \rightarrow D^{\bullet+} + Fc(0)$$
$$Fc(0) + O_2 \rightarrow \rightarrow O_2^{\bullet-}$$
$$Fc(0) + ROOH \rightarrow OH^- + RO^\bullet + Fc(+)$$
$$O_2^{\bullet-} + H^+ \rightarrow HO_2^\bullet \rightarrow \text{and } HO_2^\bullet + ROOH \rightarrow O_2 + RO^\bullet + H_2O \quad (10.13)$$

10.2.9
Coumarin/Amine/Ferrocenium Salt

In (keto)coumarin/amine/ferrocenium salt Fc(+) systems, the ferrocenium salt plays a crucial role, which is rather complex. In a combination [44, 45] consisting of coumarin C6 (46 and 47, Chapter 9) or a KC (1–5, Chapter 9), an iron arene complex such as Fc'(+) (where the isopropyl phenyl in (75, Chapter 9) is changed for chlorophenyl) and cyano-N-phenylglycine CPG as an amine, the first reaction (Eq. (10.14)) between C6 or KC and Fc'(+) is likely an electron transfer yielding Fc'(0). Owing to its radical structure, Fc'(0) reacts with CPG through a hydrogen abstraction. Therefore, in C6/Fc'(+)/CPG, no detrimental ketyl radicals are formed as the coumarin/amine interaction is not competitive. In KC/Fc'(+)/CPG, the ketyls formed in the KC/CPG interaction are likely scavenged by Fc'(+).

$$^1C6 + Fc'(+) \rightarrow C6^{\bullet+} + Fc'(0)$$
$$^3KC + Fc'(+) \rightarrow KC^{\bullet+} + Fc'(0)$$
$$Fc'(0) + CPG \rightarrow Fc'(0)_{+H} + \text{aminoalkyl radical}$$
$$KC^{\bullet} + Fc'(+) \rightarrow KC^+ + Fc'(0) \quad (10.14)$$

Changing the η^6-chlorobenzene ligand for η^6-isopropylbenzene or η^6-hexamethylbenzene rather leads to a C6/Fc'(+) or KC/Fc'(+) energy transfer. The generation of the aminoalkyl radical remains unclear.

10.2.10
Dye/Ferrocenium Salt/Amine/Hydroperoxide

The complexity of the mechanisms involved in dye (RB)/ferrocenium salt Fc(+) (75, Chapter 9)/amine AH/hydroperoxide ROOH system [43, 46] increases. This system is particularly efficient in the curing of thick and heavily pigmented paints in an industrial UV drying line. Final coating thicknesses of 300 µm (white), 400 µm (red), 300 µm (sky blue), and 150 µm (yellow, green, beige, or black) were obtained [47].

The direct and sensitized photolysis of the dye and the salt in the presence of oxygen, amine, and hydroperoxide has been investigated (Eq. (10.15)). In a general way, as above, the primary process is an electron transfer between the excited dye and Fc(+) or inside a ground-state intramolecular ion pair complex [RB...Fc(+)]. Then, the reactions of the Fc(0) radical structure (formed *in situ* on the ferrocenium salt and which behaves as a 17, 18, or 19 electron complex) with the amine, oxygen, and hydroperoxide lead to radicals. Oxygen is thus eliminated. Thermal as well as dark reactions participate in the polymerization. This allows explaining the body cure of thick samples where the light cannot deeply penetrate.

$$RB \rightarrow\,^1RB\,(h\nu) \text{ and } ^1RB \rightarrow\,^3RB$$
$$RB + Fc(+) \rightarrow [RB\ldots Fc(+)] \text{ and } [RB\ldots Fc(+)] \rightarrow RB^{\bullet+} + Fc(0)\,(h\nu)$$
$$^{1,3}RB + Fc(+) \rightarrow RB^{\bullet+} + Fc(0)$$
$$^{1,3}RB + AH \rightarrow RB^{\bullet-} + A^{\bullet} + H^+$$
$$^3RB + O_2 \rightarrow RB +\,^1O_2$$
$$Fc(+) +\,^1O_2 \rightarrow\rightarrow Fe^{3+}$$
$$Fc(0) + ROOH \rightarrow RO^{\bullet} + OH^- + Fc(+)$$
$$Fc(0) + AH \rightarrow\rightarrow A^{\bullet}$$
$$Fc(0) \rightarrow\rightarrow Fe^{2+}\,(\Delta)$$
$$Fe^{2+} + O_2 \rightarrow Fe^{3+} \text{ and } Fe^{2+} + ROOH \rightarrow RO^{\bullet} + OH^- + Fe^{3+}$$
$$Fe^{3+} + ROOH \rightarrow ROO^{\bullet} + H^+ + Fe^{2+}$$
$$O_2^{\bullet-} + RH \rightarrow HO_2^{\bullet} + R^-$$
$$HO_2^{\bullet} + ROOH \rightarrow RO^{\bullet} + H_2O + O_2 \quad (10.15)$$

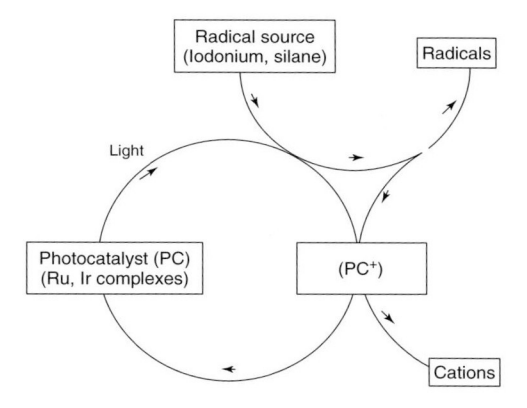

Scheme 10.4

10.2.11
Organometallic Compound/Silane/Iodonium Salt

A new concept summarized in Scheme 10.4 was recently introduced for the design of efficient dual radical/cationic photoinitiating systems on soft and visible light irradiations under air. The first step corresponds to the excitation of a suitable starting PI, called here a photocatalyst (PC) (usually a ruthenium or an iridium complex) by a visible light. The excited PI then reacts with a radical source to generate a radical and an oxidized form PC^+. In the second step, an interaction between $R^\bullet$ and PC^+ leads to a cation Cat^+ and regenerates the PC (PC^+ must be a good oxidation agent). If $R^\bullet$ and Cat^+ are efficiently and rapidly formed, the system should be operative for free radical photopolymerization (FRP) using an acrylate or FRPCP using an epoxide as well as for the manufacture of an acrylate/epoxide network or IPN (Section 12.5.2.2).

This concept was illustrated in three-component systems where PC is (i) tris(bipyridine)ruthenium(II) $Ru(bpy)_3^{2+}$ [48], $Ru(phen)_3^{2+}$ (where phen stands for a phenanthroline ligand) [49], tris(2-phenylpyridine)iridium $Ir(ppy)_3$ [50], and new iridium complexes $Ir(ligand)_3$ [51] as PIs and (ii) a silane (TTMSS) and a diphenyl iodonium salt as the radical sources. More details are provided in the section titled Photoinitiator/Silane/Iodonium Salt in Chapter 12.

10.3
Type II Photoinitiator/Silane: Search for New Properties

As in Type I PI/additive systems (Section 9.3.15), a strong improvement is noted under air when a silane or a germane derivative is added to a Type II PI [52]. Figure 10.2 shows the additive effect of TTMSS in a BP/amine system for the polymerization of TMPTA under air. A significant enhanced efficiency is observed using this three-component system. The Si–H conversion is high. Similar behaviors are also encountered in other Type II PIs based on ITX, BPSK (**21**, Chapter 8),

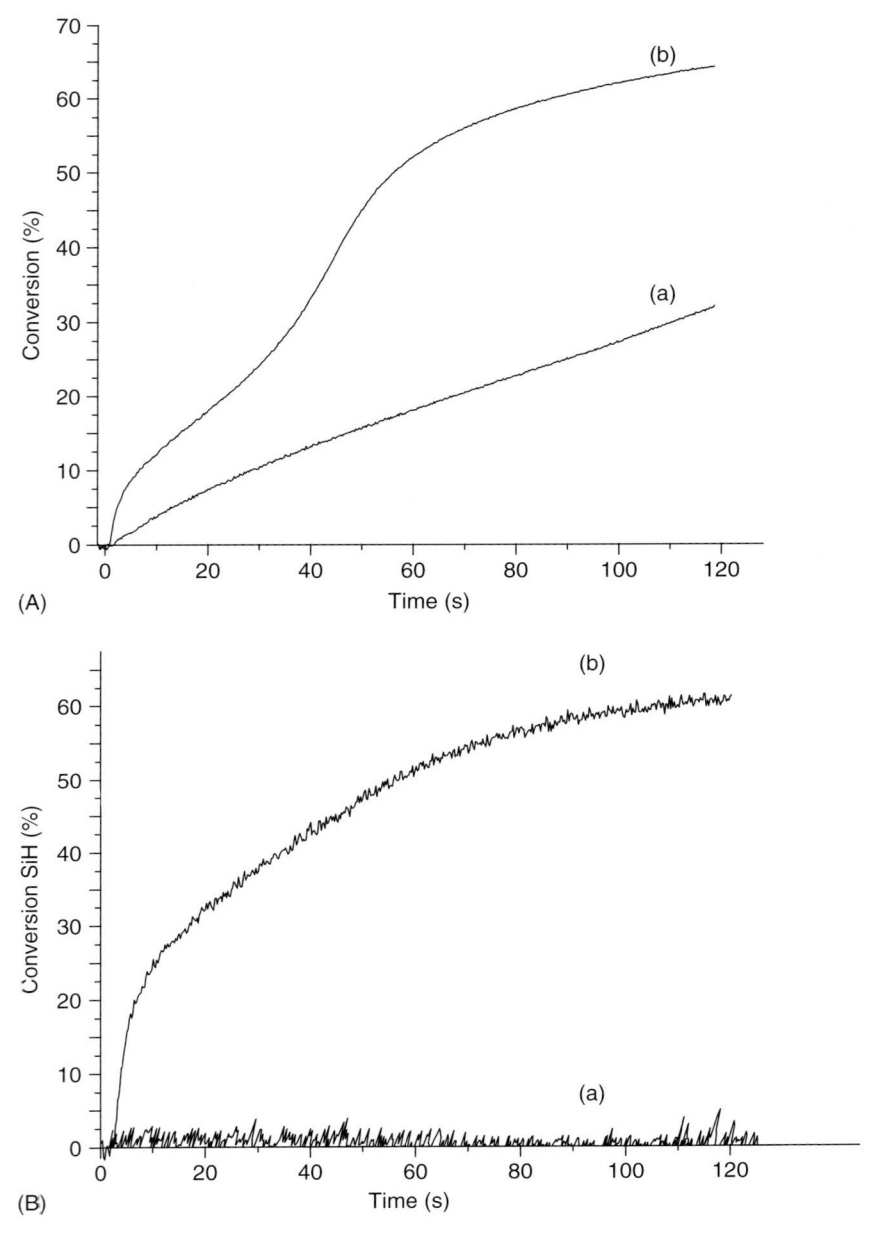

Figure 10.2 (A) Conversion versus time curves for the photopolymerization of TMPTA under air in the presence of (a) BP/ethyldimethylaminobenzoate (EDB) (1%/3% w/w) and (b) BP/EDB/TTMSS (**22**, Chapter 9) (1%/1%/2% w/w). (B) Conversion of the Si–H functions using (a) and (b). $I_0 = 44$ mW cm^{-2}; sample thickness = 20 μm. (Source: From [52].)

benzil, camphorquinone, and eosin-Y. These results demonstrate the interest of the amine/R_3XH combination in Type II PI where (i) the amine is used to get a highly efficient hydrogen abstraction and favors the initiation process and (ii) R_3XH behaves as a high-performance additive as discussed in Eq. 9.31 for Type I PI (conversion of peroxyls into new initiating structures).

10.4
Miscellaneous Multicomponent Systems

Violanthrone/silylamine/iodonium salt [53] can be relatively efficient in photopolymerization under red lights. Phosphine oxide/borane (L–BH_3) in the presence or absence of a silane can also be interesting systems for photopolymerization reactions in aerated conditions (J. Lalevée, unpublished data).

Other complex four-component systems have also been designed. Interesting results using, for example, eosin/titanocene/triazine/CBr_4 [54] were obtained. No mechanistic support was proposed.

References

1. Fouassier, J.P. and Lalevée, J. (2012) RSC Advances, ASAP. doi: 10.1039/c2ra00892k
2. Urano, T., Ito, H., and Yamaoka, T. (1999) Polym. Adv. Technol., **10**, 321–328.
3. Harada, M., Kawabata, M., and Takimoto, Y. (1989) J. Photopolym. Sci. Technol., **2**, 199.
4. Fouassier, J.P., Erddalane, A., Morlet-Savary, F., Sumiyoshi, I., Harada, M., and Kawabata, M. (1994) Macromolecules, **27**, 3349–3356.
5. Fouassier, J.P., Ruhlmann, D., Takimoto, Y., Harada, M., and Kawabata, M. (1993) J. Polym. Sci., Part A: Polym. Chem., **31**, 2245–2248.
6. Fouassier, J.P. and Chesneau, E. (1991) Makromol. Chem., **192**, 1307–1315.
7. Padon, K.S. and Scranton, A.B. (2000) J. Polym. Sci., Part A: Polym. Chem., **38**, 3336–3346.
8. Kim, K. and Scranton, A. (2004) J. Polym. Sci., Part A: Polym. Chem., **42**, 5863–5871.
9. Gomez, M.L., Avila, V., Montejano, H.A., and Previtali, C.M. (2003) Polymer, **44**, 2875–2881.
10. Gomez, M.L., Montejano, H.A., del Valle Bohorquez, M., and Previtali, C.M. (2004) J. Polym. Sci., Part A: Polym. Chem., **42**, 4916–4920.
11. Erddalane, A., Fouassier, J.P., Morlet-Savary, F., and Takimoto, Y. (1996) J. Polym. Sci., Part A: Polym. Chem., **34**, 633–642.
12. Gomez, M.L., Previtali, C.M., and Montejano, H.A. (2007) Polymer, **48**, 2355–2361.
13. Hassoons, S. and Neckers, D.C. (1995) J. Phys. Chem., **99**, 9416–9421.
14. Timpe, H.J., Ulrich, S., Decker, C., and Fouassier, J.P. (1994) Eur. Polym. J., **30** (11), 1301–1307.
15. Cavitt, T.B., Phillips, B., Hoyle, C.E., Pan, B., Hait, S.B., Viswanathan, K., and Joensson, E.S. (2004) J. Polym. Sci. Part A: Polym. Chem., **42**, 4009–4015.
16. Cavitt, T.B., Hoyle, C.E., Kalyanaraman, V., and Jonsson, S. (2004) Polymer, **45** (4), 1119–1123.
17. Hoyle, C.E., Viswanathan, K., Clark, S.C., Miller, C.W., Nguyen, C., Jonsson, S., and Shao, L. (1999) Macromolecules, **32**, 2793–2795.
18. Mullings, M.E., Walker, C.D., McDonough, A., and Cavitt, T.B. (2007) J. Coat. Technol. Res., **4**, 255–261.
19. Fouassier, J.P., Allonas, X., Lalevée, J., and Visconti, M. (2000) J. Polym. Sci., Part A: Polym. Chem., **38**, 4531–4541.

20. Grotzinger, C., Burget, D., Jacques, P., and Fouassier, J.P. (2001) *Macromol. Chem. Phys.*, **202**, 3513–3522.
21. Grotzinger, C., Burget, D., Jacques, P., and Fouassier, J.P. (2003) *Polymer*, **44**, 3671–3677.
22. Fouassier, J.P. (1995) *Photoinitiation, Photopolymerization, and Photocuring*, Hanser, Münich.
23. Tarzi, O.I., Allonas, X., Ley, C., and Fouassier, J.P. (2010) *J. Polym. Sci., Polym. Chem.*, **48**, 2594–2603.
24. Kabatc, J., Zasada, M., and Paczkowski, J. (2007) *J. Polym. Sci., Part A: Polym. Chem.*, **45**, 3626–3636.
25. Kabatc, J., Czech, Z., and Kowalczyk, A. (2011) *J. Photochem. Photobiol., A: Chem.*, **219**, 16–25.
26. Jockusch, S., Timpe, H.J., Schnabel, W., and Turro, N.J. (1997) *J. Phys. Chem.*, **101**, 440–445.
27. Yang, J. and Neckers, D.C. (2004) *J. Polym. Sci., Part A: Polym. Chem.*, **42**, 3836–3841.
28. Paczkowski, J. and Neckers, D.C. (1991) *Macromolecules*, **24**, 3013–3019.
29. Kabatc, J. (2010) *Polymer*, **51**, 5028–5036.
30. Kabatc, J. and Paczkowski, J. (2009) *J. Polym. Sci., Part A: Polym. Chem.*, **47**, 576–588.
31. Murphy, S.T., Chaofeng, Z., Miers, J.B., Ballew, R.M., Dlott, D.D., and Schuster, G.B. (1993) *J. Phys. Chem.*, **97**, 13152–13157.
32. Liu, A.D., Trifunac, A.D., and Krongrauz, V. (1992) *Chem. Phys. Lett.*, **198**, 200–205.
33. Urano, T., Tsurutani, Y., Ishikawa, M., and Itoh, H. (2000) *J. Photopolym. Sci. Technol.*, **13**, 83–88.
34. Urano, T. (2003) *J. Photopolym. Sci. Technol.*, **16**, 129–156.
35. Allonas, X., Fouassier, J.P., Kaji, M., and Murakami, S. (2003) *Photochem. Photobiol. Sci.*, **2**, 224.
36. Allonas, X., Fouassier, J.P., Kaji, M., Miyasaka, M., and Hidaka, T. (2001) *Polymer*, **42**, 7627–7634.
37. Allonas, X., Fouassier, J.P., Obeid, H., Kaji, M., and Ichihashi, Y. (2004) *J. Photopolym. Sci. Technol.*, **17**, 35–40.
38. Lalevée, J., Allonas, X., Zadoina, L., and Fouassier, J.P. (2007) *J. Polym. Sci., Polym. Chem.*, **45**, 2494–2502.
39. Llalevée, J., Tehfe, M.A., Gigmes, D., Blanchard, N., and Fouassier, J.P. (2010) *Macromolecules*, **43**, 6608–6615.
40. Zhu, Z. and Espenson, J.H. (1994) *Organometallics*, **13**, 1893–1898.
41. Allonas, X., Fouassier, J.P., Kaji, M., and Miasaka, M. (2000) *J. Photopolym. Sci. Technol.*, **13**, 237.
42. Allonas, X., Obeid, H., Fouassier, J.P., Kaji, M., Ichihashi, Y., and Murakami, Y. (2003) *J. Photopolym. Sci. Technol.*, **16**, 123–128.
43. Burget, D. and Fouassier, J.P. (1998) *J. Chem. Soc., Faraday Trans.*, **94**, 1849–1854.
44. Fouassier, J.P., Morlet-Savary, F., Yamashita, K., and Imahashi, S. (1996) *J. Appl. Polym. Sci.*, **62**, 1877–1885.
45. Fouassier, J.P., Morlet-Savary, F., Yamashita, K., and Imahashi, S. (1997) *Polymer*, **38**, 1415–1421.
46. Catilaz-Simonin, L. and Fouassier, J.P. (2001) *J. Appl. Polym. Sci.*, **79**, 1911–1923.
47. Catilaz, L., Fouassier, J.P., and Navergoni, P. (1995) *J. Radiat. Curing*, **21**, 10–12.
48. Lalevée, J., Blanchard, N., Tehfe, M.A., Chany, A.C., Morlet-Savary, F., and Fouassier, J.P. (2010) *Macromolecules*, **43**, 10191–10195.
49. Lalevée, J., Blanchard, N., Tehfe, M.A., Peter, M., Morlet-Savary, F., Gigmes, D., and Fouassier, J.P. (2011) *Polym. Chem.*, **2**, 1986–1991.
50. Lalevée, J., Peter, M., Blanchard, N., Tehfe, M.A., Peter, M., Morlet-Savary, F., and Fouassier, J.P. (2011) *Polym. Bull.*
51. Lalevée, J., Peter, M., Tehfe, M.A., Dumur, F., Gigmes, D., Morlet-Savary, F., and Fouassier, J.P. *Chem. Eur. J.*, **17**, 15027–15031.
52. El Roz, M., Lalevée, J., Allonas, X., and Fouassier, J.P. (2009) *Macromolecules*, **42**, 8725–8732.
53. Tehfe, M.A., Lalevée, J., and Fouassier, J.P., *Macromolecules*, in press.
54. Decker, C. and Elzaouk, B. (1997) *J. Appl. Polym. Sci.*, **65**, 833–844.

11
Other Photoinitiating Systems

11.1
Photoinitiator-Free Systems or Self-Initiating Monomers

The direct excitation of an acrylate-based formulation as a photoinitiator (PI)-free system requires a UV exposure around 200–220 nm and is rather a low-efficiency process.

On the contrary, charge-transfer systems allow an excitation at longer wavelengths (Section 4.2.6). For example, a monomer donor (MD) (e.g., styrene) undergoes an electron transfer with a monomer acceptor (MA) (e.g., maleic anhydride). Then, the donor radical cation and the acceptor radical anion can recombine, and a biradical is thus formed: the radical polymerization can start (Eq. (11.1)). In maleimide/dodecylvinyl ether, the electron transfer might be followed by a proton transfer that yields a radical on both compounds. These couples are designed as self-initiating monomers. The addition of a PI, however, often improves the polymerization efficiency.

$$MD + MA \rightarrow [MD\ldots MA]^* \, (h\nu)$$
$$[MD\ldots MA]^* \rightarrow MD^{\bullet+} + MA^{\bullet-} \rightarrow {}^\bullet MD - MA^\bullet \quad (11.1)$$

A review has been provided in [1]. In the past, a lot of work has been devoted to PI-free systems involving maleimide/vinyl ether, acrylate/vinyl ether, divinyl fumarates, diacryl amides, and thiol-containing monomers [2–12].

Recently, di- and tri-acryloylated hydroxylamine derivatives (**1**) [13] appear as a new class of self-initiated monomers having a N–O bond that can be easily cleaved under UV light at acceptable wavelengths around 300 nm. This kind of cleavage has been observed in keto-oxime (Scheme 8.10) and (thio)hydroxamic ester PI derivatives Eq. (8.10).

Photoinitiators for Polymer Synthesis: Scope, Reactivity and Efficiency, First Edition.
Jean-Pierre Fouassier and Jacques Lalevée.
© 2012 Wiley-VCH Verlag GmbH & Co. KGaA. Published 2012 by Wiley-VCH Verlag GmbH & Co. KGaA.

11.2
Semiconductor Nanoparticles

Irradiation of a titanium dioxide pigment leads [14] to an electron e^-–hole p^+ pair (Eq. (11.2)). Anatase, but not rutile, can sensitize the decomposition of a photoinitiating system such as the anthraquinone (fluorenone, eosin, benzil)/amine combinations. In fact, this can occur if the energy of the e^-–p^+ pair E_{ep} (anatase: 54 kcal mol^{-1}, rutile: 34 kcal mol^{-1}) is higher than that of the excited triplet state of the PI E_T. The energy of the e^-–p^+ pair must be sufficient to generate radical ions having an energy E_{ions}.

$$TiO_2 \rightarrow e^- + p^+ \; (h\nu)$$
$$e^- + PI \rightarrow PI^{\bullet -} \quad \text{and} \quad p^+ + PI \rightarrow PI^*$$
$$PI^* + AH \rightarrow PI^{\bullet -} + AH^{\bullet +} \rightarrow PIH^\bullet + A^\bullet$$
$$p^+ + AH \rightarrow AH^{\bullet +} \tag{11.2}$$

The redox reactions in the semiconductor pigment (e.g., TiO_2 or CdS) can be photosensitized by a dye. A further electron transfer reaction between the photoreduced semiconductor species and a nitro compound (e.g., *p*-nitrobenzyl halide) yields a radical anion that fragments to a *p*-nitrobenzyl radical [15]. The photopolymerization of methylmethacrylate in the presence of nanosized TiO_2 particles [16, 17] and colloidal CdS [18] has been also discussed.

Quantum-sized ZnO colloids that absorb at 335 nm are able to initiate the photopolymerization of methylmethacrylate according to Eq. (11.3)[19]. Iron oxides such as Fe_3O_4 can lead to the formation of a radical on an acrylic monomer, presumably through a similar mechanism [20].

$$ZnO \rightarrow e^- + p^+ \; (h\nu)$$
$$e^- + M \rightarrow M^{\bullet -} \quad \text{and} \quad p^+ + ROH \rightarrow RO^\bullet + H^+$$
$$M^{\bullet -} + H^+ \rightarrow M_{+H}^\bullet \text{ and } M_{+H}^\bullet + M \rightarrow M_{+H}M^\bullet \tag{11.3}$$

11.3
Self-Assembled Photoinitiator Monolayers

Various UV-light-cleavable structures that have been introduced onto silicon wafers or gold surfaces (Scheme 11.1) as reviewed in [21] work as supported PIs. This allows the generation of a radical on the photoinitiating surface assembly and a further grafting of a suitable monomer M (grafting from). Self-assembled monolayers are thus achieved on the surface and lead to surface brushes. When using a dithiocarbamate-photoiniferter-functionalized silicon, a good control of the grafted layer thickness up to 100 µm is obtained. Type II PI-based systems can also be used but the coinitiator-type structure has to be obviously linked to the substrate.

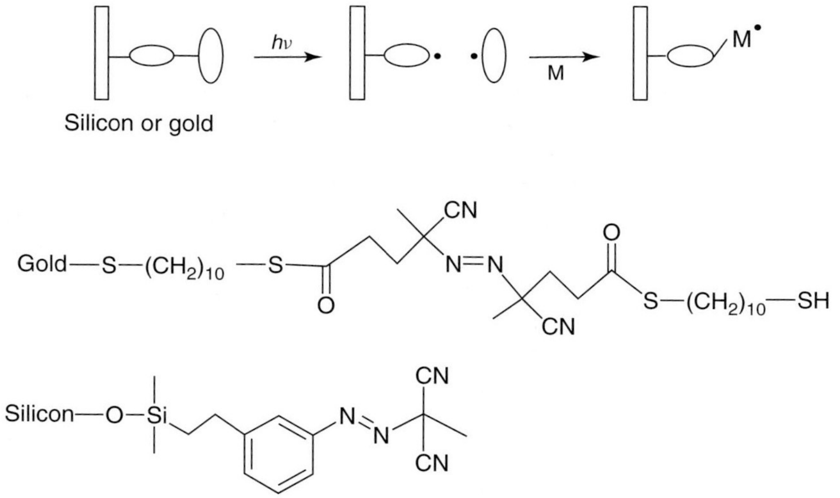

Scheme 11.1

References

1. Jonsson, E.S. and Hoyle, C.E. (2006) in *Photochemistry and UV Curing: New Trends* (ed. J.P. Fouassier), Research Signpost, Trivandrum, p. 165.
2. Hoyle, C.E., Joensson, E.S., Shimoze, M., Owens, J., and Sundell, P.E. (1997) *ACS Symp. Ser.*, **673**, 133–149.
3. Von Sonntag, J., Beckert, D., Knolle, W., and Mehnert, R. (1999) *Radiat. Phys. Chem.*, **55**, 609–613.
4. Pandey, R.B., Yang, D., Liu, Y., Hoyle, C.E., Jonsson, S., and Whitehead, J.B. (1999) *Polym. Prepr.*, **40**, 932–933.
5. Huang, X., Huang, Z., and Huang, J. (2000) *J. Polym. Sci., Part A: Polym. Chem.*, **38**, 914–920.
6. Cavitt, T.B., Hoyle, C.E., Nguyen, C., Viswanathan, K., and Jonsson, S. (2000) Proceedings of the RadTech, Baltimore, pp. 785–794.
7. Joensson, E.S., Lee, T.Y., Viswanathan, K., Hoyle, C.E., Roper, T.M., Guymon, C.A., Nason, C., and Khudyakov, I.V. (2005) *Prog. Org. Coat.*, **52**, 63–72.
8. Lee, T.Y., Guymon, C.A., Jonsson, E.S., Hait, S., and Hoyle, C.E. (2005) *Macromolecules*, **38**, 7529–7531.
9. Bongiovanni, R., Sangermano, M., Malucelli, G., and Priola, A. (2005) *Prog. Org. Coat.*, **53**(1), 46–49.
10. Zhang, X., Yang, J., Zeng, Z., Wu, Y., Huang, L., Chen, Y., and Wang, H. (2007) *Polym. Eng. Sci.*, **47**, 1082–1090.
11. Buchmeiser, M.R. (2008) *J. Polym. Sci., Part A: Polym. Chem.*, **46**, 4905–4916.
12. Karasu, F., Dworak, C., Kopeinig, S., Hummer, E., Arsu, N., and Liskra, R. (2008) *Macromolecules*, **41**, 7953–7958.
13. Dworak, C., Kopeinig, S., Hoffmann, H., and Liska, R. (2009) *J. Polym. Sci., Part A: Polym. Chem.*, **47**, 392–403.
14. Pappas, S.P. (1986) *UV-Curing: Science and Technology*, Technology Marketing Corporation, Stamford; Plenum Press, New-York (1992).
15. Eaton, D.F. (1986) in *Advances in Photochemistry*, vol. **13** (eds D.H. Hammond and G.S. Gollnick), John Wiley & Sons, Inc., New York, pp. 427–487.
16. Dong, C. and Ni, X. (2004) *J. Macromol. Sci., Pure Appl. Chem.*, **41**, 547–563.
17. Becker-Willinger, C., Schmitz-Stoewe, S., Bentz, D., and Veith, M. (2010) Micromachining and microfabrication

process technology XV, Proceedings of the SPIE, Vol. 7590, pp. 759001-1–759001-11.
18. Stroyuk, A.L., Sobran, I.V., Korzhak, A.V., Raevskaya, A.E., and Kuchmiy, S.Y. (2008) *Colloid Polym. Sci.*, **286**, 489–498.
19. Huang, Z., Barber, T., Mills, G., and Morris, M.B. (1994) *J. Phys. Chem.*, **98**, 12746–12751.
20. Sangermano, M., Vescovo, L., Pepino, N., Chiolerio, A., Allia, P., Tiberto, P., Coisson, M., Suber, L., and Marchegiani, G. (2010) *Macromol. Chem. Phys.*, **211**, 2530–2535.
21. Muftuogli, A.E., Tasdelen, M.A., and Yagci, Y. (2010) in *Photochemistry and Photophysics of Polymer Materials* (ed. N.S. Allen), John Wiley & Sons, Inc., Hoboken, pp. 509–540.

Part III
Nonradical Photoinitiating Systems

Apart the widely used radical photopolymerization reactions, ionic photopolymerizations (cationic and anionic) are also encountered. On an industrial ground, the anionic photoinitiation process is actually rather seldom described. Cationic photopolymerization is much more important, although noticeably less used than radical photopolymerization. Useful cationic photoinitiators (PIs) belong to three main classes: diazonium salts, onium salts, and organometallic complexes. Anionic PIs involves, for example, inorganic complexes or ferrocene derivatives. Acid and base photo-cross-linking reactions have also interesting applications in microelectronics and radiation curing. They require ionic and nonionic molecules that are sometimes designed as photolatent compounds as the acid or the base only appears under light exposure. They involve, for example, onium salts, iminosulfonates, or O-acyloximes.

Information can be found in many books and review papers [1–26]. This part presents all these ionic PIs, photoacids, and photobases and discusses their photochemical and chemical reactivity.

12
Cationic Photoinitiating Systems

12.1
Diazonium Salts

They generate a Lewis acid [27], for example, BF_3 (Eq. (12.1)). In the absence of hydrogen donor, BF_3 adds to the oxygen atom of the epoxy and starts the ring-opening polymerization reaction. In the presence of a hydrogen donor (e.g., an alcohol ROH), a proton and a $ROBF_3^-$ anion are formed. In that case, the proton is the initiating species. Diazonium salts suffer from severe drawbacks: poor shelf life, poor thermal and photochemical stability, and release of nitrogen (formation of bubbles and pinholes in the film).

$$Ar-N_2^+ \; BF_4^- \rightarrow Ar-F + N_2 + BF_3 \; (h\nu) \quad (12.1)$$

The efficiency of the diazonium salts depends on the substitution on the phenyl ring (which obviously controls the absorption of the molecule) and the nature of the anion (BF_4^-, PF_6^-, AsF_6^-, SbF_6^-, $SbCl_6^-$, and $FeCl_4^-$). Their photolysis has been largely discussed [28–30]. The sensitized decomposition of diazonium salts in the presence of various electron donors such as ketones, dyes, and hydrocarbons (HCs) [31] as well as the primary steps of the processes [32] and the generation of the aryl cations [33] have been presented.

12.2
Onium Salts

12.2.1
Iodonium and Sulfonium Salts

Synthesis, properties, and efficiency of iodonium and sulfonium salts are continuously reviewed [18–20, 22, 34–36].

12.2.1.1 Compounds

Iodonium and sulfonium salts represent the main and largely used class of cationic PIs [37, 38]. They consist of (i) a cationic moiety where a positively charged central atom such as iodine (in iodonium salts) or sulfur (in sulfonium salts) is usually linked to aromatic groups and (ii) a counter anion. Examples of commercially available products are shown in **1–7**. The thioxanthenium [39] and arylthianthrenium [40] structures are newly introduced compounds. Various anions such as BF_4^-, PF_6^-, SbF_6^-, and $B(C_6H_5)_4^-$ have been proposed. These salts are known as stable, nonhygroscopic, and efficient photoinitiators, acting as latent sources of cation radicals and Bröndstedt acids on light exposure [38, 41].

1–7

Both decreasing the migration of the unreacted PI and the photodecomposition products and keeping a high reactivity have led to the proposal of new structures in the sulfonium series [(R1)$_3$-Ar][(R2)$_3$-Ar][(R3)$_3$-Ar] S$^+$ where the R1 and R2 substituents are responsible for good light absorption and suitable solubility in organic matrixes (OrMs), respectively, and R3 is selected to avoid further harmful photoproducts after the S–C bond cleavage [42].

12.2.1.2 Photopolymerization Reaction

A cationic epoxide photopolymerization (cyclohexene oxide, for example) in the presence of an onium salt On$^+$ X$^-$, for example, occurs through (Scheme 12.1): (i)

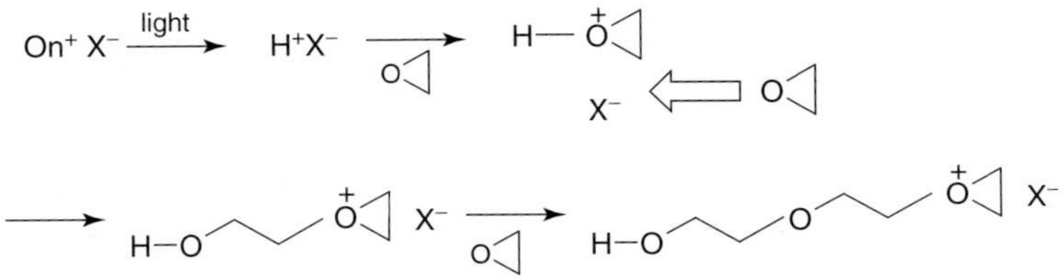

12.2 Onium Salts

Scheme 12.1

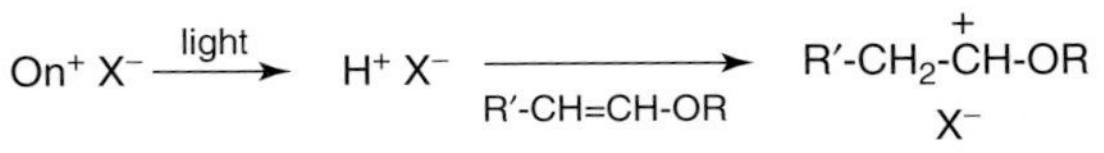

Scheme 12.2

generation of an acid from the decomposition of the onium salt, (ii) protonation of a monomer unit to form a secondary oxiranium ion, (iii) addition of a monomer to form a tertiary oxiranium ion, and then (iv) further addition of a monomer into the propagating ion pair (oxiranium cation/anion). This propagation step requires that the monomer enter into the ionic pair. As in thermal cationic polymerization, the nucleophilicity of the anion (which affects the electrostatic charge interaction in the ionic pair) is therefore expected to be a decisive factor for fast development of the propagation reaction.

In the cationic photopolymerization of vinyl ethers, the initiation step corresponds to the addition of the proton to the double bond, the carbon-centered cation being the propagating species (Scheme 12.2).

The overall polymerization rate is driven by the slowest reaction. In the case of epoxides, the kinetic behavior is strongly affected by the monomer structure. Cyclohexene oxide is highly reactive. Neopentylglycol diglycidyl ether exhibits a long induction period and then a rapid conversion. Glycidyl phenyl ether shows a shorter induction time and a slow polymerization rate. Glycidyl and glycidic esters have very poor reactivity. In the vinyl ether series, triethyleneglycol divinyl ether is highly reactive. These behaviors have recently been discussed and correlated [43] with the stability (and consequently, the reactivity) of the oxiranium ion intermediates: the highest reactivity is observed when these ions are strained and possess no other nucleophilic groups.

Termination of the polymerization occurs by reaction of the growing chain with the counteranion or nucleophiles present in the medium. Intramolecular cyclization or chain transfer reaction terminates a polymer chain but generates a new oxiranium center. The kinetics of a cationic photopolymerization reaction in fluid media has been described, and the derived models were able to fit the experimental data [44]. The termination/trapping rate constants for iodonium salts in the photopolymerization of glycidyl ether have been determined [45].

A typical feature of cationic photopolymerization is the presence of a substantial postcure effect [46] due to the well-known living character of this kind of reaction that has long been recognized for thermally initiated cationic polymerization. Compared

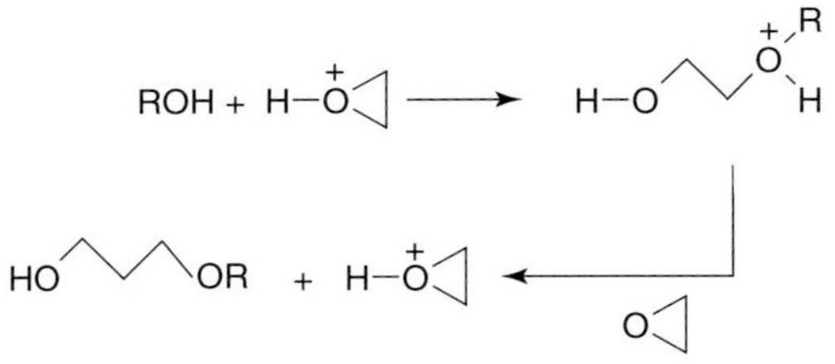

Scheme 12.3

with radical polymerization reactions, the temperature affects the polymerization rate [47]. In addition, photoinitiated cationic polymerizations are not affected by the presence of oxygen and radical impurities (except in free-radical-promoted cationic polymerization (FRPCP); see Section 12.5).

Inhibiting and retarding effects, chain transfer reactions, and termination reactions with bases, water [48, 49], or alcohols [50, 51] have been mentioned. These recent papers also discuss the role of water or alcohols (such as various benzyl alcohols) that can be sometimes beneficial in the photopolymerization of epoxide monomers (fivefold increase in the polymerization rates and final conversions increasing from 60 to 95%). Indeed, it was suggested that alcohols are responsible for an activated monomer mechanism (Scheme 12.3) in which they react with the propagating cation and generate a new oxonium cation that will add to a monomer unit. The apparent acceleration rate partly results from this chain transfer.

This effect might not be general as it depends on the initiating system. More particularly, the water or alcohol concentration has a dramatic effect, that is, acceleration of the polymerization rate at a relatively low content and inhibition at a high concentration. The optimum depends on the initiating system but can lie in the 0.05–0.5% range (in w/w) [52].

12.2.1.3 Role of the Anion

The progress of a cationic epoxide polymerization reaction is strongly affected by the anion. The degree of separation in the propagating ion pair (oxonium cation/anion) is dependent on both the size and the electron density of the anion. The larger the size, the lower is the nucleophilicity and the higher the propagation rate of the polymerization becomes. Nonnucleophilic anions allow excellent rates of polymerization, for example, the hexafluoroantimonate anion (which is the least nucleophilic) [47]; this anion (as well as AsF_6^-) is, however, toxic (see below). The reactivity order is typically: $SbF_6^- > AsF_6^- > PF_6^- > BF_4^-$ (e.g., the rates of polymerization of cyclohexene oxide in bulk are in an ~80/20/5/2 ratio). Salts bearing highly nucleophilic anions such as I^- and Cl^- are far less reactive. The low nucleophilic tetrafluoroborate or hexafluorophosphate anions are very often used, in particular, with low-reactivity monomers such as epoxides. When considering very highly reactive cationic monomers such as vinyl ethers, even highly nucleophilic triflate $CF_3SO_3^-$ or perchlorate anions can be used.

A real progress has been made with the development of a new bulky anion (tetrakis (pentafluorophenyl) borate) $B(C_6F_5)_4^-$, which exhibits low nucleophilicity

Table 12.1 Effect of the anion on the cure speed of cationic formulations (diphenyl iodonium salt; 15 μm thin layer on a polypropylene film; exposure energy: 54 mJ cm^{-2}).

Anion	Cure speed (m min^{-1}) Epoxide matrix	Cure speed (m min^{-1}) Epoxy-modified silicone matrix
B(C$_6$F$_5$)$_4^-$	$\geq 60^a$	$\geq 60^a$
SbF$_6^-$	$\geq 60^a$	20
AsF$_6^-$	20	5
PF$_6^-$	15	5
BF$_4^-$	5	2.5

^{a}One pass at 60 m min^{-1} is enough to get a tack-free coating.

and leads to a high propagation rate constant [53]. As a consequence (see Table 12.1), higher cure speeds are attained using Ph$_2$I$^+$ B(C$_6$F$_5$)$_4^-$ instead of Ph$_2$I$^+$SbF$_6^-$, which is already more efficient than Ph$_2$I$^+$PF$_6^-$ for the photopolymerization of both a usual epoxide cationic OrM and an epoxy-modified silicone (EMS) [54] under a mercury lamp fitted with a commercial conveyor. In an industrial UV line (120 W bulb Hg lamp), the cure speeds of EMS (45 μm layer on a PET film or a glassine paper) are > 200 m min^{-1} with Ph$_2$I$^+$B(C$_6$F$_5$)$_4^-$ versus <160 m min^{-1} with Ph$_2$I$^+$SbF$_6^-$. The propagation rate constants of OrM and EMS using these salts are in a 2.2 : 2.5 ratio. Oxidation potentials and reorganization energies of tetrakisaryl borates are known [55]. Tetrakis(pentafluorophenyl) gallate was also proposed as a counteranion [56–58].

12.2.1.4 Absorption Properties
In a general way, sulfonium salts exhibit a more intense absorption in the near-UV region than iodonium salts. The absorption maximum is located at 260 nm for the triphenyl sulfonium salt (and decreases until 300 nm) and at 230 nm only for the diphenyl iodonium salt (Figure 12.1).

Adequate structural modifications, however, give rise to red-shifted and enhanced absorptions. For example, the maximum absorption wavelengths and the molar extinction coefficients are change from 227 nm and 17 800 M^{-1} cm^{-1} to 265 nm and 55 000 M^{-1} cm^{-1}, respectively, when changing the diphenyliodonium salt for a phenyl fluorenone iodonium salt (whose extinction coefficient is still 745 M^{-1} cm^{-1} at 366 nm). The same behavior was observed for a lot of variously substituted triarylsulfonium salts [38]. Improved sensitivity to the near-UV lights was noted through the introduction of a 4-thiophenoxy chromophore onto triarylsulfonium salts (and further extension of this chemistry to di- or trisulfonium salts) [59], a 3-trimethylsilyl propoxy phenyl group [60], or a 9-oxo-9H-fluoren-2-yl-phenyl moiety [61]. Bistriarylsulfonium salts have an interesting absorption around 300–360 nm. Bisiodonium salts have been also proposed

[62]. Alkyl phenyl (9-phenyl thioanthracenyl)-10-sulfonium salts and arylbenzylmethyl sulfonium salt derivatives are sensitive up to 500 nm [63]. The absorption of thianthrenium or thioxanthenium structures **1–7** spreads up to the visible part of the spectrum.

12.2.1.5 Decomposition Processes of Iodonium Salts

12.2.1.5.1 Mechanism

The decomposition process is now well established and involves a cleavage of the C–I or C–S bond [6, 64]. The cleavage could be either heterolytic or homolytic and would form, for example, a singlet phenyliodide/phenyl cation pair or a phenyliodinium cation/phenyl radical, respectively. Then, an in-cage or/and an out-of-cage process can arise. In the latter process, something in the formulation must act as the hydrogen donor DH. In any case, a strong acid is generated.

Original studies first suggested a singlet state mechanism [65]. Further pulsed picosecond laser investigations also gave evidence for a bond scission process occurring after the formation of a homolytically dissociative triplet state on the picosecond scale [21, 66] in agreement with the potential energy diagram correlating the various excited states with the primary photolysis products. A dissociative charge transfer (CT) state involving an electron transfer between the cation and the counteranion in the singlet excited state (see below) was also considered. ESR and GCMS experiments [67] confirm that both singlet and triplet pathways are operative in the photolysis of iodonium salts.

A detailed plausible reaction mechanism is shown in Eq. (12.2) [19, 68–70]. In the diphenyl iodonium salt, the primary singlet state process can be described by a heterolytic or a homolytic cleavage. Both pairs have been considered as exhibiting

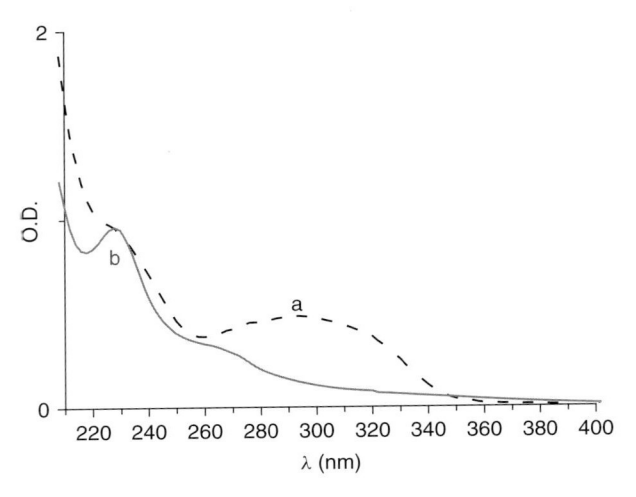

Figure 12.1 Typical absorption spectra of onium salts: (a) Ph−S−PhS$^+$Ph$_2$ in acetonitrile and (b) Ph$_2$I$^+$. (Source: Author's work.)

a comparable stability, and therefore, they can interconvert through an intrapair electron transfer. They are in equilibrium. However, this picture has been disputed [21] as the ionization potential of PhI is ~0.5 eV less than the electron affinity of Ph^+. When radical pairs are involved, the in-cage recombination process occurs in singlet pairs (where the two free electrons have opposite spins) but not in triplet pairs (spins are parallel). As a consequence, the out-of-cage process in the presence of RH that yields a proton, a radical, and phenyliodide is mostly observed when occurring from triplet pairs.

$$Ph_2I^+ \ X^- \rightarrow {}^1(Ph_2I^+ \ X^-)^* \ (h\nu)$$
$$^1(Ph_2I^+ \ X^-)^* \rightarrow {}^1[PhI^{+\bullet} \ Ph^\bullet] \rightarrow {}^1[PhI \ Ph^+]$$
$$^1[PhI^{+\bullet} \ Ph^\bullet] \rightarrow {}^3[PhI^{+\bullet} \ Ph^\bullet]$$
$$^1[PhI^{+\bullet} \ Ph^\bullet] \text{ or } {}^1[PhI \ Ph^+] \rightarrow Ph - PhI + H^+$$
$$^1[PhI^{+\bullet} \ Ph^\bullet] + DH \rightarrow PhI^{+\bullet} + Ph^\bullet + DH \rightarrow PhI + D^\bullet + Ph^\bullet + H^+$$
$$^1[PhI \ Ph^+] + DH \rightarrow PhI + Ph^+ + DH \rightarrow PhI + PhD + H^+$$
$$^3[PhI^{+\bullet} \ Ph^\bullet] \rightarrow {}^1[PhI^{+\bullet} \ Ph^\bullet]$$
$$^3[PhI^{+\bullet} \ Ph^\bullet] + DH \rightarrow PhI^{+\bullet} + Ph^\bullet + DH \rightarrow PhI + D^\bullet + Ph^\bullet + H^+$$
$$^1(Ph_2I^+ \ X^-)^* \rightarrow {}^3(Ph_2I^+ \ X^-)^* \rightarrow {}^3[PhI^{+\bullet} \ Ph^\bullet] \quad (12.2)$$

The singlet pairs generate a proton through (i) an out-of-cage process in the presence of a hydrogen donor DH (with the concomitant formation of a phenyl and a $D^\bullet$ radical) or (ii) an in-cage process that does not require the presence of DH and also forms PhPhI.

Intersystem crossing also occurs as a competitive pathway to the cleavage process. It leads to the formation of a triplet state followed by a homolytic cleavage: a triplet phenyliodinium radical/phenyl radical pair is thus formed. Interconversion of the triplet pair into a singlet pair is also feasible. This well-supported overall mechanism explains (i) the generation of H^+ even in the absence of DH and (ii) the excess of acid in the photolysis of diphenyl iodonium salts (the in-cage reactivity is ~20%) [20].

Scheme 12.3 was established in protic medium and aqueous acetonitrile [19]. In nonaqueous solvent such as in dry acetonitrile, the photodecomposition of the iodonium salt was proposed to occur through a CT mechanism, leading to a radical pair (Eq. (12.3)).

$$Ph_2I^+ \ X^- \rightarrow (Ph_2I^+ \ X^-)^* \ (h\nu)$$
$$(Ph_2I^+ \ X^-)^* \rightarrow [Ph_2I^\bullet \ X^\bullet] \rightarrow [PhI \ Ph^\bullet \ X^\bullet]$$
$$[PhI \ Ph^\bullet \ X^\bullet] \rightarrow PhI + PhX$$
$$[PhI \ Ph^\bullet \ X^\bullet] + CH_3CN \rightarrow PhI + PhH + XCH_2CN \quad (12.3)$$

The production of the phenyl radicals can play a role in free radical photopolymerization (see in Section 12.2 and three-component systems).

12.2.1.5.2 Photolysis

Decomposition quantum yields of iodonium salts on light exposure are usually high (e.g., 0.7 for the photolysis of $(C_6H_5)_2I^+AsF_6^-$ in acetonitrile) [38]. The H^+ formation is slightly affected by the solvent. For diphenyliodonium chloride, the quantum yield at $\lambda = 254$ nm is 0.76 in water and 0.73 in acetonitrile-water (3 : 7 v/v) [71].

The anion has almost no effect on the decomposition rate of the PI and the H^+ generation. The photolysis quantum yields of a diphenyl iodonium salt having Cl^-, Br^-, I^-, HSO_4^-, and BF_4^- as anions have been evaluated as 0.77, 0.7, 0.5, 1, and 0.51, respectively, (in acetonitrile/water 3 : 7 v/v) at $\lambda = 254$ nm [71]. An intrapair electron transfer between the cationic moiety and the halide (supported by the generation of haloacetonitrile in a nonaqueous solvent) has been invoked to account for this minor effect [72]. The anion effect on the cage-escape ratio is, however, much more pronounced. For example, in the photolysis of diphenyl iodonium salts at $\lambda = 254$ nm, the iodobiphenyls/iodobenzene ratio depends on both the anion and the solvent: 0.29 and 0.021 for PF_6^- and I^-, respectively, in acetonitrile but 0.35 and 0.21 for the same ions in acetonitrile/water [73]. New data have been reported in recent papers [74].

The substitution effects on the phenyl ring of diphenyl iodonium hexafluoroarsenate (e.g., by methyl groups in 3-, 4-, 3'-, 4'-, a methoxy group in 4- or nitro groups in 3-, 3'-position on the phenyl rings) introduce only slight changes in the acidity [75]. The relative efficiency of iodonium salts (containing a gallate anion) was also investigated [76].

12.2.1.5.3 Reaction of the Phenyl Iodinium Cation Radical

In acetonitrile, the phenyl iodinium cation radical absorbs around 660 nm and has a half lifetime ~ 6 μs (the decay is second order). It reacts with nucleophiles: the reaction rate constants with phenyliodide, water, and methanol are 75, 0.18, and 0.3×10^6 mol^{-1} l s^{-1}, respectively [41]. The interaction with phenyliodide explains the formation of an excess of acid relative to iodobenzene in acetonitrile (Scheme 12.4).

Monomer interaction with the cation radical results in the generation of a proton and a cation on the epoxide (Scheme 12.5) with rate constants of 5200 and 7.2×10^6 mol^{-1} l s^{-1} for butyl vinyl ether and cylcohexene oxide, respectively. Nothing is known about a possible physical quenching regenerating the ground state species.

Scheme 12.6 shows the possible role of the phenyl iodinium cation radical in acid production in alcoholic or hydrogen-donating solvents. Kinetic grounds (see

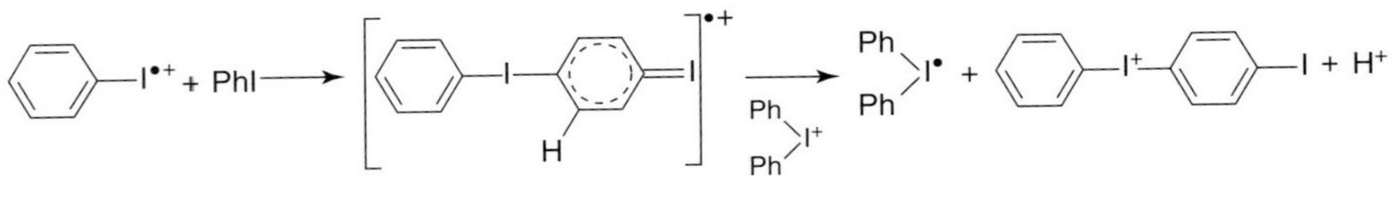

Scheme 12.4

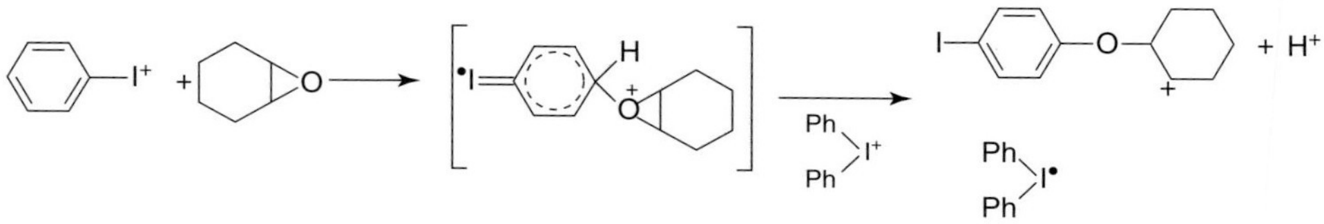

Scheme 12.5

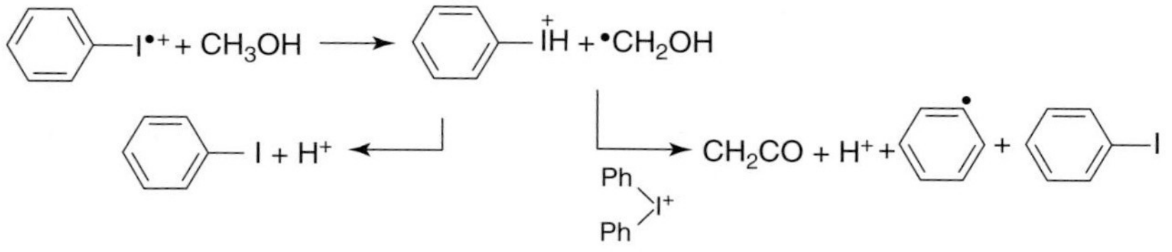

Scheme 12.6

above), however, show that monomer addition is more efficient: the quenching rate constants by methanol and cyclohexene oxide are in a 1 : 25 ratio.

12.2.1.5.4 Role of the Phenyl Radical

As shown in Eq. (12.4), an alcohol (e.g., the benzyl alcohol) can react with the phenyl radical formed in Eq. (12.4) and contribute to acid production. As the presence of the alcohol leads to the incorporation of an ether end group into the growing polymer (Scheme 12.3), a similar reaction should also occur with a R-CH_2O-polymer chain instead of R-CH_2OH. In air-equilibrated conditions, the acid quantum yield in methanol is ~3 times lower than in argon saturated conditions as the phenyls are quenched by oxygen and cannot participate in the chain reaction; in acetonitrile, there is almost no change [77]. This suggests that, in addition to the activated monomer mechanism shown in Scheme 12.3, alcohols or water should also have a noticeable positive effect on the initiation step.

$$Ph^\bullet + R\text{-}CH_2OH \rightarrow R\text{-}C(^\bullet)HOH + PhH$$
$$R\text{-}C(^\bullet)HOH + Ph_2I^+ \rightarrow Ph_2I^\bullet + R\text{-}C(^+)HOH$$
$$Ph_2I^\bullet \rightarrow Ph^\bullet + PhI$$
$$R\text{-}C(^+)HOH \rightarrow R\text{-}CHO + H^+ \quad (12.4)$$

The presence of the phenyl radical explains why iodonium salts can initiate a radical photopolymerization. An iodonium salt can therefore behave as a dual PI. This has led to the design of novel salts for the hybrid radical/cationic curing of coatings and adhesives, for example (i) sulfonium salts Ar-S^+-R(-CH_3) where, for example, Ar = phenyl, naphthyl; R = 2-indanyl, 1-ethoxycarbonylethyl [78] and (ii) N-methylbenzothiazolinium salts (see below) [79].

This has also led to the synthesis of new cationic monomers bearing a propenyl ether function 8 [80].

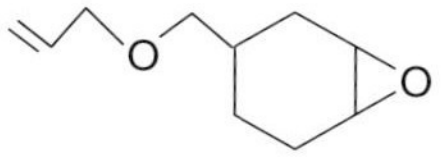

8

12.2.1.5.5 Photoinitiation Step
The photoinitiation step of the cationic polymerization of epoxides (and partly for vinyl ethers) by iodonium salts can be described according to the following reactions:

1) Addition of H^+ to a monomer unit as described above (Scheme 12.1 or 12.2).
2) Addition of the phenyl iodonium radical cation to the monomer (Scheme 12.5).
3) Quenching of the phenyl iodonium radical cation by H-donors (Scheme 12.5) is not expected to compete. In the absence of efficient H-donors at a suitable concentration (e.g., alcohols or water), this process should be disregarded. Indeed, H-abstraction with alkyl chains is still less efficient than with alcohols.

The initiation quantum yield (Scheme 12.7) is a function of the quantum yield in acid species Φ_{acsp} and the yield in monomer cations Φ_{M+}.

12.2.1.5.6 Behavior of Iodonium Salts in a Photopolymerization
The dependence of the polymerization rate of cationic monomers as a function of the experimental conditions and the iodonium salt has been deeply investigated in the early days of development of the onium salt chemistry and photochemistry [81–84]. A recent paper discussed the role of many parameters in the photopolymerization of vinyl monomers [85].

12.2.1.5.7 Role of the Iodonium Salt/Additive Combination
Iodonium Salt/Zinc Halide The effect of the addition of zinc halides in the photopolymerization of vinylethers has been checked. An adduct between the monomer (e.g., isobutyl vinyl ether) and HX (where the anion can even be highly nucleophilic) is formed. In the presence of zinc chloride, the generation of a suitable weakly nucleophilic $[X \ldots ZnCl_2]^-$ counteranion stabilizes the growing carbocation [86] and allows the development of the polymerization. A living character is observed. It results from the equilibrium between the free adduct and the zinc halide/adduct complex, the former acting as a dormant species.

Iodonium Salt/Ytterbium Triflate The drawback of the above reaction is the use of a rigorously dry medium as Lewis acids are very sensitive to traces of moisture. To

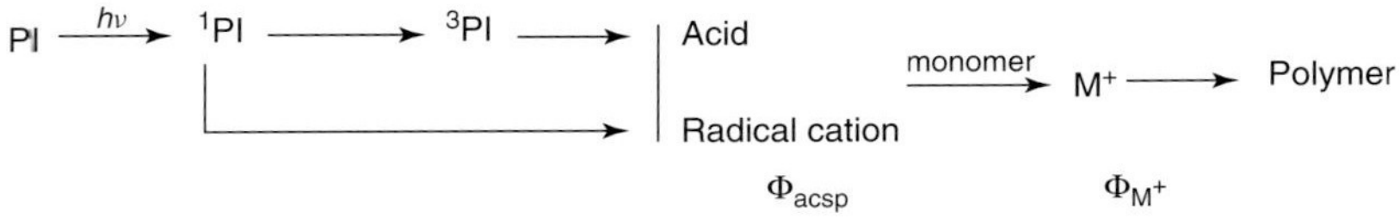

$$PI \xrightarrow{h\nu} {}^1PI \longrightarrow {}^3PI \longrightarrow \begin{array}{l} \text{Acid} \\ \text{Radical cation} \end{array} \xrightarrow{\text{monomer}} M^+ \longrightarrow \text{Polymer}$$

Φ_{acsp} Φ_{M^+}

Scheme 12.7

12.2 Onium Salts

overcome this problem, ytterbium triflate Yb(OTf)$_3$ was proposed as a water-tolerant Lewis acid additive [87]. The mechanism is similar to that discussed where ZnX$_2$ is changed for Yb(OTf)$_3$. Interestingly, this system can work in a heterogeneous medium containing 25% of water. The iodonium salt and the ytterbium triflate migrate from the aqueous phase to the organic phase where the photopolymerization occurs. This allowed for the first time a cationic polymerization in the presence of high water content, the remaining problem being the optimization of the experimental conditions.

Iodonium Salt/Alcohol The addition of an alcohol to an iodonium salt allows to use the phenyl radical (formed in the photodecomposition of the iodonium salt) for the production of new protons (Section 12.2.1.5.3) [51]. This effect on the initiation step contributes to the polymerization improvement, beside the activated monomer mechanism already shown in Scheme 12.3.

Taking advantage of the role of alcohols has led to the synthesis of new cationic monomers bearing an ether function and α-abstractable hydrogen atoms (**9** and **10**) [88, 89]. Acceleration of the cationic-ring-opening photopolymerization is noted. It results from (i) a H-abstraction by the phenyl radical at the α-carbon and (ii) an oxidation by the iodonium salt that forms a carbon-centered cation (new initiating species) and a new phenyl radical. The oxidation of these radicals gives a partial free-radical-promoted cationic polymerization character to this process. This chain process obviously allows the production of a large amount of initiating species for one photon absorbed.

12.2.1.6 Decomposition Processes of Sulfonium Salts

12.2.1.6.1 Mechanism

In triphenylsulfonium salts, a schematic mechanism consistent with all experimental data [66, 71, 72] is depicted as resulting from the direct excitation of either a $\pi-\sigma^*$ transition (in the C–S bond) or a $\pi-\pi^*$ transition followed by an intramolecular electron transfer to the σ^* MO of the C–S bond (Figure 12.2).

The primary cleavage occurs in the singlet state. In agreement with the expected higher stability of the phenyl radical pair (by ~0.8 eV) compared to the phenyl

Figure 12.2 HOMO and LUMO for Ph$_3$S$^+$ (DFT Level). (Source: Unpublished data.)

cation pair in the sulfonium salts, a heterolytic cleavage was proposed: it yields a singlet diphenylsulfide/phenyl cation pair. Then, an electron transfer within this pair produces a singlet diphenylsulfinium radical/phenyl radical pair, which can interconvert into a triplet pair. As in iodonium salts, in-cage and out-of-cage processes [20] lead to the formation of the acid (Eq. (12.5)).

$$Ph_3S^+ \ X^- \rightarrow {}^1(Ph_3S^+ \ X^-)^* \ (h\nu)$$
$${}^1(Ph_3S^+ \ X^-)^* \rightarrow {}^1[Ph_2S \ Ph^+] \ X^- \rightarrow {}^1[Ph_2S^\bullet + Ph^\bullet] \ X^-$$
$${}^1[Ph_2S^{\bullet+} \ Ph^\bullet] \rightarrow {}^3[Ph_2S^{\bullet+} \ Ph^\bullet]$$
$${}^1[Ph_2S^{\bullet+} \ Ph^\bullet] \rightarrow Ph-PhS-Ph + H^+$$
$${}^1[Ph_2S^{\bullet+} \ Ph^\bullet] \ (\text{or } {}^3[Ph_2S^{\bullet+} \ Ph^\bullet]) + DH \rightarrow Ph_2S^{\bullet+} + Ph^\bullet + DH$$
$$\rightarrow Ph_2S + Ph^\bullet + D^\bullet + H^+$$
$${}^1(Ph_3S^+ \ X^-)^* \rightarrow {}^3(Ph_3S^+ \ X^-)^* \rightarrow {}^3[Ph_2S^{\bullet+} \ Ph^\bullet] \qquad (12.5)$$

The singlet radical pair mostly undergoes an in-cage process. The homolytic dissociation in the triplet state has also been proved on some analogs [90]. Intersystem crossing followed by the formation of a triplet radical pair can also occur. The identification of the diphenylsulfinium radical cation and other species was supported by LFP [91]. The decomposition of sulfonium compounds has been also investigated by *ab initio* quantum mechanical calculations [92]. The acid quantum yield in the triarylsulfonium salt is higher than that of the diphenyliodonium salt (0.66 vs. 0.49 (BF_4^- as anion)) on photolysis at 254 nm [77]).

The photolysis of aryl dialkyl sulfonium salts follows the same mechanism. Photogeneration of acid through the conversion of the sulfonium salt to a sulfide via a 1,3 sigmatropic rearrangement was demonstrated [93].

In 5-arylthianthrenium salts [40], the cleavage occurs at the C(-aryl)–S bond (which generates the thianthrene cation) or at the C–S bond of the thianthrene ring (yielding a cation radical/radical pair). The latter cleavage is considered as an energy wasting process as the pair should easily recombine. The quantum yield of the photoacid release is lower than that of triaryl sulfonium salts (approximately twofold factor). The photoinitiating ability of both salts is, however, comparable either in vinyl or in ring-opening polymerization. This is mostly due to the better light absorption of 5-arylthianthrenium salts, which compensates their lower acid yields.

12.2.1.6.2 Photolysis

The careful investigation of the photolysis of these salts extensively studied 20 years ago (see a review in [20]) explains the generation of H^+ (even in the absence of RH) and the excess of acid in the photolysis of triphenylsulfonium salts (in-cage efficiency: 70%). In addition, the ratio of in-cage biphenyl derivatives B to cage-escape diphenylsulfide D (B/D) photoproducts increases with increasing the solvent viscosity because the diffusion of the initially formed fragments becomes limited. For example, this B/D ratio changes from 1.28 to 5.63 in the photolysis of triphenylsulfonium triflate at $\lambda = 254$ nm when going from methanol to glycerol.

The cage-escape B/D ratio exhibits dramatic changes when considering different environments. For example, the B/D ratio for the photolysis of triphenylsulfonium hexafluoroantimonate at $\lambda = 254$ nm is 1.13, 3.51, and 4.10 in acetonitrile, a PMMA film and in the solid state, respectively [71]. Under light exposure at 254 or 313 nm, the overall quantum yields of H^+ formation in acetonitrile are almost the same for the diphenyl and the triphenyl sulfonium hexafluoroarsenate (~ 0.7) [41].

12.2.1.6.3 Photoinitiation Step
As in the iodonium salt photoinitiated polymerization, the proton is still the main initiating species (Scheme 12.1). Other species can participate to some extent, such as the diphenylsulfinium cation radical (easily observed at 750 nm by LFP [94]). This species can add to an epoxy monomer (cyclohexene oxide: $k = 1.3 \times 10^5$ mol^{-1} l s^{-1}; butylvinyl ether: $k = 3.2 \times 10^5$ mol^{-1} l s^{-1}) and slowly reacts with water ($k = 2.3 \times 10^3$ mol^{-1} l s^{-1}) or methanol ($k = 5.4 \times 10^3$ mol^{-1} l s^{-1}) [94]. The rate constants are notably lower than for the phenyl iodinium cation radical PhI$\cdot^+$ ($k = 7.2 \times 10^6, 5.2 \times 10^9, 1.8 \times 10^5$; and 2.7×10^5 for the interaction with cyclohexene oxide, butylvinyl ether, water, and methanol, respectively). This reaction forms an intermediate adduct cation radical structure, which then reacts on another sulfonium salt to produce the oxiranium propagating cation and a proton (in a way similar to that described in Scheme 12.5). The qualitative picture for the photoinitiation step is similar to that of Scheme 12.7.

12.2.1.7 Acylsulfonium Salts
The highly sensitive and potentially low-cost acylsulfonium salts containing aromatic moieties (where the benzene moiety is replaced by an anthracene, a pyrene, etc.) allows the possibility of a long wavelength excitation through a noticeable shift of the absorption from the UV to the near-UV-visible range. The photolysis of dialkylphenacyl sulfonium salts has been shown [38, 81, 95] to proceed through a Norrish II hydrogen abstraction followed by the formation of an ylide (Scheme 12.8). New derivatives have been proposed [96–100]. Some of them appear as long wavelength absorbing compounds. Regeneration of the salt through a back thermal reaction can occur if the protons are not consumed.

12.2.1.8 Substituted Iodonium and Sulfonium Salt Derivatives
Other compounds are continuously synthesized and studied. They include:

1) Various sulfonium salts and substituted iodonium salts [28, 101–108] such as p-benzoyl diphenyl iodonium salt [109, 110], phenacyldimethyl sulfonium ylides [111], alkynyl phenyl iodonium salt [112], fused sulfonium salts [113], and arylcycloalkyl sulfonium salts [114]. Dialkyl arylsulfonium salts [115] exhibit a

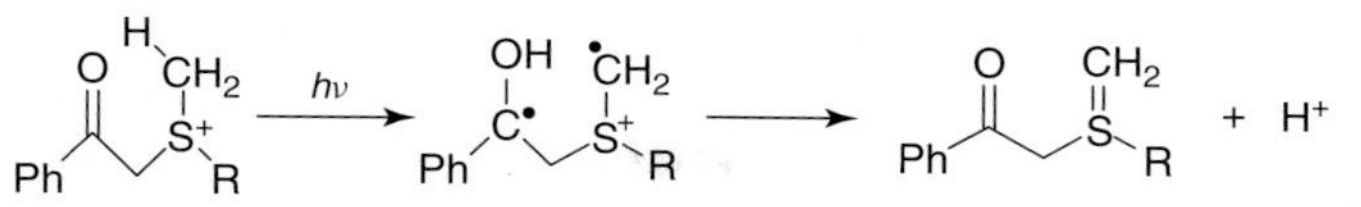

Scheme 12.8

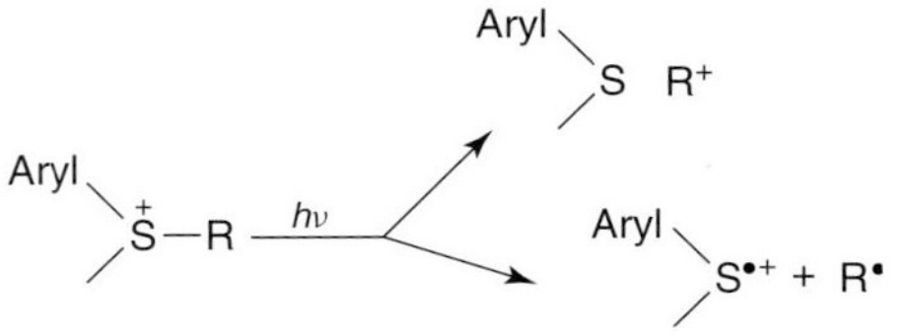

Scheme 12.9

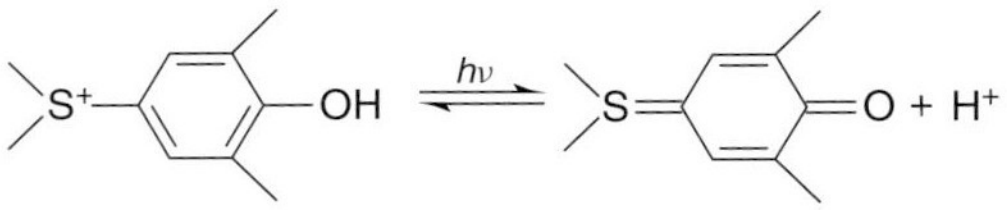

Scheme 12.10

dual radical/cationic activity as a homolytic bond breaking coexists with the heterolytic cleavage as shown in Scheme 12.9.

2) Hydroxyphenyl dialkyl sulfonium salts [82, 116, 117]. Contrary to the other sulfonium or thianthrenium salts that undergo an irreversible cleavage, these compounds photochemically cleave and form an ylide and an acid that thermally recombine (Scheme 12.10). Therefore, the cleavage becomes reversible and almost no posteffect occurs.
3) S-aryl-S,S-cycloalkyl sulfonium salts [118].
4) Triarylamine dialkylsulfonium salts [119] **11–13**. The incorporation of the triarylamine group stabilizes the radical cation formed in the homolysis of the C–S bond and favors the cleavage process. Indeed, in the corresponding phenyldimethyl sulfonium salt series, a low cleavage is observed.

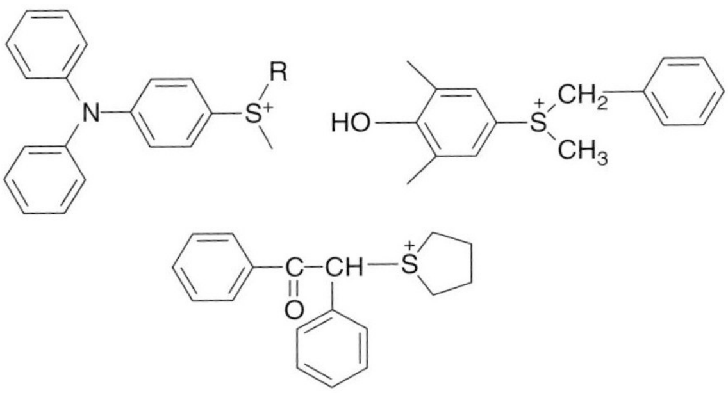

11–13

5) Polymeric iodonium salts. Their manufacture has long been achieved in systems in which the salt is incorporated either in the main chain or as a pendant group. Polyiodonium salts, iodonium salts grafted onto a polymer backbone (e.g., a silicone or a polystyrene chain), have been developed [83]. Other examples include the introduction of a O-$(CH_2)_2$-OH group, a long alkyl or alkyl oxy chain (derivatives with chains bearing 8–18 carbon atoms

become soluble in toluene [120]). The substitution by alkoxy groups increases the spectral sensitivity in the near-UV region [121].
6) Phenylethynyl onium salts. These compounds lead to an interesting red-shifted absorption, a better solubility, a change of the redox properties, and less benzene release (Scheme 12.11) [122].

12.2.2
Other Onium Salts

Although the classical iodonium and sulfonium salts shown above are mostly used, the development of other onium salts continuously received a strong attention [35]. Examples of new compounds are shown in **14–23**.

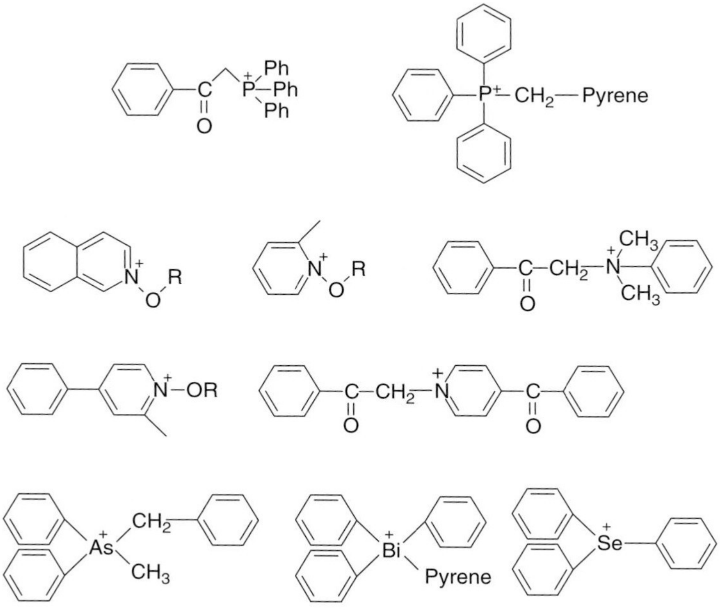

14–23

In a general way, all these new compounds are based on different central atoms such as:

1) Nitrogen as in nitronium and quinolinium salts [123], tetraalkyl ammonium [124], phenacylanilinium salts [125], and pyridinium salts [126–137]. The diethoxy-azobis (pyridinium) salt [138] exhibits a cis–trans isomerization and allows a wavelength tunability as both isomers can lead to a radical cation and a Lewis acid.
2) Phosphor as in phosphonium salts [139–141].
3) Oxygen (in pyrylium salts [142]).
4) Sulfur (in thiopyrylium [143] salts, alkyl thiobenzothiazolinium salts [144]).

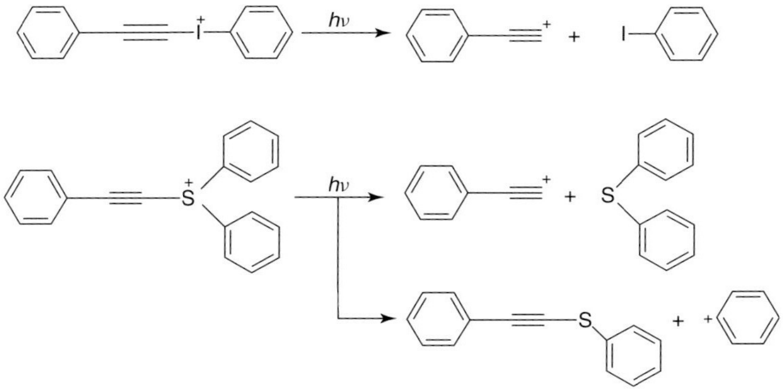

Scheme 12.11

5) Silicon [145], arsenium [146], selenium, bismuth [147, 148], carbon (in trimethoxytrityl salts [149, 150]), and chlorine or bromine [70].

The direct photolysis of these other onium salts as well as the photoacid generation was investigated, for example, in diarylchloronium and diarylbromonium salts [70], aryl ammonium, aryl phosphonium [151–153], arylarsonium [154], and triphenylselenonium [155]. The general trends look like those observed for diphenyl iodonium and triphenyl sulfonium salts: involvement of S_1 and T_1 states, cleavage of the C-onium atom bond, and in-cage and cage-escape reactivity. Examples of mechanisms [22] have been nicely described for the alkoxypyridinium R_3N^+-OR, phenacyl benzoyl pyridinium $R_3N^+-C(=O)Ph$, phosphonium R_4P^+, phenacylammonium $R_2(-Ph)N^+-CR_2 C(=O)R$, and thiobenzothiazolinium salts $R_2N^+=C(-S)-SR$. The primary cleavage occurs at the N–O, P–R, N–C, and S–R bond, respectively.

12.2.3
Search for New Properties

As seen above, many onium salts have been synthesized primarily in order (i) to achieve a broader spectral response and, in this way, extends the use of diaryl iodonium salts, triarylsulfonium salts, and some other onium salts or (ii) to increase the reactivity.

12.2.3.1 Absorption
Representative typical wavelengths of maximum absorption and molar extinction coefficients are summarized in Tables 12.2 [38] and 12.3 [38, 156]. They help to illustrate the effect of the substitution on the iodonium and sulfonium skeletons and the role of the structure in various onium salts.

Table 12.2 Substitution effects on the absorption maxima and molar extinction coefficients of some iodonium and sulfonium salts.

Compound	λ_{max} (nm)	Molar extinction coefficient (mol^{-1} l cm^{-1})
Ph_2I^+	227	17 800
$[2-NO_2Ph]_2I^+$	215	35 000
$[4-OCH_3 Ph]PhI^+$	218	43 800
$[2,4-CH_3Ph]_2I^+$	240	18 750
$Ph_3 S^+$	230	17 500
$[p-CH_3Ph]_3 S^+$	243	24 700
	278	4 900
$[p-OCH_3Ph]_3 S^+$	225	21 740
	280	10 100
$[3-CH_3, 4-OH, 5-CH_3 Ph]_3S^+$	263	25 200
	280	22 400
	316	7 700
$Ph-O-Ph-S^+ Ph_2$	263	14 000
$Ph-S-Ph-S^+ Ph_2$	225	23 400
	300	19 500

^aSource: From [38].

12.2.3.2 Solubility

From a practical point of view, in the largely used iodonium- and sulfonium-salts-derived compounds, the research studies dealing with the development of the onium salt chemistry and the applications aim at designing diaryl iodonium and triarylsulfonium salts exhibiting a better solubility in cationically polymerizable monomer matrixes, especially in highly nonpolar silicon-containing epoxy resins. Indeed, being ionic in nature, onium salts are scarcely soluble in nonpolar media, which requires the use of a polar solvent for their dissolution in multifunctional monomers: in practical UV curing applications, this procedure is not convenient. Increased solubility is observed, for example, in polymeric iodonium salts, iodonium salts introduced on a polymer backbone (e.g., polystyrene, silicone), and iodonium salts carrying long alkyl, alkyloxy, or hydroxyalkyl chain.

12.2.3.3 Stability

Onium salts, especially diphenyl iodonium salts, are not very heat stable. This might be a problem when they are used in an operating process where a heating of the formulation is required before light exposure. Salts based on the usual diphenyliodonium moiety and the tetrakis(pentafluorophenyl) borate anion exhibit higher melting points T_m. Examples of T_m for a diphenyliodonium structure

Table 12.3 Examples of structural effects on the absorption maxima and molar extinction coefficients of onium salts.

Compound	λ_{max} (nm)	Molar extinction coefficient (mol^{-1} l cm^{-1})
HO-C₆H₂(CH₃)₂-S⁺(CH₃)₂	252 300	9 300 3 000
HO-naphthyl-S⁺(CH₃)₂	248 353	16 400 6 500
C₆H₅-C(O)-CH₂-S⁺(CH₃)₂	250 290	6 900 4 100
naphthyl-C(O)-CH₂-S⁺(tetrahydrothiophene)	300 360	21 900 1 000
diphenyliodonium	264	17 300
phenyl-(fluorenone)-iodonium	265	55 000
thioxanthene-S⁺-phenyl	227 232	15 300 3 100
phenoxathiine-S⁺-phenyl	238 292	19 900 5 000

Table 12.3 (continued)

Compound	λ_{max} (nm)	Molar extinction coefficient (mol^{-1} l cm^{-1})
dibenzothiophenium phenyl (S+)	225 260 318	18 600 4 100 600
methylpyridinium N-OEt	266	5 900
isoquinolinium N-OEt	337	4 200
phenylpyridinium N-OEt	310	21 440

^aSource: From Refs. [38, 156].

are 175, 136, 140, and 58 °C for B(C$_6$F$_5$)$_4^-$, BF$_4^-$, PF$_6^-$, and the SbF$_6^-$ anion, respectively.

12.2.3.4 Benzene Release
The photolysis of the first generation of iodonium and sulfonium salts obviously yielded benzene (or toluene in the case of the ditolyliodonium salt). Because of the toxicity concerns, this has induced the design of new compounds that do not present this weak point (Section 12.2.1.1).

12.2.3.5 Odor
The formation of sulfur-containing photolysis products when sulfonium salts are used leads to a residual odor in the UV-cured coatings. A decreased odor was claimed when, on light exposure, low-molecular-weight compounds are formed to a lesser extent.

12.2.3.6 Toxicity
The toxicity is usually presented as being due to the anion (but this is not totally true), and therefore, As-based anions must be avoided. The SbF$_6^-$ anion is not suitable for application, for example, in silicone release papers where Sb-containing residue will be formed during the subsequent ageing of the UV-cured coating so that the recycling treatments and the further destruction of the cured materials become complicated. A decrease in toxicity has been mentioned when an alkoxy chain is introduced on the phenyl ring. For example, (4-octyloxyphenyl) phenyliodonium

hexafluoroantimonate is less toxic than the parent unsubstituted compound: $LD_{50} = 5000$ mg kg^{-1} versus 40 mg kg^{-1} [120]) (the oral toxicity is expressed by the LD_{50} test; for NaCl, $LD_{50} = 3750$ mg kg^{-1}). For isopropyl diphenyl iodonium tetrakis(pentafluorophenyl) borate, LD_{50} is > 2000 mg kg^{-1} [157].

12.2.3.7 Amphifunctionality
As stated before, the ability of onium salts to generate both acidic species and free radicals on photolysis can also allow their use as convenient amphifunctional PIs, that is, being able to initiate a simultaneous cationic and free radical polymerization.

12.3
Organometallic Derivatives

Many attempts have been made to develop cationic PIs that are sensitive in the visible part of the spectrum. As shown in [158, 159] or reviewed in [17, 160] and references therein, a lot of organometallic derivatives have been proposed. The photochemistry of organometallic compounds (OMCs) is well documented in the literature, for example, the reactivity of cyclopentadienyl arene complexes [161] or bisarene iron dications [162], the picosecond spectroscopy of transition metal complexes [163], and the mechanism of the photoinduced iron–cyclopentadienyl bond cleavage [164].

12.3.1
Transition Organometallic Complexes

Examples of transition organometallic complexes include [17, 160], for example, (i) ferrocenium salts (see below), (ii) metal carbonyl (based on Mn, Re, Ru, Os, Rh), (iii) metal carbonyls such as η^5 cyclohexadienyl iron tricarbonyl RFe(CO)$_3^+$, (iv) half sandwich metal cabonyls (CpFe(CO)$_2$ L$^+$, ArRe(CO)$_3^+$), and (v) titanocene dichloride. The photoinitiation mechanism in the presence of R–Fe(CO)$_3^+$ is shown in Eq. (12.6) where EPOX stands for an epoxy monomer.

$$\begin{aligned} RFe(CO)_3^+ + EPOX &\rightarrow RFe(CO)_3 EPOX^+ \ (h\nu) \\ \text{or } RFe(CO)_3^+ &\rightarrow\rightarrow RFe(CO)_2^+ + CO \ (h\nu) \\ \text{followed by } RFe(CO)_2^+ + EPOX &\rightarrow RFe(CO)_2 EPOX^+ \\ \text{or } RFe(CO)_2^+ &\rightarrow\rightarrow RFe(CO)_2 + H^+ \text{ and } H^+ \\ + EPOX &\rightarrow H-EPOX^+ \end{aligned} \quad (12.6)$$

Novel architectures have been proposed, for example, a Pd complex possessing a bichromophoric moiety (consisting in a naphthyl and Ru (polypyridyl) chromophores) for the photopolymerization of styrene [165]. On excitation, a $R_2-(-solvent)Pd^+-CH_3$ species is generated and added to styrene to form a $R_2-(-solvent)Pd^+-CH(-Ph)-CH_2(-CH_3)$ structure. The further insertion of styrene monomer units into this structure leads to polystyrene.

The most famous example of organometallic transition metal compounds that were developed for industrial cationic photopolymerization reactions in the radiation curing area is concerned with the ferrocenium salt derivatives having anions with low nucleophilicity [166]. They absorb in the visible spectral range (up to 500–550 nm with molar extinction coefficients depending on the arene moiety [22]). A well-known representative is (η^6-cumene) (η^5-cyclopentadienyl) iron (II), denoted here as Fc(+). Its absorption maxima are located at 249, 297, 375, and 440 nm with molar extinction coefficients around 12 400, 3000, 75, and 60 mol^{-1} l cm^{-1}, respectively. Changing the cumene moiety for naphthalene or pyrene shifts the absorption toward the red with new maxima at 350 ($\varepsilon = 1000$ mol^{-1} l cm^{-1}) and 480 nm ($\varepsilon = 200$ mol^{-1} l cm^{-1}) for the former and 450 nm ($\varepsilon = 1000$ mol^{-1} l cm^{-1}) and 550 nm ($\varepsilon = \sim80$ mol^{-1} l cm^{-1}) for the latter.

On photolysis in aprotic solvents, ferrocene and an iron salt are formed [167] while a crystalline crown ether complex is obtained in the presence of an excess of ethylene oxide [168] in methylene chloride. More complex reactions occur in chlorinated solvents [169] where a further photolysis of ferrocene is observed.

When using Fc(+) in the presence of an epoxide [159], the ring-opening polymerization on irradiation is assumed to take place (Scheme 12.12) through a ligand transfer reaction that expels the arene moiety, which is replaced by three molecules of epoxide (this structure is the active Lewis acid species). An arene ring slippage with a η^6 to η^4 coordination change supports the possibility of introducing three epoxides. The initiating carbocation is obtained through a thermal cleavage of the epoxy C–O single bond. Then, the polymerization proceeds outward from the coordination sphere of the iron atom. As expected, the order of increasing reactivity follows that of nucleophilicity decrease of the anions (SbF$_6^-$ is better than PF$_6^-$). The formation of cyclopentadienyl radicals has also been reported [170], and the addition of this radical to the double bond of the PBN spin trap has been followed by ESR [171]. This may partly explain the photopolymerization of acrylic monomers using Fc(+).

Addition of cumene hydroperoxide ROOH allows the generation of a stronger Lewis acid (because of the oxidation of the iron (II) species to a more electrophilic iron (III) species) and increases [159] the rate of the epoxide polymerization. This redox reaction also yields an oxygen-centered radical RO$^\bullet$ (which can initiate a radical polymerization) and a hydroxyl anion OH$^-$ (Part 2).

Photophysical studies [170] showed that these organometallic complexes can be efficiently quenched by classical redox quenchers such as methyl viologen MV^{2+} that converts into MV$^{+\bullet}$ (the complex plays the role of an electron donor) or

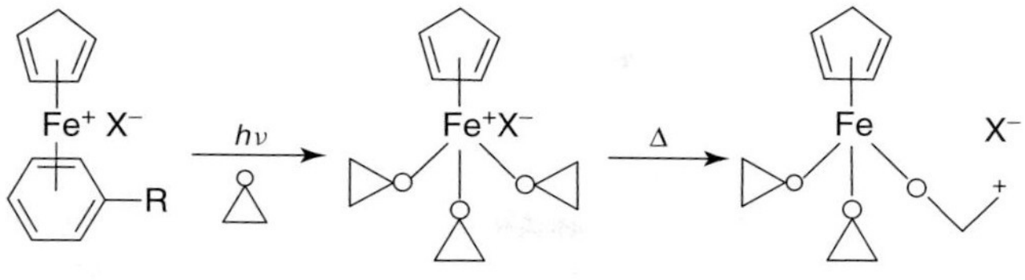

Scheme 12.12

dimethyl aniline (in that case, the electron is transferred from the amine: the complex acts as an electron acceptor).

The photolysis of ferrocenium salts can be photosensitized by polynuclear HC [172]. The mechanism seems by now to be unclear. Addition of an amine has been claimed to enhance the reaction.

Other arene–iron complexes with various arene moieties such as carbazole, anisole, diphenylether, aniline, aminonaphthalene, benzophenone (BP) (**24** and **25**), and so on have been recently synthesized [173–177] and the associated polymerization reactions investigated [178].

12.3.2
Inorganic Transition Metal Complexes

They are exemplified by (i) titanium or vanadium tetrachloride, (ii) vanadyl chloride, and (iii) ammonium salt ($FeCl_4^-$; silver-, copper-, or thalium-atom-based salts [17, 160]). In Eq. (12.7), M, THF, VC, and S stand for a vinyl monomer, tetrahydrofuran, vinylcarbazole, and solvent, respectively.

$$M + TiCl_4 \rightarrow M^{\bullet+} + TiCl4^{\bullet-} \ (h\nu) \quad (1)$$
$$THF + Ag^+ \rightarrow THF^{\bullet+} + Ag \ (h\nu)$$
$$\text{and } THF^{\bullet+} + THF \rightarrow THF^{\bullet} + H - THF^{\bullet+} \quad (2)$$
$$VC + S \rightarrow VC^{\bullet+} + S^{\bullet-} \ (h\nu) \text{ followed by } VC^{\bullet+} + VC \rightarrow (VC\text{-}VC)^{\bullet+}$$
$$\text{and } (VC\text{–}VC)^{\bullet+} + Ag^+ \rightarrow \text{initiating species} \quad (3)$$
$$AuX_4^- \rightarrow AuX_3^{\bullet-} + X^{\bullet} \ (h\nu) \text{ followed by } X^{\bullet} + VC \rightarrow VC^{\bullet+} + X^-$$
$$\text{and } VC^{\bullet+} \rightarrow VC^{\bullet} + H^+ \quad (4) \tag{12.7}$$

12.3.3
Non-Transition-Metal Complexes

Non-transition-metal complexes (Eq. (12.8)) such as (i) alkyldiarylsilyl methyl ketone/aluminum chelate or (ii) arylsilylperoxide/diketone chelate of B, Si, Ge, P, or Al centered complex [179] can cure epoxy-based systems [17].

$$Ar_2RSiC(=O)R' + R''OH \rightarrow Ar_3SiOH$$
$$Ar_3SiO\text{–}OR \rightarrow Ar_3SiO^{\bullet} + {}^{\bullet}OR \rightarrow Ar_3SiOH$$
$$\text{and Al chelate} + Ar_3SiOH \rightarrow \text{High acidity complex} \tag{12.8}$$

12.4
Onium Salt/Photosensitizer Systems

Successful photosensitization of the onium salts On^+ decomposition using both UV wavelengths and near-visible lights should clearly eliminate the handicap of these salts, that is, a poor absorption in the 300–450 nm region (see above) [180, 181]. This drawback severely limited their efficiency when irradiated with the commercially available medium- and high-pressure mercury arc lamps, which provide a substantial portion of their emission in that region.

The possible sensitized decomposition of On^+ in the presence of a photosensitizer (PS) was shown to occur through energy or electron transfer processes that are governed by energetic considerations or by thermodynamical requirements (Section 5.5). The values proposed for the triplet state T_1 energy levels E_T of the diphenyliodonium and the triphenylsulfonium salts are $E_T = 74$ and 64 kcal mol^{-1}. The ^{3}PS level should be higher than these values to get an efficient energy transfer. Free energy change ΔG calculated from the Rehm–Weller equation for various PS/onium salt couples are reported in Table 12.4 [41, 182]. The commonly used reduction potentials E_{red} of the diphenyliodonium and the triphenylsulfonium salts are $E_{red} \sim -0.2$ and ~ -1.2 eV, respectively.

An additional mechanism involving radicals can also be operative in FRPCP. This last way avoids the synthesis of visible-light-absorbing onium salts (which might be expensive) and the search for efficient PSs (that can be limited). Recent works outline the potential of FRPCP (see below and Section 12.5).

Table 12.4 Free energy change ΔG calculated from the Rehm–Weller equation for various photosensitizer PS/onium salt couples. PS oxidation potential E_{ox}^{PS} and PS excited state energy E_T used for the calculations.

Compound	E_{ox}^{PS}(eV)	E_T(kcal mol^{-1})	ΔG(kcal mol^{-1})[a]	ΔG (kcal mol^{-1})[b]	ΔG (kcal mol^{-1})[c]
Thioxanthone	1.69	66	+1	−22.2	−12.5
Xanthone	–	–	−4.8	−28	–
Benzophenone	2.39	69	+21	−2	+0.5
Acetophenone	2.69	74	+20.8	−2.4	+2.5
Anthracene	1.35	76	−30	−46.2	−30.5
Perylene	1	66	−17.5	−40.7	−28.5
Phenothiazine	0.26	57	−15.3	−38.5	−36.5
Pyrene	1.35	76.9	–	–	−30.4
Coronene	1.22	66.7	–	–	−24.2
Tetracene	0.95	60.7	–	–	−24.2

[a]Triphenylsulfonium salt.
[b]Diphenyliodonium salt.
[c]Dialkylphenacylsulfonium salt ($E_{red} = -0.63$ eV).
Source: From Refs. [41, 182].

Photosensitized photolysis has been explored using a lot of techniques such as CIDNP [183], laser flash photolysis, and ESR.

12.4.1
Photosensitization through Energy Transfer

In this process, an excited diphenyl iodonium salt is formed through a triplet–triplet energy transfer as the occurrence of a singlet–singlet energy transfer is totally excluded. This excited iodonium salt, in its triplet state, can only undergo a homolytic cleavage [66] (because of spin conservation), yielding a phenyl radical/phenyl iodinium radical pair. The diphenyl sulfinium radical cation was detected in photosensitized experiments just as in the direct photolysis [94]. The initiation step is therefore similar to that encountered in the direct photolysis of the onium salts.

This energy transfer can occur with a ketone (Ket) exhibiting a high-energy-lying triplet state (Eq. (12.9)). This is the case with acetone (triplet state energy $E_T = 80$ kcal mol^{-1}, acetone/triphenylsulfonium salt interaction rate constant: 2×10^9 mol^{-1} l s^{-1}), acetophenone, and 1-indanone [184]. The red shift of the absorption of the two-component photoinitiating system is, however, obviously small as the triplet energy of a molecule decreases when the ground state absorption is going to the red. This way of energy transfer is not really interesting.

$$\text{Ket} \rightarrow \text{ket}^*(h\nu) \text{ and } \text{Ket}^* + \text{Ph}_2\text{I}^+ \rightarrow \text{ket} + \text{Ph}_2\text{I}^{+*} \qquad (12.9)$$

12.4.2
Photosensitization through Electron Transfer

12.4.2.1 Photosensitizers
HCs (anthracene, pyrene, perylene, and N-vinyl carbazole), polymers bearing pendent HC moieties or poly(N-vinyl carbazole) [185–189], phenothiazine derivatives [190–193], dyes D, ketones Ket (camphorquinone (CQ), benzil, anthraquinone, and thioxanthone) [194], or metal complexes (such as [CpFe(CO)$_2$]$_2$) can sensitize the photolysis of cationic PIs through an electron transfer process (see, e.g., Eq. (12.10)) in the case of the diphenyliodonium salt). Such a mechanism is also observed in dye–onium salt intraion pair. As already stated, the formed diphenyl iodide radical cleaves into iodobenzene and a phenyl radical.

$$\text{HC} \rightarrow \text{HC}^*(h\nu) \text{ and } \text{HC}^* + \text{Ph}_2\text{I}^+ \rightarrow \text{HC}^{\bullet+} + \text{Ph}^\bullet + \text{PhI}$$
$$\text{D} \rightarrow \text{D}^*(h\nu) \text{ and } \text{D}^* + \text{Ph}_2\text{I}^+ \rightarrow \text{D}^{\bullet+} + \text{Ph}^\bullet + \text{PhI}$$
$$\text{Ket} \rightarrow \text{Ket}^*(h\nu) \text{ and } \text{Ket}^* + \text{Ph}_2\text{I}^+ \rightarrow \text{Ket}^{\bullet+} + \text{Ph}^\bullet + \text{PhI} \qquad (12.10)$$

The introduction of 1-pyrenemethanol or 3-perylenemethanol that appears as a PS (due to the pyrene moiety), a chain transfer agent, and a free radical accelerator (the alcohol moiety) allows fast curing of cyclohexene oxide (with a dramatic increase of R_p and a final conversion going from 50 to 70%) [64].

The quenching reaction of anthracene by a diphenyliodonium salt [38, 41] primarily involves the singlet state [195] (rate constant: 2×10^{10} mol^{-1} l s^{-1}). The diphenyliodo radical has been detected (lifetime: a few hundred picoseconds). The quenching of the triplet state occurs with a very low rate constant ($\sim 10^7$ mol^{-1} l s^{-1}) [196]. The calculated ΔG values support the view of a fast electron transfer mechanism. Back electron transfer drastically decreases the generation of the anthracene radical cation. This work has been extended to a series of various onium tetrakis(pentafluorophenyl) borates [197]. Anthracene-bound sulfonium salts appeared as highly efficient PIs [198].

Efficient electron transfer reactions were shown to occur both in the singlet and the triplet states of dyes [38, 41]. Many dyes were tested, and some of them such as xanthene [199], thioxanthene and merocyanine [200], acridone [84], tetrabenzoporphyrine [201], flavin and acridine derivatives [84], titanocenes [202], acridinediones (ADDs) [77], and Rose Bengal [203] exhibit a good sensitization activity leading to the decomposition of the diaryliodonium salts. As recently studied, the photooxydation of ADD by iodonium, sulfonium, and thianthrenium salts yields the ADD radical cation (on the nitrogen atom) and a derived enol radical cation (as the nitrogen bears a positive charge) clearly observed at 550 nm. Both species leads to a carbon-centered radical and a proton [77]. The photosensitized photolysis at 366 nm leads to acid release quantum yields lower than those obtained in the direct photolysis at 254 nm (e.g., 0.15 vs. 0.49, 0.50 vs. 0.66, and 0.25 vs. 0.30 for the diphenyl iodonium salt, the triarylsulfonium salt, and the thianthrenium salt, respectively). A fast electron transfer process between the dye and the onium salt does not necessarily parallel a strong ability to initiate the polymerization reaction. The efficiency of the photoinitiation process sharply depends on the structure of the dye. A fast back electron transfer process also decreases the efficiency of the sensitization process.

The interaction between ketones and onium salts can occur through electron transfer in the singlet and the triplet states. The former being generally (very) short-lived, the transient state responsible for the quenching reaction is primarily the triplet state. The reaction was shown to occur with ketones having a long-lived and low-energy triplet state such as thioxanthones ($E_T \sim 60-63$ kcal mol^{-1}), for example, 2-isopropylthioxanthone (ITX), ketocoumarins (~ 58 kcal mol^{-1}), and CQ (~ 50 kcal mol^{-1}). In ITX/Ph$_2$I$^+$ and CQ/Ph$_2$I$^+$, the interaction rate constants are 9.6×10^9 and $< 10^8$ mol^{-1} l s^{-1}, respectively [204].

The cation radical of thioxanthone derivatives has been directly observed through time-resolved laser spectroscopy [181, 205]. In the BP series ($E_T \sim 69$ kcal mol^{-1}), the problem was much more complex as no definitive statement of the transfer mechanism (energy vs. electron transfer) was proposed for a long time [41] until a paper adopted a definitive position in favor of an electron transfer mechanism (70% of the reaction) [68].

Typical electron transfer rate constants in selected systems have been measured as a function of the ketones, the solvent, the anions, the onium salts (centered on I$^+$, S$^+$, P$^+$, N$^+$, As$^+$, and Se$^+$) together with the ketone triplet state quenching rate constant by an epoxide monomer. The efficiency of the electron transfer is

(i) higher in acetonitrile than in methanol, (ii) almost similar whatever the ketone considered, (iii) independent of the counter anion except for benzophenone, and (iv) sharply dependent on the introduction of substituents on the onium salts. The solvent dependence in the steady state photolysis has been studied [206]. Monomer quenching of the ketone triplet state usually very efficiently compete with the onium salt interaction (e.g., 2.8×10^6 and 4.7×10^8 mol^{-1} l s^{-1} for BP/cyclohexeneoxide and BP/iodonium salt, respectively in acetonitrile). Using ITX, the situation is better: $k = 3 \times 10^5$ and 9.6×10^9 mol^{-1} l s^{-1} for ITX/EPOX and ITX/iodonium salt, respectively, in acetonitrile [204]. The electron transfer rate constants of Cl-thioxanthone (CTX) with various oniums salts (iodonium, triaryl sulfonium, dialkyl hydroxyphenyl sulfonium, dialkylphenacyl sulfonium, pyrydinium, phosphonium, arsenonium, and selenonium) in methanol have been determined (Table 12.5) [181].

In Table 12.6 [181, 196, 204, 207], electron transfer rate constants of the triplet states of CTX, BP, ketocoumarin, or dyes with diphenyliodonium tetrafluoroborate in acetonitrile as well as the monomer quenching of these PSs are reported. A fast singlet state quenching of CTX ($\sim 10^9$ mol^{-1} l s^{-1}) also occurs: it will not have, however, any important role in bulk monomer where the rate constants are diffusion limited. A study using TX, CTX, xanthone, BP, pyrene, benzyl, and phenothiazine as PS together with two substituted iodonium and sulfonium salts bearing different anions (tetrafluoroborate, hexafluoroantimonate, and tosyl) in methanol, acetonitrile, and heptane reveals a more or less noticeable PS, solvent, and anion effect on the rate constants [208].

The relative values of k_e [onium salt] and k_q [Monomer] allow a large panel of situations for the yield of electron transfer (and the production of the radical cation on PS) both in monomer solution and bulk media, for example, 0.5 with the iodonium salt in bulk 1,2 epoxybutane (11.5 M) and 0.19 with the sulfonium salt in bulk cyclohexene oxide (9.7 M) using [onium salt]= 0.05 M. Yields $\sim$ 1 can be obtained when considering dialkylphenacyl sulfonium salts.

In order to decrease the toxicity of the HC-based PS, new compounds in which perylene, phenothiazine, or N-vinyl carbazole (**26**) are linked to a cationically polymerizable group have been proposed [209]. They allow a photosensitizing effect, and they are incorporated into the matrix during the photopolymerization reaction.

26

New PSs based on thiophene (ThP) derivatives for the cationic photopolymerization of epoxides, vinylethers, N-vinylcarbazole, and styrene in the presence of iodonium salts have been proposed, for example, dithienothiophene [210] or highly conjugated diphenyldithienothiophene (DDT) derivatives **27** [211]. These

Table 12.5 Electron transfer rate constants of the Cl-thioxanthone triplet state with various oniums salts in methanol.

Compound	$10^{-7}\, k_e\,(\text{mol}^{-1}\,\text{l}\,\text{s}^{-1})$
4-HO-3,5-dimethylphenyl dimethylsulfonium	35
(phenylthio)phenyl diphenylsulfonium	4
2-naphthoylmethyl tetrahydrothiophenium	570
9-fluorenyl triphenylphosphonium	110
phenanthridinium	100
diphenyliodonium	24
bis(4-nitrophenyl)iodonium	400
bis(3,5-dimethylphenyl)iodonium	36
benzyl diphenylarsonium	38
triphenylselenonium	15

Source: From [181].

Table 12.6 Electron transfer rate constants k_e in various photosensitizers/diphenyliodonium tetrafluoroborate couples. For comparison, $k_e = 14 \times 10^7$ mol^{-1} l s^{-1} in CTX/triarylsulfonium salt.

Compound	$10^{-7} k_e$ (mol^{-1} l s^{-1})	$10^{-7} k_q$ (mol^{-1} l s^{-1})
Cl-thioxanthone	24	0.1 (1)
		0.3 (2)
Benzophenone	47	0.7 (1)
Ketocoumarin	6.4	0.04 (2)
Acridinedione[a]	440	0.027 (3)
ITX	960	0.03 (3)
Camphorquinone	<10	<0.01 (3)

[a] Acridinedione with methyl and H substitution on C and N, respectively.
Interaction rate constants k_q of various photosensitizers with a cationic monomer: (1) 1,2 epoxybutane, (2) cyclohexene oxide, and (3) EPOX. In methanol (except in acetonitrile for EPOX with diphenyliodonium hexafluorophosphate).
Source: From Refs. [204, 207].

last compounds are sensitive under visible lights (480 nm). Structurally related compounds such as quinoxaline [212] or benzotriazole derivatives [213] have been further developed. The primary process is an electron transfer from DDT to the iodonium salt. The resulting radical cation on the ThP DDT$^\bullet$ leads to the initiation as explained below.

27

In the same way, a polythiophene film (that is an extensively investigated structure of organic conjugated polymers) can be synthesized using DDT and an iodonium salt (but not with a sulfonium salt) [211]. After the formation of DDT$^\bullet$ the reaction proceeds as a photoinduced step-growth photopolymerization involving (i) formation of $^+$DDT $-$ DDT$^+$ (by addition of two DDT$^{\bullet+}$) and then DDT$_{-H}$–DDT$_{-H}$ (by elimination of two protons), (ii) further DDT$_{-H}$–DDT$_{-H}$/iodonium salt light-induced reaction, and so on. This mechanism is similar to that proposed for the cationic photopolymerization of ThP using iodonium salts (in that case, ThP$^{\bullet+}$ is formed from the interaction between ThP and the PhI$^{\bullet+}$ cation radical obtained by irradiation of Ph$_2$I$^+$). The mechanism of the photoinitiation step by the phenyliodinium and diphenylsulfinium radical cations with ThP was discussed in [214]. Interaction rate constants of PhI$^{\bullet+}$ and Ph$_2$S$^{\bullet+}$ with ThP are 1.2×10^{10} and 1.7×10^5 M^{-1} s^{-1}, respectively.

Tricyclohexylphosphine tetrafluoroborate or tributylphosphine tetrafluoroborate in combination with ITX can also be efficient photoinitiating systems for epoxide polymerization on UV light exposure (J. Lalevée, M.A. Tehfe, and J.P. Fouassier, unpublished data). This demonstrates the ability of the phosphoniumyl cation radical $R_3P^{+\bullet}$ to initiate a cationic polymerization (Eq. (12.11)). The known reaction of this species with H_2O traces generating an acid might also contribute to some extent.

$$ITX + R''_3P^+\text{-}H \rightarrow R''_3P^{+\bullet} + ITX\text{-}H^{\bullet} \; (h\nu)$$
$$R''_3P^{+\bullet} + H_2O \rightarrow R''_3P^{\bullet}\text{-}OH + H^+ \quad (12.11)$$

12.4.2.2 Photoinitiation Step

When using a photosensitization process by electron transfer, the cation radical formed on the PS (Eq. (12.12)) adds to the monomer unit. It can also react with a hydrogen donor, water, or a phenyl radical or dimerize [35, 36]), the produced cations being the initiating species.

$$PS + On^+ \rightarrow PS^{+\bullet} + On^{\bullet} \; (h\nu)$$
$$PS^{+\bullet} + H_2O \rightarrow PS\text{-}OH + H^+$$
$$PS^{+\bullet} + M \rightarrow {}^{\bullet}PS\text{-}M^+$$
$$PS^{+\bullet} + Ph^{\bullet} \rightarrow {}^+PS\text{-}Ph$$
$$PS^{+\bullet} + PS^{+\bullet} \rightarrow {}^+PS\text{-}PS^+ \quad (12.12)$$

PS/monomer hydrogen abstraction followed by monomer radical/onium salt electron transfer has been also reported as a possible competitive route to the PS/On$^+$ interaction (Eq. (12.13)). A proton and a positively charged monomer unit are formed.

$$PS + M \rightarrow PS\text{-}H^{\bullet} + M_{-H}{}^{\bullet} \; (h\nu)$$
$$PS\text{-}H^{\bullet} + On^+ \rightarrow PS + H^+ + On^{\bullet}$$
$$M_{-H}{}^{\bullet} + On^+ \rightarrow On + M_{-H}{}^+ \quad (12.13)$$

In solution, the PS/On$^+$ electron transfer process (e.g., from ITX (9.6×10^9 mol^{-1} l s^{-1}), which promotes the decomposition of the iodonium salt) is much more efficient than the dicycloaliphatic epoxide monomer quenching (0.3×10^6 mol^{-1} l s^{-1}). However, in bulk (monomer viscosity: 250 cp at 25 °C; calculated diffusion rate constant: 2.3×10^7 M^{-1} s^{-1}), the rate constants level off (see Section 5.7.3 and Chapter 17) and the situation might be in favor of a PS/monomer interaction. The balance between the two routes should be dependent on the PS.

Scheme 12.13 (where PI stands for the cationic PI) summarizes the different routes that lead to initiation. The initiation quantum yield is dependent on the quantum yield of electron or energy transfer Φ_{eTorET} and the yields in cations Φ_{as} and monomer cations Φ_{M+}.

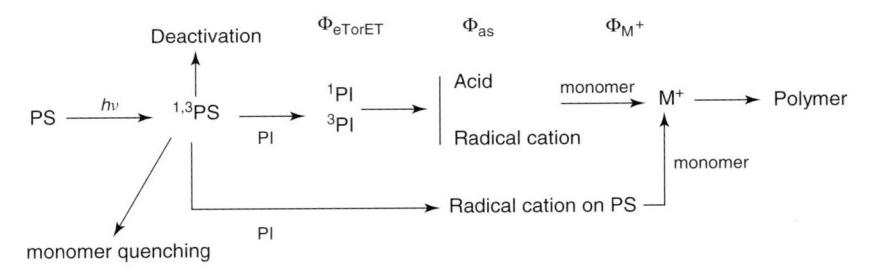

```
                          Φ_eTorET        Φ_as          Φ_M+
         Deactivation
              ↑                         Acid
              |          ¹PI                    monomer
  PS  --hν--> ¹,³PS  -->                       -------->  M+  ------>  Polymer
                         ³PI                                ↑
                    PI          Radical cation              |
              \         \                                monomer
               \         _____> Radical cation on PS __|
                \           PI
                 ↓
         monomer quenching
```

Scheme 12.13

12.5
Free-Radical-Promoted Cationic Photopolymerization

12.5.1
Radical/Onium Salt Interaction

Photosensitization is also feasible through an electron transfer between a radical that can be generated from a usual radical PI and the iodonium salt (Eq. (12.14)). This way is referred to as *leading to a free-radical-promoted cationic photopolymerization (FRPCP)*. The basic interest is related to the obvious possibility of using a lot of radical generators that are able to absorb in the visible range. The problem is to create a radical exhibiting a high reactivity toward the iodonium salt. Changing Ph_2I^+ for Ph_3S^+ should lead to the same reaction scheme. The reduction reaction by the sulfonium salt Ph_3S^+, however, is probably as inefficient as the reduction potential of Ph_3S^+ is $-1.1/-1.35$ V, respectively (Section 12.5.2.4).

$$\text{Radical source} \rightarrow R^\bullet \ (h\nu)$$
$$R^\bullet + Ph_2I^+ \rightarrow R^+ + PhI + Ph^\bullet \tag{12.14}$$

The main drawback is the oxygen sensitivity of the system as the addition of radicals to oxygen (peroxidation process) is usually highly efficient and close to the diffusion limit. The generated peroxyl radicals are electrophilic structures and can hardly be oxidized. Therefore, the cationic polymerization, which normally presents the great interest to be oxygen insensitive, is noticeably dropped down under air when using the various historically proposed radical source/iodonium salt systems.

12.5.2
Radical Source/Onium Salt Based Systems

12.5.2.1 Radical Source/Iodonium Salt Two-Component Systems

12.5.2.1.1 Ketone/Iodonium Salt
The FRPCP mechanism was originally observed with Type I PIs where radicals are formed through a fast α-cleavage process from a short-lived triplet state (Scheme 12.14). Benzoin ethers [215, 216], dialkoxyacetophenones [217], or benzoyl

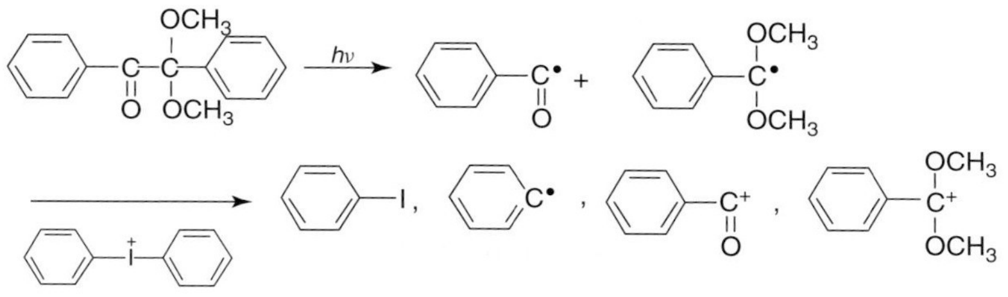

Scheme 12.14

phosphine oxides [218–220] have been reported as interesting sources of benzoyl or phosphinoyl radicals.

The rate constants measured for the interaction between the benzoyl or the phosphinoyl radicals and the iodonium salts ($< 10^6$ mol^{-1} l s^{-1}) make the reaction rather unable to efficiently compete with the monomer quenching, for example, with butyl vinyl ether (4×10^6 mol^{-1} l s^{-1}) for the diphenylphosphinoyl radical [218]. The situation is better with the hydroxyisopropyl radical (6×10^7 mol^{-1} l s^{-1} [221]) formed from a hydroxyl alkyl acetophenone derivative. The ketyl radical K$^\bullet$ of BP as well as the aminoalkyl radical A$^\bullet$ that can be formed in dye/iodonium salt/amine AH systems is also reduced by the iodonium salt [222]. The rate constants for the K$^\bullet$ (or A$^\bullet$)/Ph$_2$I$^+$ interaction are usually very high ($> 10^7$ mol^{-1} l s^{-1}) as discussed in Section 19.

Cyclohexene oxide, N-vinylcarbazole, and n-butyl vinyl ether can be polymerized by FRPCP. When considering cyclic monomers, the initiation step corresponds to a ring-opening reaction by the carbocations. When using vinyl ethers, the adduct formed by addition of the radicals (generated upon photolysis of a radical PI) to the double bond can be oxidized by the iodonium salt, the resulting cation being the initiating species (Scheme 12.15).

12.5.2.1.2 Polysilane/Iodonium Salt

The photolysis of catena-poly-phenyl-4-phenylphenyl silicon at 365–400 nm presumably yields free radicals that are oxidized by a diphenyliodonium salt: the

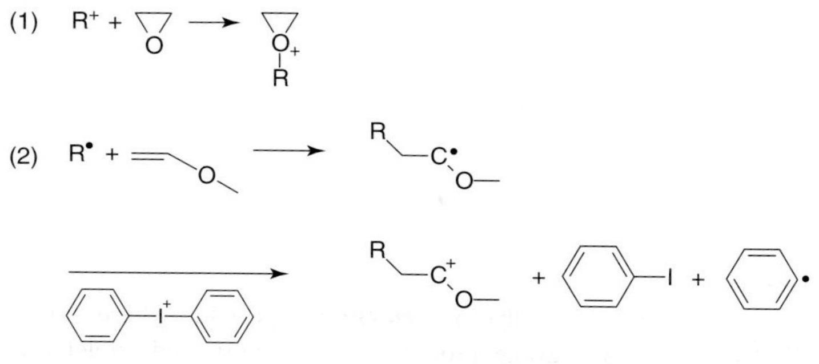

Scheme 12.15

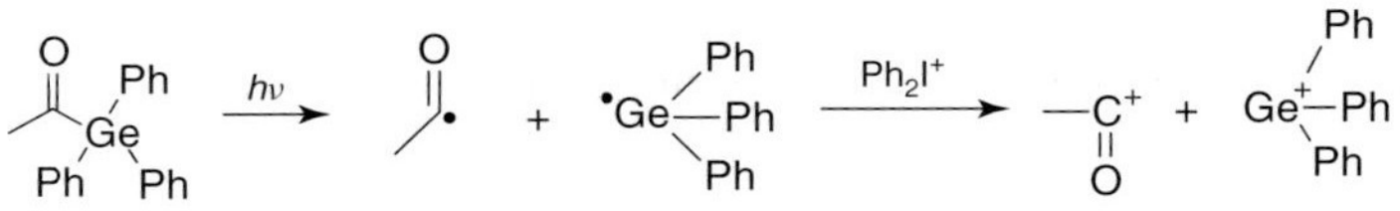

Scheme 12.16

photoinitiation of the cationic polymerization of THF, n-butyl vinyl ether, cyclohexene oxide, or vinylcarbazole is feasible [223]. Polysilane/pyridinium salt [224] and polysilane/pyridinium salt/benzaldehyde systems [225] have been mentioned. The efficiency of the cationic UV polymerization involving silicon polymer radicals appears, however, as rather low.

12.5.2.1.3 Germane/Iodonium Salt

The benzoyl trimethylgermane/diphenyliodonium salt combination was very recently reported as an interesting system that absorbs up to 450 nm and undergoes an α-cleavage process yielding a benzoyl and a germyl radical that are oxidized by the onium salt [226]. A detailed investigation of the processes involved in the acetyltriphenylgermane/diphenyliodonium salt system is detailed in [227] (Scheme 12.16).

12.5.2.1.4 Benzodioxinone/Iodonium Salt

Benzodioxinone (BD)/iodonium salt is an interesting system that works under a stepwise two-photon absorption [228]. The first photon excites BD, which generates in situ BP (through a reaction already depicted in Scheme 9.11); the second photon excites the benzophenone, which further works as described in Eq. (12.15).

$$\begin{aligned} &BD \rightarrow BD^*(h\nu) \text{ and } BD^* \rightarrow BP \\ &BP \rightarrow BP^*(h\nu) \\ &BP^* + DH \rightarrow BP-H^\bullet + D^\bullet \\ &BP-H^\bullet + Ph_2I^+ \rightarrow BP-H^+ \rightarrow BP + H^+ \end{aligned} \quad (12.15)$$

12.5.2.1.5 Vinyl Halide/Iodonium Salt

The irradiation of an aryl-substituted vinyl bromide leads to a C–Br bond breaking, which results in free radicals and ions [229]. The vinyl cation–bromide ion pair cannot initiate the polymerization as the bromide is highly nucleophilic. Addition of an onium salt $On^+MX_n^-$ generates a vinyl cation/MX_n^- ion pair through an attack on the free vinyl radical (the anion must obviously have a low nucleophilicity), which further reacts with a vinyl ether monomer.

12.5.2.2 Radical Source/Iodonium Salt Three-Component Systems

The use of carefully selected radical source/onium salt based three-component systems in which the radical source is a Type II PI allows a longer wavelength photosensitivity (large choice of PI) and/or a better efficiency as well as a strong decrease of the oxygen inhibition.

12.5.2.2.1 Photoinitiator/Amine/Iodonium Salt

Visible-light-photosensitive three-component systems have been proposed for a long time. For example, the xanthenic dye (Xan)/amine/iodonium salt system [230] works according to Eq. (12.16). A Xan–H$^\bullet$/Ph$_2$I$^+$ interaction reaction can also contribute. This system has a low stability.

$$\text{Xan} \rightarrow \text{Xan}^*(h\nu) \text{ and } \text{Xan}^* + \text{AH} \rightarrow \text{Xan–H}^\bullet + \text{A}^\bullet$$
$$\text{A}^\bullet + \text{Ph}_2\text{I}^+ \rightarrow \text{A}^+ \rightarrow \text{initiation} \quad (12.16)$$

The ketyl radicals produced, for example, in benzophenone/dimethylaniline are also efficient. The same behavior holds true in thioxanthone fluorene carboxylic acid (or thioxanthone carbazole)/hydrogen donor/iodonium salt [231].

12.5.2.2.2 Photoinitiator/Alkylhalide/Iodonium Salt

The dimanganese decacarbonyl/alkyl halide/iodonium salt three-component system [232] works according to Eq. (12.17).

$$\text{Mn}_2(\text{CO})_{10} \rightarrow 2\,\text{Mn}^\bullet(\text{CO})_5\,(h\nu)$$
$$\text{Mn}^\bullet(\text{CO})_5 + \text{CH}_2\text{Cl}_2 \rightarrow \text{Mn(CO)}_5\text{Cl} + \text{CH}_2\text{Cl}^\bullet$$
$$\text{CH}_2\text{Cl}^\bullet + \text{Ph}_2\text{I}^+ \rightarrow \text{CH}_2\text{Cl}^+ \rightarrow \text{initiation} \quad (12.17)$$

12.5.2.2.3 Photoinitiator/Silane/Iodonium Salt

Recent progress [233] has been made using a three-component photoinitiating system based on the specific input of the silyl radical chemistry (see Section 9.3.15 and Chapter 18) and the change of the amine for a silane. The absence of an amine avoids the dark reactions that occur in iodonium salt/amine mixture. The PI/silane/iodonium salt system has four advantages, three of them being already schematically presented in the case of PI/silane systems for radical photopolymerization (Section 9.1.2.7): (i) a suitable selection of the PIs obviously allows to easily tune the absorption from the near-UV range to the visible wavelength range and leads in any case to an active silyl radical R$_3$Si$^\bullet$, (ii) the silyls consume oxygen and scavenge the peroxyls, (iii) new silyls are regenerated, and (iv) the same cationic structure R$_3$Si$^+$ is formed (Eq. (12.18)). Therefore, oxygen plays a positive role in the cationic photopolymerization reaction due to the remarkable reactivity of the silyl radicals toward oxygen (Section 9.3.15). A minor contribution of the R$_1^\bullet$, R$_2^\bullet$, and PI–H$^\bullet$ radicals with Ph$_2$I$^+$ can be also expected.

$$\text{PI} \rightarrow \text{PI}^*(h\nu)$$
$$\text{PI}^* + \text{R}_3\text{SiH} \rightarrow \text{R}_3\text{Si}^\bullet + \text{PI–H}^\bullet$$
$$\text{PI}^* \rightarrow \text{R1}^\bullet + \text{R2}^\bullet \rightarrow\rightarrow \text{R}_3\text{Si}^\bullet$$
$$\text{R}_3\text{Si}^\bullet \rightarrow \text{R}_3\text{SiOO}^\bullet \rightarrow \text{R}_3\text{Si}^\bullet$$
$$\text{R}_3\text{Si}^\bullet + \text{Ph}_2\text{I}^+ \rightarrow \text{R}_3\text{Si}^+ + \text{Ph}_2\text{I}^\bullet \quad (12.18)$$

Moreover, it appears that (i) PI should not be necessarily a compound used as a conventional radical PI (indeed, it has only to absorb the light and react with the silane), (ii) the onium salt does not play any role except to oxidize the silyl radical (due to the values of the redox potentials, at present, the iodonium salts are among the best candidates, for example, $Ph_2I^+PF_6^-$; other oxidation agents such as the silver salts can be used), and (iii) the very high efficiency of the PI/silane/onium salt system (the counter anion usually being hexafluorophosphate) avoids the use of an onium salt containing a counteranion more suitable for a fast propagation of the cationic polymerization (as it is necessary in a PI/onium salt system to enhance the efficiency; this more suitable anion is also more expensive, e.g., $B(C_6F_5)_4^-$ or toxic, e.g., SbF_6^-). In the following examples, the cationic monomer is an epoxide.

Ketone/Silane/Ph_2I^+ Among PIs, CQ, ITX, and eosin [204] as well as ADD [207] appear as suitable compounds. For example, (Figure 12.3), when introducing a silane R_3SiH into CQ/Ph_2I^+, both the polymerization rates and final conversions on an exposure with $\lambda > 400$ nm are higher under air. In the same conditions, the 2,2'-dimethoxyphenyl acetophenone/Ph_2I^+ and bisphosphine oxide BAPO/Ph_2I^+ reference systems do not work.

The overall role of the silane is qualitatively explained on the basis of Eq. (12.19) [204]. The formed silyl radicals are efficiently oxidized by the iodonium salt ($10^5 - 10^7$ mol^{-1} l s^{-1}), albeit the rate constants are lower than for the aminoalkyl radical of EDB. They are also strongly quenched by oxygen ($3-4 \times 10^9$ mol^{-1} l s^{-1}). In this last case, the generated $R_3Si-O-O^\bullet$ peroxyl radical is reduced by the onium salt ($\sim 10^6$ mol^{-1} l s^{-1}). This oxidation ability of the silyl or silylperoxyl radicals as well as some other subsequent side reactions involving oxygen (such as a possible silyl/oxygen redox process yielding a Si cation and an oxygen radical

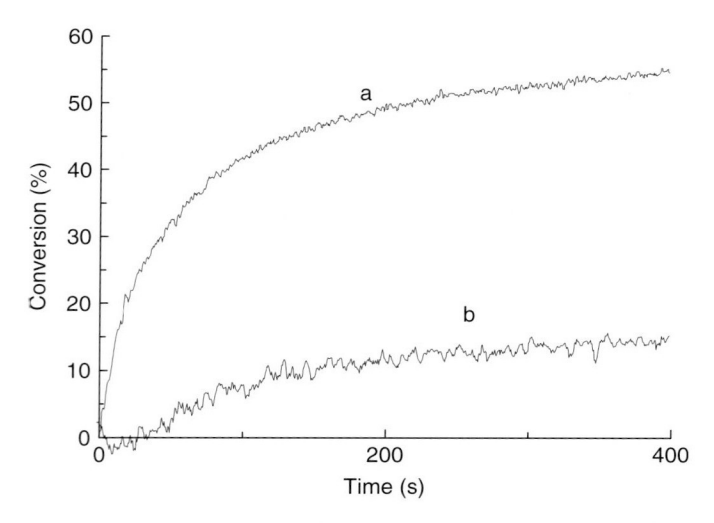

Figure 12.3 Polymerization profiles of an epoxide EPOX under air in the presence of (a) CQ/tris(trimethylsilyl)silane/Ph_2I^+ (3%/3%/1%) and (b) CQ/Ph_2I^+ (3%/1%). Irradiation with a Xenon lamp ($\lambda > 400$ nm). (Source: From [204].)

anion) can contribute to the enhanced polymerization reaction under air. The hydrogen abstraction reaction between a silyl radical and an epoxide monomer is low ($< 10^4$ mol^{-1} l s^{-1}). When changing the PIs, the overall behavior is probably a strong interplay between the efficiencies of both the hydrogen transfer process and the Ph$_2$I$^+$ reduction reaction. Under air, the silyl radical/O$_2$ interaction is decisive.

$$^3\text{PI} + \text{Ph}_2\text{I}^+ \rightarrow \text{PhI} + \text{Ph}^\bullet + \text{PI}^{\bullet+} \text{ or/and } ^3\text{PI} + \text{R}_3\text{Si-H}$$
$$\rightarrow \text{R}_3\text{Si}^\bullet + \text{PI-H}^\bullet$$
$$\text{Ph}^\bullet + \text{R}_3\text{Si-H} \rightarrow \text{Ph-H} + \text{R}_3\text{Si}^\bullet$$
$$\text{PI}^{\bullet+} + \text{R}_3\text{Si-H} \rightarrow \text{R}_3\text{Si}^\bullet + \text{H}^+ + \text{PI}$$
$$\text{R}_3\text{Si}^\bullet + \text{Ph}_2\text{I}^+ \rightarrow \text{Ph}_2\text{I}^\bullet + \text{R}_3\text{Si}^+$$
$$\text{R}_3\text{Si}^\bullet + \text{O}_2 \rightarrow \text{R}_3\text{SiOO}^\bullet \text{ and } \text{R}_3\text{SiOO}^\bullet + \text{Ph}_2\text{I}^+$$
$$\rightarrow \text{R}_3\text{SiOO}^+ + \text{PhI} + \text{Ph}^\bullet \quad (12.19)$$

Ketone/Germane/Ph$_2$I$^+$ The photosensitizing efficiency of different germanes-based R$_3$GeH photoinitiating systems has also been tested [234]. The most outstanding result is still the acceleration of the polymerization rate and the final conversion when using CQ/triphenylgermane/iodonium salt for a polymerization carried out under air. This enhancement was ascribed, as in Eq. (12.19), to (i) the competitive oxidation of the germyl radicals that have low ionization potentials (5–6 eV) and (ii) the oxidation of the formed peroxyl radicals R$_3$Ge–O–O$^\bullet$ by the onium salt.

Thiopyrylium Salt/Silane/Ph$_2$I$^+$ The striking interests of using thiopyrylium salts as PIs [235] are both the high efficiency and the long wavelength absorption (good matching with the emission spectrum of the Xe lamp or the visible laser lines). Compared to CQ/silane/diphenyliodonium salt, the 2,4,6(4-methoxyphenyl) thiopyrylium salt/tris(trimethylsilyl)silane/diphenyliodonium salt system [236] exhibits a very similar efficiency under the same experimental conditions. The primary step involves an electron transfer process between the thiopyrylium salt TP$^+$ and the silane. A 2,4,6-tri methoxyphenyl thiabenzene radical TP$^\bullet$, and a silane radical cation is formed. A subsequent hydrogen transfer between these two species yields a silyl radical, which further reacts with the iodonium salt (Eq. (12.20)).

$$^3\text{TP}^+ + \text{R}_3\text{Si-H} \rightarrow \text{TP}^\bullet + \text{R}_3\text{Si-H}^{\bullet+} \rightarrow \rightarrow \rightarrow \text{R}_3\text{Si}^\bullet$$
$$\text{R}_3\text{Si}^\bullet + \text{Ph}_2\text{I}^+ \rightarrow \text{Ph}_2\text{I}^\bullet + \text{R}_3\text{Si}^+ \quad (12.20)$$

A similar electron transfer mechanism occurs when using dimetals (e.g., disilanes) instead of silanes [237] (Eq. (12.21)). In the presence of an iodonium salt, the R$_3$M$^\bullet$ radical is oxidized, leading to the same electrophilic R$_3$M$^+$ structure (M=Si or Ge) that is able to initiate the cationic photopolymerization. The decomposition of Ph$_2$I$^+$ after reduction with TP$^\bullet$ can also occur. As already observed in other thiopyrylium salt/iodonium salt systems, a direct ^{3}TP$^+$ (or ^{1}TP$^+$)/Ph$_2$I$^+$ interaction should exist, the cation radical TP$^{•+}$ being the initiating species.

$$^3TP^+ + R_3M\text{--}MR_3 \rightarrow TP^• + (R_3M\text{--}MR_3)^{•+} \rightarrow TP^• + R_3M^• + R_3M^+$$

$$TP^• + Ph_2I^+ \rightarrow TP^+ + Ph_2I^• \rightarrow TP^+ + PhI + Ph^•$$

$$^3TP^+ \text{ (or } ^1TP^+) + Ph_2I^+ \rightarrow TP^{2+} + Ph_2I^• \rightarrow TP^{2+} + PhI + Ph^•$$

(12.21)

Metal Carbonyl Compound/Silane/Ph$_2$I$^+$ When using OMCs such as Mn$_2$CO$_{10}$, Re$_2$(CO)$_{10}$, Cp$_2$Fe$_2$CO$_4$ [238], ArCr(CO)$_3$, Cp$_2$Mo$_2$(CO)$_6$, Fe$_2$(CO)$_9$, and Cp$_2$Ru$_2$(CO)$_4$ [239], the conversions reached after 400 s are always very low (<10%). The addition of a silane (tris(trimethylsilyl)silane) TTMSS dramatically improves the polymerization process for both the laser diode (405 nm) and the Xe lamp irradiation. In the presence of Mn$_2$CO$_{10}$/TTMSS/Ph$_2$I$^+$, the polymerization rates (Figure 12.4) are better than those found with the titanocene/TTMSS/Ph$_2$I$^+$ system (about 5- to 10-fold factor), but the final conversions are quite similar (~60%). Moreover, it can also initiate the cationic photopolymerization of a renewable monomer (epoxidized soybean oil) under visible lights in aerated conditions. Very interestingly, an outdoor exposure of the sample for about 1 h under air was enough to obtain a final tack-free coating. The addition of TTMSS to Cp$_2$Mo$_2$(CO)$_6$ (or Fe$_2$(CO)$_9$, Cp$_2$Ru$_2$(CO)$_4$)/Ph$_2$I$^+$ also enhances the polymerization (using ArCr(CO)$_3$; the observed lack of polymerization is ascribed to a detrimental quenching of the formed ArCr(CO)$_2$ structure by the epoxide

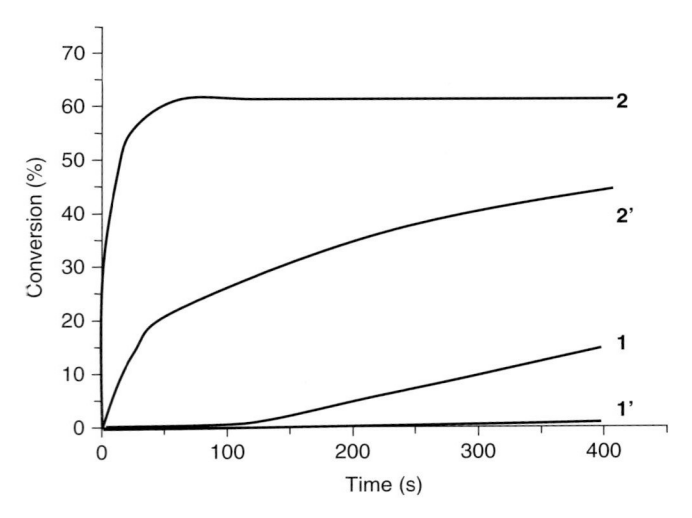

Figure 12.4 Photopolymerization profiles of an epoxide EPOX under air on a Xe lamp irradiation ($\lambda > 390$ nm) in the presence of (1) Mn$_2$CO$_{10}$/Ph$_2$I$^+$ (1%/1% w/w), (2) Mn$_2$CO$_{10}$/TTMSS/Ph$_2$I$^+$ (1%/3%/1% w/w) and (1') Cp$_2$Fe$_2$CO$_4$/Ph$_2$I$^+$ (0.2%/1% w/w), (2') Cp$_2$Fe$_2$CO$_4$/TTMSS/Ph$_2$I$^+$ (0.2%/3%/1% w/w). (Source: From [238].)

($\sim$2 × 10^6 mol^{-1} l s^{-1})). The Cp$_2$Mo$_2$(CO)$_6$/TTMSS/Ph$_2$I$^+$ system is particularly efficient on a laser diode exposure at 405, 457, 473, and 532 nm.

The low efficiency of the OMC/Ph$_2$I$^+$ systems suggests that (i) the radicals generated from the transition metal carbonyls are not easily oxidized by Ph$_2$I$^+$ and/or (ii) the generated cationic species does not efficiently initiate the cationic polymerization process. The oxidation of Mn$^\bullet$(CO)$_5$ by Ph$_2$I$^+$ is rather hard (<10^6 mol^{-1} l s^{-1}). On the contrary, a very high rate constant is found (1.5 × 10^9 mol^{-1} l s^{-1}) for Re$^\bullet$(CO)$_5$, supporting the easy formation of a cationic species (Re$^+$(CO)$_5$). However, the low efficiency of the Re$_2$(CO)$_{10}$/Ph$_2$I$^+$ system probably reflects the inability of this cation to initiate the cationic polymerization. The better oxidation property of Re$^\bullet$(CO)$_5$ compared to Mn$^\bullet$(CO)$_5$ is ascribed to its lower ionization potential (7.3 vs. 7.8 eV).

In the OMC/TTMSS/Ph$_2$I$^+$ three-component system, the OMC/TTMSS interaction is the primary decisive step. Indeed, the oxidation rate constant of (TMS)$_3$Si$^\bullet$ by Ph$_2$I$^+$ (2.6 × 10^6 mol^{-1} l s^{-1}) is considerably higher than that obtained in the case of the OMC$^\bullet$ radical. The silylium cation is a nice polymerization initiating structure. This behavior is in full agreement with the excellent photopolymerization ability of Mn$_2$(CO)$_{10}$/TTMSS/Ph$_2$I$^+$. In Re$_2$(CO)$_{10}$/TTMSS/Ph$_2$I$^+$, the oxidation of Re$^\bullet$(CO)$_5$ is faster than the hydrogen transfer with TTMSS (k[Ph$_2$I$^+$] > k'[TTMSS] : 5.6 × 10^5 vs. > 2.4 × 10^4 mol^{-1} l s^{-1}). As a consequence, in the presence of Ph$_2$I$^+$, few silyl radicals are generated, in line with the observed low polymerization efficiency.

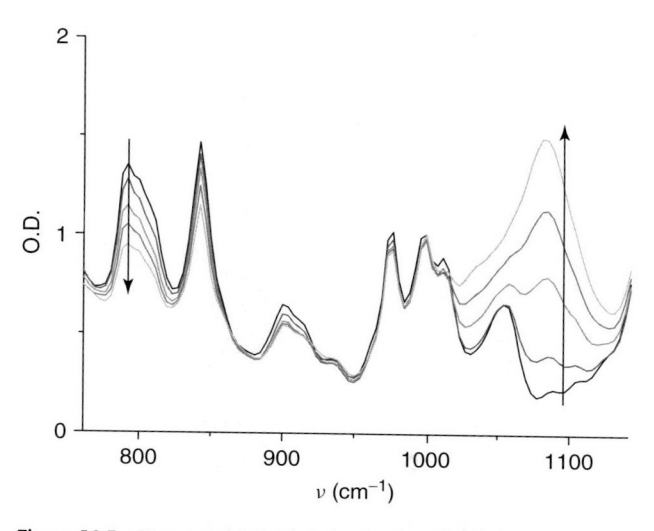

Figure 12.5 IR spectra recorded during the photopolymerization of an epoxide EPOX in the presence of W$_{10}$O$_{32}$$^{4-}$/TTMSS/Ph$_2I^+$ (1%/3%/2% w/w), under sunlight at $t = 0$ to 45 min. The conversion of EPOX can be followed at 790 cm^{-1} and the formation of the polyether network at 1080 cm^{-1}. (Source: From [240].)

Decatungstate/Silane/Ph$_2$I$^+$ The recently proposed decatungstate W$_{10}$O$_{32}$$^{4-}$/silane/ iodonium salt system [240] (decatungstate is well known in organic chemistry, see, e.g., [241]) leads to a noticeably efficient FRPCP of an epoxide under a usual fluorescent bulb or sun (Figure 12.5); this could still be a step to a green polymer photochemistry.

Decatungstate leads to an intermediate (designed as wO; lifetime = ~60 ns) within 30 ps [242], which further reacts with the silane to generate the necessary silyl radical (Eq. (12.22)).

$$W_{10}O_{32}^{4-} \rightarrow \rightarrow \rightarrow wO \;(h\nu)$$
$$wO + R_3SiH \rightarrow H^+W_{10}O_{32}^{5-} + R_3Si^\bullet \quad (12.22)$$

Ruthenium Salt (or/and Violanthrone)/Silane/Ph$_2$I$^+$ In addition to their promising ability under fluorescence bulb, LED lamp, or sun [243], new systems based on ruthenium salts and violanthrone (Vi) derivatives, usable in cationic photopolymerization even under air, have opened a new way for polychromatic light irradiation [244] and red or green laser diode exposure inaccessible up to now [244]. This allows to introduce a cationic matrix with a shrinkage lower than that encountered with the usually employed radical formulations. It could be helpful for applications in optics under laser excitation above 488 nm (e.g., holographic recording). Examples are encountered in the Vi/silane/Ph$_2$I$^+$ or Vi/ruthenium trisbipyridine Ru/silane/Ph$_2$I$^+$ system for irradiation at 635 nm and Ru/silane/Ph$_2$I$^+$ or Ru/Vi/silane/Ph$_2$I$^+$ system for exposure at 532 nm.

The mechanism observed in Ru/silane/Ph$_2$I$^+$ is basically described by Eq. (12.23) where the silane is TTMSS and Ru stands for Ru(bpy)$_3$$^{2+}$ [243] and Ru(phen)$_3$$^{2+}$ [245]. The primary electron transfer is favorable according to the redox properties of these reactants (E_{ox}(Ru) = 1.23 V, E_{red}(Ph$_2$I$^+$) ~ −0.2 V, $E(^3$Ru) = 2.07 eV, free energy change $\Delta G <$ 0 eV, and rate constants are in the 10^8 – 10^9 mol^{-1} l s^{-1} range). The quenching of ^{3}Ru(bpy)$_3$$^{2+}$ by TTMSS is slow ($k <$ 10^7 mol^{-1} l s^{-1}). The positive role of oxygen is explained as discussed above. The overall redox reaction cycles have been presented in Section 10.2.11.

$$Ru(bpy)_3^{2+} \rightarrow {}^3Ru(bpy)_3^{2+} \;(h\nu)$$
$$^3Ru(bpy)_3^{2+} + Ph_2I^+ \rightarrow Ru(bpy)_3^{3+} + Ph^\bullet + Ph\text{–}I$$
$$Ph^\bullet + (TMS)_3Si\text{–}H \rightarrow Ph\text{–}H + (TMS)_3Si^\bullet$$
$$(TMS)_3Si^\bullet + Ph_2I^+ \rightarrow (TMS)_3Si^+ + Ph^\bullet + Ph\text{–}I$$
$$(TMS)_3Si^\bullet + Ru(bpy)_3^{3+} \rightarrow (TMS)_3Si^+ + Ru(bpy)_3^{2+}$$
$$(TMS)_3Si^+ + M \rightarrow (TMS)_3Si\text{–}M^+ \quad (12.23)$$

The mechanism of formation of cations and radicals in the presence of Vi has been investigated [244]. A fast interaction of the Vi excited singlet state (lifetime: a few nanoseconds) with Ph$_2$I$^+$ occurs (k = 3–5 × 10^9 mol^{-1} l s^{-1}). The interaction rate constant with TTMSS is slower (k = 1.5 × 10^8 mol^{-1} l s^{-1}). The formation of Ph$^\bullet$ and R$_3$Si$^\bullet$ is supported by ESR. These results are consistent with the mechanism proposed in Eq. (12.24).

$$Vi + Ph_2I^+ \rightarrow Vi^{\bullet+} + Ph^\bullet + Ph\text{–}I \ (h\nu)$$
$$Ph^\bullet + R_3Si\text{–}H \rightarrow Ph\text{–}H + R_3Si^\bullet \text{ and } R_3Si^\bullet + Ph_2I^+$$
$$\rightarrow R_3Si^+ + Ph^\bullet + Ph\text{–}I \qquad (12.24)$$

When Ru is added [244], more photons are absorbed by the Ru/Vi/TTMSS/Ph$_2$I$^+$ system under a polychromatic excitation (such as sunlight). The mechanism is based on the reactions of Eq. (12.25). After absorption of the light both by Ru and Vi, ^{3}Ru/Vi and ^{1}Vi/Ru electron transfer reactions are quite feasible (E_{ox}(Vi) = 0.88 V, E_{red}(Ru) = −1.33 V, $E(^1V-79) \sim 1.95$ eV, $E(^3$Ru$) \sim 2.1$ eV, and $\Delta G_{et} = +0.11$ and $+0.26$ eV). The phenyl radical reacts as in Eq. (12.24). These reactions contribute to the enhancement of the polymerization efficiency (rate and conversion).

$$^3Ru(bpy)_3{}^{2+} + Ph_2I^+ \rightarrow Ru(bpy)_3{}^{3+} + Ph^\bullet + Ph\text{–}I$$
$$\text{and } {}^3Ru(bpy)_3{}^{2+} + Vi \rightarrow Ru(bpy)_3{}^+ + Vi^{\bullet+} \quad (1)$$
$$^1Vi + Ru(bpy)_3{}^{2+} \rightarrow Ru(bpy)_3{}^+ + Vi^{\bullet+}$$
$$\text{and } Ru(bpy)_3{}^+ + Ph_2I^+ \rightarrow Ru(bpy)_3{}^{2+} + Ph^\bullet + Ph\text{–}I \quad (2) \qquad (12.25)$$

The performance for the FRPCP of an EPOX under a usual Xe–Hg lamp and green/red diode laser exposure is shown in Figure 12.6. Exposure to sunlight, LED, and fluorescent bulbs is also possible as well as the photopolymerization of renewable monomers. Other lasers or diode laser lines operating in the 420–670 nm range should likely be successfully used. Other PSs might be proposed to extend the photosensitivity up to 800–900 nm.

Iridium Complexes/Silane/Ph$_2$I$^+$ A series of Ir(Ligand)$_3$ complexes allow an interesting tunable absorption [246, 247] **(28–33)**. The observed mechanisms are roughly similar to that of Eq. (12.23). For the oxidative cycle, as in the case of Ru(L)$_3{}^{2+}$, the key parameters are the absorption properties and the oxidation ability of the Ir(Ligand)$_3$ excited states by Ph$_2$I$^+$. These two last Ru- and Ir-based systems can also work as efficient dual radical/cationic PIs (Section 10.2.9). The iridium complex is really suitable under soft irradiations provided by household fluorescent bulbs and household LED lamps [248]. Moreover, due to the already described catalytic cycle, the iridium complexes are not consumed and are incorporated into the final polymerized material; the film exhibits a photoluminescence that can be used in organic light emitting diode (OLED) applications.

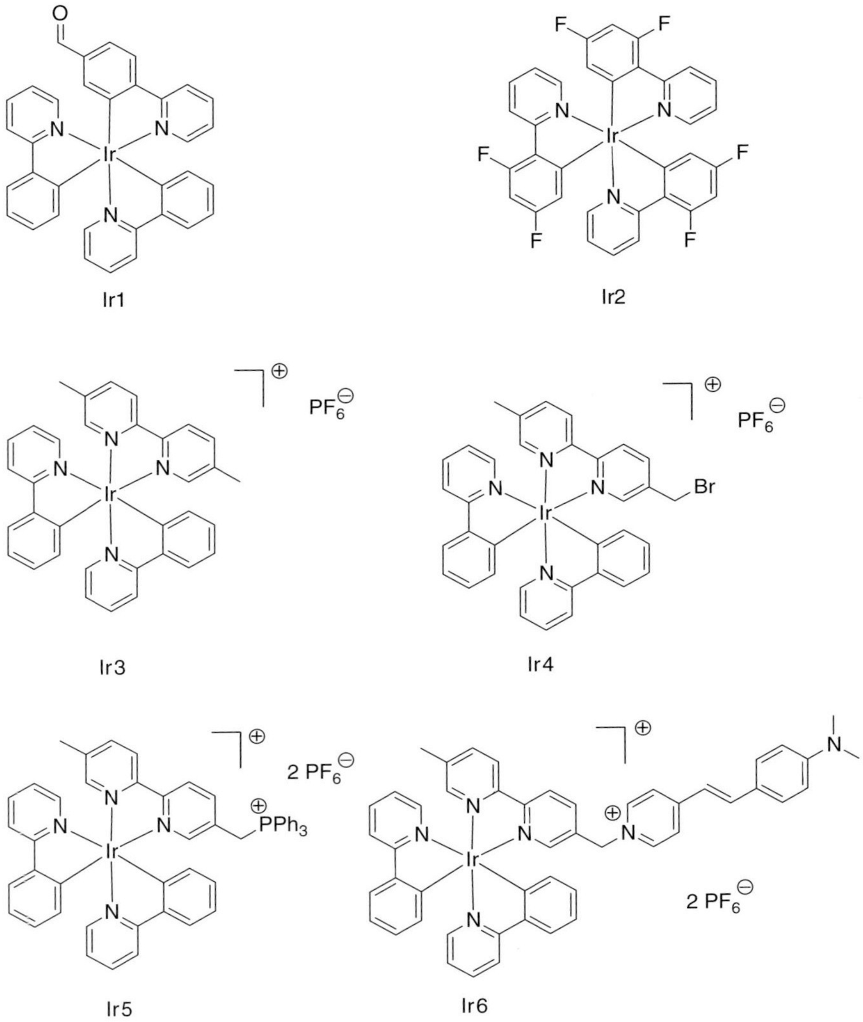

328 | *12 Cationic Photoinitiating Systems*

Ir1, Ir2, Ir3, Ir4, Ir5, Ir6

28–33

12.5.2.2.4 Photoinitiator/Alcohol/Iodonium Salt

Introducing a PI in the iodonium salt/alcohol combination allows an efficient visible-light-induced FRPCP [249]. As above, this three-component system was claimed to avoid the presence of an amine and more costly silanes (but its reactivity is lower and it does not really work under air). Scheme 12.17 summarizes the observed reactions using CQ.

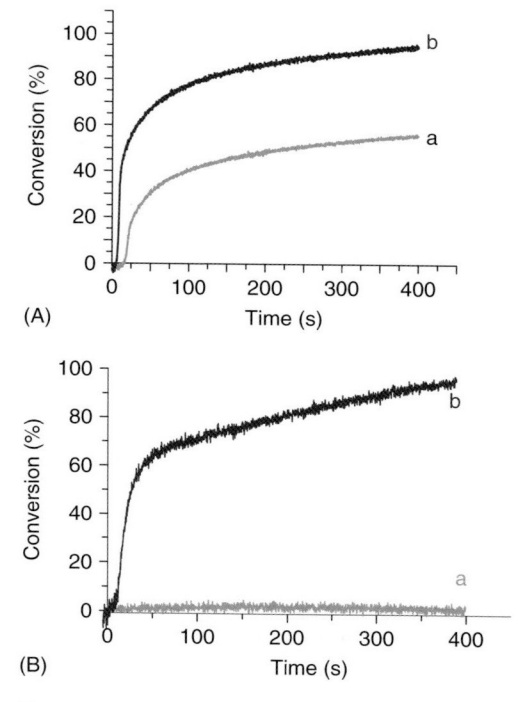

Figure 12.6 Photopolymerization profiles of an epoxide EPOX (a) on xenon lamp irradiation ($\lambda > 390$ nm) in the presence of (a) Ru(phen)$_3^{2+}$/Ph$_2$I$^+$ (0.2%/2% w/w) and (b) Ru(phen)$_3^{2+}$/TTMSS/Ph$_2$I$^+$ (0.2%/3%/2% w/w) and (B) on a diode laser irradiation (532 nm) in the presence of (a) Vi/Ph$_2$I$^+$ (0.2%/2% w/w) and (b) Ru/Vi/Ph$_2$I$^+$/TTMSS (0.2%/0.2%/2%/3% w/w). Under air. (Source: From [245].)

CQ + PhCH$_2$OH $\xrightarrow{h\nu}$ CQ$\overset{\bullet}{}$H + PhC$\overset{\bullet}{}$HOH

$\xrightarrow{\text{Ph}_2\text{I}^+}$ Ph$\overset{+}{\text{C}}$HOH
CQH$^+$ $\longrightarrow$ H$^+$ + CQ + PhCHO

Scheme 12.17

A similar behavior was noted using the titanium complex/benzylalcohol/iodonium salt system [250] in deaerated media. The primary process is a hydrogen abstraction generating the PhC($\bullet$)HOH radical, which further reduces the iodonium salt.

12.5.2.3 Role of the Onium Salt

As said above, the most efficient onium salt in FRPCP is clearly the iodonium salt. Triaryl sulfonium salts have strongly negative reduction potentials ($E_{1/2}^{\text{red}} \sim -1.1$ to -1.35 eV) that prevent such a kind of decomposition. On the contrary, an allylsulfonium salt (e.g., 2-ethoxycarbonyl-2-propenylthiophenium hexafluoroantimonate) reacts with benzophenone to form a ketyl radical that further reduces another allyl sulfonium salt ($E_{1/2}^{\text{red}} = -0.58$ eV) and generates

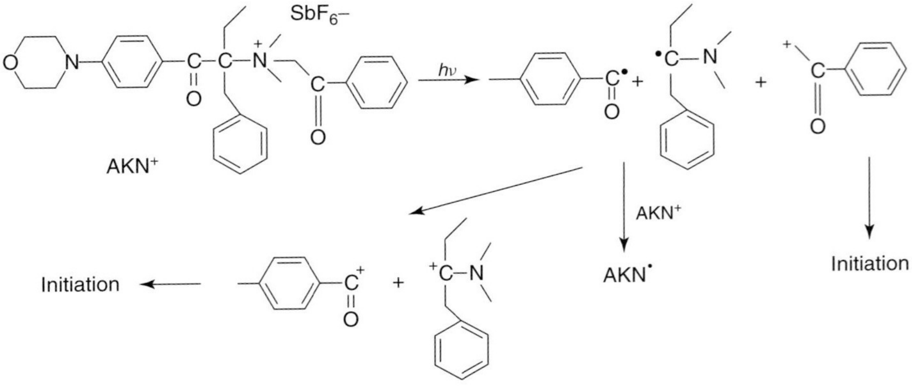

Scheme 12.18

a proton [251]. The reaction was also carried out with phosphonium salts [87, 252]. N-ethoxy-2-methylpyridinium (EMP$^+$) is characterized by lower reduction potentials than Ph$_3$S$^+$ (−0.58 V) and can also be used in FRPCP processes [253].

12.5.2.4 One-Component Radical Source-Onium Salt System

A new way has been recently explored [254] consisting of synthesizing a radical PI modified by an onium salt moiety. This was illustrated by AKN$^+$SbF$_6^-$ that is based on a commercial aminoketone PI MPPK moiety (Scheme 12.18) and appears as an ammonium salt. In this salt, the absorption maximum is shifted to 361 nm ($\varepsilon = 40400$ mol^{-1} l cm^{-1}). After α-cleavage as in the parent MPPK compound, the electron transfer occurring between the two radicals and the starting compound leads to three carbon-centered cations (the phenacyl cation being due to a secondary cleavage of the $^{\bullet}$C(R$_2$)–N$^+$(R$_3$) radical). A concomitant mechanism might be a homolytic (followed by electron transfer) or heterolytic N–C cleavage yielding MPPK and the phenacyl cation.

12.5.2.5 Photoinitiation Step

The photoinitiation step in FRPCP using a radical source RS and an onium salt On$^+$ is depicted in Scheme 12.19. The initiation quantum yield is a function of the

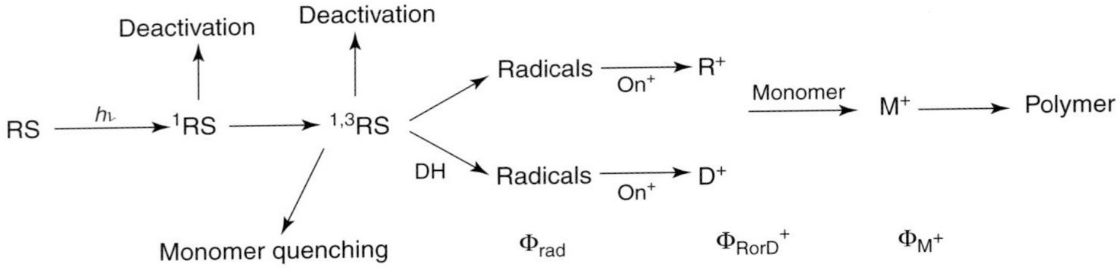

Scheme 12.19

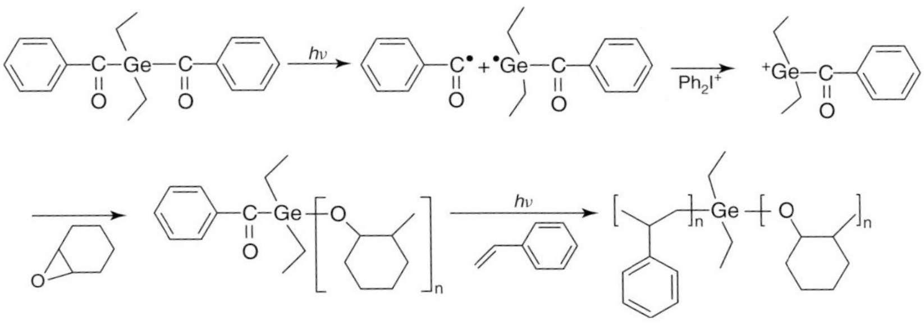

Scheme 12.20

quantum yield in free radicals Φ_{rad} and the yields in cations Φ_{R+} and monomer cations Φ_{M+}.

12.5.2.6 Recent Applications

In addition to its use in the cationic polymerization area, the FRPCP reaction was elegantly carried out for the synthesis of block copolymers [255] as seen, for example, in Scheme 12.20. The irradiation of dibenzoyl diethyl germane in the presence of an iodonium salt and cyclohexene oxide leads to a polyether. Further irradiation creates a germyl radical that initiates a radical polymerization, thereby producing, for example, a polystyrene/polyether copolymer.

12.5.3
Radical/Metal-Salt-Based Systems

The mechanism shown in Scheme 12.21 [256] for the 2,2-dimethoxy-2-phenyl acetophenone/iodonium salt/silver salt system [257] is based on the generation of (i) a carbon-centered cation from the initially formed radicals (it is assumed that mainly the dimethoxybenzyl radical is oxidized) and (ii) *in situ* silver nanoparticles Ag(0)NP. Using a Type II PI, the carbon-centered radical (e.g., the ketyl radical of CQ) can be easily oxidized by the metal salt. This procedure was also extended to gold nanocomposites [258].

12.5.4
Addition/Fragmentation Reaction

Another FRPCP way suggests the use of an addition/fragmentation reaction involving a radical source and a cationic salt (see e.g., in [259]). For example,

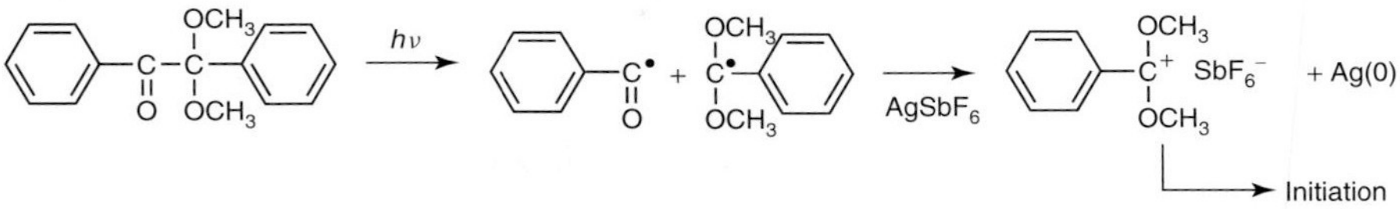

Scheme 12.21

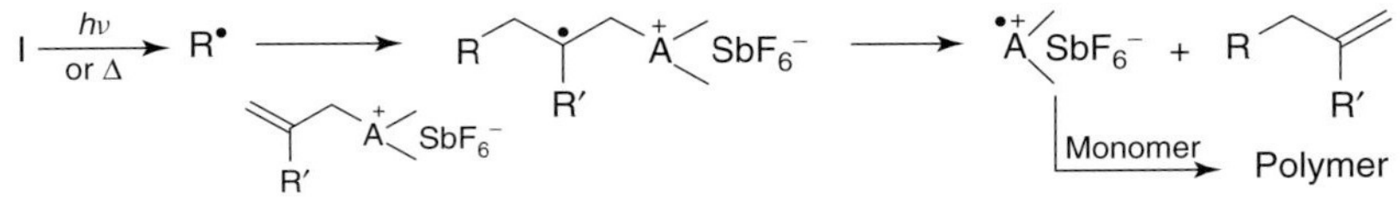

Scheme 12.22

(Scheme 12.22), the radical adds to a suitable allyl onium salt (pyridinium, sulfonium, phosphonium, and ammonium). The formed adduct further cleaves into a cation radical structure and an unsaturated compound. The radical source and the cationic salt can also be linked in a single molecule such as in the morpholinoketone ethyl acrylate ammonium salt or the benzophenone allyl ammonium salt. A careful selection of the radical source allows the tuning of the absorption to higher wavelengths [260–266].

12.6
Miscellaneous Systems

Using vinyl halides Vi-Br and ZnI_2 allows to initiate a cationic photopolymerization. The primary heterolytic cleavage of Vi-Br on light absorption forms an ion pair Vi^+ Br^- that adds on a vinyl ether double bond. The Br... ZnI_2 association favors the propagation of the vinyl ether polymerization (Scheme 12.23) [267] as the C–Br bond is polarized. The carbocation is the propagating species.

12.7
Photosensitive Systems for Living Cationic Polymerization

Living cationic polymerization has been evoked in Section 4.2.11. Photosensitive systems for such a process still lie on onium salts. The key point when using vinyl ethers is the stabilization of the carbocation. This could be achieved by (i) introducing highly nucleophilic counteranions (bromide or chloride) [268] or (ii) adding nucleophilic agents such as zinc chloride (Section 12.2.1.5.7) [86] or sulfides (Scheme 12.24) [269].

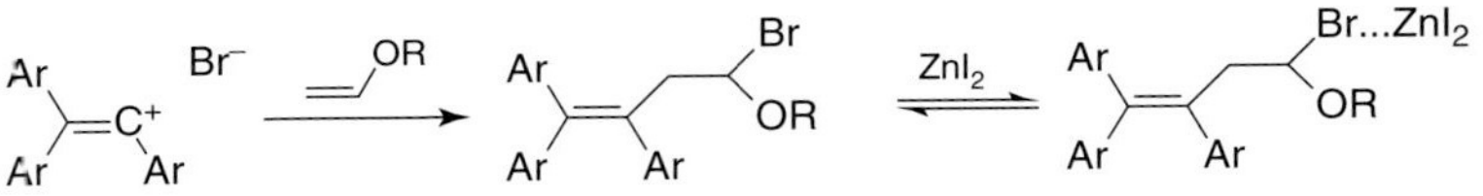

Scheme 12.23

Scheme 12.24

Scheme 12.25

The living cationic photopolymerization of epoxides can be conducted using the same strategy, for example, by adding sulfides (Scheme 12.25) [270].

12.8
Photosensitive Systems for Hybrid Cure

As shown in Section 4.2.8, the hybrid cure technology requires either a one- or a two-step irradiation using two PIs: a radical PI and a cationic PI. Only one PI might also be used as, for example, iodonium and sulfonium salt derivatives release radicals on photolysis (Section 12.2.1.5.1 and 12.2.1.6.1). Radical and cationic PIs, unfortunately, present a strong difference in practical efficiency (partly due to their absorption spectra) and reactivity. UV lights have to be used for the excitation with, for example, a benzoin ether (or a BAPO derivative) and an iodonium (or a sulfonium) salt. This drawback becomes still more important in the few encountered examples when using a visible-light-absorbing radical PI (e.g., a titanocene) and an available UV cationic PI under a visible light source (Hg doped, Xe lamp, or sun) where less UV light is available. The network characteristics are affected by the formation kinetics (fast for the acrylate and slow for the epoxide) and the final conversion of the polymer (higher for the acrylate than for the epoxide). In other words, almost similar fast polymerization rates and conversions can be achieved only under UV lights.

To overcome this problem and use visible lights, a new concept has been recently proposed for the design of a dual radical/cationic PI (Sections 10.2.11 and 12.5.2.2.3). This has opened the way for the remarkable simultaneous generation of radicals and cations (i) under visible lights and soft irradiation conditions under air and (ii) together with a catalytic behavior of the PI (allowing a low product consumption) and an access to a large range of photopolymerizable formulations (including natural and renewable products). For example, using a Ru(phenanthroline)$_3^{2+}$/silane/iodonium salt system has allowed an efficient hybrid cure of an acrylate/epoxide formulation on a fluorescent bulb irradiation as both monomers exhibit a high conversion at the same exposure time (acrylate and epoxide conversions are ~65 and ~45%, respectively) [245].

References

1. Hoyle, C.E. and Kinstle, J.F. (eds) (1990) *Radiation Curing of Polymeric Materials*, ACS Symposium Series, Vol. 417, American Chemical Society.
2. Fouassier, J.P. and Rabek, J.F. (eds) (1990) *Lasers in Polymer Science and Technology: Applications*, CRC Press, Boca Raton, FL.
3. Pappas, S.P. (1986) *UV-Curing: Science and Technology*, Technology Marketing Corporation, Stamford; Plenum Press New York (1992).

4. Fouassier, J.P. and Rabek, J.F. (eds) (1993) *Radiation Curing in Polymer Science and Technology*, Chapman & Hall, London.
5. Reiser, A. (1989) *Photoreactive Polymers: The Science and Technology of Resists*, John Wiley & Sons, Inc., New York.
6. Fouassier, J.P. (1995) *Photoinitiation, Photopolymerization, Photocuring*, Hanser Publishers, Münich.
7. Scranton, A.B., Bowman, A., and Peiffer, R.W. (eds) (1997) *Photopolymerization: Fundamentals and Applications*, ACS Symposium Series, Vol. 673, American Chemical Society, Washington, DC.
8. Olldring, K.H. (ed.) (1997) *Chemistry and Technology of UV and EB Formulation for Coatings, Inks and Paints*, vol. I–VIII, John Wiley & Sons, Ltd and Sita Technology, Ltd, London.
9. Davidson, S. (1999) *Exploring the Science, Technology and Application of UV and EB Curing*, Sita Technology, Ltd, London.
10. Neckers, D.C. (1999) *UV and EB at the Millenium*, Sita Technology, Ltd, London.
11. Crivello, J.V. and Dietliker, K. (1999) in *Photoinitiators for Free Radical Cationic and Anionic Photopolymerization* (ed. G Bradley), John Wiley & Sons, Inc., pp. 125–141.
12. Belfied, K.D. and Crivello, J.V. (eds) (2003) *Photoinitiated Polymerization*, ACS Symposium Series, Vol. 847, American Chemical Society, Washington, DC.
13. Fouassier, J.P. (ed.) (2006) *Photochemistry and UV Curing*, Research Signpost, Trivandrum.
14. Schwalm, R. (2007) *UV Coatings: Basics, Recent Developments and New Applications*, Elsevier, Oxford.
15. Allen, N.S. (ed.) (2010) *Photochemistry and Photophysics of Polymer Materials*, John Wiley & Sons, Inc., Hoboken.
16. Fouassier, J.P. and Allonas, X. (eds) (2010) *Basics of Photopolymerization Reactions*, Research Signpost, Trivandrum.
17. Cunningham, A.F. and Desobry, V. (1993) in *Radiation Curing in Polymer Science and Technology*, vol. 2 (eds J.P. Fouassier and J.F. Rabek), Elsevier, Barking, pp. 323–374.
18. Timpe, H.J., Jockush, S., and Korner, K. (1993) in *Radiation Curing in Polymer Science and Technology*, vol. 2 (ed. J.P. Fouassier), Elsevier, Barking, pp. 575–602.
19. Crivello, J.V. (1993) in *Radiation Curing in Polymer Science and Technology*, vol. 2 (eds J.P. Fouassier and J.F. Rabek), Elsevier, Barking, pp. 435–472.
20. Hacker, N.P. (1993) in *Radiation Curing in Polymer Science and Technology*, vol. 2 (eds J.P. Fouassier and J.F. Rabek), Elsevier, Barking, pp. 473–504.
21. Sahyun, M.R., De Voe, R.J., and Olofson, P.M. (1993) in *Radiation Curing in Polymer Science and Technology*, vol. 2 (ed. J.P. Fouassier), Elsevier, Barking, pp. 505–529.
22. Kahveci, M.U., Gilmaz, A.G., and Yagci, Y. (2010) in *Photochemistry and Photophysics of Polymer Materials* (ed. N.S. Allen), John Wiley & Sons, Inc., Hoboken, pp. 421–478.
23. Ivan, M.G. and Scaiano, J.C. (2010) in *Photochemistry and Photophysics of Polymer Materials* (ed. N.S. Allen), John Wiley & Sons, Inc., Hoboken, pp. 479–508.
24. Lalevée, J., El Roz, M., Allonas, X., and Fouassier, J.P. (2009) in *Organosilanes: Properties, Performance and Applications*, Chapter 6 (eds E. Wyman and M.C. Skief), Nova Science Publishers, Hauppauge, pp. 174–198.
25. Lalevée, J., Tehfe, M.A., Allonas, X., and Fouassier, J.P. (2010) in *Polymer Initiators*, Chapter 8 (ed. W.J. Ackrine), Nova Science Publishers, Hauppauge, pp. 201–229.
26. Lalevée, J. and Fouassier, J.P. (2011) *Polym. Chem.*, **2**, 1107–1113.
27. Muller, U., Utterodt, A., Morke, W., Deubzer, B., and Herzig, C. (2003) *Photoinitiated Polymerization*, ACS Symposium Series, Vol. 847 (eds K.D. Belfied and J.V. Crivello), American Chemical Society, pp. 202–212.
28. Park, J., Kihara, N., Ikeda, T., and Endo, T. (1993) *J. Polym. Sci. Part A: Polym. Chem.*, **31**, 1083–1085.
29. Cao, W., Zhao, C., and Cao, J. (1998) *J. Appl. Polym. Sci.*, **69**, 1975–1982.

30. Grimshaw, J. (2004) *CRC Handbook of Organic Photochemistry and Photobiology*, 2nd edn (eds A. Griesbeck, J. Cook, F. Ghetti), CRC Press, pp. 43 / 1–43/17.
31. Timpe, H.J. (1986) *Sitzungsber. Akad. Wiss. DDR (Leipzig)*, **13**, 1.
32. Ambroz, H.B., Przybytniak, G.K., Stradowski, C.Z., and Wolszczak, M. (1990) *J. Photochem. Photobiol., A: Chem.*, **52**, 369–374.
33. Milanesi, S., Fagnoni, M., and Albini, A. (2005) *J. Org. Chem.*, **70**, 603–610.
34. Crivello, J.V. (2002) *Des. Monomers Polym.*, **5**, 141–154.
35. Aydogan, B., Gacal, B., Yildirim, A., Yonet, N., Yuksel, Y., and Yagci, Y. (2006) in *Photochemistry and UV Curing* (ed. J.P. Fouassier), Research Signpost, Trivandrum, pp. 187–201.
36. Sangermano, M. and Crivello, J.V. (2003) in *Photoinitiated Polymerization*, ACS Symposium Series, Vol. 847 (eds K.D. Belfied and J.V. Crivello), American Chemical Society, Washington, DC, pp. 242–252.
37. Crivello, J.V. and Lam, J.H. (1979) *J. Polym. Sci., Part A: Polym. Chem.*, **17**, 977–982.
38. Crivello, J.V. (1984) *Adv. Polym. Sci.*, **62**, 1–48.
39. Norcini, G., Casiraghi, A., Visconti, M., and Li Bassi, G. (2003) PCT Int. Appl. WO 03 8 404.
40. Crivello, J.V., Ma, J., and Jiang, F. (2002) *J. Polym. Sci., Part A: Polym. Chem.*, **40**, 3465–3480.
41. Pappas, S.P. (1985) *Prog. Org. Coat.*, **13**, 35–64.
42. Studer, K. (2009) Proceedings of the RadTech Europe Conference.
43. Crivello, J.V. (2010) in *Basics of Photopolymerization Reactions*, vol. 2 (eds J.P. Fouassier and X. Allonas), Research Signpost, Trivandrum, pp. 101–118.
44. Corcione, C.E., Frigione, M.E., Greco, A., and Maffezzoli, A. (2003) in *Quantitative Level of Chemical Reaction* (eds G.E. Zaikov and A. Jimenez), Nova Science Publischers, New York, pp. 21–39.
45. Sipani, V. and Scranton, A.B. (2003) *J. Photochem. Photobiol., A: Chem.*, **159**, 189–195.
46. Crivello, J.V. (1983) *Ann. Rev. Mater. Sci.*, **13**, 173–190.
47. Crivello, J.V., Lam, J.H.W., and Volante, C.N. (1977) *J. Radiat. Curing*, **4**, 2–16.
48. Huang, R., Ficek, B.A., Glover, S.O., and Scranton, A.B. (2007) *Radtech Rep.*, **21**, 30, 32–35.
49. Kim, D. and Jessop, J.L.P. (2006) Proceedings of the RadTech USA.
50. Yagci, Y. and Schnabel, W. (1999) *Angew. Makromol. Chem.*, **270**, 38–41.
51. Crivello, J.V. and Ortiz, R.A. (2002) *J. Polym. Sci., Part A: Polym. Chem.*, **40**, 2298–2309.
52. Souane, R., Tehfe, M.A., Lalevée, J., Gigmes, D., and Fouassier, J.P. (2010) *Macromol. Chem. Phys*, **211**, 1441–1445.
53. Castellanos, F., Fouassier, J.P., Priou, D., and Cavezzan, A. (1997) US Patent 5, 668, 192, Sep. 16, 1997, Rhone Poulenc.
54. Castellanos, F., Fouassier, J.P., Priou, C., and Cavezzan, J. (1996) *J. Appl. Polym. Sci.*, **60**, 705–713.
55. Murphy, S.T., Zou, C., Miers, J.B., Ballew, R.M., Dlott, D.D., and Schuster, G.B. (1993) *J. Phys. Chem.*, **97**, 13152–13157.
56. Komarova, E.Y., Ren, K., and Neckers, D.C. (2002) *Langmuir*, **18**, 4195–4197.
57. Li, H., Ren, K., Zhang, W., Malpert, J.H., and Neckers, D.C. (2001) *Macromolecules*, **34**, 2019.
58. Ren, K., Malpert, J.H., Gu, H., Li, H., and Neckers, D.C. (2002) *Tetrahedron*, **58**, 5267–5273.
59. Crivello, J.V., Lee, J.L., and Conlon, D.A. (1988) *Makromol. Chem., Macromol. Symp.*, **13/14**, 145.
60. Gu, H., Zhang, W., Feng, K., and Neckers, D.C. (2000) *J. Org. Chem.*, **65**, 3484–3490.
61. Hartwig, A., Harder, A., Luhring, A., and Schroder, H. (2001) *Eur. Polym. J.*, **37**, 1149–1455.
62. Koser, G.F. and Carman, C.S. (1985) U.S. Patent 4, 513, 137.
63. Saeva, F.D., Breslin, D.T., and Martic, P.A. (1989) *J. Am. Chem. Soc.*, **111**, 1328–1330.

64. Crivello, J.V., Ma, J., Jiang, F., Hua, H., Ahn, J., and Ortiz, R.A. (2004) *Macromol. Symp.*, **215**, 165–177.
65. Pappas, S.P. (1985) *Prog. Org. Coat.*, **13**, 35–42.
66. de Voe, R.J., Sahyun, M.R.V., Serpone, N., and Sharma, D.K. (1987) *Can. J. Chem.*, **65**, 2342–2349.
67. Jancovicova, V., Brezova, V., Ciganek, M., and Cibulkova, Z. (2000) *J. Photochem. Photobiol., A*, **136**, 195–202.
68. Dektar, J.L. and Hacker, N.P. (1990) *J. Org. Chem.*, **55**, 639–647.
69. Hacker, N.P. and Welsh, K.M. (1991) *Macromolecules*, **24**, 2137.
70. Dektar, J.L. and Hacker, N.P. (1991) *J. Org. Chem.*, **56**, 1838–1844.
71. Dektar, J.L. and Hacker, N.P. (1989) *J. Photochem. Photobiol., A: Chem.*, **46**, 233–238.
72. Timpe, H.J. and Schikowsky, V. (1989) *J. Prakt. Chem.*, **331**, 447–460.
73. Dektar, J.L. and Hacker, N.P. (1991) *J. Org. Chem.*, **56**, 2280–2282.
74. Gu, H., Ren, K., Grinevich, O., Malpert, J., and Neckers, D.C. (2001) *J. Org. Chem.*, **66**, 4161–4164.
75. Crivello, J.V., Burr, D., and Fouassier, J.P. (1994) *Macromol. Sci., Pure Appl. Chem., A*, **31**, 677–684.
76. Ren, K., Serguievski, P., Gu, H., Grinevich, O., Malpert, J.H., and Neckers, D.C. (2002) *Macromolecules*, **35**, 898–904.
77. Selvaraju, C., Sivikumar, A., and Ramamurthy, P. (2001) *J. Photochem. Photobiol., A: Chem.*, **138**, 213–226.
78. Takahashi, E., Sanda, F., and Endo, T. (2003) *J. Polym. Sci., Part A: Polym. Chem.*, **41**, 3816–3827.
79. Takahashi, E., Sanda, F., Yamamoto, T., and Endo, T. (2003) *J. Polym. Sci., Part A: Polym. Chem.*, **41**, 3828–3837.
80. Rajaraman, S.K., Mowers, W.A., and Crivello, J.V. (1999) *Macromolecules*, **32**, 36–47.
81. Crivello, J.V. and Lam, J.H.W. (1979) *J. Polym. Sci., Polym. Chem. Ed.*, **17**, 2877.
82. Crivello, J.V. and Lam, J.H.W. (1980) *J. Polym. Sci., Polym. Chem. Ed.*, **18**, 1021.
83. Crivello, J.V. and Lam, J.H.W. (1979) *J. Poym. Sci., Polym. Chem. Ed.*, **17**, 3845.
84. Crivello, J.V. and Lam, J.H.W. (1978) *J. Polym. Sci., Polym. Chem. Ed.*, **16**, 2441.
85. Chen, S., Cook, W.D., and Chen, F. (2009) *Macromolecules*, **42**, 5965–5975.
86. Kahveci, M.U., Tasdelen, M.A., and Yagci, Y. (2008) *Macromol. Rapid Commun.*, **29**, 202–206.
87. Kahveci, M.U., Tasdelen, M.A., Cook, W.D., and Yagci, Y. (2008) *Macromol. Chem. Phys.*, **209**, 1881–1886.
88. Crivello, J.V. and Acosta Ortiz, R. (2001) *J. Polym. Sci., Part A: Polym. Chem.*, **39**, 2385–2392.
89. Crivello, J.V. and Acosta Ortiz, R. (2001) *J. Polym. Sci., Part A: Polym. Chem.*, **39**, 3578–3586.
90. Welsh, K.M., Dektar, J.L., Hacker, N.P., and Turro, N.J. (1989) *Polym. Mater. Sci. Eng.*, **61**, 181.
91. Welsh, K.M., Dektar, J.L., Garcia-Garibaya, M.A., and Turro, N.J. (1992) *J. Org. Chem.*, **57**, 4179–4184.
92. Ohmori, N., Nakazono, Y., Hata, M., Hoshino, T., and Tsuda, M. (1998) *J. Phys. Chem. B*, **102**, 927–930.
93. Saeva, F.D., Morgan, B.P., and Luss, H.R. (1985) *J. Org. Chem.*, **50**, 4360–4362.
94. Tilley, M., Pappas, B., Pappas, S.P., Yagci, Y., Schnabel, W., and Thomas, J.K. (1989) *J. Imaging Sci.*, **33**, 62.
95. Crivello, J.V. and Lee, J.L. (1983) *Macromolecules*, **16**, 864.
96. Crivello, J.V. and Kong, S. (2000) *Macromolecules*, **33**, 825–832.
97. Crivello, J.V. and Kong, S. (2000) *J. Polym. Sci., Part A: Polym. Chem.*, **38**, 1433–1442.
98. Kasapoglu, F., Aydin, M., Arsu, N., and Yagci, Y. (2003) *J. Photochem. Photobiol., A: Chem.*, **159**, 151–159.
99. Yagci, Y., Durmaz, Y.Y., and Aydogan, B. (2007) *Chem. Rec.*, **7**, 78–90.
100. Kong, S., Jang, M., Jiang, F., Gomurashvili, Z., Hua, and Y., Crivello, J.V. (2004) Proceedings of the RadTech USA.
101. Crivello, J.V. and Fan, M. (1992) EP 476426.

102. Ohmori, N., Nakazono, Y., Tsujino, A., Hata, M., Oikawa, S., and Tsuda, M. (1995) *J. Photopolym. Sci. Technol.*, **8**, 653–656.
103. Park, J., Kihara, N., Ikeda, T., and Endo, T. (1997) *Macromolecules*, **30**, 3414–3416.
104. Hoyle, C.E., Thames, S.F., Mullen, L.R., and Schmidt, D.L. (2001) *J. Polym. Sci., Part A: Polym. Chem.*, **39**, 571–584.
105. Takahashi, E., Sanda, F., and Endo, T. (2004) *J. Appl. Polym. Sci.*, **91**, 589–597.
106. Kim, J.Y., Patil, P.S., Seo, B.J., Kim, T.S., Kim, J., and Kim, T.H. (2008) *J. Appl. Polym. Sci.*, **108**, 858–862.
107. Saeva, F.D. and Breslin, D.T. (1989) *J. Org. Chem.*, **54**, 712.
108. Fernandez-Francos, X., Salla, J.M., Cadenato, A., Morancho, J.M., Mantecon, A., Serra, A., and Ramis, X. (2006) *J. Polym. Sci., Part A: Polym. Chem.*, **45**, 16–25.
109. He, J.H. and Mendoza, V.S. (1996) *J. Polym. Sci., Part A: Polym. Chem.*, **34**, 2809–2816.
110. Crivello, J.V. and Bulut, U. (2006) *J. Polym. Sci., Part A: Polym. Chem.*, **44**, 6750–6764.
111. Bhatnagar, U. and Srivastava, A.K. (1991) *J. Photochem. Photobiol. A: Chem.*, **59**, 393–396.
112. Kitamura, T., Tanaka, T., and Taniguchi, H. (1992) *Chem. Lett.*, **11**, 2245.
113. Sundell, P.E., Skolling, O., Williamson, S., Hoyle, C., and Jonsson, S. (1993) Proceedings of the RadTech Europe Conference, pp. 430–436.
114. Sanrame, C.N., Brandao, M.S.B., Coenjarts, C., Scaiano, J.C., Pohlers, G., Suzuki, Y., and Cameron, J.F. (2004) *J. Photochem. Photobiol. Sci.*, **3**, 1052–1059.
115. Yagci, Y. and Reetz, I. (1998) *Prog. Polym. Sci.*, **23**, 1485–1495.
116. Crivello, J.V. and Ahn, J. (2003) *J. Polym. Sci., Part A: Polym. Chem.*, **41** (16), 2570–2587.
117. Auhn, J. and Crivello, J.V. (2002) *Polym. Prepr.*, **43**, 918–919.
118. Crivello, J.V. (2006) *J. Macromol. Sci., Part A: Pure Appl. Chem.*, **43**, 1339–1353.
119. Zhou, W., Kuebler, S.M., Carrig, D., Perry, J.W., and Marder, S.R. (2002) *J. Am. Chem. Soc.*, **124**, 1897–1901.
120. Crivello, J.V. and Lee, J.L. (1989) *J. Polym. Sci.: Part A: Polym. Chem.*, **27**, 3951–3968.
121. Akhtar, S.R., Crivello, J.V., Lee, J.L., and Schmitt, M.L. (1990) *Chem. Mater.*, **2**, 732–737.
122. Hoefer, M. and Liska, R. (2009) *J. Polym. Sci., Part A: Polym. Chem.*, **47**, 3419–3430.
123. Yagci, Y., Kornowski, A., and Schnabel, W. (1992) *J. Polym. Sci., Part A: Polym. Chem.*, **30**, 1987–1991.
124. Gomez, M.L., Palacios, R.E., Previtali, C.M., Montejano, H.E., and Chesta, C.A. (2002) *J. Polym. Sci.: Part A: Polym. Chem.*, **40**, 901–913.
125. Kasapoglu, F., Onen, A., Bicak, N., and Yagci, Y. (2002) *Polymer*, **43**, 2575–2579.
126. Crivello, J.V., Suh, D.H., and Carter, A.M. (1992) *Polym. Mater. Sci. Eng.*, **67**, 258–259.
127. Yagci, Y. and Denizligil, S. (1995) *J. Polym. Sci., Part A: Polym. Chem.*, **33**, 1461–1464.
128. Dossow, D., Zhu, Q.Q., Hizal, G., Yagci, Y., and Schnabel, W. (1996) *Polymer*, **37**, 2821–2826.
129. Kozlecki, T. and Wilk, K.A. (1996) *J. Phys. Org. Chem.*, **9**, 645–651.
130. Yagci, Y. and Endo, T. (1997) *Adv. Polym. Sci.*, **127**, 59–86.
131. Arsu, N., Oenen, A., and Yagci, Y. (1996) *Macromolecules*, **29**, 8973–8974.
132. Monecke, P., Schnabel, W., and Yagci, Y. (1997) *Polymer*, **38**, 5389–5395.
133. Schnabel, W. (2000) *Macromol. Rapid Commun.*, **21**, 628–642.
134. Bhowmik, P.K., Burchett, R.A., Han, H., and Cebe, J.J. (2001) *Macromolecules*, **34**, 7579–7581.
135. Bhowmik, P.K., Burchett, R.A., Han, H., and Cebe, J.J. (2002) *Polymer*, **43**, 1953–1958.
136. Yonet, N., Bicak, N., and Yagci, Y. (2006) *Macromolecules*, **39**, 2736–2738.

137. Czech, Z., Gasiorowska, M., and Soroka, J. (2007) *J. Appl. Polym. Sci.*, **106**, 558–561.
138. Durmaz, Y.Y., Zaim, O., and Yagci, Y. (2008) *Macromol. Rapid Commun.*, **29**, 892–896.
139. Yagci, Y., Kminek, I., and Schnabel, W. (1992) *Eur. Polym. J.*, **28**, 387–390.
140. Schnabel, W., Yagci, Y., Kornowski, A., and Massonne, K. (1992) EP 498194.
141. Timpe, H.J. and Rautschek, H. (1991) *Angew. Makromol. Chem.*, **193**, 135–145.
142. Ledwith, A. (1982) *Polym. Prepr.*, **23**, 323–326.
143. Tsao, J.T. (1979) US Patent 4, 139, 655.
144. Takahashi, E., Sanda, F., and Endo, T. (2002) *J. Polym. Sci., Part A: Polym. Chem.*, **40**, 1037–1046.
145. Celler, J.A., Schwabacher, A.W., and Schultz, A.R. (1983) *Ind. Eng. Chem.*, **22**, 20.
146. Bi, Y. and Neckers, D.C. (1992) *Tetrahedron Lett.*, **33**, 1139–1142.
147. Matano, Y., Shinokura, T., Yoshikawa, O., and Imahori, H. (2008) *Org. Lett.*, **10**, 2167–2170.
148. Matano, Y. and Imahori, H. (2009) PCT Int. Appl. WO 2009113217.
149. Abu-Abdoun, I.I. and Ledwith, A. (2007) *J. Polym. Res.*, **14**, 99–105.
150. Abu-Abdoun, I.I. (2008) *Int. J. Polym. Mater.*, **57**, 584–593.
151. Neckers, D.C. and Abu-Abdoun, I.I. (1984) *Macromolecules*, **17**, 2468.
152. Alonso, E.O., Johnston, L.J., Scaiano, J.C., and Toscano, V.G. (1992) *Can. J. Chem.*, **70**, 1784.
153. Takata, T., Takuma, K., and Endo, T. (1993) *Makromol. Chem. Rapid Commun.*, **14**, 203.
154. Breslin, D.T. and Saeva, F.D. (1988) *J. Org. Chem.*, **53**, 713.
155. Hacker, N.P. and Dektar, J.L. (1989) *Polym. Mater. Sci. Eng.*, **61**, 76.
156. Durmaz, Y.Y., Yilmaz, G., Tasleden, M.A., Aydogan, B., Koz, B., and Yagci, Y. (2010) in *Basics of Photopolymerization Reactions*, vol. 1 (eds J.P. Fouassier and X. Allonas), Research Signpost, Trivandrum, pp. 7–22.
157. Priou, C., Soldat, A., Cavezzan, J., and Fouassier, J.P. (1995) *J. Coat. Technol.*, **67**, 71–78.
158. Lohse, F. and Zweifel, H. (1986) *Adv. Polym. Sci.*, **78**, 62–69.
159. Klingert, B., Riediker, M., and Roloff, A. (1988) *Comments Inorg. Chem.*, **7**, 109–115.
160. Kunding, E.P., Xu, L.H., Kondratenko, M., Cunningham, A.F. Jr., and Kuntz, M. (2007) *Eur. J. Inorg. Chem.*, **2007**, 2934–2943.
161. Ronco, S., Ferraudi, G., Roman, E., and Hernandez, S. (1989) *Inorg. Chim. Acta*, **161**, 183.
162. Lehmann, R.E. and Kochi, J.K. (1991) *J. Am. Chem. Soc.*, **113**, 501.
163. Serpone, N. and Jamieson, M.A. (1989) *Coord. Chem. Rev.*, **93**, 87–92.
164. Leong, N.S. and Manners, I. (2008) *J. Organometal. Chem.*, **693**, 802–807.
165. Murata, K., Ito, M., Inagaki, A., and Akita, M. (2010) *Chem. Lett.*, **39**, 915–917.
166. Meier, K. and Zweifel, H. (1985) Proceedings of the RadCure Europe, Basle, Technical Paper FC 85-417.
167. Gill, T.P. and Mann, K.R. (1983) *Inorg. Chem.*, **22**, 1986.
168. Meier, K. and Rhis, G. (1985) *Angew. Chem.*, **97**, 879.
169. Rabek, J.F., Linden, L.A., and Wu, S.K. (1991) Proceedings of the RadTech Europe, pp. 618–624.
170. Roman, E., Barrera, M., Hernandez, S., and Giannotti, C. (1989) *NATO ASI Ser. C*, **257**, 327.
171. Criqui, A., Lalevée, J., Allonas, X., and Fouassier, J.P. (2008) *Macromol. Chem. Phys.*, **209**, 2223–2231.
172. Gaube, G. (1986) Proceedings of the Radcure '86, pp. 15–22.
173. Valis, J. and Weidlich, T. (2003) *Adv. Colour Sci. Technol.*, **6**, 73–82.
174. Wang, T., Chen, J.W., Li, Z.Q., and Wan, P.Y. (2007) *J. Photochem. Photobiol. A: Chem.*, **187**, 389–394.
175. Li, Z.Q., Li, M., Li, G.L., Chen, Y., Wang, X.N., and Wang, T. (2009) *Int. J. Photoenergy*, 6, Article ID 981065, www.hindawi.com.
176. Wang, T., Li, Z.Q., Zhang, Y., and Lu, M.B. (2009) *Prog. Org. Coat.*, **65**, 251–256.

177. Li, M., Chen, Y., Zhang, H., and Wang, T. (2010) *Prog. Org. Coat.*, **68**, 234–239.
178. Chen, X. and Chen, Y. (1997) *J. Appl. Polym. Sci.*, **66**, 2551–2554.
179. Hayase, S., Onighi, Y., Suzuki, S., and Wada, M. (1986) *Macromolecules*, **19**, 968–973.
180. Sangermano, M. and Crivello, J.V. (2003) in *Photoinitiated Polymerization*, ACS Symposium Series, Vol. 847 (eds K.D. Belfied and J.V. Crivello), American Chemical Society, pp. 242–252.
181. Hua, Y. and Crivello, J.V. (2003) in *Photoinitiated Polymerization*, ACS Symposium Series, Vol. 847 (eds K.D. Belfied and J.V. Crivello), American Chemical Society, pp. 219–230.
182. Manivannan, G., Fouassier, J.P., and Crivello, J.V. (1992) *J. Polym. Sci., Polym. Lett.*, **30**, 1999.
183. Eckert, G. and Goez, M. (1999) *J. Am. Chem. Soc.*, **121**, 2274–2280.
184. Dektar, J.L. and Hacker, N.P. (1988) *J. Org. Chem.*, **53**, 1833–1837.
185. Hua, Y., Jiang, F., and Crivello, J.V. (2002) *Chem. Mater.*, **14**, 2369–2377.
186. Yagci, Y. (1998) *Macromol. Symp.*, **134**, 177–188.
187. Toba, Y. (2000) *J. Polym. Sci., Part A: Polym. Chem.*, **38**, 982–987.
188. Kura, H., Fujihara, K., Kimura, A., Ohno, T., Matsumura, M., Hirata, Y., and Okada, T. (2001) *J. Polym. Sci., Part B: Polym. Phys.*, **39**, 2937–2946.
189. Crivello, J.V. and Jang, M. (2003) *J. Photochem. Photobiol. A: Chem.*, **159**, 173–188.
190. Rodrigues, M.R. and Neumann, M.G. (2000) *J. Polym. Sci., Part A: Polym. Chem.*, **39**, 46–55.
191. Crivello, J.V. (2001) *J. Polym. Sci., Part A: Polym. Chem.*, **39**, 1187–1197.
192. Gomurashvili, Z. and Crivello, J.V. (2002) *Macromolecules*, **35**, 2962–2969.
193. Crivello, J.V. (2008) *J. Polym. Sci., Part A: Polym. Chem.*, **46**, 3820–3829.
194. Rodrigues, M.R. and Neumann, M.G. (2001) *Macromol. Chem. Phys.*, **202**, 2776–2782.
195. De Voe, R.J., Sahyun, M.R.V., Schmidt, E., Serpone, N., and Sharma, D.K. (1988) *Can. J. Chem.*, **66**, 319.
196. Fouassier, J.P., Burr, D., and Crivello, J.V. (1989) *J. Photochem. Photobiol., A: Chem.*, **49**, 318.
197. Toba, Y., Saito, M., and Usui, Y. (1999) *Macromolecules*, **32**, 3209–3215.
198. Pappas, S.P., Tilley, M.G., and Pappas, B.C. (2003) *J. Photochem. Photobiol., A: Chem.*, **159**, 161–171.
199. Neckers, D.C. (1989) *J. Photochem. Photobiol., A: Chem.*, **47**, 1.
200. Kawabata, M. and Takimoto, Y. (1990) *J. Photopol. Sci. Tech.*, **3**, 147.
201. Yasuike, M., Sakuragi, M., and Ichimura, K. (1992) *Photochem. Photobiol. A: Chem.*, **64**, 115.
202. Degirmenci, M., Onen, A., Yagci, Y., and Pappas, S.P. (2001) *Polym. Bull.*, **46**, 443–449.
203. Linden, S.M. and Neckers, D.C. (1988) *J. Amer. Chem. Soc.*, **110**, 1257–1262.
204. Lalevée, J., El Roz, M., Allonas, X., and Fouassier, J.P. (2008) *J. Polym. Sci. Polym. Chem.*, **46**, 2008–2014.
205. Manivannan, G. and Fouassier, J.P. (1991) *J. Polym. Sci., Polym. Chem. Ed.*, **29**, 1113–1118.
206. Eckert, G., Goez, M., Maiwald, B., and Muller, U. (1996) *Ber. Bunsenges. Phys. Chem.*, **100** (7), 1191–1198.
207. Tehfe, M.A., Morlet-Savary, F., Lalevée, J., Louerat, F., Graff, B., Allonas, X., and Fouassier, J.P. (2010) *Eur. Polym. J.*, **46**, 2138–2144.
208. Kunze, A., Muller, U., Tittes, K., Fouassier, J.P., and Morlet-Savary, F. (1997) *J. Photochem. Photobiol., A: Chem.*, **110**, 115–122.
209. Hua, Y. and Crivello, J.V. (2001) *Macromolecules*, **34**, 2488–2495.
210. Aydogan, B., Gundogan, A.S., Ozturk, T., and Yagci, Y. (2008) *Macromolecules*, **41**, 3468–3471.
211. Aydogan, B., Gunbas, G.E., Durmus, A., Toppare, L., and Yagci, Y. (2009) *Macromolecules*, **43**, 101–106.
212. Bulut, U., Gunbas, G.E., and Toppare, L. (2010) *J. Polym. Sci., Part A: Polym. Chem.*, **48**, 209–213.
213. Bulut, U., Balan, A., and Caliskan, C. (2011) *J. Polym. Sci., Part A: Polym. Chem.*, **49**, 729–733.
214. Yagci, Y., Jockusch, S., and Turro, N.J. (2007) *Macromolecules*, **40**, 4481–4485.

215. Ledwith, A. (1979) *Makromol. Chem.* **3**, 348–358.
216. Schnabel, W., Bottcher, A., Hasebe, K., Hizal, G., Yagci, Y., and Stellberg, P. (1991) *Polymer*, **32**, 2289.
217. Yagci, Y. and Ledwith, A. (1988) *J. Polym. Sci. Polym. Chem. Ed.*, **26**, 1911.
218. Yagci, Y. and Schnabel, W. (1987) *Makromol. Chem., Rapid Commun.*, **8**, 209.
219. Dursun, C., Degirmenci, M., Yagci, Y., Jockusch, S., and Turro, N.J. (2003) *Polymer*, **44**, 7389–7396.
220. Atmaca, L., Kayihan, I., and Yagci, Y. (2000) *Polymer*, **41**, 6035–6041.
221. Yagci, Y., Pappas, S.P., and Schnabel, W. (1987) *Z. Naturforsch.*, **42a**, 1425.
222. Bi, Y. and Neckers, D.C. (1994) *Macromolecules*, **27**, 3683–3693.
223. Yagci, Y., Kminek, I., and Schnabel, W. (1993) *Polymer*, **34**, 426–428.
224. Guo, H.Q., Kajiwara, A., Morishima, Y., and Kamachi, M. (1996) *Polym. J.*, **28**, 960–964.
225. Guo, H.Q., Kajiwara, A., Morishima, Y., Kamachi, M., and Schnabel, W. (1997) *Polym. J.*, **29**, 446–449.
226. Durmaz, Y.Y., Moszner, N., and Yagci, Y. (2007) *Macromolecules*, **41**, 6714–6718.
227. Lalevée, J., Allonas, X., and Fouassier, J.P. (2009) *Chem. Phys. Lett.*, **469**, 293–303.
228. Tasleden, M.A., Kumbaraci, V., Jockusch, S., Turro, N.J., Tallini, N., and Yagci, Y. (2008) *Macromolecules*, **41**, 295–297.
229. Johnen, N., Schnabel, W., Kobayashi, S., and Fouassier, J.P. (1992) *J. Chem. Soc., Faraday Trans.*, **88**, 1385–1389.
230. Bi, Y. and Neckers, D.C. (1994) *Macromolecules*, **27**, 3683–3693.
231. Yilmaz, G., Beyazit, S., Yagci, Y., and Part, A. (2011) *J. Polym. Sci., Polym. Chem.*, **49**, 1591–1596.
232. Yagci, Y. and Hepuzer, H. (1999) *Macromolecules*, **32**, 6367–6370.
233. Lalevée, J., Tehfe, M.A., Morlet-Savary, F., Graff, B., Allonas, X., and Fouassier, J.P. (2010) *Prog. Org. Coat.*, **70**, 23–31.
234. Lalevée, J., Dirani, A., El Roz, M., Allonas, X., and Fouassier, J.P. (2008) *J. Polym. Sci., Polym. Chem.*, **46**, 3042–3047.
235. Lalevée, J., El Roz, M., Tehfe, M., Allonas, X., and Fouassier, J.P. (2009) *Macromolecules*, **42**, 8669–8674.
236. El Roz, M., Morlet-Savary, F., Lalevée, J., Allonas, X., and Fouassier, J.P. (2008) *J. Polym. Sci., Polym. Chem.*, **46**, 7369–7375.
237. El Roz, M., Lalevée, J., Morlet-Savary, F., Allonas, X., and Fouassier, J.P. (2009) *Macromolecules*, **42**, 4464–4469.
238. Lalevée, J., Tehfe, M.A., Gigmes, D., Blanchard, N., and Fouassier, J.P. (2010) *Macromolecules*, **43**, 6608–6615.
239. Tehfe, M.A., Lalevée, J., Gigmes, D., and Fouassier, J.P. (2010) *J. Polym. Sci., Polym. Chem.*, **48**, 1830–1837.
240. Lalevée, J., Blanchard, N., Tehfe, M.A., and Fouassier, J.P. (2011) *Macromol Rapid Commun.*, **32**, 838–843.
241. Angioni, S., Ravelli, D., Emma, D., Dondi, D., Fagnoni, M., and Albini, A. (2008) *Adv. Synth. Catal.*, **350**, 2209–2214.
242. Kothe, T., Martschke, R., and Fischer, H. (1998) *J. Chem. Soc., Perkin Trans. 2*, 503–507.
243. Lalevée, J., Blanchard, N., Tehfe, M.A., Chany, A.C., Morlet-Savary, F., and Fouassier, J.P. (2010) *Macromolecules*, **43**, 10191–10195.
244. Tehfe, M.A., Lalevée, J., and Fouassier, J.P., to be published.
245. Lalevée, J., Blanchard, N., Tehfe, M.A., Peter, M., Morlet-Savary, F., Gigmes, D., and Fouassier, J.P. (2011) *Polym. Chem.*, **2**, 1986–1991.
246. Lalevée, J., Blanchard, N., Tehfe, M.A., Peter, M., Morlet-Savary, F., and Fouassier, J.P. (2011) *Macromol. Rapid Commun.*, **32**, 917–920.
247. Lalevée, J., Peter, M., Dumur, F., Gigmes, D., Blanchard, N., Tehfe, M.A., Fouassier, J.P., to be published.
248. Lalevée, J., Blanchard, N., Tehfe, M.A., Peter, M., Morlet-Savary, F., and Fouassier, J.P. (2012) *Polym. Bull.*, **68**, 341–347.
249. Crivello, J.V. and Part, A. (2009) *J. Polym. Sci., Polym. Chem.*, **47**, 866–875.

250. Crivello, J.V. and Part, A. (2009) *J. Macromol. Sci., Pure Appl. Chem.*, **46**, 474–483.
251. Denizligil, S., Resul, R., Yagci, Y., Mc Ardle, C., and Fouassier, J.P., (1996) *Macromol. Chem. Phys.*, **197**, 1233–1240.
252. Toneri, T., Watanabe, K., Sanda, F., and Endo, T. (1999) *Macromolecules*, **32**, 1293–1299.
253. Denizligil, S., Yagci, Y., and Mc Ardle, C. (1995) *Polymer*, **36**, 3093–3098.
254. Kreutzer, J., Demir, K.D., and Yagci, Y. (2011) *Eur. Polym. J.*, **47**, 792–799.
255. Durmaz, Y.Y., Kukut, M., Moszner, R., Yagci, Y., and Part, A. (2009) *J. Polym. Sci., Polym. Chem.*, **47**, 4793–4799.
256. Abdul-Rasoul, F.A.M., Ledwith, A., and Yagci, Y. (1978) *Polym. Bull.*, **1**, 1–4.
257. Sangermano, M., Yagci, Y., and Rizza, G. (2007) *Macromolecules*, **40**, 8827–8829.
258. Yagci, Y., Sangermano, M., and Rizza, G. (2008) *Macromolecules*, **41**, 7268–7270.
259. Colak, D., Yurteri, S., Kiskan, B., and Yagci, Y. (2006) in *Photochemistry and UV Curing* (ed. J.P. Fouassier), Research Signpost, Trivandrum, pp. 175–186.
260. Reetz, I., Bacak, V., and Yagci, Y. (1997) *Polym. Int.*, **43**, 27–32.
261. Yagci, Y., Yildirim, S., and Onen, A. (2001) *Macromol. Chem. Phys.*, **202**, 527–531.
262. Onen, A. and Yagci, Y. (2001) *Polymer*, **42**, 6681–6685.
263. Onen, A. and Yagci, Y. (2001) *Macromol. Chem. Phys.*, **202**, 1950–1954.
264. Yurteri, S., Onen, A., and Yagci, Y. (2002) *Eur. Polym. J.*, **38**, 1845–1850.
265. Degirmenci, M., Hepuzer, Y., and Yagci, Y. (2002) *J. Appl. Polym. Sci.*, **85**, 2389–2395.
266. Gupta, M.K., Mehare, R., and Singh, R.P. (2010) *Polym. Bull.*, **65**, 25–34.
267. Kahveci, M.U., Uygun, M., Tasdelen, M.A., Schnabel, W., Cook, W.D., and Yagci, Y. (2009) *Macromolecules*, **42**, 4443–4448.
268. Kwon, S., Chun, H., and Mah, S. (2006) *J. Appl. Polym. Sci.*, **101**, 3581–3586.
269. Percec, V. and Tomazos, D. (1987) *Polym. Bull.*, **18**, 239–246.
270. Falk, B., Zonca, M.R., and Crivello, J.V. (2005) *J. Polym. Sci., Polym. Chem.*, **43**, 2504–2519.

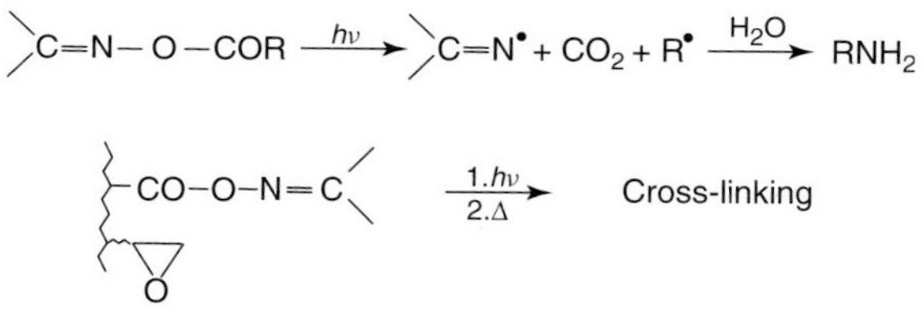

Scheme 15.1

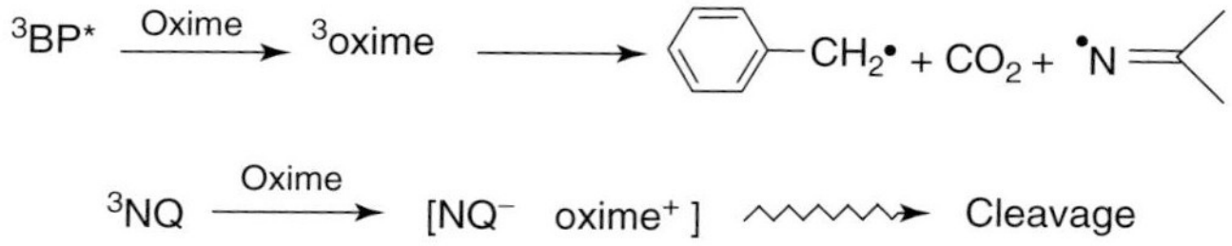

$$\text{C=N-O-COR} \xrightarrow{h\nu} \text{C=N}^\bullet + CO_2 + R^\bullet \xrightarrow{H_2O} RNH_2$$

$$\text{-CO-O-N=C} \xrightarrow[2.\Delta]{1.h\nu} \text{Cross-linking}$$

Scheme 15.2

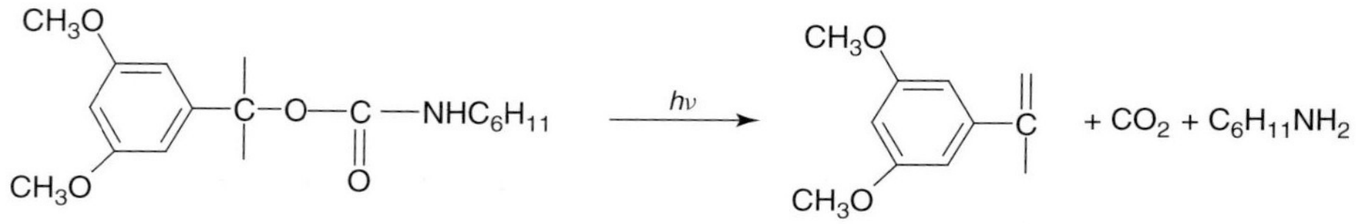

$$^3BP^* \xrightarrow{\text{Oxime}} {}^3\text{oxime} \longrightarrow \text{Ph-CH}_2^\bullet + CO_2 + {}^\bullet N=$$

$$^3NQ \xrightarrow{\text{Oxime}} [NQ^- \text{ oxime}^+] \rightsquigarrow \text{Cleavage}$$

Scheme 15.3

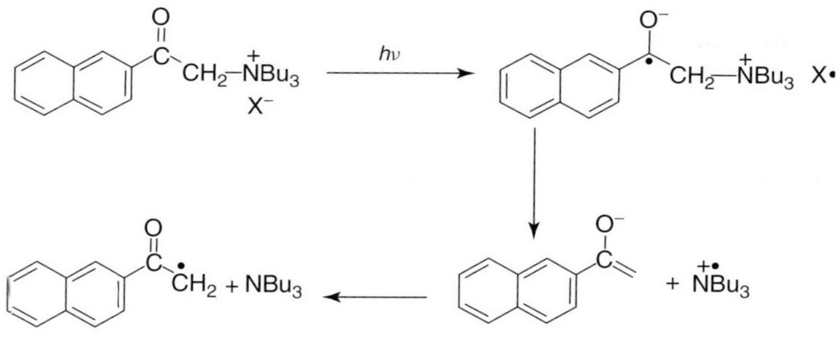

$$\text{(3,5-(CH}_3O)_2\text{C}_6\text{H}_3)\text{-C-O-C(O)-NHC}_6\text{H}_{11} \xrightarrow{h\nu} \text{(3,5-(CH}_3O)_2\text{C}_6\text{H}_3)\text{-C=} + CO_2 + C_6H_{11}NH_2$$

Scheme 15.4

Naphthyl-CO-CH$_2$-$\overset{+}{N}$Bu$_3$ X$^-$ $\xrightarrow{h\nu}$ Naphthyl-C(O$^-$)$^\bullet$-CH$_2$-$\overset{+}{N}$Bu$_3$ X$^\bullet$

$\longrightarrow$ Naphthyl-C(O$^-$)=CH$_2$ + $\overset{+\bullet}{N}$Bu$_3$

$\longrightarrow$ Naphthyl-CO-$^\bullet$CH$_2$ + NBu$_3$

15.4
N-Benzylated-Structure-Based Photobases

Very recently, new photobases were proposed in several patents. They certainly present an interesting starting point for practical applications [16]. One example consists in an N-benzylated structure that yields, on cleavage, an amidine base (Scheme 15.5). The primary or secondary amines generated in other photobases are not very strong bases but rather good nucleophiles. Moreover, in these systems, one amine chemically participates in the reaction and forms one cross-link (or at

13
Anionic Photoinitiators

Anionic photoinitiators are considerably less investigated. Patents do exist, but rather few papers were published in the past (e.g., [1–5]). Examples of the polymerization of cyanoacrylates (CAs) include the following.

13.1
Inorganic Complexes

Inorganic complexes are able to release negatively charged nucleophiles. Such compounds are exemplified by (i) chromium amine thiocyanate complexes [6] (Scheme 13.1) where the initiating anion is formed by a photoinduced cleavage of the Cr–NCS bond) and (ii) platinum acetylacetonate complexes.

13.2
Organometallic Complexes

Organometallic complexes [7] have been considered, for example, (i) pentacarbonyl $M(CO)_5L$ complexes (where M is Cr or W and L a pyridine), (ii) Schiff bases [8], (iii) ferrocenes $FeCp_2$ (where Cp is the cyclopentadienyl moiety) [9–11], (iv) metallocenes [12], (v) alkylaluminum porphyrins [13], and (vi) p-chlorophenyl o-nitrobenzylether/titanium tetraisopropoxide [14].

Ferrocene (Scheme 13.2) forms a ground-state complex with the CA monomer; on excitation, an electron transfer occurs and leads to the ferricenium cation and a monomer radical anion [12].

In benzoyl-substituted ferrocene, the formation of a benzoyl-substituted cyclopentadienide anion after a metal–ring bond cleavage (Scheme 13.3) is suggested as the initiating species [11].

UV light exposure of silicon-bridged ferrocenophanes (which contain a ferrocene moiety linked to one silicon atom) in the presence of NaC_5H_4R results in a ring-opening polymerization (ROP) that proceeds through an iron–cyclopentadienyl bond cleavage (Scheme 13.4) and generates an anion [15]. The living polymer chain can be capped with methanol. The same procedure is usable for the synthesis of block copolymers with a controlled architecture by a sequential photocontrolled ROP.

Photoinitiators for Polymer Synthesis: Scope, Reactivity and Efficiency, First Edition.
Jean-Pierre Fouassier and Jacques Lalevée.
© 2012 Wiley-VCH Verlag GmbH & Co. KGaA. Published 2012 by Wiley-VCH Verlag GmbH & Co. KGaA.

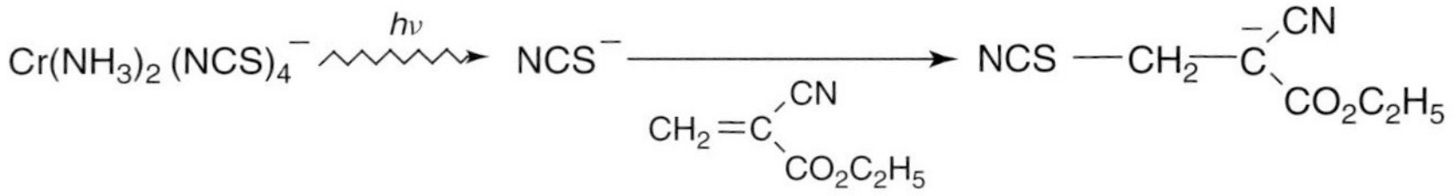
Cr(NH3)2 (NCS)4−
hν
NCS−
CH2=C(CN)CO2C2H5
NCS —CH2—C−(CN)CO2C2H5

Scheme 13.1

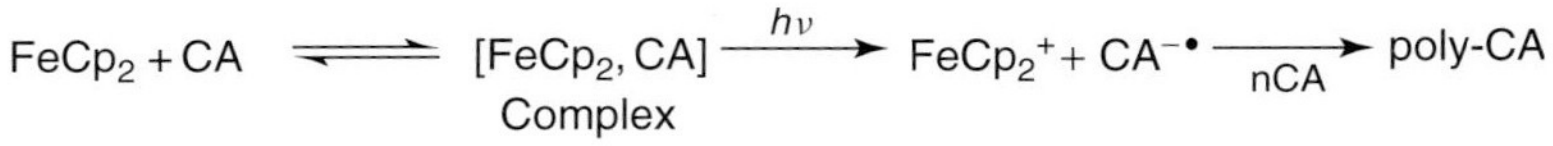
FeCp2 + CA
[FeCp2, CA]
Complex
hν
FeCp2+ + CA−•
nCA
poly-CA

Scheme 13.2

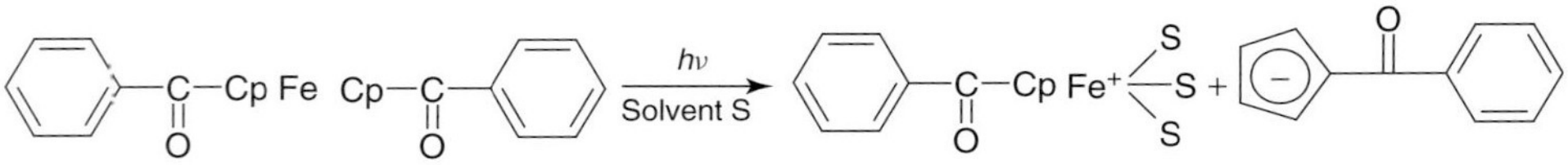
C—Cp Fe Cp—C
O
O
hν
Solvent S
C—Cp Fe+
S
S
S
+
−

Scheme 13.3

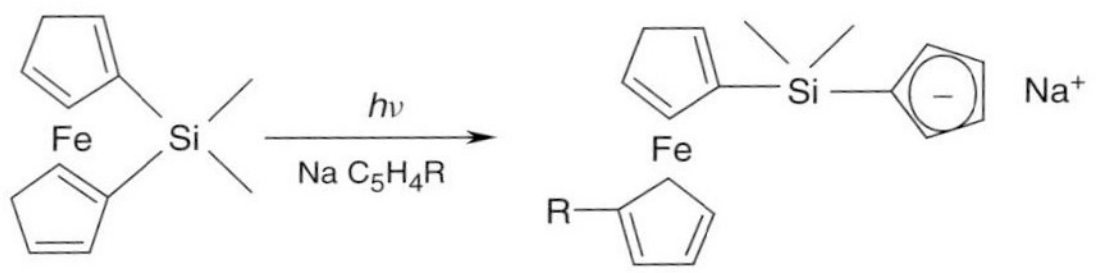
Fe
Si
hν
Na C5H4R
Fe
Si
−
Na+
R

Scheme 13.4

Irradiation of an alkylaluminum porphyrin and an acrylate results in the addition of the methyl aluminate to the acrylate and leads to an aluminum enolate [16]. Anionic photopolymerization of an epoxide in the presence of zinc propylthio *N*-methyl tetraphenylporphyrin involves the formation of a zinc alkoxide [17].

13.3 Cyano Derivative/Amine System

A suitable photoinduced electron transfer between an aromatic cyano derivative and *N,N,N′,N′*-tetramethylbenzidine [18] leads to a further efficient C–C bond cleavage and the generation of a radical and an anion.

13.4 Photosensitive Systems for Living Anionic Polymerization

Few examples of living anionic photopolymerization have been reported, for example, using phosphorus-bridged ferrocenophanes [19] or fluorinated-silicone-bridged ferrocenophanes [20].

References

1. Cunningham, A.F. and Desobry, V. (1993) in *Radiation Curing in Polymer Science and Technology*, vol. 2 (eds J.P. Fouassier and J.F. Rabek), Elsevier, Barking, pp. 323–374.
2. Palmer, B.J., Kutal, C., Billing, R., and Hennig, H. (1995) *Macromolecules*, 28, 1328–1329.
3. Gamble, G., Grutsch, P.A., Lavallee, R.J., Palmer, B.J., and Kutal, C. (1995) *Proc. Conf. Coord. Chem.*, 239–244.
4. Yamaguchi, Y. and Kutal, C. (1999) *Monogr. Ser. Int. Conf. Coord. Chem.*, 4, 209–214.
5. Kutal, C. (2001) *Coord. Chem. Rev.*, 211, 353–368.
6. Kutal, C., Grutsch, P.A., and Yang, D.B. (1991) *Macromolecules*, 24, 26–31.
7. Paul, R.B., Kelly, J.M., Pepper, D.C., and Long, C. (1997) *Polymer*, 38, 2011–2014.
8. Wang, C., Wei, Y., Gruber, H., and Wendrinsky, J. (1997) *Polym. Prepr.*, 38, 219–220.
9. Yamaguchi, Y., Palmer, B.J., Kutal, C., Wakamatsu, T., and Yang, D.B. (1998) *Macromolecules*, 31, 5155–5157.
10. Yamaguchi, Y., Palmer, B.J., Wakamatsu, T., Yang, D.B., and Kutal, C. (1998) *Polym. Mater. Sci. Eng.*, 79, 1–2.
11. Yamaguchi, Y. and Kutal, C. (2000) *Macromolecules*, 33, 1152–1156.
12. Sanderson, C.T., Palmer, B.J., Morgan, A., Murphy, M., Dluhy, R.A., Mize, T., Amster, I.J., and Kutal, C. (2002) *Macromolecules*, 35, 9648–9652.
13. Kuroki, M., Aida, T., and Inoue, S. (1987) *J. Am. Chem. Soc.*, 109, 4737–4741.
14. Fukuchi, Y., Takahashi, T., Nogushi, H., Saburi, M., and Uchida, Y. (1987) *Maromolecules*, 20, 2316–2322.
15. Herbert, D.E., Gilroy, J.B., Chan, W.Y., Chabanne, L., Staubitz, A., Lough, A.J., and Manners, I. (2009) *J. Am. Chem. Soc.*, 131, 14958–14968.
16. Kodaira, T. and Mori, K. (1990) *Makromol. Chem. Rapid Commun.*, 23, 2612–2616.
17. Watanabe, Y., Aida, T., and Inoue, S. (1991) *Macromolecules*, 24, 3970–3976.
18. Maslak, P., Kula, J., and Chateauneuf, J.E. (1991) *J. Am. Chem. Soc.*, 113, 2304–2308.
19. Patra, S.K., Whittell, G.R., Nagiah, S., Ho, C.-L., Wong, W.-Y., and Manners, I. (2010) *Chem.-A Eur. J.*, 16, 3240–3250.
20. Smith, G.S., Patra, S.K., Vanderark, L., Saithong, S., Charmant, J.P.H., and Manners, I. (2010) *Macromol. Chem. Phys.*, 211, 303–312.

14
Photoacid Generators (PAG) Systems

Photoacid generators (PAGs) are encountered in the imaging area (see a review in [1]). They involve ionic systems such as onium salts [2, 3] (see Section 12.2) or nonionic compounds. The nonionic-photoacid-initiating systems that belong to photolatent compounds [4–10] have a large range of solubility in organic matrices and generate sulfinic acid, sulfonic acid, carboxylic acid, and phosphoric acid. They are based on several kinds of structures: iminosulfonates; esters of sulfonic-, sulfinic-, or carboxylic acids; nitrobenzyl esters; arene sulfonate derivatives; hydroxyimide sulfonates; oxime sulfonates; phenacyl sulfone derivatives; triaryl phosphate derivatives; selenides; organo-silane-based compounds; and naphthalimides. Other proposals include 1-chloro-substituted thioxanthones [11], phenols [12], and thianthrene skeletons [13]. Polymeric PAGs have been developed [14]. Most of these systems are used in organic photopolymers.

Sensitizer/PAG systems containing dispersed photopolymers have been developed for micron-scale photolithography [15]. Two-photon acid generation has been achieved in thin polymer films [16].

In a first photochemical step, photoacids liberate a proton that leads, in a second thermal step, to a catalyzed cross-linking reaction. They are useful for the design of chemically amplified photoresists. Chemical amplification means that the reactions induced by the light can continue on a thermal treatment or a postcure process once the light has been switched off.

14.1
Iminosulfonates and Oximesulfonates

On irradiation, iminosulfonates (Scheme 14.1) form a sulfonic acid. Then, a thermal process favors the cross-linking reactions of, for example, an epoxide. In this procedure, a negative image is obtained under solvent development.

The cleavage of iminosulfonates occurs in the singlet state [17]. The use of a triplet photosensitizer shows that the N–O bond breaking also occurs in the triplet state. Examples include the compounds shown in **1** and **2** [10].

Photoinitiators for Polymer Synthesis: Scope, Reactivity and Efficiency, First Edition.
Jean-Pierre Fouassier and Jacques Lalevée.
© 2012 Wiley-VCH Verlag GmbH & Co. KGaA. Published 2012 by Wiley-VCH Verlag GmbH & Co. KGaA.

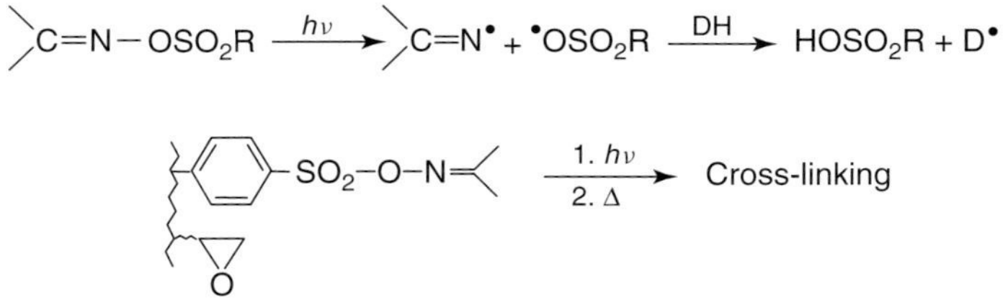

Scheme 14.1

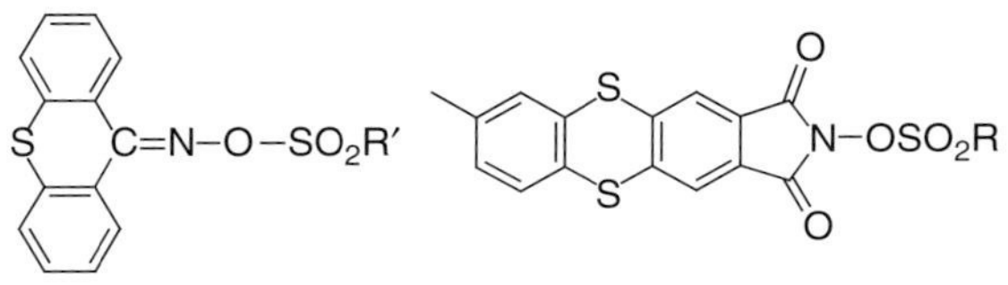

$$\text{>C=N-OSO}_2\text{R} \xrightarrow{h\nu} \text{>C=N}^\bullet + {}^\bullet\text{OSO}_2\text{R} \xrightarrow{DH} \text{HOSO}_2\text{R} + \text{D}^\bullet$$

Ar–SO$_2$–O–N< $\xrightarrow{\text{1. } h\nu}_{\text{2. } \Delta}$ Cross-linking

1: thioxanthene-C=N–O–SO$_2$R′

2: thianthrene-fused naphthalimide N–OSO$_2$R

The sensitized decomposition of oxime derivatives has been studied [18–20].

14.2
Naphthalimides

The photophysics, photochemistry, and photopolymerization activity and applications of naphthalimides alone or in the presence of photosensitizers has been largely explored [21–25].

In 1,8-naphthalimide PAGs, the singlet excited state exhibits a low polarity and is strongly deactivated by an efficient intersystem crossing process (closely lying singlet and triplet states). In a polar solvent, a homolytic singlet cleavage of the N–O bond occurs and leads to the production of a naphthalimide radical and a sulfonyl radical (Scheme 14.2 for N-(trifluoromethanesulfonyloxy)-1,8-naphthalimide (NIOTf)) [26, 27] and references therein). The former radical then undergoes a ring-opening reaction to generate an isocyanate radical intermediate, which is hydrolyzed in the presence of water; the formation of a carbamic acid derivative is then induced. Hydrogen abstraction between the radicals is followed by CO_2 release, benz[cd]indol-2(1H)-one (bione), and acid formation.

The photodecomposition of a naphthalimide derivative (e.g., NIOTf) in the presence of a pyrromethene dye (PYR) leads to trifluorosulfonic acid (Scheme 14.3 for PYR = Py). This reaction involves an electron transfer process from the excited Py singlet state to NIOTf. The triplet state of the Py dyes is unreactive toward the naphthalimide [28]).

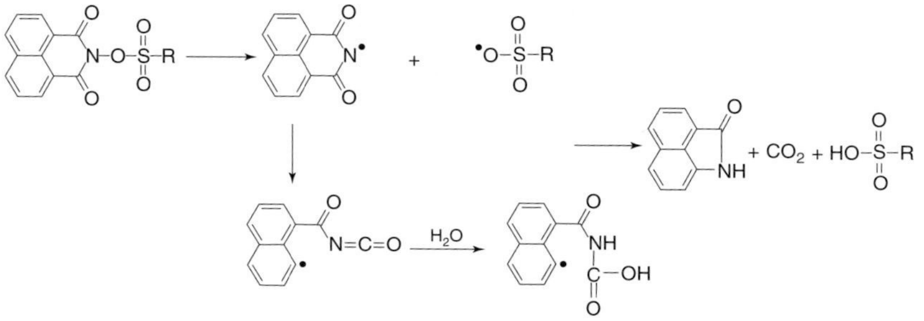

Scheme 14.2

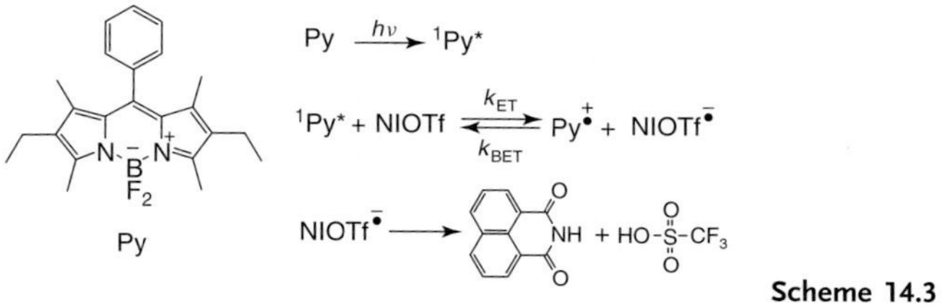

Scheme 14.3

14.3
Photoacids and Chemical Amplification

A chemistry based on a photochemical event that requires one photon to produce one acid has obviously a limited efficiency. The first chemical amplification process ([29] and references therein) where one photon is able to indirectly induce several reactions was evidenced in the depolymerization of polyphthaldehyde. This acid-catalyzed reaction produces a positive relief image. Such a positive image can also be formed by starting with an alkali insoluble polymer P where an acidic function (e.g., P–OH) is protected by, for example, a *t*-BOC group (*t*BOC = *N*-*tert*-butoxycarbonyl group; the starting polymer is thus P-O-*t*-BOC). Under irradiation with an acid generator, the liberated acid is used to deprotect the acidic function of the polymer and form P–OH. Therefore, the polymer becomes soluble in an alkali. A postbaking treatment favors the acid diffusion and the chemical amplification. Negative images can be formed through cross-linking of epoxy resins. The chain-amplified photoacid generation has been studied, for example, in [30–32] and reviewed in [1].

References

1. Ivan, M.G. and Scaiano, J.C. (2010) in *Photochemistry and Photophysics of Polymer Materials* (ed. N.S. Allen), John Wiley & Sons, Inc., Hoboken, pp. 479–508.
2. Ito, H. (2008) *J. Photopolym. Sci. Tech.*, **21**, 475–491.
3. Crivello, J.V. (2008) *J. Photopolym. Sci. Technol.*, **21**, 493–497.
4. Kahveci, M.U., Gilmaz, A.G., and Yagci, Y. (2010) in *Photochemistry and Photophysics of Polymer Materials* (ed. N.S. Allen), John Wiley & Sons, Inc., Hoboken, pp. 421–478.

5. Shirai, M. and Tsunooka, M. (1998) *Bull. Chem. Soc. Jpn.*, **71**, 2483–2507.
6. Shirai, M. and Tsunooka, M. (1996) *Prog. Polym. Sci.*, **21**, 1–9.
7. Okamura, H., Sakai, K., Tsunooka, M., and Shirai, M. (2003) *J. Photopolym. Sci. Technol.*, **16**, 701–706.
8. Shirai, M. and Tsunooka, M. (1998) *Bull. Chem. Soc. Jpn.*, **71**, 2483–2487.
9. Nishimae, Y., Yamato, H., Asakura, T., and Ohwa, M. (2008) *J. Photopolym. Sci. Technol.*, **21**, 377–381.
10. Shirai, M. and Okamura, H. (2009) *Prog. Org. Coat.*, **64**, 175–181.
11. Shah, M., Allen, N.S., Salleh, N.G., Corrales, T., Edge, M., Catalina, F., Bosch, P., and Green, A. (1997) *J. Photochem. Photobiol. A: Chem.*, **111**, 229–232.
12. Hino, T. and Endo, T. (2004) *Macromolecules*, **37**, 1671–1673.
13. Okamura, H., Naito, H., and Shirai, M. (2008) *J. Photopolym. Sci. Technol.*, **21**, 285–288.
14. Jung, M.S., Choi, T.L., Huh, N., Ko, C., and Jung, H.T. (2008) *Polym. Adv. Technol.*, **19**, 237–243.
15. Ichimura, K. (2007) *J. Mater. Chem.*, **17**, 632–641.
16. Billone, P.S., Park, J.M., Blackwell, J.M., Bristol, R., and Scaiano, J.C. (2010) *Chem. Mater.*, **22**, 15–17.
17. Lalevée, J., Allonas, X., Fouassier, J.P., Shirai, M., and Tsunooka, M. (2003) *Chem. Lett.*, **32**, 178–181.
18. Tehfe, M.A., Morlet-Savary, F., Lalevée, J., Louerat, F., Graff, B., Allonas, X., and Fouassier, J.P. (2010) *Eur. Polym. J.*, **46**, 2138–2144.
19. Takahara, S., Nishizawa, N., and Tsumita, T. (2009) *J. Photopolym. Sci. Technol.*, **22**, 289–294.
20. Takahara, S., Suzuki, S., Allonas, X., Fouassier, J.P., and Yamaoka, T. (2008) *J. Photopolym. Sci. Technol.*, **21**, 499–504.
21. Noppakundilograt, S., Suzuki, S., Urano, T., Miyagawa, M., Takahara, S., and Yamaoka, T. (2002) *Polym. Adv. Technol.*, **13**, 527–533.
22. Andraos, J., Barclay, G.G., Medeiros, D.R., Baldovi, M.V., Scaiano, J.C., and Sinta, R. (1998) *Chem. Mater.*, **10**, 1694–1699.
23. Ortica, F., Coenjarts, C., Scaiano, J.C., Liu, H., Pohlers, G., and Cameron, J.F. (2001) *Chem. Mater.*, **13**, 2297–2304.
24. Coenjarts, C., Garca, O., Llauger, L., Palfreyman, J., Vinette, A.L., and Scaiano, J.C. (2003) *J. Am. Chem. Soc.*, **125**, 620–621.
25. Iwashima, C., Imai, G., Okumara, H., Tsunooka, M., and Shirai, M. (2003) *J. Photopolym. Sci. Technol.*, **16**, 91–96.
26. Malval, J.P., Morlet-Savary, F., Allonas, X., Fouassier, J.P., Suzuki, S., Takahara, T., and Yamaoka, S. (2007) *Chem. Phys. Lett.*, **443**, 323–327.
27. Suzuki, S., Allonas, X., Fouassier, J.P., Urano, T., Takahara, T., and Yamaoka, S. (2006) *J. Photochem. Photobiol. A: Chem.*, **181**, 60–66.
28. Malval, J.P., Morlet-Savary, F., Allonas, X., Fouassier, J.P., Suzuki, S., Takahara, T., and Yamaoka, S. (2008) *J. Phys. Chem. A*, **112**, 3879–3885.
29. Yun Moon, S., Kim, J.M., and Photochem, J. (2007) *Photobiol. C: Photochem. Rev.*, **8**, 157–173.
30. Scaiano, J.C., Barra, M., and Sinta, R. (1996) *Chem. Mater.*, **8**, 161–166.
31. Sakamizu, T., Shiraishi, H., and Ueno, T. (1995) *Micro-electronics Technology*, ACS Symposium Series, Vol. **614**, American Chemical Society, pp. 124–136.
32. Takahara, S. and Yamaoka, T. (2004) *J. Photopolym. Sci. Technol.*, **17**, 719–725.

15
Photobase Generators (PBG) Systems

The available structures of photoinitiating systems, photobase generators (PBGs), that are able to liberate a base on light exposure and usable in photo-cross-linking reactions have been reviewed in several papers [1–5]. The following directions have been explored.

15.1
Oxime Esters

On light excitation, oxime esters [6–8] release carbon dioxide, a radical, and an iminyl radical. Hydrolysis of the photoproducts leads to an amine (Scheme 15.1). The cleavage likely occurs in the singlet state. Polymeric amines can be prepared by photolysis of a polymer bearing such oxime ester moieties. The thermal cross-linking reaction corresponds to the addition of the amine to the epoxide.

Decomposition of an O-acyloxime derivative used as a photobase occurs through energy transfer when benzophenone (BP) is used and electron transfer in the presence of naphthoquinone (NQ) (Scheme 15.2) [9].

15.2
Carbamates

Carbamates (Scheme 15.3) cleave, release carbon dioxide, and liberate an amine [10–12]. As in oximes, one photon yields one base molecule.

15.3
Ammonium Tetraorganyl Borate Salts

Ammonium tetraorganyl borate salts [13, 14] directly cleave through an intraion-pair electron transfer between the cationic moiety and the borate followed by a C–N bond scission. The amine that is then released (Scheme 15.4, where X stands for the borate anion) is suitable for the thermal curing of epoxides where a nucleophilic attack is favored. Free radicals are also liberated. The design of a bicyclic guanidinium tetraphenylborate [15] allows generation of a stronger base usable for (i) the cross-linking of PMMA with a hydroxyl alkyl cellulose or (ii) the living anionic ring-opening photopolymerization of cyclic esters.

Photoinitiators for Polymer Synthesis: Scope, Reactivity and Efficiency, First Edition.
Jean-Pierre Fouassier and Jacques Lalevée.
© 2012 Wiley-VCH Verlag GmbH & Co. KGaA. Published 2012 by Wiley-VCH Verlag GmbH & Co. KGaA.

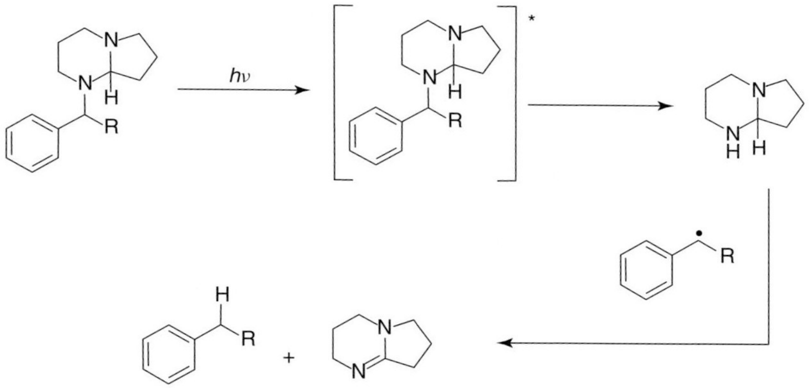

Scheme 15.5

most two cross-links if a diamine is used). On the contrary, the amidine derivatives are strong and nonnucleophilic amines. As a consequence, they act as true catalysts in cross-linking reactions that require strong bases, for example, the addition of an alcohol on an isocyanate function during the formation of a polyurethane from a polyol and a polyisocyanate.

15.5
Other Miscellaneous Systems

Examples [13] are found in tertiary-amine-releasing systems [17], aromatic formamides, cobalt–amine complexes, amine–imides [18, 19], and aminoketones [20]. All these compounds yield amines on irradiation, and the photo-cross-linking of epoxides or isocyanates is feasible.

A careful design of a tertiary amine source elaborated from usual α-amino acetophenones for application in car refinishing has been done. A change of the chromophore (to avoid a yellowing of the clear coat by the photolysis products) and the amine (to modify the steric shielding) seems promising for the development of efficient systems [21].

Other nucleophiles can also be formed, for example, in ketoprofene derivatives (Scheme 15.6). On irradiation, decarboxylation occurs and the strong base is liberated [22].

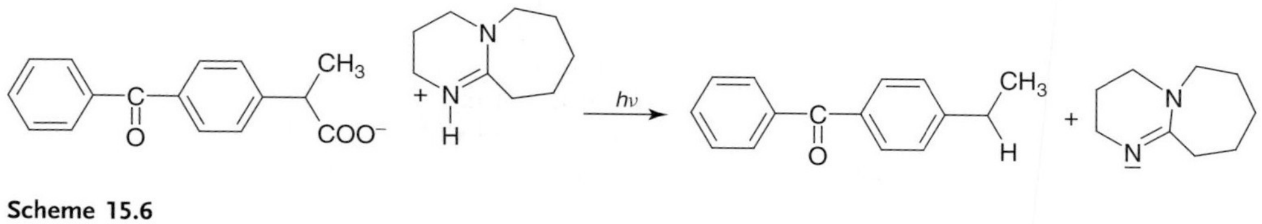

Scheme 15.6

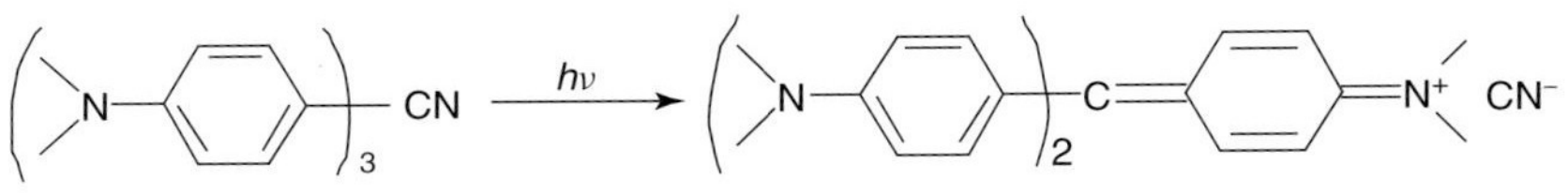

Scheme 15.7

Phosphazene bases [23, 24], phosphonium and pyridinium salts (which liberate a phosphine or a substituted pyridine, respectively) were also mentioned. In dye leuconitriles [25] such as in Scheme 15.7, a cyanide anion is released on irradiation. Other nucleophilic leaving groups can also be obtained in leucocarbinols and leucoethers.

15.6
Photobases and Base Proliferation Processes

The design of new compounds (e.g., 9-fluorenyl carbamate) that can undergo a base-catalyzed decomposition to liberate a base has led to the concept of a base amplifier [26]. An enhancement of the overall photoreactivity is noted as one photon allows creation of more than one base molecule.

Advances in ionic photopolymerization have been recently reviewed in [27].

References

1. Ken-Ichi, I., Masamichi, N., Minoru, S., and Part, A. (1994) *J. Polym. Sci., Polym. Chem.*, **32**, 2177–2185.
2. Harkness, B.R., Takeuchi, K., and Tachikawa, M. (1998) *Macromolecules*, **31**, 4798–4805.
3. Shirai, M., Suyama, K., and Tsunooka, M. (1999) *Trends Photochem. Photobiol.*, **5**, 169–180.
4. Tsunooka, M., Suyama, K., Okamura, H., and Shirai, M. (2006) *J. Photopolym. Sci. Technol.*, **19**, 65–70.
5. Suyama, K. and Shirai, M. (2009) *Prog. Polym. Sci.*, **34**, 194–209.
6. Ken-Ichi, I., Masamichi, N., Minoru, S., Masahiro, T., and Part, A. (1994) *J. Polym. Sci., Polym. Chem.*, **32**, 1793–1796.
7. Tsunooka, M., Matsuoka, T., Miyamoto, Y., and Suyama, K. (1998) *J. Photopolym. Sci. Technol.*, **11**, 123–124.
8. Crivello, J.V., Fan, M., and Part, A. (1992) *J. Polym. Sci., Polym. Chem.*, **30**, 31–39.
9. Lalevée, J., Allonas, X., Fouassier, J.P., Shirai, M., Tachi, H., Izumitani, A., and Tsunooka, M. (2002) *J. Photochem. Photobiol. A: Chem.*, **151**, 27–32.
10. Camperon, J.F., Willson, C.G., and Frechet, J.M.J. (1997) *J. Chem. Soc., Perkin Trans.*, **1**, 2429–2442.
11. Tachi, H., Yamamoto, T., Shirai, M., Tsunooka, M., and Part, A. (2001) *J. Polym. Sci., Polym. Chem.*, **39**, 1329–1341.
12. Tsunooka, M., Tachi, H., Yamamoto, T., and Shirai, M. (2003) in *Photoinitiated Polymerization*, ACS Symposium Series (eds K.D. Belfied and J.V. Crivello), American Chemical Society, pp. 351–362.
13. Yu, X., Chen, J., Yang, J., Zeng, Z., and Chen, Y. (2005) *Polym. Int.*, **54**, 1212–1219.
14. Yu, X., Chen, J., Yang, J., Zeng, Z., and Chen, Y. (2006) *J. Appl. Polym. Sci.*, **100**, 399–405.

15. Sun, X., Gao, J.P., and Wang, Z.Y. (2008) *J. Am. Chem. Soc.*, **130**, 7542–7549.
16. Dietliker, K., Hüsler, R., Birbaum, J.L., Ilg, S., Villeneuve, S., Studer, K., Jung, T., Benkhoff, J., Kura, H., Matsumoto, A., and Oka, H. (2007) *Prog. Org. Coat.*, **58**, 146–157.
17. Jensen, K.H. and Hanson, J.E. (2002) *Chem. Mater.*, **14**, 918–923.
18. Katogi, S. and Yusa, M. (2001) *J. Photopolym. Sci. Technol.*, **14** (2), 151–152.
19. Katogi, S., Yusa, M., Shirai, M., and Tsunooka, M. (2003) *Chem. Lett.*, **32**, 418–419.
20. Kura, H., Oka, H., Birbaum, J.-L., and Kikuchi, T. (2000) *J. Photopolym. Sci. Technol.*, **13**, 145–152.
21. Dietliker, K., Braig, A., and Ricci, A. (2010) *Photochemistry*, **38**, 344–368.
22. Arimitsu, K. and Endo, R. (2010) *J. Photopolym. Sci. Technol.*, **23**, 135–136.
23. Boerner, H.G. and Heitz, W. (1998) *Macromol. Chem. Phys.*, **199**, 1815–1820.
24. Aslan, F., Demirpence, Z., Tatsiz, R., Turkmen, H., Ozturk, A., and Arslan, M. (2008) *Z. Anorg. Allg. Chem.*, **634**, 1140–1144.
25. Jarikov, V.V. and Neckers, D.C. (2000) *Macromolecules*, **33**, 7761–7764.
26. Arimitsu, K., Miyamoto, M., and Ichimura, K. (1999) *Polym. Mater. Sci. Eng.*, **81**, 93–94.
27. Yagci, Y., Jockusch, S., and Turro, N.J. (2010) *Macromolecules*, **65**, 6245–6260.

Part IV
Reactivity of the Photoinitiating System

This part is concerned with a general discussion of the reactivity versus efficiency relationships in photoinitiating systems based on both experimental and theoretical considerations. It shows how typical data on the transient states are obtained and how the driving factors of the processes involved (cleavage, hydrogen abstraction with H-donors, monomer quenching, oxygen quenching, radical addition, and radical quenching) can be understood. By using both these information and the polymerization results, it becomes possible to derive interesting (photo)chemical reactivity/practical efficiency relationships. For example, the rate constants and/or the different quantum yields of the primary events associated with the excited states and the initiating radicals can be tentatively correlated to the polymerization characteristics (polymerization rates Rp, conversions, and behavior under aerated conditions). The huge progress made in the theoretical approach really improves the interpretation of the observed practical efficiency compared to what was done in the past. The understanding of the chemical reactivity is a fascinating and tempting challenge. Everybody likes to understand what happens in a reaction. The knowledge of the factors that drive a reaction rate constant should also allow the design of more suitable systems.

More than 30 years of research with laser techniques have made possible the proposals of many diagrams of excited-state processes in photoinitiators (PIs) based on kinetic measurements. In very recent years, a series of works involving laser-induced photocalorimetry, ESR, and quantum mechanical calculations (in addition to the more classic laser flash photolysis (LFP)) noticeably improves the panel of available data. At present, such diagrams are largely known [1–10] for (i) type I PI, for example, benzoin ethers, hydroxyalkyl ketones, dialkoxyacetophenones, α-keto-oxime esters, phosphine oxides, sulfonyl ketones, and morpholino ketones, (ii) type II photoinitiatiors, for example, ketone (benzophenone, thioxanthone, camphorquinone, and ketocoumarin)/hydrogen donors (amine, thiol, silane, borane, and germane), dyes (Rose Bengal, methylene blue, and coumarin)/amine, and (iii) some multicomponent photoinitiating systems, for example, dye (or ketone)/H-donor/iodonium salt, and coumarin (or ketone)/HABI/amine (or thiol). The role of the structure and the substitution in PIs as well as the

Photoinitiators for Polymer Synthesis: Scope, Reactivity and Efficiency, First Edition.
Jean-Pierre Fouassier and Jacques Lalevée.
© 2012 Wiley-VCH Verlag GmbH & Co. KGaA. Published 2012 by Wiley-VCH Verlag GmbH & Co. KGaA.

particular features observed in multifunctional, copolymerizable, hydrophilic, and water-soluble PIs has been largely explored.

Most of the reactivity/efficiency relationships were discussed in fluid media (monomers dissolved in an inert solvent or, more recently, low-viscosity monomers). For a long period (in the 1980s), many authors (Parts 2 and 3) [11–16] tried to fit the measured polymerization rates $Rp = f$ (rate constants, reactant concentrations, light intensity, and quantum yields) with a model predicted from the investigation of the excited-state processes. A lot of interesting conclusions on the general behavior were therefore deduced, but, in many cases (except in homogeneous series of photoinitiators), it was difficult to derive complete and quantitative predicted laws of polymerization in agreement with the experimental results. More recently, in the past 10 years, attempts were also conducted to have an idea of these relationships in viscous photopolymerizable media close (or, ideally, similar) to those used in practical applications.

Many data on radical photoinitiating systems have been gathered (see in Part 2 and throughout Part 4). Considerably less kinetic, thermodynamical, or theoretical information related to Schemes 12.7, 12.13, and 12.19 are available on cationic photoinitiators (Part 3). The same holds true for anionic, neutral photoacid, and photobase photoinitiators (Part 3). On the other hand, the results on photoinitiating systems usable in free-radical-promoted cationic photopolymerization (FRPCP) came to be well documented (see this last point in Part 4). In the following, we focus our attention on radical PIs.

16
Role of the Experimental Conditions in the Performance of a Radical Photoinitiator

The PI efficiency is strongly connected with the experimental conditions: high-/low-viscosity monomer-based formulations, laminate/air film photopolymerization reactions, polychromatic/monochromatic light, and high/low light intensity exposure. Indeed (Section 5.7.3), monomer viscosities can range, for example, from 14 500 cP for an epoxy acrylate oligomer/monomer matrix (e.g., bisphenol A epoxy diacrylate oligomer diluted with 25% of tripropyleneglycol diacrylate monomer; EpAc) to ~100 cP for a trifunctional monomer such as trimethylolpropane triacrylate (TMPTA) (lower viscosities can still be reached for mono- or difunctional acrylates). The polymerization reaction can be carried out in a film that is either in contact with the surrounding aerated atmosphere (e.g., for a coating) or protected from oxygen (through the use of an inert film or a glass sandwich, nitrogen inerting). According to the applications, a lot of very different light sources are used (Section 2.3): high-power polychromatic Hg lamps (UV and visible emission spectrum; power density typically >1.2 W cm^{-2} in the 280–445 nm range), xenon lamps or Xe–Hg lamps (visible light), LEDs (quasi-monochromatic light; band pass ~40 nm; output >10 mW cm^{-2}), diode lasers (monochromatic lights; output 10–100 mW cm^{-2}; operating at selected wavelengths such as 405, 457, 473, 532, 635, and 780 nm), sun (white light with near-UV emission, typical power density <5 mW cm^{-2} in the 350–800 nm range) and more recently very soft irradiation sources such as fluorescence or LED bulbs (<5 mW cm^{-2} in the 400–800 nm range) [17].

The overall quantum yield in radicals as well as the radical reactivity toward the starting matrix (Section 5.7) will obviously govern the efficiency of the initiation step and as a consequence the efficiency of the polymerization reaction (Rp and final conversion). Both aspects are related to the kinetics of the excited state processes and the thermodynamics of the involved reactions. In reactions where high-viscosity monomer/oligomer formulations and high-intensity light sources are used (as in the radiation curing area), many radical PIs are operative even under air. On thecontrary, the viscosity of the formulation, the intensity and the spectral distribution of the light source, and the presence of oxygen play a decisive role and appear of prime importance, for example, when the goal consists in photopolymerizing a low-viscosity matrix on a visible light excitation with a very

Photoinitiators for Polymer Synthesis: Scope, Reactivity and Efficiency, First Edition.
Jean-Pierre Fouassier and Jacques Lalevée.
© 2012 Wiley-VCH Verlag GmbH & Co. KGaA. Published 2012 by Wiley-VCH Verlag GmbH & Co. KGaA.

low intensity source under air. The film thickness is also important, that is, for thin samples, the polymerization in aerated conditions is more difficult.

16.1
Role of Viscosity

Let us take an example to illustrate the viscosity effects. As already stated (Section 5.7.3), the diffusion rate constants evaluated from the Stokes–Einstein equation are $\sim 4 \times 10^5$ and 9×10^7 mol^{-1} l s^{-1} for a viscous epoxy acrylate (EpAc) (see above) and TMPTA, respectively. In EpAc, all bimolecular rate constants higher than the diffusion limit of this medium will level off at $\sim 4 \times 10^5$ mol^{-1} l s^{-1}, whereas, in TMPTA, rate constants $k < 9 \times 10^7$ mol^{-1} l s^{-1} will be discriminated. As a consequence, the k values (i.e., that reflect the true reactivity of the concerned states or species) will have a more important effect in low-viscosity matrices where the bimolecular reactions (e.g., the initiating radical production vs the monomer or oxygen quenching of the excited state) can become competitive to decrease the initiation quantum yield. For example, in laminates under low-intensity irradiation [18], using BP/silane instead of BP/amine in fluid TMPTA leads to a lower relative efficiency (i.e., the amine is a better co-initiator) than that found in EpAc where the silane behaves now as a better co-initiator than the amine (Figure 16.1).

As seen in Scheme 16.1, the triplet quenching of ^{3}BP by TMPTA ($k_q = 7 \times 10^7$ mol^{-1} l s^{-1}) appears in strong competition with the ^{3}BP/silane hydrogen abstraction reaction ($k_H = \sim 10^6 - 10^8$ mol^{-1} l s^{-1} depending on the silane structure). Taking into account the respective monomer and silane concentrations, the ^{3}BP/monomer interaction is expected as the major pathway (k_q [TMPTA] $\sim 3 \times 10^8$ s$^{-1} \gg k_H$ [silane] $\sim 5 \times 10^5$ s^{-1}). It is less important in ^{3}BP/EDB, that is, k_H being higher than for silane (k_H [EDB] $\sim 4.5 \times 10^6$ s^{-1}). In TMPTA, the k_Hs are lower than the diffusion rate constant and the reduced efficiency of silanes

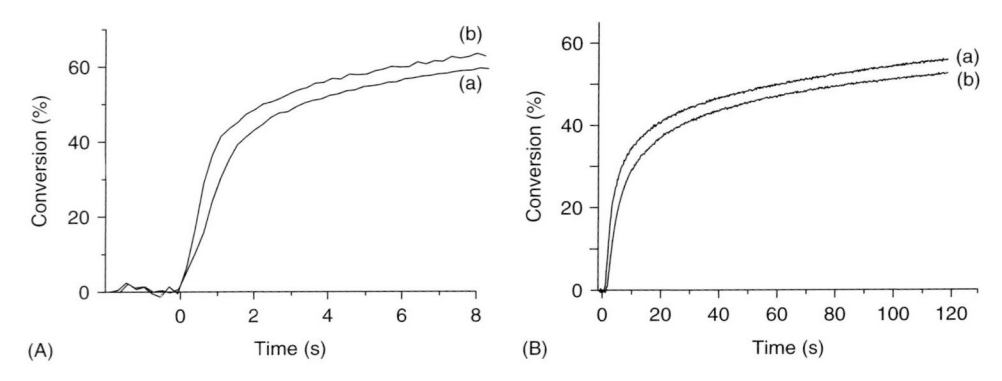

Figure 16.1 Conversion versus time curves for the photopolymerization of (A) EpAc and (B) TMPTA under laminated conditions. Hg–Xe lamp. $I_0 = 44$ mW cm^{-2}. Photoinitiating systems: benzophenone/ethyldimethylaminobenzoate (EDB) (a) and benzophenone/tris(trimethylsilyl)silane (b) (1%/1% w/w). (Source: From [18].)

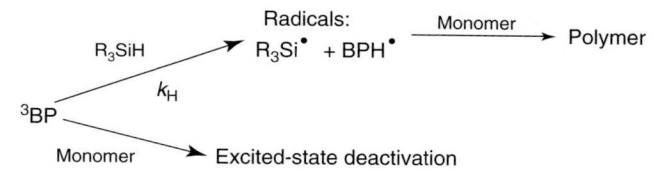

```
                          Radicals:        Monomer
          R₃SiH    ↗  R₃Si• + BPH•  ————————→  Polymer
                 k_H
³BP  ↗
     ↘
     Monomer  ——→  Excited-state deactivation
```

Scheme 16.1

compared to EDB is explained by a decrease in the hydrogen abstraction yield. Contrary to TMPTA, the hydrogen abstraction in EpAc more efficiently competes with the triplet-state/monomer quenching (as both the k_{HS} of silane and EDB are now leveled off by the diffusion). The difference between the two pathways decreases (k_q [monomer] $\sim 1 \times 10^6$ s^{-1} > k_H [silane] $\sim 2 \times 10^4$ s^{-1}, and = k_H [EDB] $\sim 2 \times 10^4$ s^{-1}). As a consequence, the BP/silane system is now more efficient in EpAc. Indeed, with similar quantum yields in initiating radicals (silane vs amine), the ability of the silyl radicals to initiate the polymerization is higher than that of the EDB-derived aminoalkyl radical (the initiation rate constants are given in Section 18.9).

16.2
Role of the Surrounding Atmosphere

Oxygen inhibition, which refers to the detrimental interaction of oxygen with the formulation, might be a serious drawback in free radical photopolymerization (FRP) and FRPCP (Section 7.5). According to their lifetimes, the excited states of PIs and PSs are more or less quenched by O_2. Both the initiating R• and propagating radicals R-M_n• are scavenged by O_2. This leads to highly stable peroxyl radicals ROO•(R-M_nOO•), which cannot participate in any further polymerization initiation reactions but yield hydroperoxides ROOH (R-M_nOOH) through hydrogen abstraction with a polymer chain or a hydrogen donor (see Section 18.11 for a discussion of the peroxyl reactivity). The radical/O_2 interaction is a nearly diffusion-controlled reaction. The polymerization starts only when oxygen is consumed [19]. The reoxygenation process is slow in highly viscous or thick samples but fast in very low viscosity or thin samples: this explains the difference in the monomer conversion–time curves in Figure 16.2. Accordingly, a low oxygen inhibition in FRP is observed in thick and viscous samples contrary to the very strong inhibition noted in thin- and low-viscosity samples.

16.3
Role of the Light Intensity

The light source intensity I_0 governs the energy absorbed I_{abs} by the photosensitive system (Section 5.6) and the amount of the initiating radicals generated. The search

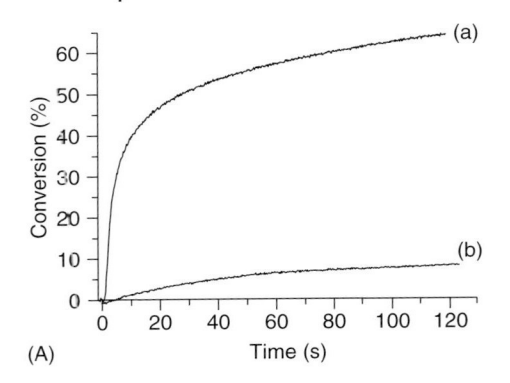

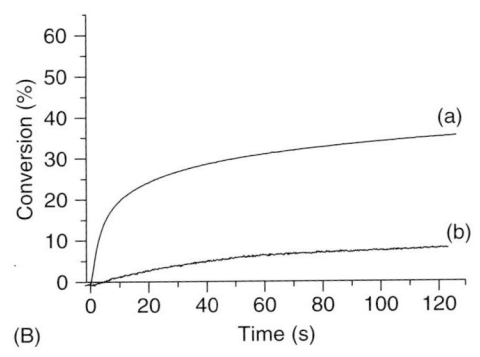

Figure 16.2 (A) Conversion versus time curves for the photopolymerization of (a) EpAc and (b) TMPTA under air. Photoinitiating system: benzophenone/ethyldimethylaminobenzoate (1%/1% w/w). Hg–Xe lamp. $I_0 = 44$ mW cm^{-2} (film thickness: 20 µm). (B) Conversion versus time curves for the photopolymerization of TMPTA under air. Photoinitiating system: benzophenone/ethyldimethylaminobenzoate (1%/1% w/w): film thickness: (a) 50 µm and (b) 20 µm. (Author's work).

for initiating systems working under relatively low light intensity and/or visible lights (e.g., a xenon lamp, LED, and diode lasers) has been the subject of many efforts ([20–45] and references therein).

Considering Eq. (7.2), it clearly appears that I_{abs} drives the polymerization rate Rp. It is also responsible for the more or less pronounced oxygen effect. When I_0 is reduced, the initial O_2 consumption decreases and the amount of initiating radicals becomes lower. Much more initiating radicals are thus scavenged by O_2, and, as a consequence, the oxygen inhibition has a larger dramatic effect on the polymerization profile. Coming back to our example, in a TMPTA film exposed to a relatively low light intensity (44 mW cm^{-2}) where a significant O_2 effect is expected, BP/silane is better than BP/EDB in agreement with the ability of silyl radicals to overcome the oxygen inhibition (Section 9.3.15). When the irradiation intensity is increased (220 mW cm^{-2}), BP/EDB is more efficient in TMPTA (Figure 16.3). This is likely ascribed to the lower oxygen effect at high intensity associated with a fast O_2 consumption by the aminoalkyl radicals.

For a given I_0, increasing the PI concentration allows an increase in I_{abs} and the radical concentration, which partly overcomes the problem. In fact, at high [PI], the polymerization can be better at the surface (because of a higher I_{abs}). It might be worse, however, in the core of the coating (as the penetration of the light into the film is less important).

Using high light intensity sources as in the UV curing area (Hg lamps, doped Hg lamps, Hg bulb lamps), suitable PIs and high PI concentrations allow satisfactory curing of high- and (to a lesser extent) low-viscosity monomer/oligomer films under air for conventional coating applications on an industrial ground (see below). The situation is not the same if one considers applications where (i) only low intensities (e.g., with sunlight, LED, diode lasers, fluorescence and LED bulbs) are available or (ii) quite low-viscosity monomers (particular acrylates in FRP or cationic monomers

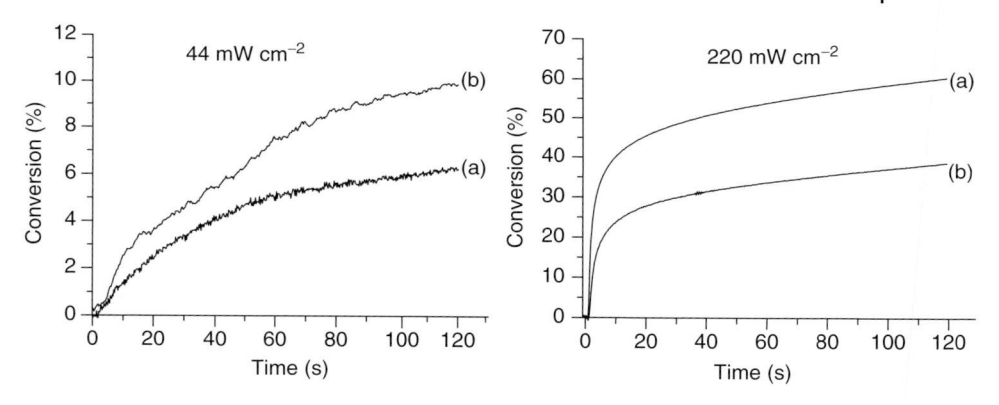

Figure 16.3 Conversion versus time curves for the photopolymerization of TMPTA under air and on different light intensities: $I_0 = 44$ mW cm^{-2} and $I_0 = 220$ mW cm^{-2} (Hg–Xe lamp). Photoinitiating systems: (a) BP/EDB and (b) BP/tris(trimethylsilyl)silane (1%/1% w/w). (Source: From [18].)

in FRPCP) or thin films have to be employed. This is the reason why (i) the use of sunlight in the radiation curing field, despite efforts during the past 30 years, still remains a great challenge and (ii) the research for high-power diode laser and LED arrays now knows a very rapid development.

Actually, upon low light intensity (sunlight: $I_0 < 10$ mW cm^{-2} in the 300–800 nm range), Type I PIs based on phosphine oxides (TPO, BAPO) lead to good to excellent polymerization profiles under air in viscous media such as EpAc [17]. In very viscous polymer matrices (acrylates dispersed in styrene-butadiene rubber) [46], the oxygen inhibition is still more easily overcome even for a low I_0. In Figure 16.4, the addition of tris(trimethylsilyl)silane (TTMSS) to BAPO significantly improves the polymerization of EpAc, that is, tack-free coatings are obtained after 12 min versus 1 h of sunlight irradiation using BAPO/TTMSS (2%/3% w/w) instead of BAPO alone (2% w/w), respectively [17]. The same holds true with TPO/TTMSS versus TPO (12 vs 45 min). The ability of the silyl radical chemistry to overcome the oxygen inhibition is presented in detail in Sections 9.3.15 and 18.9. Up to now, tack-free coatings cannot be obtained under sunlight exposure with low-viscosity monomers such as TMPTA or ethoxylated pentaerythritol tetraacrylate (EPT) [17, 46].

In conclusion, the observed photopolymerization efficiency is a strong interplay between light intensity, medium viscosity, surrounding atmosphere, and chemistry of the photoinitiating systems. The monomer used can also affect the polymerization in aerated conditions, that is, the propagation reaction being in competition with the radical/oxygen quenching. For acrylates exhibiting higher propagation rate constants than methacrylates, the propagation reaction is favored (Section 18.18). Under similar polymerization conditions (light intensity, sample viscosity, atmosphere, and PIs), the polymerization under aerated conditions is more efficient for acrylates than for methacrylates.

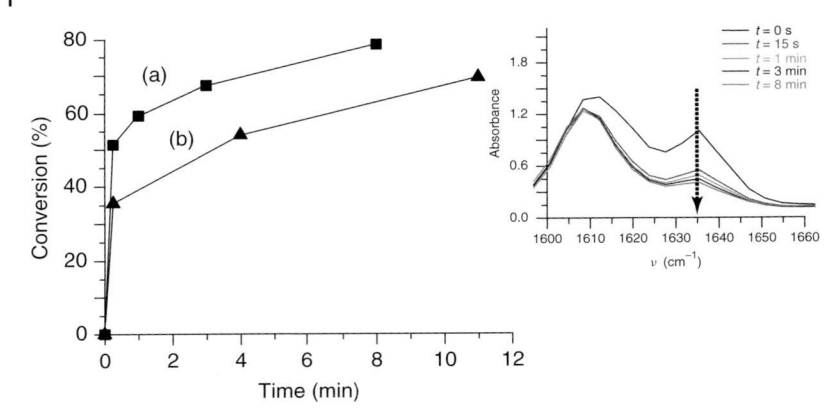

Figure 16.4 Photopolymerization of EpAc under sunlight and air. Photoinitiating systems: (a) BAPO/TTMSS (2%/3% w/w) and (b) BAPO (2% w/w). Insert: evolution of the acrylate IR band (followed by real-time Fourier transform infrared (FTIR) spectroscopy). (Source: From [17].)

References

1. Pappas, S.P. (1986) *UV-Curing: Science and Technology*, Technology Marketing Corporation, Stamford, CT; Plenum Press, New York (1992).
2. Fouassier, J.P. and Rabek, J.F. (eds) (1993) *Radiation Curing in Polymer Science and Technology*, Chapman & Hall, London.
3. Fouassier, J.P. (1995) *Photoinitiation, Photopolymerization, Photocuring*, Hanser, Münich.
4. Neckers, D.C. (1999) *UV and EB at the Millennium*, Sita Technology, London.
5. Dietliker, K. (2002) *A Compilation of Photoinitiators Commercially Available for UV Today*, Sita Technology, Ltd, London.
6. Mishra, M.K. and Yagci, Y. (eds) (2009) *Handbook of Vinyl Polymers*, CRC Press.
7. Allen, N.S. (ed.) (2010) *Photochemistry and Photophysics of Polymer Materials*, John Wiley & Sons, Inc., Hoboken, NJ.
8. Fouassier, J.P. and Rabek, J.F. (eds) (1990) *Lasers in Polymer Science and Technology*, CRC Press, Boca Raton, FL.
9. Fouassier, J.P., Ruhlmann, D., Graff, B., Morlet-Savary, F., and Wieder, F. (1995) *Prog. Org. Coat.*, **25**, 235–271.
10. Fouassier, J.P., Ruhlmann, D., Graff, B., and Wieder, F. (1995) *Prog. Org. Coat.*, **25**, 169–202.
11. Fouassier, J.P. (1990) *Prog. Org. Coat.*, **18**, 229–252.
12. Salmassi, A., Eichler, J., Herz, C.P., and Schnabel, W. (1982) *Polym. Photochem.*, **2**, 209–221.
13. Ruhlmann, D., Fouassier, J.P., and Wieder, F. (1992) *Eur. Polym. J.*, **28**, 1577–1581.
14. Schnabel, W. (1991) in *Laser in Polymer Science and Technology: Applications*, vol. 2 (eds J.P. Fouassier and J.F. Rabek), CRC Press, pp. 95–135.
15. Lalevée, J., Allonas, X., Jradi, S., and Fouassier, J.P. (2006) *Macromolecules*, **39**, 1872–1878.
16. Timpe, H.J., Jockusch, S., and Korner, K. (1993) in *Radiation Curing in Polymer Science and Technology*, vol. 2 (ed. J.P. Fouassier), Elsevier, Barking, pp. 575–601.
17. Lalevée, J. and Fouassier, J.P. (2011) *Polym. Chem.*, **2**, 1107–1113.
18. El-Roz, M., Lalevée, J., Allonas, X., and Fouassier, J.P. (2009) *Macromolecules*, **42**, 8725–8732.
19. Decker, C. and Jenkins, A.D. (1985) *Macromolecules*, **18**, 1241–1248.

20. Oster, G.K., Oster, G., and Pratti, G. (1957) *J. Am. Chem. Soc.*, **79**, 595–599.
21. Timpe, H.J. (1993) in *Radiation Curing in Polymer Science and Technology*, vol. 2 (ed. J.P. Fouassier), Elsevier, Barking, pp. 529–554.
22. Nguyen, C.K., Hoyle, C.E., Lee, T.Y., and Joensson, S. (2007) *Eur. Polym. J.*, **43**, 172–177.
23. Padon, K.S. and Scranton, A.B. (2000) *J. Polym. Sci. A: Polym. Chem.*, **38**, 3336–3346.
24. Kabatc, J., Pietrzak, M., and Paczkowski, J. (2002) *J. Chem. Soc., Perkin Trans. 2*, 287–295.
25. Jakubiak, J. and Rabek, J.F. (1999) *Polimery*, **44**, 447–461.
26. Rabek, J.F., Linden, L.A., Fouassier, J.F., Nie, J., Andrzejewska, E., Paczkowski, J., Jakubiak, J., Wrzyszczynski, A., and Sionkowska, A. (1999) in *Trends in Photochemistry and Photobiology: Photosensitive Systems for Photopolymerization Reactions*, vol. 5 (ed. J.P. Fouassier), Research Trends, Trivandrum, pp. 51–62.
27. Nie, J., Andrzejewska, E., Rabek, J.F., Linden, L.A., Fouassier, J.P., Paczkowski, J., Scigalski, F., and Wrzyszczynski, A. (1999) *Macromol. Chem. Phys.*, **200**, 1692–1701.
28. Urano, T. (2003) *J. Photopolym. Sci. Technol.*, **16**, 129–156.
29. Takahashi, T., Watanabe, H., Miyagawa, N., Takahara, S., and Yamaoka, T. (2002) *Polym. Adv. Technol.*, **13**, 33–39.
30. Riediker, M., Meier, K., and Zweifel, H. (1986) Ciba Geigy. European Patent 186626.
31. Eaton, D. (1979) *Photogr. Eng. Sci.*, **23**, 150–154.
32. Liu, A.D., Trifunac, A.D., and Krongrauz, V. (1991) *J. Phys. Chem.*, **95**, 5822–5827.
33. Yang, J. and Neckers, D.C. (2004) *J. Polym. Sci. A: Polym. Chem.*, **42**, 3836–3841.
34. Tehfe, M.A., Blanchard, N., Fries, C., Lalevée, J., El-Roz, M., Allonas, X., and Fouassier, J.P. (2009) *Macromol. Chem. Rapid Commun.*, **31**, 473–478.
35. Lalevée, J., El-Roz, M., Tehfe, M.A., Allonas, X., and Fouassier, J.P. (2009) *Macromolecules*, **42**, 8669–8674.
36. Lalevée, J., El-Roz, M., Morlet-Savary, F., Allonas, X., and Fouassier, J.P. (2009) *Macromol. Chem. Phys.*, **210**, 311–319.
37. El-Roz, M., Morlet-Savary, F., Lalevée, J., Allonas, X., and Fouassier, J.P. (2008) *J. Polym. Sci. Polym. Chem.*, **46**, 7369–7375.
38. Burget, D. and Fouassier, J.P. (1998) *J. Chem. Soc., Faraday Trans.*, **94**, 1849.
39. Fouassier, J.P., Allonas, X., and Burget, D. (2003) *Prog. Org. Coat.*, **47**, 16–36.
40. Burget, D., Grotzinger, C., Jacques, P., and Fouassier, J.P. (2001) *J. Appl. Polym. Sci.*, **81**, 2368.
41. Allonas, X., Obeid, H., Fouassier, J.P., Kaji, M., Ichihashi, Y., and Murakami, Y. (2003) *J. Photopolym. sci. Technol.*, **16**, 123.
42. Burget, D., Mallein, C., and Fouassier, J.P. (2003) *Polymer*, **44**, 7671.
43. Burget, D., Fouassier, J.P., Amat-Guerri, F., Sastre, R., and Mallavia, R. (1999) *Acta Polim.*, **50**, 337.
44. Chesneau, E. and Fouassier, J.P. (1985) *Angew. Makromol. Chem.*, **135**, 41–64.
45. Yamashita, K., Imahashi, S., Morlet-Savary, F., and Fouassier, J.P. (1996) *Polymer*, **38**, 1415–1421.
46. Decker, C., Nguyen Thi Viet, T., and Le Xuan, H. (1996) *Eur. Polym. J.*, **32**, 1319–1326.

17
Reactivity and Efficiency of Radical Photoinitiators

17.1
Relative Efficiency of Photoinitiators

Typical examples of selected photopolymerization results obtained under various conditions (in solution or in film under monochromatic or polychromatic lights) are given in several tables.

17.1.1
Photopolymerization of MMA in Organic Solvents

Examples of relative initiation quantum yields Φ_i (see Eq. (5.18)) for the methylmethacrylate (MMA) photopolymerization in benzene under a 365 nm monochromatic light and cleavage yields (or yield in initiating radicals) Φ_c of Type I photoinitiator (PI) [1–4] are shown in Table 17.1. Large differences in Φ_i and Φ_c are observed but Φ_c is not the only driving parameter. The role of the initiating radicals in the monomer addition reaction is likely important (see the values of addition rate constants in Chapter 18). A refined analysis based on a quantitative kinetic treatment is almost impossible.

17.1.2
Photopolymerization of Various Monomers in Organic Solvents and Bulk Media

The role of the medium (mono- and difunctional monomers in solvents, polymer matrices, and bulk oligomers) in polymerization reactions under polychromatic lights is illustrated in Table 17.2, where the relative initiation efficiency ϕ_{rel} of a given PI (compared to a reference PI) has been expressed using Eq. (7.5) to take into account the amount of energy absorbed by each sample. The efficiency scale defined by PI $= f$(monomer) clearly depends on PI. A discussion of these results is provided below.

Photoinitiators for Polymer Synthesis: Scope, Reactivity and Efficiency, First Edition.
Jean-Pierre Fouassier and Jacques Lalevée.
© 2012 Wiley-VCH Verlag GmbH & Co. KGaA. Published 2012 by Wiley-VCH Verlag GmbH & Co. KGaA.

Table 17.1 Examples of available data concerning cleavable photoinitiators: (i) calculated cleavage yield Φ_c in solution of methylmethacrylate MMA 7 mol l^{-1}; (ii) measured relative initiation quantum yield (Φ_i).

Compound	100 Φ_c	Φ_i	Compound	100 Φ_c	Φ_i
DMPA*	100	100		90	45
TPMK	80	60		46	12
	38	12		38	8
	35	0.1	HAP	29	100

17.1 Relative Efficiency of Photoinitiators

In benzene/MMA 7 mol l^{-1}. Under monochromatic light at 365 nm. Optical density of the PI: 0.1. Values of Φ_i are normalized to that of DMPA(*) (Φ_i = 100). Results mostly gathered from Ref. [4].

Table 17.2 Relative initiation efficiency ϕ_{rel} values as a function of the polymerizable medium.

PI	EpAc[a]	HDDA[b]	BA[b]	MA[b]	MMA[b]	BMA[b]
(structure 1)	1	1	1	1	1	1
(structure 2)	0.7	3.9	3.4	3.6	0.7	1.35
(structure 3)	0.4	–	3.5	3.5	0.85	1
(structure 4)	0.7	2.8	2.1	2.3	0.65	1
(structure 5)	0.7	–	3.2	3.7	3.5	3.5
(structure 6)	1.5	–	3.2	3.5	1.6	2.45
(structure 7)	0.15	–	1.7	–	–	–

Table 17.2 (continued)

PI	EpAc[a]	HDDA[b]	BA[b]	MA[b]	MMA[b]	BMA[b]
(structure 1)	<0.05	–	0.45	–	–	–
(structure 2)	0.4	–	0.5	–	–	–
(structure 3)	0.2	–	1.2	–	–	–
(structure 4)	<0.05	–	<0.05	–	–	–

See text. Under argon. Light intensity: ∼44 mW cm^{-2}. The error on ϕ_{rel} is typically ±0.1 except for the experiments in EpAc where the error reaches ±0.2.
[a] Ebecryl 605: bulk.
[b] HDDA, BA, MA, MMA, BMA, 1 mol l^{-1} of double bond in toluene.
From [5].

17.1.3
Photopolymerization of TMPTA in Film

Typical polymerization rates of the low-viscosity monomer trimethylolpropane triacrylate (TMPTA) in film in the presence of various Type I PIs under polychromatic irradiations (Hg–Xe or Xe lamp) under air are given in Table 17.3. A strong O$_2$ inhibition occurs with a strong decrease in the final conversions under air compared to laminated experiments. The O$_2$ effect when using Type II PI in aerated conditions is also shown in Table 17.4. Albeit that the absorbed light is different for the investigated PIs, these tables provide their relative efficiency under given irradiation conditions.

Silyl- or germyl-radical-generating PIs lead to good/excellent polymerization initiating abilities under air (Table 17.5), which are better than that of a reference

Table 17.3 Polymerization rates of TMPTA under air ($R_p/[M_0]*100$ in s^{-1}) using a Type I PI (1% w/w) except otherwise noted; the final conversions at $t = 120$ s are given in brackets.

Photoinitiator	Irradiation conditions[b]			
	Hg–Xe lamp	Xe	405 nm	473 nm
	35 (46) (70^d)	np (63^d)	np (54^d)	–
	10 (24)^a (68^d)	2.1 (6) (57^d)	np (52^d)	–
	1.4 (3.4)	np (55^d)	np	–
	–	34 (39) (69^d)	27 (28) (52^d)	–
	65 (37) (71^d)	5.4 (11) (67^d)	2.3 (3) (55^d)	–
	np^a	np	np	–
	np	np	–	–

Table 17.3 (continued)

Photoinitiator	Irradiation conditions[b]			
	Hg–Xe lamp	Xe	405 nm	473 nm
[Ti structure with fluorinated aryl pyrrole ligands]	–	np (64[d])	–	1.9 (7.3)
[benzoin-type structure with OSiSi group]	30 (43)	np	np	–
[PhC(O)-S-C(S)-OEt]	np[c] (59[d])	–	–	–

np, no polymerization observed.
Under air. The final conversions in laminated conditions are also given.
[a]A lower light intensity is used (~11 mW cm^{-2}).
[b]Irradiation sources: polychromatic lights from a Hg–Xe lamp ($I \sim 22$ mW cm^{-2} in the 300–450 nm range) or a Xe lamp (filtered light at $\lambda > 390$ nm; $I \sim 60$ mW cm^{-2} in the 390–800 nm range); monochromatic light from diode lasers at 405 nm ($I_0 \approx 12$ mW cm^{-2}) or 473 nm ($I_0 \approx 100$ mW cm^{-2}).
[c]Light intensity ~44 mW cm^{-2}.
[d]Final conversion for polymerization in laminated conditions.
From [6].

(DMPA). This table also illustrates the effect of the PI absorption spectrum. Other systems generating germyl radicals have been described [7–10].

17.1.4
Photopolymerization of Monomer/Oligomer Film

Examples of photopolymerization data for highly viscous formulations on high-intensity polychromatic light exposure under air are given in Table 17.6. The effect of phenolic derivatives (Sections 7.7 and 8.6.4) on the cure speed of a UV conveyor to get a tack-free coating is outlined by the change of the substrate

Table 17.4 Polymerization rates of TMPTA under air ($R_P/[M_0]*100$ in s^{-1}) using a Type II PI system (1% w/w); the final conversions at $t = 120$ s are given in brackets.

Photoinitiator	Irradiation conditions			
	Hg–Xe	Xe	405 nm	473 nm
isopropyl thioxanthone	—	0 (<5)	0 (<5)	—
benzil	0.2 (17)	0.07 (12)	np	—
camphorquinone	—	—	—	0.4 (7)
7-diethylamino-4-methylcoumarin	—	np^a	—	—

17.1 Relative Efficiency of Photoinitiators

np	np	np	—
13 (23)	np	np	—
—	np[a]	np[a]	—

[a] No polymerization with ethyldimethylaminobenzoate co-initiator (EDB) 3% w/w. See also legend of Table 17.3.
From [6].

Table 17.5 Photopolymerization of EpAc in the presence of silyl- or germyl-radical-generating PIs.

	$R_{p\ rel}$ (laminate)	$R_{p\ rel}$ (under air)
[dibenzodioxa-disilane structure with Ph groups]	$0.96^a\ (3.37^b)$	$0.85^a\ (2.99^b)$
[difluorenyl-disilane structure]	$0.35^a\ (1.11^b)$	$0.12^a\ (0.38^b)$
Acetyl-GePh$_3$	1.5^a	1.6^a

^{a}The R_ps are relative to DMPA.
^{b}The R_ps in brackets are corrected from the amount of light absorbed (see text).
From [11, 12].

(glass vs wood). The positive effect of the silane versus an amine in a Type II PI is revealed by the increase of the surface hardness of the coating in reactions carried out under air.

17.1.5
Photopolymerization of Acrylamide in Water

The role of ionic PIs in the photopolymerization of acrylamide in water is shown in Table 17.7. Strong differences are observed (see comments below).

17.1.6
Photopolymerization of Oil-Soluble Monomers in Direct Micelles

The photopolymerization of MMA and styrene has been carried out in oil-in-water sodium dodecyl sulfate (SDS) micelles: some results are presented in Table 17.7 (see the discussion below).

Table 17.6 Relative rates of polymerization of a polyurethane diacrylate/oxazolidone monoacrylate resin. Conversion after 1 s in parenthesis; medium pressure mercury lamp. From Ref. [13]. Number of passes to get a tack-free coating with an epoxyacrylate/HDDA matrix on glass or wood (UV conveyor equipped with an 80 W cm^{-1} Hg lamp; curing speed: 52.6 m mn^{-1}; from Ref. [14]). Hardness of a cured epoxyacrylate/HDDA coating (Xe lamp; from Ref. [15]). Under air.

Photoinitiator	R_p polyurethane diacrylate	Tack-free glass	Tack-free wood	Hardness
DMPA	80 (85%)	–	3	–
HCAP	100 (90%)	–	–	–
PDO	70 (80%)	–	–	–
BP/MDEA	12 (30%)	1	2	290[a]
BAPO	–	2	2	305
BAPO/MDEA	–	1	Sticky	–
BAPO/ITX	–	1	2	–
BP/TTMSS	–	–	–	320[b]
BAPO/TTMSS	–	–	–	320[b]

[a]EDB (1% w/w) instead of MDEA.
[b]TTMSS (1% w/w).

17.1.7
Photopolymerization of Water-Soluble Monomers in Reverse Micelles

The behavior of ionic PIs in the photopolymerization of acrylamide in reverse AOT (bis-2 ethyl hexyl sodium sulfosuccinate) micelles is also shown in Table 17.7 (see the discussion below).

17.2
Role of the Excited-State Reactivity

17.2.1
Generation of Radicals

The generation of the first monomer radical is obviously one of the driving factors for high polymerization efficiency. For a Type I PI working in its triplet state, the initiation quantum yield ϕ_i (defined by Eqs. 5.18 test and 5.19) is a function of (i) the intersystem crossing quantum yield ϕ_{isc}, (ii) the cleavage yield ϕ_c in the presence of the monomer, and (iii) the first monomer radical yield ϕ_{RM}. The real issues for a detailed analysis in fluid media are (i) the cleavage rate constant k_c and the monomer quenching rate constant of short-lived triplet states cannot be really measured (an estimate for an upper value of k_c can only be expressed as the reciprocal value of the triplet-state lifetime), (ii) the only accessible parameter (Section 5.7.2) is the dissociation quantum yield $\phi_{diss}(=\phi_{isc}\phi_c)$ in the absence of monomer, (iii) ϕ_{diss} (but not ϕ_c) takes into account the cage escape yield in radicals

Table 17.7 Relative rates of polymerization of (i) acrylamide (AA) (0.7 mol l^{-1}) in water, (ii) MMA and styrene (0.5 mol l^{-1}) in SDS (0.5 mol l^{-1}) micelles in the presence of water- or oil-soluble photoinitiators (and MDEA 0.05 mol l^{-1}): 4-benzophenone (BP$^+$, BP$^-$, BP), 4-benzil (BZ$^+$, BZ$^-$), and 2-thioxanthone (TX$^+$, TX$^-$, CTX) derivatives; DMPA. $\lambda = 365$ nm. OD = 0.1, iii) acrylamide AA (0.25 mol l^{-1}) in reverse AOT (bis-2 ethyl hexyl sodium sulfosuccinate) micelles ([AOT] = 1.1 mol l^{-1}; [decane] = 2.6 mol l^{-1}; [water] = 2 mol l^{-1}). In degassed media.

PI		R_p AA water	R_p MMA SDS	R_p Styrene SDS	R_p AA AOT [TEA] = 0	R_p AA AOT [TEA] = 0.15
BP$^+$	BP-CH$_2$N$^+$(CH$_3$)$_3$Cl$^-$	194	250	162	18	6.2
BP$^-$	BP-CH$_2$SO$_3^-$ Na$^+$	76	45	150	3.7	2.5
BP	BP-H	–	50	37	12	3
TX$^+$	TX-O(CH$_2$)$_3$N$^+$(CH$_3$)$_3$ CH$_2$SO$_3^-$	206	135	165	–	–
TX$^-$	TX-OCH$_2$COO$^-$H$^+$	–	70	156	–	–
BZ$^+$	BZ-CH$_2$N$^+$(CH$_3$)$_3$Cl$^-$	7	–	–	–	–
BZ$^-$	BZ-CH$_2$SO$_3^-$ Na$^+$	6	–	–	–	–
CTX	TX-Cl	–	80	–	–	–
DMPA[a]	–	–	95	–	–	–

[a]Without amine.
From [16].

(Section 5.7.4), and (iv) ϕ_{RM} is hardly calculated as it depends on true first-order reactions as well as second-order bimolecular reactions.

Similarly, for a Type II (PI/hydrogen donor) working in its triplet state, ϕ_i is a function of (i) ϕ_{isc}, (ii) ϕ_H (yield in hydrogen transfer (HT); in the case of systems working through electron/proton transfer, ϕ_H corresponds to the electron transfer yield ϕ_e), (iii) ϕ_R the yield in initiating radicals, and (iv) ϕ_{RM}. Unfortunately, ϕ_R, ϕ_{RM} as well as the cage escape yield in radicals is also hardly accessible or calculable. In a few cases, the quantities $\phi_{Rad}(= \phi_{isc}\phi_H\phi_R)$, $\phi_H\phi_R$, or ϕ_H can be directly measured (Section 6.1.3).

In Table 17.1, where the cleavage yields ϕ_c and the experimental relative ϕ_I are reported, noticeable differences are seen in the $\phi_{isc}\phi_{RM}(= \phi_I/\phi_c)$ values. The same holds true in Type II PIs where $\phi_i = \phi_{isc}\phi_H\phi_{RM}$ or $\phi_i = \phi_{isc}\phi_e\phi_R\phi_{RM}$. Interesting rough trends can sometimes be observed [17]. As discussed in many papers (e.g., [1–3, 5, 18–24] and references therein), it clearly appears now that it is almost impossible to go further in a clean and deeper analysis.

17.2.2
Role of Monomer Quenching

Monomer quenching is an important deactivation process in PIs, especially those having long-lived triplet states. A large panel of rate constants is given in Table 17.8.

Table 17.8 Adiabatic ionization potentials IPa of the studied monomers and quenching rate constants (in 10^5 mol^{-1} l s^{-1}) of the PI triplet states of BP, TX, CTX by monomers. In benzene or acetonitrile.

	IPa (eV)	Benzophenone	Thioxanthone	2-Chlorothioxanthone
2,3-Dimethyl-2-butene	8.27	90 000	–	–
NVP	8.49	4 000	400	300
BVE	8.58	1 160	10	10
2-Methyl-2-butene	8.68	3 600	–	–
Cyclohexene	8.94	570	–	–
EVE	8.8	880	–	64
VA	9.2	50	2	0.2
MMA	9.7	700	150	31
MA	9.9	220	–	1
2-Chloroethylvinylether	9.97	800	–	–
AN	10.9	300	40	4
HDDA	–	460	–	13
TMPTA	–	710	–	21

From [25], and references therein.

Several mechanisms have been proposed for the ketone triplet-state/acrylate (or MMA) monomer interaction:

1) **Energy transfer.** The energetic requirement ($E_T^{donor} > E_T^{acceptor}$) must be fulfilled. It was proposed that energy transfer mostly occurs between electron-poor monomers and ketones having $E_T > 68$ kcal mol^{-1} [26]. In the case of acrylates, because of their high triplet energy (>72 kcal mol^{-1}), energy transfer is ruled out. On the contrary, in the benzophenone/styrene couple, energy transfer occurs (Figure 17.1 curve b).
2) **Hydrogen abstraction.** This process (Section 5.7.1) is also ruled out as no ketyl radicals are detected by laser flash photolysis (LFP) in most usual monomers.
3) **Electron transfer.** The quenching of nπ^* carbonyl triplet states (e.g., BP) by olefins is well known. A charge-transfer (CT) mechanism has been proposed on the basis of the dependence of the quenching rate constant k_q on the ionization potential (IP) (Section 6.4) and its decrease with the solvent polarity [26]. This mechanism was considered as true for electron-rich monomers and ketones (e.g., BP) having $E_T < 75$ kcal mol^{-1}. On the contrary, in $\pi\pi^*$ triplet states (e.g., in xanthone (XT) or thioxanthone (TX) derivatives), k_q decreases when the polarity increases. This was tentatively attributed to a change in the relative position of the closely lying nπ^* and $\pi\pi^*$ states [27, 28]. An investigation of the triplet-state quenching of a large set of PIs by electron-rich and electron-poor double bonds [25] do not show clear correlations between k_q and IP on the whole range of IP values. This might be attributed to a CT reaction from (i) the olefin to the ketone for IP < 9 eV (π/n interaction) and (ii) the ketone to

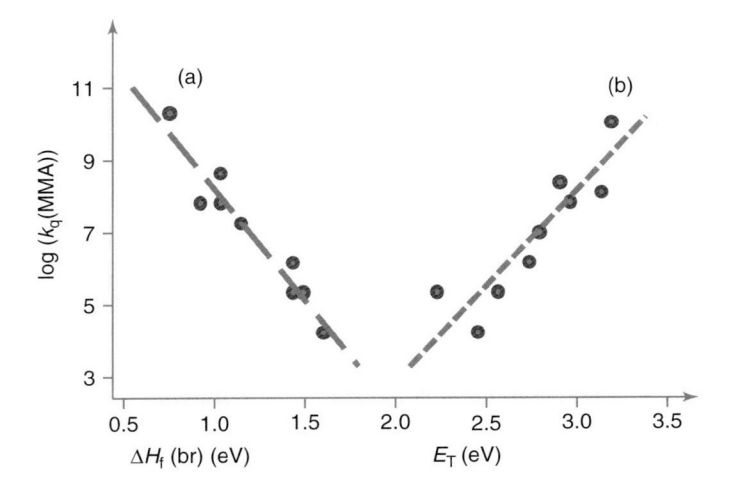

Figure 17.1 Log k_q (MMA) versus the biradical formation enthalpy ΔH_f (br) (curve a) and the experimental values of the triplet-state energies of ketones (curve b). (Source: From [25].)

the olefin for IP > 9 eV (π^*/π^* interaction). The calculated free enthalpy changes according to Eq. (6.4) are not, however, in favor of a CT mechanism.

4) **Formation of a biradical (resulting from the addition of the excited ketone to the double bond) and then an oxetane.** This mechanism has been largely suspected in the case of MMA [29] through the evaluation of the oxetane formation enthalpy ΔH_f (ox) using Eq. (17.1), the rate constant k_q becoming linearly dependent on E_T (assuming that the formation enthalpy of the C=O and C—O bonds are independent of the ketone). Moreover, the photopolymerzation of MMA in the presence of BP alone is effective (but R_p is very low): as no hydrogen abstraction is observed using LFP, the plausible mechanism is an initiation by this biradical.

$$\Delta H_f \text{ (ox)} = -\Delta H_f \text{ (C=O)} + \Delta H_f \text{ (C—O)} - \Delta H_f \text{ (C=C)} - E_T \tag{17.1}$$

This is still exemplified in [25] where the number of ketones has been increased: k_q correlates with E_T in the case of MMA. The same holds true for acrylonitrile (AN), vinylpyrrolidone (NVP), and butylvinylether (BVE). However, the oxetane is not detected, which suggests that the biradical transient rather leads to the recovery of the starting compounds. The biradical formation enthalpy ΔH_f (br) in the case of MMA was therefore recalculated using Eq. (17.2). A quite good correlation between k_q and ΔH_f (br) was thus obtained [25] (Figure 17.1 curve a).

$$\Delta H_f \text{ (br)} = H_f \text{ (ox)} + \Delta H_f \text{ (C—C)} \tag{17.2}$$

This supports the involvement of the 1,4 biradical for both electron-rich (BVE, NVP) and electron-poor (AN, MMA) monomers. A CT might be present in the formation of this biradical, which would explain the observations reported in (iii). No other treatment has been proposed in the literature so far.

17.3
Role of the Medium on the Photoinitiator Reactivity

As recalled above, the investigation of the photophysical processes involved in PIs as well as the polymerization reactions are often explored in fluid solution. Other works (e.g., for industrial applications) require the evaluation of the practical efficiency in polymerization experiments carried out in bulk formulations. The question concerning the possibility of extrapolating the behavior of a PI from solution to bulk is fascinating.

17.3.1
Reactivity in Solution

In this paragraph, the comments are concerned with the general PI behavior in organic solvents where the competitive reactions are not diffusion limited. They are often based on works carried out more than 15 years ago on largely used commercial PIs (see in [1–3, 17, 19, 30, 31] and references therein). All these comments still remain valid.

As expected, the monomer interacts with the PI triplet state. The absence of a chemical quenching (i.e., the absence of a significant amount of photolysis products) reveals a predominant physical quenching process, leading to the deactivation of the triplet state without significantly producing any free radicals or biradicals.

In photoinduced electron-transfer reactions in the presence of amines, back electron transfer, and/or free ion formation arise in the primary radical pair (RP) formed after electron transfer. The electron transfer rate constant k_e depends on the IP of the electron donor. In the case of amines, the n (lone pair) or π (double bond) nature of the involved electron donor should be taken into account [32–34]. This is the reason why an electron donor with the lowest IP is not necessarily the best one. The information available on the subsequent proton transfer step is rather scanty. Tertiary aliphatic or aromatic amines are better than the others. Sometimes, fair correlations can exist between the photopolymerization activity and IP but this is not systematic [35–38]. Studies of the ketone/amine interactions and their effect on R_ps show that the mechanisms remain rather complex.

Regarding the competition between cleavage and quenching by electron donors and monomers, two comments can be made. First, cleavable PIs possessing a short-lived triplet state are almost not affected by the monomer quenching. This holds true, for example, for benzildimethyl ketal, benzoin ethers, hydroxy alkyl or cyclohexyl acetophenones, benzoyl phosphine oxides, benzoyl oxime esters, dialkoxy acetophenones. In α-morpholino ketones and α-amino para morpholino benzoyl ketones, the slower α-cleavage process is balanced by a lower monomer quenching. Second, in systems working through electron transfer, the rate constants are usually high, for example, in BP (or CTX)/aliphatic (or aromatic) amine couples. Monomer quenching is reasonably high in BPs but quite low in TXs (Table 17.8): the difference in ϕ_i, however, does not reflect the difference in ϕ_e. In efficient cleavable PIs, amine quenching usually does not compete with the dissociation process. The practical effects that are sometimes noted on R_p when adding an amine are explained by the occurrence of secondary reactions.

The interaction of triplet states with oxygen is usually strong (Sections 9.1.3 and 9.1.4). Other compounds present in the formulation such as light stabilizers or phenols (see in Sections 7.6, 7.7, and Chapter 18) can also induce competitive reactions detrimental to the initiation step [39, 40].

In photosensitized experiments, the reaction that largely governs the efficiency is obviously the photosensitizer/PI interaction (Section 5.5). Many rate constants have been measured in a lot of couples [3].

The yield in radicals for the ketone–amine systems is dependent on the polarity of the solvent. For example, the ion pair between the ketone anion and the amine cation is relatively stable in alcohol but rapidly decays in nonpolar solvents, thus increasing the initiating radical yield. Therefore, the formation of radical ions competes with that of the free radicals when the polarity or the dielectric constant increases. The photopolymerization efficiency usually increases on a decrease in the solvent polarity. The excited-state behavior may become extremely solvent dependent. In molecules such as in para aminophenyl ketones, the intersystem crossing and the photoreduction quantum yields decrease when the solvent polarity increases because of a change in the spectroscopic nature of the excited state (mixing or inversion of the CT state and the $n\pi^*$ state). In TXs where the $n\pi^*$ and $\pi\pi^*$ states are very close, the monomer quenching rate constants k_q drastically decrease when the solvent polarity increases. Such a behavior was also observed in the ITX/silane couple where the ketyl radical quantum yields are very different in benzene or acetonitrile (e.g., 0.3 and 0.7, respectively, using N,N-diethyl-1,1-dimethylsilylamine) evidencing the role of the $n\pi^*$ vs. $\pi\pi^*$ triplet-state character on the hydrogen abstraction process [15]. On the other hand, in the BP series where the lowest lying excited state remains $n\pi^*$, k_q increases with the solvent polarity. All these considerations should be kept in mind when considering neat monomers or oligomers as photopolymerizable media. The observed differences in reactivity might also be correlated with polarity effects in the case of various acrylate matrices (Section 9.1.4).

The structure/property relationships are strongly dependent on chemistry. As described in previous works [1–3, 17, 19, 30, 31], changing the PI skeleton drastically affects the reactivity through the involved photochemical processes (cleavage or electron transfer), the spectroscopic nature of the lowest lying triplet state ($n\pi^*/\pi\pi^*$), the nature of the cleavable bond (C–C, C–N, C–O, C–S, and C–P). Substitution in a given structure usually induces changes in the reactivity. The α-cleavage process in aromatic ketones is known to be strongly affected by the nature of the transition state, for example, in phenylacetophenones and benzoin derivatives or α,α-diphenylacetophenones. Fair correlations have been established, as shown, for example, for the initiation quantum yield versus the cleavage yield of hydroxy alkyl acetophenones [2, 3]. When a methylthio group is introduced at the para position of the benzoyl moiety, the photochemistry is completely changed. The reactivity generally decreases, as compared with the parent unsubstituted compound. The same holds true when introducing an amino group. Substitution on the TX skeleton at the 2-position has little effect on the rate constants, while the change is much more significant for compounds substituted in 3-position. The mixing of the $n\pi^*$ and $\pi\pi^*$ lowest triplet states of TXs results in a substantial modification of the rate constants that depend on the solvent (its polarity seems to be one of the driving factors; see above). In para-substituted BPs used in conjunction with H-donors (such as THF), the rate of polymerization, the ratio between the H-abstraction and the monomer quenching rate constants correlate pretty well with the Hammett constant σ of the substituent.

When changing the backbone of a PI, one can expect a drastic influence of the nature of the initiating radical. In fact, the by-side reactions of the formed radicals (that may give rise to secondary reactions, especially with amines), the low efficiency for the addition to a double bond, the interaction of the excited species with O_2, and the deactivation of the RPs without radical release lead to detrimental pathways even if the ϕ_c or ϕ_e values are close to unity. For example, (i) ϕ_c is not the only key parameter for the design of a skeleton yielding an efficient Type I PI (e.g., the lower triplet-state reactivity of hydroxy alkyl phenyl ketones such as HCAP and HAP is balanced by the generation of two very active initiating radicals) and (ii) ϕ_e alone does not govern the efficiency of Type II PIs. The role of the initiating radical is discussed in Chapter 18.

The same kind of discussion has been done for ionic PIs in water ([16, 41, 42] and references therein). Few data are available (see examples in Section 9.1.11) but fair descriptions of the photoinitiation step were achieved. In addition, specific processes related to polarity and PH effects can occur. Acid–base equilibrium exits in some radicals (e.g., ketyls) or PIs (e.g., xanthenic dyes). The formation of ketyl or dye radical anions as well as the ketyl/amine interaction (Eq. (17.1)) renders the mechanisms still more complicated.

$$\begin{aligned} BPH^\bullet &\longleftrightarrow BP^{\bullet-} + H^+ \\ BPH^\bullet + AH &\longleftrightarrow BP^{\bullet-} + AH_2^+ \\ DyeH^\bullet &\longleftrightarrow Dye^{\bullet-} + H^+ \end{aligned} \qquad (17.3)$$

Recently, a water-soluble carbene-borane (2,4-dimethyl-1,2,4-triazol-3-ylidene borane) has been proposed as an efficient co-initiator for the polymerization of 2-hydroxyethylacrylate in water [43]. Interestingly, the associated boryl radicals exhibit the same reactivity in the absence or presence of water.

In the past five years, works based on new experimental and theoretical investigations have been carried out on Si–C, Si–Si, C–Ge cleavable bond containing PIs, the use of other H-donors (silanes, germanes, and boranes) and the formation of novel silyl-, germyl-, or boryl-initiating radicals. These works, which give a new impetus to the knowledge of the processes involved, are discussed in Chapter 18.

17.3.2
Reactivity in Microheterogeneous Solution

Relatively few works on the chemical reactivity in microheterogeneous media have been carried out since the former times (20/30 years ago), when the interest of the photochemists was captured by the specific processes arising in such media, for example, for ketones [44–48]. In that period, similar works have been devoted to the behavior of PIs in micelles [16]. Various examples of photopolymerization reactions in direct and reverse micelles, microemulsions, emulsions, vesicles, or dispersion can be found (Section 4.1). Most of the recent works refer to applications for the manufacture of new polymer materials rather than to the role of the PI [49].

The radical photopolymerization reactions in microheterogeneous media (micelle, microemulsion, emulsion) generally exhibit both a high rate and a high degree of polymerization. Fundamental questions arise as to the specific role played by the PI: (i) interaction with the environment, (ii) localization in the surfactant assembly, (iii) exit efficiency of the radicals from the micelle, (iv) micropolarity and microbasicity effects (in the bulk and the microphase), (v) charges of both the PI and the micelle, (vi) acid–base equilibrium, and (vii) local concentration effects.

For example [16], in positively and negatively charged BPs in direct oil/water micelles, neither the excitation intensity (which increases the number of excited molecules) nor the concentration effects (which vary the occupancy) are detected, which suggests that, owing to compartmentalization of the reactants, the bimolecular reactions that take place in homogeneous solution (self-quenching and triplet–triplet annihilation) are avoided in micelles. In the presence of an electron or a hydrogen donor, a radical pair (RP) is formed in the triplet state. It converts into a singlet RP (through intersystem crossing). Free radicals appear after decorrelation and exit. The exit depends on the nature of the solute (neutral or ionic compounds), the surfactant (anionic or cationic character), or the counter-radical. The mobility of the partners is a predominant factor for the escape of the radical. High monomer and amine quenching are noted (Table 17.9).

Higher yields of exit are obtained in the presence of a magnetic field (Table 17.9), which decreases the intersystem crossing between the triplet and the singlet RP and, thereby, indirectly favors the yield of escape [50–53] (Section 5.7.4).

The reactivity in these organized media is fascinating (Table 17.9). Water-soluble PIs, compared to the parent oil-soluble PIs, do not systematically exhibit the best efficiency. Typical results in the presence of charged BPs and TXs in direct micelles show that [54, 55] (i) a lower quenching of water-soluble PI triplet states by styrene occurs, whereas a very strong quenching is observed in homogeneous solution, (ii) high rates of polymerization in SDS micelles are obtained, (iii) the introduction of a water-solubilizing group is not enough to guarantee a high efficiency (Table 17.7), and (iv) the cationic derivatives are the best PIs whatever the charge of the surfactant (anionic for SDS vs cationic for DTAC (dodecyl trimethylammonium chloride)). The higher performance of cationic PIs in SDS micelles (a fivefold ratio between BP^+ and BP^-) is not connected with the amine electron transfer and monomer-quenching efficiency.

The water affinity is not a driving (but can be a contributing) parameter: (i) BP^- is located in a region of high polarity similar to bulk water, whereas BP^+ is encapsulated in the SDS micelle (as a result of two antagonist forces: a water/BP^+ hydrophilic character interaction and a BP^+/sulfate group coulombic attraction) and (ii) the DMPA or chlorothioxanthone (CTX) oil-soluble PIs are more efficient

17.3 Role of the Medium on the Photoinitiator Reactivity

Table 17.9 Examples of available data concerning radical photoinitiators in micelles.

PI	$R = R_p(B=800\,G)/R_p(B=0)$		$10^{-9}\,k_q$ (mol^{-1} l s^{-1}) MMA	$10^{-9}\,k_q$ (mol^{-1} l s^{-1}) styrene	$10^{-9}\,k_e$ (mol^{-1} l s^{-1}) MDEA	% exit (SDS)		$10^{-9}\,k_q$ (mol^{-1} l s^{-1}) AA
	SDS	DTAC	SDS	SDS	SDS	B = 0 G	B = 2000 G	AOT
BP$^+$	1.8	1	0.7	0.14	0.4	6	41	4
BP$^-$	1.1	1.5	1	0.4	1.5	45	62	13
BP	2.4	2.6	1.5	0.2	1.5	7	60	—
TX$^+$	65	—	—	0.077	0.02	—	—	8.3
TX$^-$	63	—	10^{-3}	7 × 10^{-3}	0.04	—	—	—
CTX	—	—	1.4 × 10^{-3}	0.095	0.34	—	—	—

4-Benzophenone (BP$^+$, BP$^-$, BP) and 2-thioxanthone (TX$^+$) derivatives. Effect of a magnetic field: $R = R_p(B = 800\,G)/R_p(B = 0\,G)$ for the photopolymerization of MMA in SDS (sodium dodecyl sulfate) and DTAC (dodecyl trimethylammonium chloride) micelles in PI/TEA 0.005 mol l^{-1}. Exit fraction (%) of the radicals in SDS micelles 0.05 mol l^{-1}. Electron transfer rate constant k_e (with MDEA) and MMA bimolecular quenching rate constant k_q in SDS 0.5 mol l^{-1}. Acrylamide quenching k_q in AOT (1.1 mol l^{-1})-decane (2.6 mol l^{-1})-water reverse micelles.
From [16].

than BP^-. In DTAC micelles, BP^+ is located close to the water phase and the relative R_p (BP^+/BP^-) is still in a sevenfold ratio.

The achieved performance can be ascribed to the effects of the micelle environment: (i) localization of PI in the surfactant assembly, (ii) close contact of the different partners, (iii) crossing of the micellar interface and radical exit, (iv) compartmentalization as in thermally initiated microheterogeneous polymerization, and (v) termination reactions that are influenced by the presence of the polymer chain trapping ketyl radical. A quantitative kinetic treatment still appears as rather hard.

In reverse micelles based on AOT (bis-2 ethyl hexyl sodium sulfosuccinate), decane, and water, the photopolymerization rate of acrylamide is higher with BP^+ than with BP or BP^-. Initiating radicals can be formed on decane, AOT, amine, and monomer. In reverse microemulsion, additional radicals are created on cyclohexanol. A complete explanation for the high BP^+ performance has not been provided yet [56].

17.3.3
Reactivity in Bulk

The reactivity in bulk (where the diffusion limits the mobility of the transient species) compared to that in model conditions in solution was more deeply investigated in recent years [5, 57]. Photopolymerization experiments clearly show the huge effect of the viscosity of the medium on the efficiency scale of different PIs (Table 17.2). One has to be careful when extrapolating data from solution to bulk experiments (see above and in Section 5.7.3). Second-order interaction rate constants are very high in solution but are often diffusion controlled in bulk, thereby decreasing the values determined in solution by several orders of magnitude. For example, in the media used in Table 17.2, the rate constants are leveled off to about $\sim 10^{10}, 4 \times 10^8, 9 \times 10^7$, and 4×10^5 mol^{-1} l s^{-1} in benzene, hexane diol diacrylate (HDDA) or tripropyleneglycol diacrylate (TPGDA), TMPTA, and a high viscosity medium (e.g., EpAc; see above), respectively. It means that absolute rate constants of 10^9 and 10^6 mol^{-1} l s^{-1} in solution are leveled off to the same value (about 4×10^5 mol^{-1} l s^{-1}) in EpAc. As a consequence, the apparent higher reactivity in solution completely disappears in bulk. This effect explains the difficulty when extrapolating efficiency/reactivity relationships from solution to bulk.

As the two partners have to move before interacting, the viscosity effect on the generation of initiating radicals is particularly important with Type II PIs. This is obviously not the case with Type I PIs. The viscosity, however, has a direct effect on the escape of the radicals from the radical pair (RP) formed from the cleavage process. This does not change the cleavage rate constant but modifies the dissociation quantum yield ϕ_{diss}. Indeed, the photolysis of a PI in the liquid phase or in bulk leads to the generation of two radicals that exist for a given time as an RP keeping the spin memory (Section 5.7.4). The surrounding solvent molecules (or monomer/oligomer units) create a cage around RP where recombination and intersystem crossing occur [15]. The cage effect is strongly dependent on the solvent viscosity and the temperature. The same also holds true in RPs formed in Type II systems. During the course of the polymerization reaction, the PI should not have a constant efficiency as a result of a ϕ_{diss} change with the viscosity [57].

It was shown (Sections 8.1.5 and 8.5.4) that many cleavable PIs where the fast cleavage process is obviously the main pathway are also able to undergo an intermolecular hydrogen abstraction (e.g., occurring between two PI molecules or a PI and a co-initiator). In photopolymerization experiments in film, the PI concentration is high and this last reaction might be favored. In a viscous medium, however, the efficiency of the bimolecular process levels off. Therefore, the monomolecular reaction still remains the largely dominant process. For slow cleavable systems in a viscous matrix, both processes can be likely operative.

Few works have been devoted to quantitatively evidence the effect and the role of the viscosity when using various PIs [58]. This was described in detail in a recent work [5] that allows to fit the experimental ϕ_i (expressed as a relative value ϕ_{rel} using DMPA as a reference) as a function of ϕ_{diss} in several media (Figure 17.2): bulk monomer/oligomer (EpAc), monomer (butylacrylate (BA), or HDDA) in PMMA and also monomer (BA or MMA 1M) in an inert solvent (toluene). A quite good correlation exits between the relative efficiency of a set of cleavable PIs in viscous EpAc and the ϕ_{diss} measured in solution. This is owing to the fact that (i) the cleavage is very fast (the monomer quenching of the triplet state is inefficient) and (ii) the addition rate constants of the generated radicals to the monomer are leveled off. The correlation remains valid (not for all compounds) in less viscous or fluid

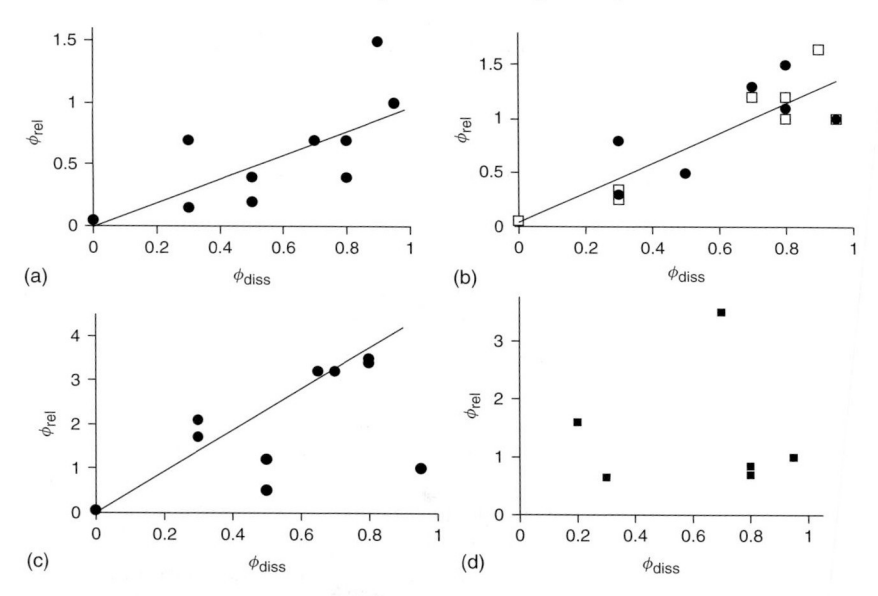

Figure 17.2 Plot of ϕ_{rel} versus ϕ_{diss} for the photopolymerization of (a) bulk EpAc; (b) BA/PMMA 75/25% w/w (open square) and HDDA/PMMA 75/25% w/w (plain circles); (c) BA (1M) in toluene; and (d) MMA (1M) in toluene. See text. (Source: From [5].)

media provided the monomer quenching of the PI triplet state is weak (i.e., no change of ϕ_{diss}) and the radical addition to the monomer is efficient. This is the case for most PIs: they have rather short-lived triplet states and most usual initiating radicals exhibit a similar (and high) reactivity toward acrylates and methacrylates (except the benzoyl radical; see Chapter 18). The dependence is more complicated in the case of methacrylates (lack of correlation) because of the high values of the monomer-quenching rate constants.

This analysis shows that a deep knowledge of the photophysics and photochemistry of cleavable PIs in solution can serve as a good basis for the prediction of the polymerization efficiency in film experiments. Several approaches also suggest that a direct investigation on the excited-state processes in bulk media or in high-speed photopolymer coating layers may lead to a deeper insight into the understanding of the practical efficiency ([5, 59] and references therein).

17.4
Structure/Property Relationships in Photoinitiating Systems

The interpretation of the PI reactivity has made considerable progress in the past years. This part discusses the most relevant and recent considerations that provide useful information on two specific aspects: (i) the role of the involved bond dissociation energies (BDEs) in Type I PIs and co-initiators (H-donors) and (ii) the reactivity of the initiating radicals. At present, the dependence of k_c, k_e, k_q, k_H, ϕ_H, ϕ_R on the PI molecular structures is not totally understood as these parameters are not always experimentally accessible. Side reactions can also hardly be investigated. The access to a quantitative evaluation of the cage effects that change ϕ_{diss} or ϕ_i is restricted to a few examples [60].

17.4.1
Role of the Bond Dissociation Energy in Cleavable Systems

BDEs are accessible by molecular orbital calculations. The calculated values, assuming an adapted level of theory (e.g., density functional theory (DFT) or *ab initio* procedures), are often in very good agreement with the experimental data extracted by time-resolved photoacoustic spectroscopy [61, 62]. A common statement suggests that the dissociation efficiency of a cleavable PI should increase with the BDE decrease. However, the ϕ_{diss}s of a set of PIs do not correlate [5] with the dissociation enthalpy $\Delta H_{diss}(= BDE - E_T$ where E_T is the triplet energy). As a consequence, the enthalpy is not the unique driving factor for the cleavage process.

As discussed in Section 6.3, the cleavage of several acetophenone derivatives is mainly governed by the pre-exponential factor of the Arrhenius' law for the rate constant evidencing an entropy contribution. The participation of polar effect with a CT in the transition state structure was also involved although a complete quantification remains hard.

17.4 Structure/Property Relationships in Photoinitiating Systems

In ketosulfone derivatives where a β-cleavage process occurs in the triplet state, BDE(C–S) calculations effectively show that (i) a bond breaking occurs when the triplet state is higher than the corresponding BDE and (ii) the reaction is strongly exothermic although a slow cleavage process is observed in some cases (see the explanation in Section 8.2.5).

In a series of light-sensitive alkoxyamines, the BDEs do not govern the cleavage efficiency, the distance between the involved C–O bond and the chromophore being an important parameter (see also Section 8.30) [63].

17.4.2
Role of the Bond Dissociation Energy in Noncleavable Systems

In Type II systems, the hydrogen abstraction reaction either corresponds to a PHT process or an electron transfer followed by a proton transfer (EPHT process) (Scheme 17.1). The competition between these two mechanisms is related to the redox properties of the reactants and the free-energy changes ΔG_{Et} for the electron transfer is calculated from (Eq. (6.4)) [64].

For amines and thiols, the EPHT process occurs, whereas a pure H-transfer is usually expected for the other H-donors. Typical interaction rate constants k_H of the BP triplet state (^{3}BP) with different representative hydrogen donors are given in Table 17.10. For triethylamine (TEA) and mercaptobenzoxazole (that are good representatives of amines and thiols), it was shown that (i) ΔG_{Et} is < 0 supporting the EPHT mechanism and (ii) the k_Hs lie close to the diffusion limit and are higher than the corresponding k_Hs with tBuO$^\bullet$. In the ^{3}BP reaction with alcohols, silanes, germanes, stannanes, or ligated boranes, ΔG_{Et} is > 0 and the EPHT mechanism

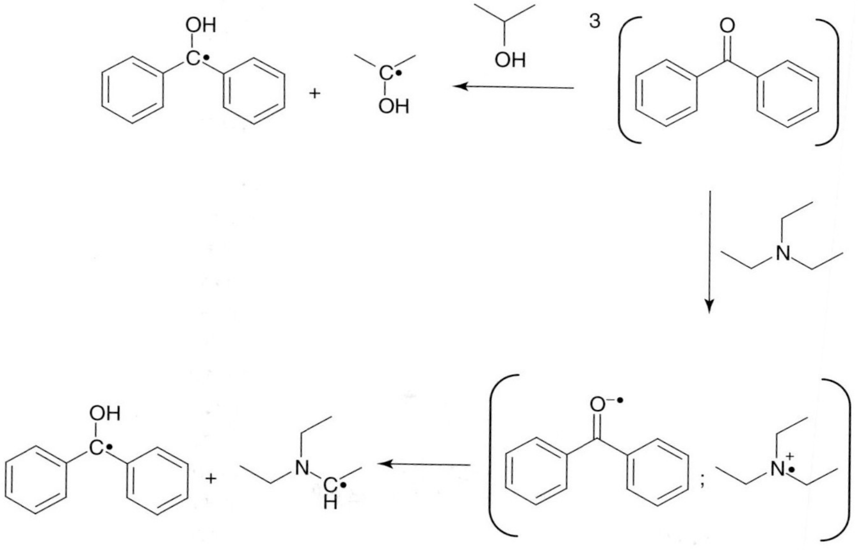

Scheme 17.1

Table 17.10 Interaction rate constants k_H of typical hydrogen donors with the benzophenone triplet state ^{3}BP. The ketyl radical BPH$^\bullet$ quantum yields for the ^{3}BP/hydrogen donor reaction are given in brackets (column 2) ($\bar{e} - H^+$ ≡ EPHT; HT is a pure hydrogen transfer noted also PHT).

Hydrogen donors	k_H (^{3}BP) in 10^7 (mol^{-1} l s^{-1})	BDE(X–H) kcal mol^{-1}	Mechanism	Generated radicals
Triethylamine	310 (0.94)	~91	$\bar{e} - H^+$	BPH$^\bullet$, Et$_2$NC($^\bullet$)HCH$_3$
Mercaptobenzoxazole	150 (0.5)	–	$\bar{e} - H^+$	BPH$^\bullet$, RS$^\bullet$
Isopropanol	0.16	–	HT	BPH$^\bullet$, 2 – hydroxypropyl
Tris(trimethylsilyl)silane (TMS)$_3$Si–H^c	10.2 (0.95)	~80	HT	BPH$^\bullet$, (TMS)$_3$Si$^\bullet$
Tris(trimethylsilyl)germane (TMS)$_3$GeH^c	105 (0.81)	~75	HT	BPH$^\bullet$, (TMS)$_3$Ge$^\bullet$
Ph$_3$Sn–H	46 (0.74)	~73	HT	BPH$^\bullet$, Ph$_3$Sn$^\bullet$
Et$_3$N → BH$_3$	9.1 (0.73)	~100	HT	BPH$^\bullet$, Et$_3$N → BH$_2^\bullet$

Extracted from [65].

is not favorable. The measured k_Hs are quite similar to those obtained with tBuO$^\bullet$. Therefore, ^{3}BP behaves like an alkoxyl radical.

Co-initiators leading to the EPHT sequence are preferred as higher interaction rate constants with the PI excited state are expected. This reduces the by-side reactions (monomer or oxygen quenching) and ensures a high initiating radical production. Other co-initiators involving the PHT process were also proposed (silanes and ligated boranes). One expects that the BDE of the H-donors (for both the EPHT or PHT mechanisms) plays a significant role as the more stable radical (which is the easiest to form) should derive from the starting compound exhibiting the lowest BDE. These BDEs partly govern the PHT rate constants; the highest rate constants are found for compounds exhibiting the lowest BDEs [65].

Many works devoted to the determination of a large set of BDEs for primary, secondary, and tertiary aliphatic amines show that there is no satisfactory k_H(^{3}BP/amine) versus BDE correlation. One explanation is that the BDE governs a direct hydrogen abstraction rather than a EPHT transfer.

General comments on BDEs are the following [66]: (i) the α(C–H) BDEs are almost similar (around 90 kcal mol^{-1}) for unconstrained amines; (ii) the evolution of N–H and α(C–H) bond strength allows underlining of the slight influence of C- and N-alkylation on the BDEs; (iii) the effect of the N-alkylation on the α(C–H) BDEs is ascribed to a three electron, two-orbital interaction (Section 18.5); (iv) the C-alkylation has clearly no effect on the α(C–H) BDEs; (v) the stabilization observed in the effect of the N-alkylation on the N–H BDEs occurs through C–C hyperconjugation; and (vi) in constrained amines, the role of a through-bond stabilization is clearly observed.

The regioselectivity of the hydrogen abstraction reaction, particularly for the EPHT process is often unclear. This point is discussed in Sections 18.5 and 18.12.

17.4.3
Role of the Initiating Radical in the Initiation Step

17.4.3.1 Generation of the First Monomer Radical

The initiating radical R$^\bullet$ plays a decisive role, which is reflected by the addition rate constant k_i to the monomer M unit and the production yield ϕ_{RM} of RM$^\bullet$ (Eq. (5.18)). In the past years, although the direct detection of radicals in LFP experiments was often quite hard because of their low absorptions and sometimes their short lifetimes (μs – ms domain), a lot of radicals have been studied according to various experimental strategies [67–92], for example, hydroxyalkyl, benzoyl, benzyl, alkylamine, thiyl, xanthone ketyl, benzyl peroxyl, phenacetyl, sulfydryl, benzodithiazolyl, phenyl selenyl, benzoyl peroxyl, cumyloxyl, alkoxy, diphenylpentadienyl, phenoxyl, (benzoyl phenyl) diphenyl methyl, hydroxyisopropyl, phosphinoyl, aryloxyl, acrylate, vinyl radicals and AIBN, and DMPA and dye-derived radicals.

Clean radical production methods have also been recently developed to characterize the reactivity of R$^\bullet$ with different monomer substrates; many k_i values became accessible (Chapter 18). Obviously, ϕ_{RM} strongly depends on the side reactions (hydrogen abstraction, fragmentation, recombination, reaction with stabilizers, and

O_2; pseudo first-order rate constant: k_r) that can trap the radical before the addition to the double bond; k_r is usually unknown or hardly accessible. However, for a given structure, it can be noted that k_i strongly affects ϕ_{RM}. A better knowledge of the dependence of k_i on the PI and monomer structures is required (see above).

17.4.3.2 Addition Rate Constants to Monomer

Many interaction rate constants of various initiating radicals with monomer double bonds as well as with other compounds such as oxygen, amines, thiols, phenols, halides, spin traps, and so on, have therefore been determined. A general presentation for each class of radicals is given in Chapter 18. The approach using the state correlation diagram described in Section 6.6 has been applied to the reactivity of a large series of carbon-centered radicals (including the benzoyl, hydroxyl alkyl, and aminoalkyl radicals) toward various electron-rich (vinylether (VE), and vinylacetate (VA)) or electron-poor monomer units (acrylate MA, MMA, and AN) [93].

17.4.3.3 Description of the Radical/Monomer Reactivity

A detailed investigation of the behavior of three specific radicals $(CH_3)_2NCH_2^\bullet$, $CH_3^\bullet$, and $NC\text{-}CH_2^\bullet$ exhibiting an initiating ability has been reported in [94, 95] within an approach based on experimental and theoretical grounds (Section 6.6). Calculating E_a (barrier) and ΔE_{pol} (stabilization ascribed to the polar effect) yields E_{enth} (barrier for a given addition enthalpy ΔH_r). Linear relationships between E_{enth} and ΔH_r are generally found, evidencing the strong influence of the reaction exothermicity.

For example, Figure 17.3 shows the enthalpy effect as represented by the evolution of the enthalpy (calculated as the difference between the energy of the products and the reactant-optimized structures) versus the monomer electronegativity for the three investigated radicals. Electron-rich monomer (VE and VA) are characterized by a low electronegativity and a low exothermicity. In electron-poor monomers (MMA and AN), the electronegativity increases as well as the exothermicity. On addition of the aminoalkyl radical, the enthalpy increases when going from VE (12.6 kcal mol^{-1}) to AN (21.6 kcal mol^{-1}). On the other side, the barrier decreases from VE (12.9 kcal mol^{-1}) to AN (1.5 kcal mol^{-1}). As a consequence, it is obvious that the enthalpy partly governs the barrier.

The evolution of the amount of charge transfer (CT) δ^{TS} in the transition state directly reveals the polar effects. Electron-deficient monomers in the presence of a nucleophilic radical ($R_2NCH_2^\bullet$) are characterized by high δ^{TS} (from the radical to the monomer), whereas in electron-rich monomer/electrophilic radical ($NCCH_2^\bullet$) couples, the net radical to monomer CT increases ($\delta < 0$). The absolute value of δ^{TS} increases from VE to AN for $R_2NCH_2^\bullet$ and from AN to VE for $NCCH_2^\bullet$, thereby revealing the nucleophilic character of $R_2NCH_2^\bullet$ and the electrophilic character of $NCCH_2^\bullet$. The methyl radical exhibits an amphiphilic behavior, that is, nucleophilic toward MA or AN ($\delta > 0$) and electrophilic toward VE or VA ($\delta < 0$). A detailed calculation shows that the polar effect in the addition of the aminoalkyl radical is less important for VE (0.33 kcal mol^{-1}) than for AN (4.2 kcal mol^{-1}). This

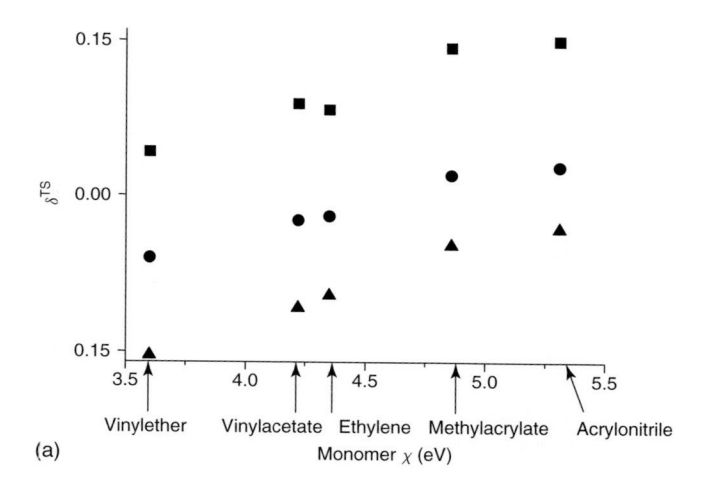

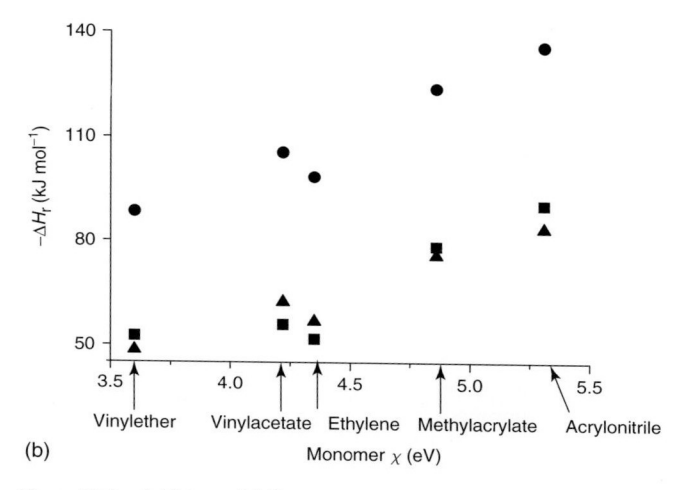

Figure 17.3 Addition of different representative initiating radicals $(CH_3)_2NCH_2^\bullet$ (square), $CH_3^\bullet$ (circle), $NC-CH_2^\bullet$ (triangle) to double bonds (vinylether, vinylacetate, ethylene, methylacrylate, acrylonitrile). Evolution of (a) the amount of charge transfer δ^{TS} and (b) the reaction enthalpy ΔH_r versus the absolute electronegativity of the monomer χ_{DB}. (Source: From Refs. [94, 95].)

outlines the lowering of the barrier by the polar effect, that is, more particularly for electron-deficient monomers. It thus appears that the proposed theoretical treatment adequately describes the reactivity trends and can be confidently used to predict the efficiency evolution for the addition of a given radical to different monomer double bonds.

The reversibility of the addition reaction is an interesting subject. In a general way, exothermic reactions (Scheme 17.2) are not reversible as the barrier $E_{a_{add}}$ is higher for the fragmentation process $E_{a_{frag}}$ (contrary to the case of an endothermic reaction where the barrier is lower). This situation is encountered with most of

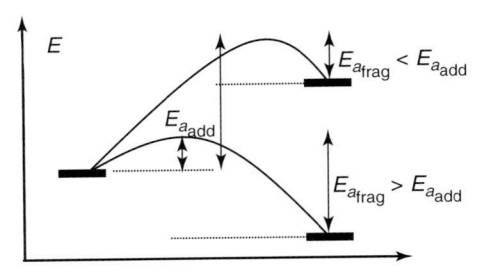

Scheme 17.2

radicals (alkyls, aminoalkyls, and benzoyls). Some radicals (e.g., thiyls) or particular radical/monomer couples (NHC-boryl/VE where NHC = N-heterocyclic carbene) lead to a significant back reaction as $E_{a_{frag}} < E_{a_{add}}$ (see below). In the case of the benzoyl radical itself [96], it was experimentally shown that the back reaction is a very minor pathway.

Substantial progress has been made in the recent past as most of the experimental k_i values are now explained on the basis of a theoretical approach (Chapter 18). All these works help to (i) describe the transition state, (ii) explain the role of the enthalpy/polar effects in the addition properties in a very large set of radical/monomer couples, (iii) interpret the large scale of values for the addition rate constants, (iv) study the reversibility of the addition process (more particularly for the thiyl or boryl radical addition), and (v) design radicals exhibiting both a high reactivity and a low selectivity (e.g., the tetrazole-derived thiyl radical). This latter property can be highly worthwhile as nucleophilic radicals (aminoalkyl and hydroxy isopropyl) usually exhibit significant addition rate constants only to electron-poor monomers, whereas electrophilic radicals (thiyl and cyano isopropyl) are mainly reactive toward electron-rich monomers.

References

1. Fouassier, J.P. (1995) *Photoinitiation, Photopolymerization, Photocuring*, Hanser, Münich.
2. Fouassier, J.P., Ruhlmann, D., Graff, B., Morlet-Savary, F., and Wieder, F. (1995) *Prog. Org. Coat.*, **25**, 235–271.
3. Fouassier, J.P., Ruhlmann, D., Graff, B., and Wieder, F. (1995) *Prog. Org. Coat.*, **25**, 169–202.
4. Fouassier, J.P. (2000) *Rec. Res. Dev. Photochem. Photobiol.*, **4**, 51–74.
5. Lalevée, J., Allonas, X., Jradi, S., and Fouassier, J.P. (2006) *Macromolecules*, **39**, 1872–1878.
6. Lalevée, J., Tehfe, M.A., Morlet-Savary, F., Graff, B., and Fouassier, J.P. (2011) *Prog. Org. Coat.*, **70**, 83–90.
7. Ganster, B., Fischer, U.K., Moszner, N., and Liska, R. (2008) *Macromol. Rapid Commun.*, **29**, 57–62.
8. Hayashi, H. and Mochida, K. (1983) *Chem. Phys. Lett.*, **101**, 307–311.
9. Wakasa, M., Mochida, K., Sakaguchi, Y., Nakamura, J., and Hayashi, H. (1991) *J. Phys. Chem.*, **95**, 2241–2246.
10. Tehfe, M.A., Blanchard, N., Fries, C., Lalevée, J., Allonas, X., and

Fouassier, J.P. (2010) *Macromol. Rapid Commun.*, **31**, 473–478.
11. Lalevée, J., Allonas, X., and Fouassier, J.P. (2009) *Chem. Phys. Lett.*, **469**, 298–303.
12. Lalevée, J., El-Roz, M., Morlet-Savary, F., Graff, B., Allonas, X., and Fouassier, J.P. (2007) *Macromolecules*, **40**, 8527–8530.
13. Decker, C. and Moussa, K. (1988) *Makromol. Chem.*, **189**, 2381–2394.
14. Obeid, H., Dossot, M., Allonas, X., Jacques, P., Fouassier, J.P., Dumarcay, S., and Merlin, A. (2003) Proceedings of the RadTech Europe International Conference Berlin, pp. 887–893.
15. Lalevée, J., Dirani, A., El-Roz, M., Allonas, X., and Fouassier, J.P. (2008) *Macromolecules*, **41**, 2003–2010.
16. Fouassier, J.P. and Lougnot, D.J. (1990) in *Lasers in Polymer Science and Technology*, vol. II (eds J.P. Fouassier and J.F. Rabek), CRC Press, Boca Raton, FL, pp. 145–168.
17. Fouassier, J.P. (1990) *Prog. Org. Coat.*, **18**, 229–252.
18. Pappas, S.P. (1986) *UV-Curing: Science and Technology*, Technology Marketing Corporation, Stamford, CT; Plenum Press, New York, (1992).
19. Fouassier, J.P. and Rabek, J.F. (eds) (1993) *Radiation Curing in Polymer Science and Technology*, Chapman & Hall, London.
20. Neckers, D.C. (1999) *UV and EB at the Millenium*, Sita Technology, London.
21. Dietliker, K. (2002) *A Compilation of Photoinitiators Commercially Available for UV Today*, Sita Technology Ltd, London.
22. Mishra, M.K. and Yagci, Y. (eds) (2009) *Handbook of Vinyl Polymers*, CRC Press.
23. Allen, N.S. (ed.) (2010) *Photochemistry and Photophysics of Polymer Materials*, John Wiley & Sons, Inc., Hoboken, NJ.
24. Fouassier, J.P. and Rabek, J.F. (eds) (1990) *Lasers in Polymer Science and Technology*, CRC Press, Boca Raton, FL.
25. Lemée, V., Burget, D., Jacques, P., and Fouassier, J.P. (2000) *J. Polym. Sci. A: Polym. Chem.*, **38**, 1785–1792.
26. Wagner, P.J., Truman, R.J., and Scaiano, J.C. (1985) *J. Am. Chem. Soc.*, **107**, 7093–7098.
27. Fouassier, J.P., Jacques, P., and Encinas, M.V. (1988) *Chem. Phys. Lett.*, **148**, 309–312.
28. Scaiano, J.C. (1980) *J. Am. Chem. Soc.*, **102**, 7747–7751.
29. Timpe, H.J., Kronfeld, K.P., and Schiller, M. (1991) *J. Photochem. Photobiol. A: Chem.*, **62**, 245–259.
30. Schnabel, W. (1991) in *Laser in Polymer Science and Technology: Applications*, vol. 2 (eds J.P. Fouassier and J.F. Rabek), CRC Press, pp. 95–135.
31. Timpe, H.J., Jockusch, S., and Korner, K. (1993) in *Radiation Curing in Polymer Science and Technology*, vol. 2 (ed. J.P. Fouassier), Elsevier, Barking, pp. 575–601.
32. Jacques, P. and Allonas, X. (1995) *Chem. Phys. Lett.*, **233**, 533–537.
33. Jacques, P. and Allonas, X. (1997) *Chem. Phys.*, **215**, 371–378.
34. Jacques, P. and Allonas, X. (1994) *J. Photochem. Photobiol. A: Chem.*, **78**, 1–5.
35. Kucybala, Z., Pietrzak, M., and Paczkowski, J. (1998) *Chem. Mater.*, **10**, 3555–3561.
36. Kucybala, Z. and Paczkowski, J. (1999) *J. Photochem. Photobiol. A: Chem.*, **128**, 135–142.
37. Linden, L.A., Paczkowski, J., Rabek, J.F., and Wrzyszczyski, A. (1999) *Polimery*, **44**, 161–167.
38. Paczkowski, J. and Neckers, D.C. (2001) in *Electron Transfer in Chemistry*, vol. 5 (ed. I.R. Gould), Wiley-VCH Verlag GmbH, Weinheim, pp. 516–585.
39. Dossot, M., Obeid, H., Allonas, X., Jacques, P., Fouassier, J.P., and Merlin, A. (2004) *J. Appl. Polym. Sci.*, **92**, 1154–1164.
40. Valdebenito, A., Lissi, E.A., and Encinas, M.V. (2001) *Macromol. Chem. Phys.*, **202**, 2581–2585.
41. Corrales, T., Catalina, F., Allen, N.S., and Peinado, C. (2006) in *Photochemistry and UV Curing: New Trends* (ed. J.P. Fouassier), Research Signpost, Trivandrum, pp. 31–45.
42. Fouassier, J.P., Lougnot, D.J., Zuchowicz, I., Green, P.N., Timpe, H.J., Kronfeld, K.P., and Muller, U. (1987) *J. Photochem.*, **36**, 347–363.
43. Tehfe, M.A., Monot, J., Malacria, M., Fensterbank, L., Fouassier, J.P.,

Curran, D.P., Lacôte, E., and Lalevée, J. (2012) *ACS Macro Lett.*, **1**, 92–95.

44. Jacques, P., Lougnot, D.J., and Fouassier, J.P. (1984) in *Surfactants in Solution*, vol. **2** (eds K.L. Mittal and B. Lindman), Plenum Press, New York, p. 1177.

45. Scaiano, J.C. and Abuin, E.B. (1981) *Chem. Phys. Lett.*, **81**, 209–214.

46. Krieg, M., Pileni, M.O., Braun, A.M., and Graetzel, M. (1981) *J. Colloid Interface Sci.*, **83**, 209–215.

47. Braun, A.M., Krieg, M., Turro, N.J., Aikawa, M., Gould, I.R., Graff, G.A., and Lee, P.C. (1981) *J. Am. Chem. Soc.*, **103**, 7312–7315.

48. Sakaguchi, Y., Hayashi, H., Murai, H., and L'Haya, Y.J. (1984) *Chem. Phys. Lett.*, **110**, 275–279.

49. Pioge, S., Nesterenko, A., Brotons, G., Pascual, S., Fontaine, L., Gaillard, C., and Nicol, E. (2011) *Macromolecules*, **44**, 594–603.

50. Khudyakov, I.V., Serebrennikov, Y.A., and Turro, N.J. (1993) *Chem. Rev.*, **93**, 537–545.

51. Hayashi, H., Sakaguchi, Y., and Wakasa, M. (2001) *Bull. Chem. Soc. Jpn.*, **74**, 773–779.

52. Vink, C.B. and Woodward, J.R. (2004) *J. Am. Chem. Soc.*, **126**, 16730–16735.

53. Scaiano, J.C. and Lougnot, D.J. (1984) *J. Phys. Chem. B*, **88**, 3379–3385.

54. Fouassier, J.P. and Lougnot, D.J. (1986) *J. Appl. Polym. Sci.*, **32**, 6209–6226.

55. Fouassier, J.P. and Lougnot, D.J. (1987) *J. Appl. Polym. Sci.*, **34**, 477–488.

56. Fouassier, J.P., Lougnot, D.J., and Zuchowicz, I. (1986) *Eur. Polym. J.*, **22**, 933–938.

57. Khudyakov, I.V., Arsu, N., Jockusch, S., and Turro, N.J. (2003) *Des. Monomers Polym.*, **6**, 91–96.

58. Valdebenito, A. and Encinas, M.V. (2003) *J. Polym. Sci. A: Polym. Chem.*, **41**, 2368–2373.

59. Urano, T. (2003) *J. Photopolym. Sci. Technol.*, **16**, 129–156.

60. Khudyakov, I.V. and Turro, N.J. (2010) *Des. Monomers Polym.*, **13**, 487–496.

61. Lalevée, J., Allonas, X., and Fouassier, J.P. (2002) *J. Mol. Struct. (TheoChem)*, **588**, 233–238.

62. Allonas, X., Lalevée, J., and Fouassier, J.P. (2003) *J. Photochem. Photobiol. A: Chem.*, **159**, 127–133.

63. Versace, D.L., Guillaneuf, Y., Fouassier, J.P., Bertin, D., Lalevée, J., and Gigmes, D. (2011) *Org. Biomol. Chem.*, **9**, 2892–2898.

64. Rehm, D. and Weller, A. (1970) *Isr. J. Chem.*, **8**, 259–271.

65. Lalevée, J. and Fouassier, J.P. (2012) in *Encyclopedia of Radicals in Chemistry, Biology and Materials* (eds A. Studer and C. Chatgililoglu), John Wiley & Sons Inc., pp. 34–57.

66. Lalevée, J., Allonas, X., and Fouassier, J.P. (2002) *J. Am. Chem. Soc.*, **124**, 9613–9621.

67. Hunt, P., Worrall, D.R., Wilkinson, F., and Batchelor, S.N. (2003) *Photochem. Photobiol. Sci.*, **2**, 518–523.

68. Colley, C.S., Grills, D.C., Besley, N.A., Jockusch, S., Matousek, P., Parker, A.W., Town, N.J., Gill, P.M.W., and George, M.W. (2002) *J. Am. Chem. Soc.*, **124**, 14952–14958.

69. Hristova, D., Gatlik, I., Rist, G., Dietliker, K., Wolf, J.P., Birbaum;, J.L., Savitsky, A., Möbius, K., and Gescheidt, G. (2005) *Macromolecules*, **38**, 7714–7720.

70. Lei, X.-G., Jockusch, S., Ottaviani, M.F., and Turro, N.J. (2003) *Photochem. Photobiol. Sci.*, **2**, 1095–1100.

71. Janovsky, I., Knolle, W., Naumov, S., and Williams, F. (2004) *Chem. A Eur. J.*, **10**, 5524–5534.

72. Wenska, G., Taras-Goslinska, K., Skalski, B., Hug, G.L., Carmichael, I., and Marciniak, B. (2005) *J. Org. Chem.*, **70**, 982–988.

73. Sakamoto, M., Cai, X., Hara, M., Fujistuka, M., and Majima, T. (2005) *J. Phys. Chem. A*, **109**, 2452–2458.

74. El Dib, G., Chakir, A., Roth, E., Brion, J., and Daumont, D. (2006) *J. Phys. Chem. A*, **100**, 7848–7857.

75. Zhang, X. and Nau, W.M. (2000) *J. Phys. Org. Chem.*, **13**, 634–639.

76. Das, T.N., Huie, R.E., Neta, P., and Padmaja, S. (1999) *J. Phys. Chem. A*, **103**, 5221–5226.

77. Vlasyuk, I.V., Bagryansky, V.A., Gritsan, N.P., Molin, Y.N., Yu, M.A., Gatilov, Y.V., Shcherbukhin, V.V., and

Zibarek, A.V. (2001) *Phys. Chem. Chem. Phys.*, **3**, 409–415.

78. Beletskaya, I.P., Sigeev, A.S., Kuzmin, V.A., Tatikolov, A.S., and Hevesi, L. (2000) *J. Chem. Soc., Perkin Trans.*, **2**, 107–109.

79. Hoshino, M., Konishi, R., Seto, H., Seki, H., Sonoki, H., Yokoyama, T., and Shimamori, H. (2001) *Res. Chem. Intermed.*, **27**, 189–204.

80. Pischel, U. and Nau, W.M. (2001) *J. Am. Chem. Soc.*, **123**, 9727–9737.

81. Gross, A., Schneiders, N., Daniel, K., Gottwald, T., and Hartung, J. (2008) *Tetrahedron*, **64**, 10882–10889.

82. Miranda, M.A., Font-Sanchis, E., Perez-Prieto, J., and Scaiano, J.C. (2002) *J. Org. Chem.*, **67**, 6131–6135.

83. Dabestani, R., Ivanov, I.N., Britt, P.F., Sigman, M.E., and Buchanan, A.C. III (2002) *Prepr. Symp. – Am. Chem. Soc., Div. Fuel Chem.*, **47**, 390–392.

84. Jarikov, V.V., Nikolaitchik, A.V., and Neckers, D.C. (2000) *J. Phys. Chem.*, **104**, 5131–5140.

85. Gatlik, I., Rzadek, P., Gescheidt, G., Rist, G., Hellrung, B., Wirz, J., Dietliker, K., Hug, G., Kunz, M., and Wolf, J.P. (1999) *J. Am. Chem. Soc.*, **121**, 8332–8336.

86. Fede, J.M., Jockusch, S., Lin, N., Moss, R.A., and Turro, N.J. (2003) *Org. Lett.*, **5**, 5027–5030.

87. Weber, M., Khudyakov, I.V., and Turro, N.J. (2002) *J. Phys. Chem.*, **106**, 1938–1945.

88. Turro, N.J., Lei, X.-G., Jockusch, S., Li, W., Liu, Z., Abrams, L., and Ottaviani, M.F. (2002) *J. Org. Chem.*, **67**, 2606–2618.

89. Spichty, M., Giese, B., Matsumoto, A., Fischer, H., and Gescheidt, G. (2001) *Macromolecules*, **34**, 723–726.

90. Goumans, T.P.M., van Alem, K., and Lodder, G. (2008) *Eur. J. Org. Chem.*, **3**, 435–443.

91. Terazima, M., Nogami, Y., and Tominaga, T. (2000) *Chem. Phys. Lett.*, **332**, 503–507.

92. Stanoeva, T., Neshchadin, D., Gescheidt, G., Ludvik, J., Lajoie, B., and Batchelor, S. (2005) *J. Phys. Chem.*, **109**, 11103–11109.

93. Lalevée, J., Allonas, X., and Fouassier, J.P. (2004) *J. Phys. Chem. A*, **108**, 4326–4334.

94. Lalevée, J., Allonas, X., and Fouassier, J.P. (2005) *Macromolecules*, **38**, 4521–4524.

95. Lalevée, J., Allonas, X., Fouassier, J.P., Rinaldi, D., Lopez, R., and Rivail, J.L. (2005) *Chem. Phys. Lett.*, **415**, 202–205.

96. Griesser, M., Neshchadin, D., Dietliker, K., Moszner, R., Liska, R., and Gescheidt, G. (2009) *Angew. Chem. Int. Ed.*, **48**, 9359–9361.

18
Reactivity of Radicals toward Oxygen, Hydrogen Donors, Monomers, and Additives: Understanding and Discussion

As stated before, a large set of radicals encountered in photopolymerization are based now on $R_3C^\bullet$, $RC(=O)^\bullet$, $RS^\bullet$, $R_3Si^\bullet$, $R_3Ge^\bullet$, $R_2P(=O)^\bullet$, $R_3Sn^\bullet$, $RO^\bullet$, and $ROO^\bullet$. Their addition efficiency to various substrates is a function of their chemical reactivity, which is dependent on the involved molecular orbitals (MOs) and geometry as well as selectivity, regioselectivity, and nucleophilic/electrophilic properties [1–5]. In this chapter, all these points are detailed as a function of the considered radical.

18.1
Alkyl and Related Carbon-Centered Radicals

Primary, secondary, and tertiary alkyl radicals $R^\bullet$ are usually planar and characterized by a π structure, that is, the single electron is located in a 2p orbital. Obviously, both the geometry and the singly occupied molecular orbital (SOMO) are affected by the substituents. These radicals can be generated by (i) a halogen abstraction reaction from R-X (X = Cl, Br, I) or (ii) a decarboxylation reaction from the corresponding acyl radicals ($R-C^\bullet(=O) \rightarrow R^\bullet + CO$). Methyl radicals arise from the fragmentation of, for example, the dimethoxybenzyl radical derived from the cleavage of DMPA, tert-butoxyl, or cumyloxyl radicals (Scheme 18.1).

The formation of $R^\bullet$ from the corresponding alkanes through hydrogen abstraction is slow, as the C–H bond dissociation enthalpies (BDEs) are usually rather high [6]. As a consequence, the hydrogen abstraction rate constants from ^3BP or the alkoxyl radicals are low ($<10^7$ mol^{-1} l s^{-1}). Substituted carbon-centered radicals R_1, R_7-R_9, R_{12}, and R_{14} 1 formed in commercial Type I or II PIs (Part 2) and other radicals added for comparison have been investigated. They allow a study of the role of the substituent in the radical center (OH for R_6-R_9, CN for $R_{10}-R_{11}$, Ph for R_9 and $R_{12}-R_{14}$, C =O for $R_{15}-R_{16}$). These radicals are hardly observed by laser flash photolysis (LFP) (excepting R_9) and usually exhibit a significant absorption at $\lambda < 320$ nm. These structures can be detected by ESR, more particularly in ESR-spin trapping experiments. The kinetic data are usually obtained through radical clock indirect approaches or kinetic ESR [7, 8].

Photoinitiators for Polymer Synthesis: Scope, Reactivity and Efficiency, First Edition.
Jean-Pierre Fouassier and Jacques Lalevée.
© 2012 Wiley-VCH Verlag GmbH & Co. KGaA. Published 2012 by Wiley-VCH Verlag GmbH & Co. KGaA.

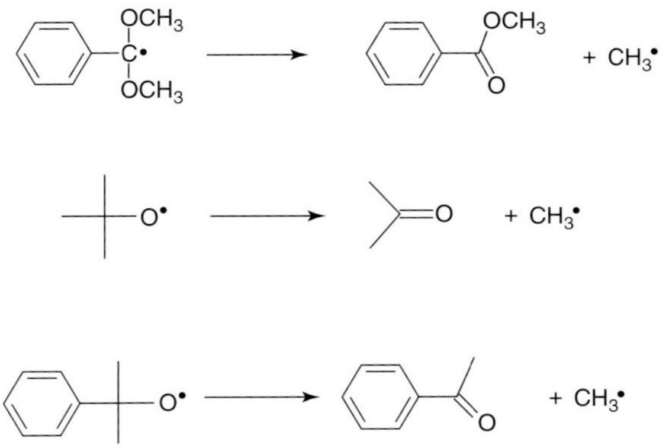

Scheme 18.1

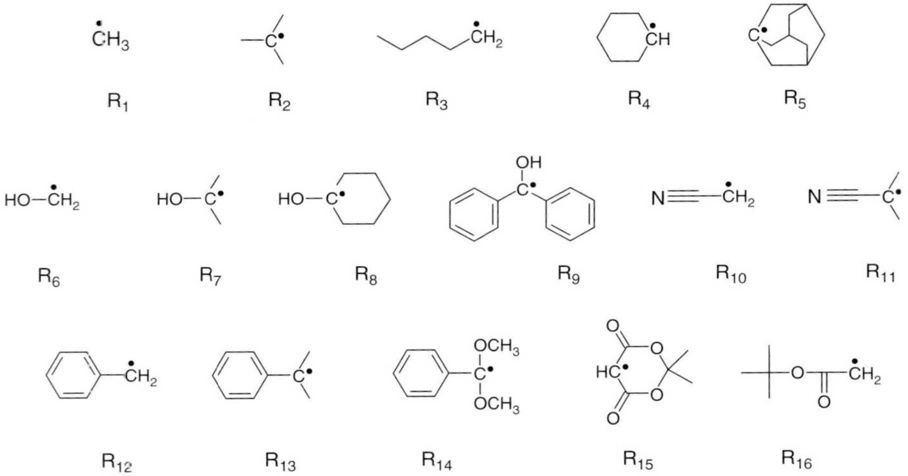

R$_1$: ĊH$_3$; R$_2$: secondary alkyl ; R$_3$: ·CH$_2$–(butyl) ; R$_4$: cyclohexyl ; R$_5$: adamantyl

R$_6$: HO–ĊH$_2$; R$_7$: HO–Ċ(tertiary) ; R$_8$: HO–Ċ(cyclohexyl) ; R$_9$: Ph$_2$Ċ–OH ; R$_{10}$: N≡C–ĊH$_2$; R$_{11}$: N≡C–Ċ(tertiary)

R$_{12}$: PhĊH$_2$; R$_{13}$: PhĊ(tertiary) ; R$_{14}$: PhĊ(OCH$_3$)$_2$; R$_{15}$: (dioxolanyl radical) ; R$_{16}$: acyloxy-CH$_2$ radical

1

Carbon-centered radicals exhibit rather high addition rate constants to acrylates ($k_i > 10^5$ M^{-1} s^{-1}), the nucleophilic character increasing in the series tertiary > secondary > primary radicals (Table 18.1). Methylacrylate (MA) and ethylvinylether (EVE) were selected as representative electron-poor and electron-rich monomers, respectively, to highlight the polar effects that strongly affect the radical reactivity: (i) for nucleophilic radicals exhibiting a OH substitution, the reactivity toward the addition to electron-deficient monomers is enhanced (for R$_6$–R$_8$, the addition rate constants to MA are significantly higher than for alkyls) and (ii) for a CN (R$_{10}$–R$_{11}$) or a C=O (R$_{15}$–R$_{16}$) group-substituted electrophilic radical, the addition to electron-rich monomers is more efficient. The stabilized ketyl radical R$_9$ is usually considered as a terminating agent of growing polymer chains. Its addition to electron-deficient monomers (MA and methylmethacrylate (MMA)) as

Table 18.1 Addition rate constants k_i of different alkyl (and derived) radicals to methylacrylate and ethylvinylether.

Radical	$k_i(10^5\ \text{mol}^{-1}\ \text{l}\ \text{s}^{-1})$ methylacrylate	$k_i(10^5\ \text{mol}^{-1}\ \text{l}\ \text{s}^{-1})$ ethylvinylether	Radical	$k_i(10^5\ \text{mol}^{-1}\ \text{l}\ \text{s}^{-1})$ methylacrylate	$k_i(10^5\ \text{mol}^{-1}\ \text{l}\ \text{s}^{-1})$ ethylvinylether
R_1	3.4	0.14	R_9	<0.09	<0.09
R_2	11	0.0039	R_{10}	1.1	0.43
R_3	6.2	–	R_{11}	0.0037	0.00108
R_4	33	–	R_{12}	0.0043	0.00014
R_5	1000	–	R_{13}	0.008	–
R_6	7.1	0.0018	R_{14}	<0.1	–
R_7	350	0.0032	R_{15}	1.1	3.0
R_8	250	–	R_{16}	4.9	–

From [9–14].

well as electron-rich monomers (vinyl acetate (VA) and EVE) has a low efficiency. This structure can also be easily oxidized (Chapter 19 and 12.5 for the different rate constants associated with these processes).

The observed modulation of the reactivity for these carbon-centered radicals can be ascribed to large polar and/or enthalpy effects. For $R_{12}-R_{14}$, the stabilization of benzyl-derived radicals can also be involved (the delocalization decreases ΔH_r).

Carbon-centered radicals are very sensitive to oxygen, with formation rate constants of peroxyls close to the diffusion limit in organic solvents ($>10^9\ \text{mol}^{-1}\ \text{l}\ \text{s}^{-1}$) [15].

Alkyl and growing alkyl-type macroradicals slowly react with phenols ($k \sim 10^4\ \text{mol}^{-1}\ \text{l}\ \text{s}^{-1}$ for primary alkyl radicals and 2,4,6-trimethylphenol) [16]. Such a reaction (e.g., with traces of a phenolic stabilizer) will not compete with the monomer addition (k [phenol] $\sim 1\ \text{s}^{-1}$ vs k_p [acrylate] $\sim 5 \times 10^4\ \text{s}^{-1}$). The situation is likely different when working on a wood substrate as a high amount of phenolic compounds can migrate into the coating formulation.

18.2
Aryl Radicals

In photoinitiating systems, the phenyl (or aryl) radicals are obtained by the reduction of iodonium salts Eq. (18.1) from (i) a photosensitizer working by electron-transfer reaction or (ii) a radical, for example, in free-radical-promoted cationic photopolymerization (FRPCP) (Section 12.5). Phenyl radicals can hardly be observed by LFP but are easily characterized by ESR-spin trapping (with phenyl-N-tertbutyl nitrone (PBN), $a_N = 14.3$ G and $a_H = 2.2$ G) [17].

$$PS^* + Ph_2I^+ \rightarrow PS^{\bullet+} + PhI + Ph^\bullet$$
$$R^\bullet + Ph_2I^+ \rightarrow R^+ + PhI + Ph^\bullet \tag{18.1}$$

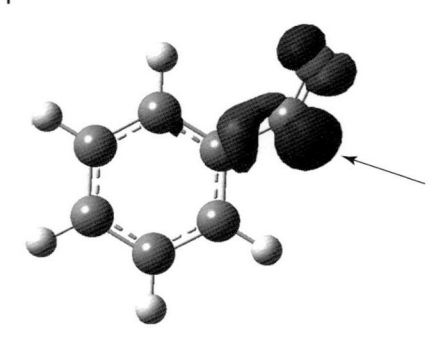

Figure 18.1 SOMO of a benzoyl radical calculated at DFT level; the radical center is indicated by an arrow. (Author's work.)

These radicals exhibit very high k_i to acrylates since the reaction is highly exothermic ($k_i = 1.9 \times 10^8$ mol^{-1} l s^{-1} for MA; $\Delta H_r \sim -160$ kJ mol^{-1} compared to $k_i = 3.4 \times 10^5$ mol^{-1} l s^{-1} and $\Delta H_r \sim -110$ kJ mol^{-1} for the addition of the methyl radical [18]). The addition of aryl radicals to O_2 is diffusion controlled. Arylperoxyl radicals are observed in the visible wavelength range ($\sim$470 nm) [19]. These structures are also very efficient in the hydrogen abstraction from H-donors (silanes and boranes) in agreement with the formation of a strong C–H bond in Ph–H (Ph$^\bullet$ + R–H → Ph–H + R$^\bullet$).

18.3
Benzoyl Radicals

Benzoyl radicals that represent a very important class of polymerization initiating structures are generated from the cleavage of Type I PIs working through a Norrish I mechanism (Section 8.1). Benzoyl radicals are considered as σ radicals as shown by the SOMO depicted in Figure 18.1. Substituents on the phenyl have almost no effect on the SOMO as the radical is located in an orbital perpendicular to the π system: this explains the low substituent effect on the reactivity (see below).

Benzoyl radicals exhibit a rather weak absorption at $\lambda < 350$ nm. They can be easily characterized by steady-state ESR ($g = 2.0008$; $a_H = 0.12$ mT) [20] or ESR-spin trapping experiments (the hyperfin coupling constants hfc being highly specific, e.g., $a_N = 14.6$ G and $a_H = 4.5$ G using PBN; these hfcs are usually not affected by the substituent at the para position for the aryloyl derivatives) [21]. The benzoyl radical reactivity was also probed by time-resolved IR spectroscopy (characteristic transient absorption between 1790 and 1820 cm^{-1} depending on the benzoyl structures (Figure 18.2); the C=O stretching frequencies of the RC(=O)$^\bullet$ radicals are higher than those of the corresponding aldehydes, RCHO) [22].

The reactivity of acyl radicals toward a variety of substrates is roughly comparable to that of small alkyl radicals (Table 18.2). Interaction rate constants of the benzoyl radical with some additives are typically: 6.0×10^4 (CCl$_4$), 4.8×10^7 (C$_6$H$_5$SH), 2.2×10^8 (CCl$_3$Br), 1.1×10^9 (2,2,6,6, tetramethylpiperidine N-oxyl radical (TEMPO)), and 1.8×10^9 mol^{-1} l s^{-1} (O$_2$). Benzoyls and alkanoyls have a

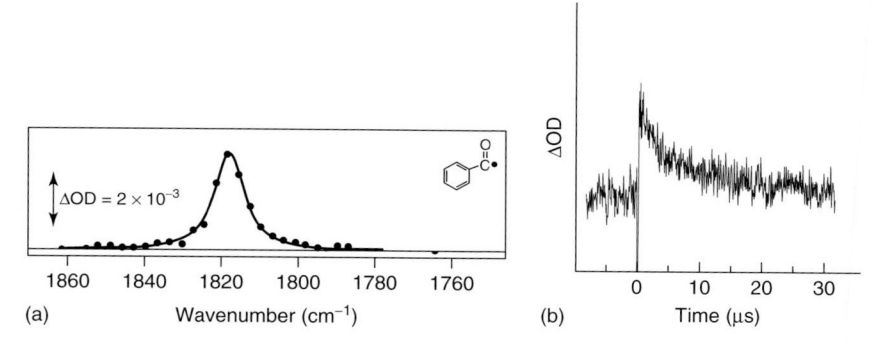

Figure 18.2 (a) Time-resolved IR spectrum of PhC(=O)•. (Source: Reproduced with permission from ACS [22].) (b) Decay of PhC(=O)• at 1820 cm^{-1}. (Source: From [23].)

similar reactivity [24]. The k_is are only slightly affected by the benzoyl substitution ($1.3–5.5 \times 10^5$ mol^{-1} l s^{-1}) [22, 25].

18.4
Acrylate and Methacrylate Radicals

Acrylate and methacrylate radicals as well as acrylic or vinylic radicals M• can hardly be observed and studied by LFP ($\lambda < 330$ nm) [26–30] contrary to the acrylate-derived radical TEA−M• (where TEA is the triethylaminoalkyl group) whose absorption is red-shifted (Figure 18.3). For M =AN, a quite similar spectrum is found for TEA−M•. The TEA−M−M• or TEA−M−M−M• radicals cannot be detected (lack of absorption; the calculated intensity of the visible transition strongly decreases [31]).

The properties of the radical center in M• and TEA−M• are very similar (for more details, see Sections 6.6 and 18.18). This unique feature of TEA−M• provides a direct access to its reactions with, for example, well-known polymerization inhibitors such as hydroquinone methylether (HQME) and TEMPO given in Table 18.3.

The reaction between oxygen and (meth)acrylate radicals [32] is nearly diffusion controlled (close to 3×10^9 mol^{-1} l s^{-1}) in line with both the usual high reactivity of carbon-centered radicals with oxygen and the strong inhibition of radical polymerization reactions under air.

The reactivity of TEA−MA• or TEA−MMA• toward TEMPO [33] is particularly worthwhile: the interaction rate constants are close to 3×10^8 mol^{-1} l s^{-1}. These values are in line with the very efficient reactions observed elsewhere between other carbon-centered radicals and TEMPO. The slightly lower reactivity noted for TEA−MMA• toward oxygen and TEMPO can probably be ascribed to a weak steric hindrance effect compared to that of TEA−MA•.

The reaction of HQME with TEA−MA• and TEA−MMA• is less favorable ($<5 \times 10^5$ mol^{-1} l s^{-1}). This highlights that, in a photopolymerization, the

Table 18.2 Rate constants characterizing the reactivity of different benzoyl radicals: addition to acrylate (k_i) and hydrogen abstraction from thiol (k_H).

Radical	k_i (10^5 mol^{-1} l s^{-1}) acrylate	k_H (10^7 mol^{-1} l s^{-1}) PhSH
benzoyl	2.7	4.8–1.8
HO-CH₂CH₂-O-C₆H₄-C(O)•	3.5	–
MeS-C₆H₄-C(O)•	5.5	–
morpholino-C₆H₄-C(O)•	3.6	5.4
2,4,6-trimethylbenzoyl	1.8	0.39
2,6-dimethoxybenzoyl	1.3	0.71

From [22, 25].

stabilizer probably reacts with the initiating radicals and less efficiently with the polymer-chain-propagating radical. Interaction rate constants with thiols are in the order of 10^5 mol^{-1} l s^{-1} [34].

18.5
Aminoalkyl Radicals

18.5.1
Reactivity

In Type II systems, the interaction of a ketone triplet state with an amine AH leads to ketyl K• and α-aminoalkyl radicals A•. Ketyl radicals are usually stabilized structures and cannot initiate a polymerization (k_is are very low, i.e., often $<10^3$ mol^{-1} l s^{-1};

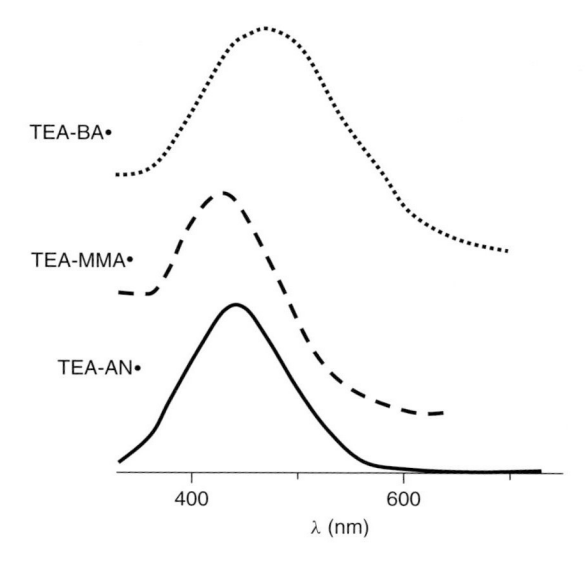

Figure 18.3 Transient absorption spectra for the TEA−BA•, TEA−MMA•, and TEA−AN• radicals.

Table 18.3 Interaction rate constants of monomer derived radicals (methylacrylate, methylmethacrylate, and acrylonitrile) toward O_2, HQME, and TEMPO.

	$k_q(O_2)$ $(10^9 \text{ mol}^{-1} \text{ l s}^{-1})$	$k_q(\text{HQME})$ $(10^6 \text{ mol}^{-1} \text{ l s}^{-1})$	$k_q(\text{TEMPO})$ $(10^8 \text{ mol}^{-1} \text{ l s}^{-1})$
TEA−MA•	3.0	<0.5	3.3
TEA−MMA•	1.9	<0.5	1.8

From [31].

Table 18.1). Therefore, the role of the α-aminoalkyl radicals is crucial, these species being the initiating radicals (Section 9.1.2).

The transient absorptions of radical structures derived from trialkylamines and aromatic amines (e.g., ethyldimethylaminobenzoate (EDB)) are located at ∼340 and ∼500 nm, respectively, but they usually remain much weaker than that of the concomitantly formed K• [35]. To investigate the reactivity, the precursor must be silent in LFP : tBuO• is usually selected (tBuO• + A−H → tBuOH + A•) as it exhibits no significant absorption at λ > 300 nm (Section 6.1.5). The A• radicals are easily characterized by ESR-spin trapping experiments (hfc being a_N ∼14.7 G and a_H ∼2.5 G for the TEA-derived aminoalkyl radical with PBN) [36].

Aminoalkyl radicals are stabilized by a two-orbital three-electron interaction through the delocalization of the 2p orbital of the carbon radical center with the nitrogen lone pair (Scheme 18.2). Accordingly, the spin density is partly located on the nitrogen (0.1 vs about 0.9 for the carbon). This stabilization quite well explains

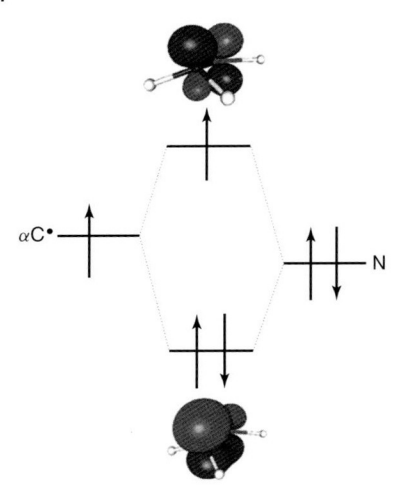

Scheme 18.2

the reduced BDE(αC−H) of the amines compared to the alkanes, that is, 90 versus 96–100 kcal mol^{-1}, respectively [6, 37]. From this orbital interaction, a low energy is expected for the SOMO, that is, aminoalkyls exhibit low ionization potentials and are characterized by a rather high nucleophilic character.

The structural effects on the reactivity of representative aminoalkyl radicals (shown in **2** and generated by hydrogen abstraction (A_1−A_{12}) or C–C bond cleavage in Type I PIs (A_{12}−A_{13}) [38]) toward different additives (oxygen, MA, and a radical trap such as TEMPO) has been investigated through LFP and quantum mechanical calculations [35].

2

Table 18.4 Interaction rate constants between aminoalkyl radicals and O_2, TEMPO, and methylacrylate (MA).

Radical	k_{O_2} (10^9 mol^{-1} l s^{-1})	k_{TEMPO} (10^8 mol^{-1} l s^{-1})	k_i^a (mol^{-1} l s^{-1})
A_1	2.9	3.3	2.0×10^7
A_2	1.5	5.0	9.0×10^6
A_3	0.9	1.3	6.0×10^5
A_4	3.0	2.0	9.0×10^5
A_5	0.04	0.002	$<4.0 \times 10^4$
A_6	0.45	0.075	$<4.0 \times 10^4$
A_7	1.5	0.5	9.7×10^5
A_8	0.6	0.35	3.0×10^5
A_9	0.5	1.0	5.0×10^5
A_{10}	0.9	0.8	6.5×10^5
A_{11}	0.8	0.8	6.2×10^5
A_{12}^b	–	–	2.9×10^7
A_{13}^b	–	–	6.1×10^6

[a] Addition to methylacrylate from [35].
[b] Addition to n-butylacrylate from [38].
From [35].

All the interaction rate constants are given in Table 18.4. A large range of values are obtained: $k_{O_2} = (0.04–3) \times 10^9$ mol^{-1} l s^{-1} for O_2, $k_{TEMPO} = (0.002–5) \times 10^8$ mol^{-1} l s^{-1} for TEMPO and $k_i = (<0.004–2) \times 10^7$ mol^{-1} l s^{-1} for MA.

MO calculations [35] provide a clear correlation of the reaction enthalpy ΔH_r with k_{O_2}. For exothermicity >140 kJ mol^{-1}, the reaction is diffusion controlled ($k_{O_2} = 3 \times 10^9$ mol^{-1} l s^{-1}). However, when the exothermicity decreases, the rate constants decrease. This result is of prime interest for the understanding of the oxygenated species formation or the design of oxygen-insensitive radicals, which should allow the finding of more efficient radical structures in polymerization reactions.

The change of log (k_{TEMPO}) versus ΔH_r has also been checked. For radicals characterized by exothermicity >80 kJ mol^{-1}, a high reactivity is observed as supported by interaction rate constants close to the diffusion (3×10^8 mol^{-1} l s^{-1}) in line with the typical values encountered for the reaction between carbon-centered radicals and TEMPO. The steric hindrance directly affects the rate constants that range between 3×10^7 and 10^8 mol^{-1} l s^{-1}: this is well exemplified in triisopropylamine, which is characterized by the lowest value. The rate constants markedly decrease with the exothermicity, particularly when this factor is lower than 40 kJ mol^{-1}. Interestingly, k_{TEMPO} decreases with k_{O_2}. The dependence of the recombination of the aminoalkyl radicals with oxygen and TEMPO on ΔH_r shows that both reactions are strongly related [35].

Noticeable differences are noted in the addition rate constants to MA between the A_1-A_{11} radicals, for example: $k_i = 2 \times 10^7$ mol^{-1} l s^{-1} for

$A_1, = 6 \times 10^5$ mol^{-1} l s^{-1} for A_3, and $<4 \times 10^4$ mol^{-1} l s^{-1} for A_5. These radicals are also less reactive than the others toward O$_2$ and TEMPO. An excellent linear relationship was found between log (k_i) and the barrier as expected from the classical Arrhenius equation. Despite k_i being roughly related to ΔH_r, some deviations are observed. As evidenced above, the radical addition reaction to an acrylate unit exhibits a strong interplay between the enthalpy and the polar contributions. These results well explain why the complete set of data does not reveal any stringent correlation between k_i and k_{O_2} or k_{TEMPO}. This lack of correlation was attributed to the fact that both polar and enthalpy effects govern the radical/acrylate interaction, whereas the enthalpy effects mostly control k_{O_2} or k_{TEMPO}.

Aminoalkyl radicals are nucleophilic. High charge transfers from the radical to MA are found in the transition state (TS) structure. The addition to electron-rich monomers (VA, vinylcarbazole (VC), and ethylvinylether VE) is slow ($k_i < 10^4$ mol^{-1} l s^{-1}). A low reactivity toward phenolic derivatives in the hydrogen transfer (HT) process (A$^\bullet$ + PO−H → A−H + PO$^\bullet$) is also noted ($<10^5$ mol^{-1} l s^{-1}) [36]. The aminoalkyl radicals can also be easily oxidized. The oxidation potential is very low (−1.12 V) for the triethylamine-derived radical [39]), that is, the oxidation rate constants by iodonium salts are usually almost diffusion controlled (Section 12.5).

18.5.2
Role of the Class of the Amine

The hydrogen abstraction rate constants k_H from amines by the benzophenone (BP) triplet state usually decrease in the series primary < secondary < tertiary. These amines can lead to aminoalkyl radicals by α(C−H) hydrogen abstraction. In primary and secondary amines, aminyl radicals can be also generated (Section 18.12). The reactivity of aminoalkyl radicals for the addition to MA is very similar in the series primary, secondary, tertiary: rate constants being $1.7 \times 10^7, 6.7 \times 10^6$, and 2×10^7 mol^{-1} l s^{-1} for the radicals derived from cyclohexylamine, diethylamine, and triethylamine, respectively [36]. Therefore, the different photoinitiating ability in this series is probably governed by the hydrogen abstraction process as well as the nature of the generated radicals, that is, the aminoalkyls are efficient initiating structures contrary to aminyls (Section 18.12).

18.5.3
N-Phenyl Glycine Derivatives

In the well-known N-phenyl glycine derivatives (NPG) (Ar-NHCH$_2$COOH), different sites for the hydrogen abstraction can be expected on (i) the nitrogen, (ii) the α(C−H) carbon, and (iii) the oxygen atom. The experimental as well as the calculated radical spectrum clearly demonstrates that the hydrogen abstraction takes place at the carbon atom. A very fast decarboxylation reaction then occurs and generates A_{10}. The addition of this radical to MA is rather slow:

$k_i = 6.5 \times 10^5 \text{ mol}^{-1} \text{ l s}^{-1}$. NPG in the presence of a ketone (e.g., BP) leads, however, to a photopolymerization ability higher than that of a classical amine. As it cannot be ascribed to k_i, this ability is attributed to the fast proton transfer (Ar−N$^{•+}$HCH$_2$COOH → Ar−NHCH$^{•}$ COOH + H$^+$) in line with (i) the relative BDEs (α(C−H) BDE of NPG = 78.5 kcal mol^{-1} vs ∼90 kcal mol^{-1} for usual amines) and (ii) a fast decarboxylation reaction that avoids the back electron-transfer reaction in the ketone/NPG CTC [35].

18.5.4
Chain Length Effect

A very similar reactivity is observed for the reaction of both the monomeric A_9 and polymeric A_{11} α-aminoalkyl radicals with oxygen, TEMPO, acrylates (e.g., MA), phenols (vitamin E (VIE), HQME), Ph$_2$I$^+$, or CBr$_4$ [40]. For example: $k_{O2} = 5 \times 10^8$ and 8×10^8 mol^{-1} l s^{-1} ($\Delta H_r = -115.7$ and -116.4 kJ mol^{-1}); $k_{TEMPO} = 1 \times 10^8$ and 0.8×10^8 mol^{-1} l s^{-1} ($\Delta H_r = -101.8$ and -102.5 kJ mol^{-1}); and $k_i = 5.0 \times 10^5$ and 6.2×10^5 mol^{-1} l s^{-1} ($\Delta H_r = -91$ and -92.1 kJ mol^{-1}). These results unambiguously demonstrate the weak influence of the polymer chain length on these reaction rate constants.

The macroradical formed here on the polymeric amine can tentatively mimic a growing low-molecular-weight polymer chain as in A_{11}; it roughly corresponds to eight to nine acrylate units. In pulsed laser polymerization (PLP) methods that allow the determination of the propagation k_p and termination k_t rate constants of the radical polymerization reaction of small monomers (Section 7.1.4), one recurrent question is the effect of the chain length on k_p or on other polymer reactions below the entangled region where long macromolecular chains are formed. This effect, referred to as the *penultimate unit effect*, has been the subject of numerous studies [41] and remains partly unclear. In a computational approach carried out on the propagation of the ethylene polymerization [42, 43], a noticeable chain length effect was evidenced through a decrease of k_p by a factor of about 10 when going from the monomeric to the corresponding macropolyethylene radical. The behavior difference observed in these experimental (above) and theoretical data might be ascribed to (i) the choice of the polymer radicals, (ii) the different possible conformations in larger systems (like A_{11}) that can reduce the contribution of the mechanical effect on the chain length, (iii) the retained hypothesis for the preexponential factor calculations, and (iv) the error bars of the determined addition rate constants (∼10%).

18.5.5
Regioselectivity of the Hydrogen Abstraction Reaction

The regioselectivity of the hydrogen abstraction reaction in primary or secondary amines (aminoalkyl vs aminyl for C–H or N–H abstraction) is an interesting subject (Section 18.12).

The regioselectivity can be also well exemplified by a carefully selected structure: N,N-diethyl-1,1-dimethylsilylamine (DEDMSA) (3). ESR-spin trapping and LFP clearly demonstrated (J. Lalevée et al., unpublished data) that the hydrogen abstraction reaction with several types of ketones and radicals leads to aminoalkyl and silyl radicals at the α(C−H) or Si–H position, respectively. It was suggested that a pure HT can lead to a selectivity different from that of the electron/proton transfer sequence EPHT.

H₃C–CH(H)–N(Si–H)(CH₃)–...

3

A high interaction rate constant (2.5×10^8 mol^{-1} s^{-1}) is obtained in tBuO$^{\bullet}$/DEDMSA together with a high selectivity ($\sim$70% for the Si–H abstraction) which supports the Si–H hydrogen abstraction PHT as the main process. This can be explained by the particular stabilized character of the produced aminosilyl radical (the stabilization is due to a hyperconjugation between the nitrogen lone pair and the π orbital of the Si radical center) and a lower BDE (BDE(Si−H) = 89.6 for DEDMSA versus 95 kcal mol^{-1} for triethylsilane; α(C−H) BDE of DEDMSA = 90.8 kcal mol^{-1}). In the ^{3}ketone/DEDMSA, the PHT versus EPHT process and the selectivity are dependent on the ketone. The regioselectivity of the lophyl radical L$^{\bullet}$ is different from that of tBuO$^{\bullet}$. This probably gives evidence for an EPHT process or at least a partial charge-transfer character for the hydrogen abstraction PHT (see also Section 18.15 for the L$^{\bullet}$/H−donor reaction).

18.5.6
Aminoalkyl Radicals and the Halogen Abstraction Reaction

The phenyl- and adamantyl-carbon-centered radical/alkyl halide interactions [44, 45] and the very high reactivity of aminoalkyl radicals toward a halogen abstraction reaction [46] were reported. The reaction (18.2) allows a change from R-X (where X is the halogen atom) to R-H (A$^{\bullet}$ being an aminoalkyl radical) and consists in a halogen abstraction followed by HT. This reaction can occur, for example, in CBr$_4$ containing multicomponent radical initiating systems (Part 2).

$$A^{\bullet} + R\text{-}X \rightarrow A\text{-}X + R^{\bullet}$$
$$R^{\bullet} + A\text{-}H \rightarrow R\text{-}H + A^{\bullet} \tag{18.2}$$

The usual driving force of the halogen abstraction is the BDE(R-X) and BDE(A-X). The carbon-centered radical/alkyl chloride interaction exhibits a rather low exothermicity (as both the cleavage and the formation of a C–Cl bond are involved) and is a rather slow reaction ($<10^5$ mol^{-1} s^{-1}). On the contrary, a high charge transfer from the radical to the alkyl halides was highlighted (an electron-transfer process can be even expected in some specific radical/RX couples). Accordingly, the reaction rate constants of radicals A_1 and A_9 **2** with CCl_4 and CBr_4 are close to the diffusion limit. They are about four to five orders of magnitude higher than those previously determined for typical alkyl radicals.

18.5.7
Reactivity under Air

It is still currently admitted that amines help in reducing the oxygen inhibition effect in free radical polymerization. A classic largely used mechanism is shown in Scheme 18.3. The benefit of amines is ascribed to the conversion of peroxyls (uneffective in propagating the polymerization) to hydroperoxydes (that can further decompose into oxyl-and hydroxyl-initiating radicals) through an HT with the amine. Therefore, amines are considered as being able to consume oxygen and then to regenerate an α-aminoalkyl radical [47–49].

However, this proposed mechanism is the subject of debate as the peroxyl radical/amine interaction was recently found to be rather weak: only 6 $mol^{-1} l\, s^{-1}$ for the $tBuOO^{\bullet}$/EDB interaction rate constant (as determined by kinetic ESR; see the discussion in Section 18.11). This result shows that the reaction is slower than formerly considered [50]. For primary to tertiary amines, the interaction with the model $tBuOO^{\bullet}$ peroxyl remains $< 25\, mol^{-1} l\, s^{-1}$.

For typical [amine] ~ 0.2–$0.3\,M$, the conversion of peroxyls is slow (k [amine] $\sim 1.2\, s^{-1}$ for EDB) in agreement with a lower ability of amine to overcome the oxygen inhibition compared to other compounds (silanes, boranes; see in Section 18.11) exhibiting a higher reactivity with peroxyls (k [TTMSS] $\sim 70\, s^{-1}$ for tris(trimethylsilyl)silane). As (i) the radical yields in (1a) of Scheme 18.3 is higher for amines than for silanes, (ii) reaction (1b) is diffusion controlled, and (iii) PI/amine is less efficient than PI/silane in photopolymerization, it should be concluded than (2) can probably be neglected.

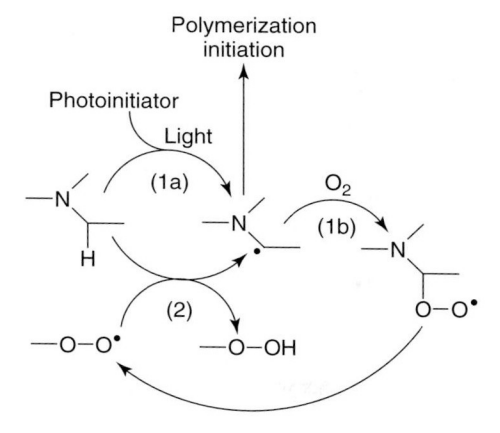

Scheme 18.3

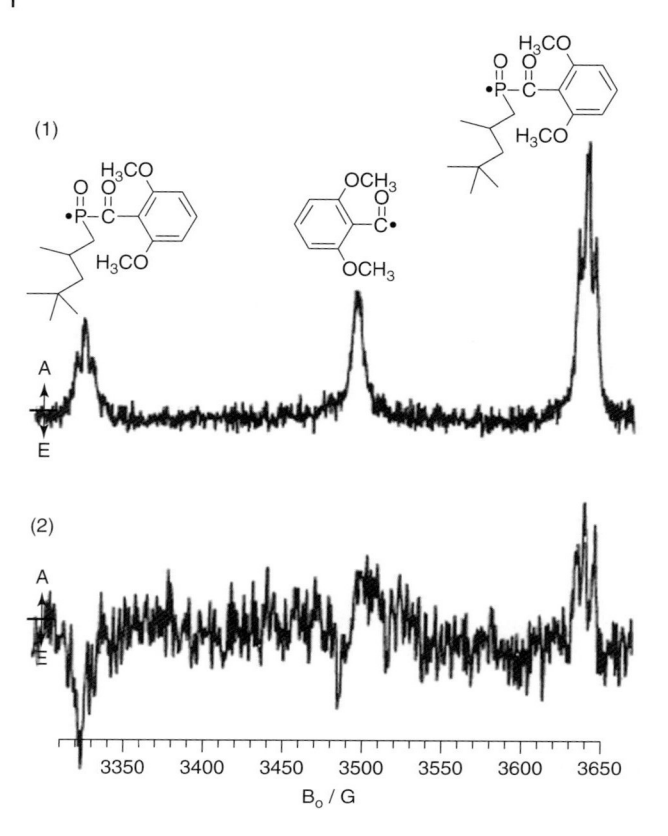

Figure 18.4 TR-ESR spectra recorded (a) 50–250 ns and (b) 750–1050 ns following the 355 nm laser excitation of the corresponding phosphine oxide in deoxygenated acetonitrile solution at 23 °C. (Source: From [51]. Reproduced with permission from ACS.)

18.6
Phosphorus-Centered Radicals

Phosphorus-centered radicals and more particularly phosphinoyls $R_1R_2P^{\bullet}(=O)$ are very attractive polymerization-initiating species. They are generated in acylphosphine oxides, that is, in commercial Type I PIs based on a C-P homolytic cleavage (Section 8.6). They can be observed by time-resolved ESR (Figure 18.4) [51, 52], LFP (absorption dependent on the phosphinoyl structures i.e., $\lambda = 340$ nm for $R_1 = R_2 = Ph$; $\lambda = 450$ nm for $R_1 = Ph$ and $R_2 = C(=O)$—mesityl [53, 54]), and ESR-spin trapping experiments (*hfc* being $a_N = 14.2$ G; $a_H = 2.94$ G; $a_P = 19.2$ G for $Ph_2P^{\bullet}(=O)$ using PBN) [53]. Phosphinoyl radicals usually exhibit a pyramidal structure as a result of a relatively high phosphorus spin localization associated with a σ-character. However, the shape and the electronic character of phosphinoyl radicals are affected by the substitution. For $R_1 = R_2 = Ph$, a σ radical is expected. For $R_1 = Ph$ and $R_2 = C(=O)$-mesityl, a significant π-character is found; this is also associated with a more planar structure [51, 52, 55].

The reactivity of various compounds toward several types of phosphonyl radicals has been already reviewed and discussed [56, 57] (Table 18.5). Quite high rate constants are found for the addition to double bonds, halogen abstraction, hydrogen abstraction from thiols, or the reaction with oxygen. Polar effects are important in these radical reactions. For example, the dependence of the reaction rate constants k_i between thiophosphinoyl radicals and various unsaturated monomers has been checked.

An interesting qualitative trend in the dependence of k_i on the chemical nature of the two partners has been reported using the Q-e scheme [56–58] used for the copolymerization of unsaturated monomers (Eq. (18.3)) where P_R and Q are the reactivity parameters, e_R and e_M the polarity parameters that take into account the influence of substituents on the electron density at the unpaired electron site of the radical R• and the double bond M•).

$$k_i = P_R Q \exp(-e_R e_M) \tag{18.3}$$

The high reactivity of the phosphinoyl radicals has been attributed to their pyramidal structure (which allows little steric constraints in the addition to double bonds) resulting from a high degree of σ-character and spin localization on the phosphorus atom [55]. As observed by the measurement of the ^{31}P hyperfine coupling in the ESR spectra, substituents that lead to a flattening of the radical reduce the reactivity in agreement with an increase of the π-character of the radical center and a more important conjugative interaction with the benzoyl moiety.

Interestingly, phosphinoyl radicals are also characterized by a rather low selectivity as supported by the high k_i for the addition both to electron-rich and electron-poor monomers, that is, in the case of $Ph_2P^•(=O)$: $k_i = 2 \times 10^7, 3.5 \times 10^7, 0.4 \times 10^7$, and 0.16×10^7 mol^{-1} l s^{-1} with AN, MA, n-butyl vinyl ether, and VA, respectively [59].

18.7
Thiyl Radicals

In Type II PIs, amines represent the widely used class of H-donors. For specific applications, thiols were also introduced [60, 61]. The reactivity of thiyl radicals RS•, for example, 4 is an important item. Thiyls can be generated from (i) a tBuO•/thiol or PI/thiol hydrogen abstraction or (ii) the photolysis of disulfides through an S-S cleavage.

Table 18.5 Rate constants characterizing the reactivity of different phosphorus-centered radicals.

Radical	$k_i{}^a$ (10^7 mol^{-1} l s^{-1}) methylacrylate	k_H (10^6 mol^{-1} l s^{-1}) PhSH	k_{O_2} (10^9 mol^{-1} l s^{-1})	k_{PhCH_2Br} (10^6 mol^{-1} l s^{-1})
Ph-P(=O)-Ph•	2.8	4.2^b	4.2	5.9^e
Ph-P(=S)-Ph•	0.2^c	–	–	–
(EtO)$_2$P(=O)•	1.6^d	–	–	–
Ph-P(=O)(OiPr)•	2.1^d	–	–	–
(2,4,6-trimethylbenzoyl)-P(=O)-Ph•	1.1	–	2.7	–
(2,6-dimethoxybenzoyl)-P(=O)-Ph•	1.25	–	3.0	–
(2,4,6-trimethylbenzoyl)-P(=O)-CH$_2$C(CH$_3$)$_3$•	–	–	2.5	–

^{a}Addition to n-butylacrylate (except otherwise indicated).
^{b}From [22].
^{c}Addition to methyl methacrylate from [58].
^{d}Addition to methylacrylate from [59].
^{e}[51].

18.7 Thiyl Radicals

[Structures S1-S9 labeled as 4]

These radicals are usually ESR silent in steady-state ESR but can be characterized by ESR-spin trapping (*hfc* being $a_N = 13.9$ G; $a_H = 1.8$ G for $p-CH_3OC_6H_4-S^\bullet$ using PBN) [62]. They usually exhibit a quite intense absorption in the visible range and can be also observed by LFP. The absorption spectrum of the thiyl of mercaptobenzothiazole (MBT) is depicted in Figure 18.5. The kinetic data associated with these structures remain scarce.

Thiyl radicals are characterized by a low sensitivity to oxygen ($k_{O_2} < 10^5$ mol^{-1} l s^{-1} (Eq. (18.4)). The reaction is reversible (which is remarkably unusual), that is, thiylperoxyls undergo a β-fragmentation process [64]. Thiyls hardly abstract a halogen atom ($k_{iodopropane} < 10^5$ mol^{-1} l s^{-1}) and are likely not oxidized by an iodonium salt ($k_{iodonium} < 10^6$ mol^{-1} l s^{-1}). They efficiently add to a monomer double bond. This reaction is also followed by a further back fragmentation of the monomer radical, the equilibrium constant depending on the thiyl/double bond couples (Eq. (18.4)) [65–70]. Some equilibrium constants

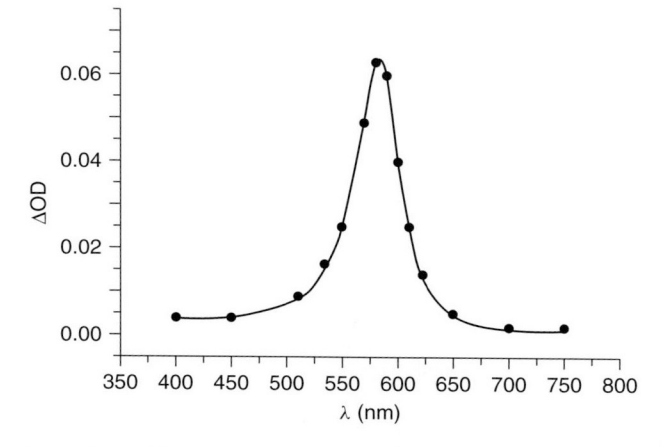

Figure 18.5 Absorption spectrum of radical S_3 (generated by a tBuO$^\bullet$/thiol hydrogen abstraction). (Source: From [63].)

Table 18.6 Addition rate constants k_i (in mol^{-1} l s^{-1}) of thiyl radicals to acrylonitrile (AN), methylacrylate (MA), vinylacetate (VA), and ethylvinyl ether (EVE).

Monomer radical	k_i			
	AN[a]	MA[a]	VA[a]	EVE[a]
S_1	6.5×10^4 $(0.55\ M^{-1})^c$	2×10^5 $(0.1\ M^{-1})^c$	5.6×10^5 $(0.004\ M^{-1})^c$	6×10^5 $(0.03\ M^{-1})^c$
S_2	$<10^4$	$<10^4$	$<10^4$	$<10^4$
S_3	3.7×10^5 $(0.4\ M^{-1})^c$	6×10^5 $(0.4\ M^{-1})^c$	8×10^5 $(0.01\ M^{-1})^c$	10×10^5 $(0.08\ M^{-1})^c$
S_4	–	1.6×10^6	–	–
S_5	–	5.6×10^5	–	–
S_6	–	$<10^4$	–	–
S_7	–	$2.7 \times 10^{5\,b}$	–	–
S_8	2.5×10^7 $(33.3\ M^{-1})^c$	1×10^8 $(16.7\ M^{-1})^c$	2.0×10^8 $(4.17\ M^{-1})^c$	4.4×10^8 $(16.7\ M^{-1})^c$
S_9	–	2×10^7	–	–

[a] In acetonitrile.
[b] From [73].
[c] The equilibrium constant ($K = k_i/k_{-i}$).
From [63, 71, 72].

(k_i/k_{-i}) are given in Table 18.6 for representative structures.

$$RS^\bullet + O_2 \rightleftarrows RS\text{-}OO^\bullet$$

$$RS^\bullet + C\text{–}C\text{-}R' \rightleftarrows RS\text{-}C\text{-}C^\bullet\text{-}R' \quad \left(\frac{k_i}{k_{-i}}\right) \tag{18.4}$$

The relative contributions of the polar and enthalpy effects as well as the factors that govern the fragmentation process are outstanding problems. In a general way, thiyl radicals exhibit a good reactivity and selectivity toward a large set of monomer double bonds, albeit the more pronounced electrophilic character generally found as the addition to electron-rich monomers (e.g., VE) is faster than that to electron-deficient monomers (e.g., acrylates). The results reported in Table 18.6 show that the k_is are generally high (the highest values are noted for S_8 and S_9 as discussed below). The worse values are quite similar to those found for the benzoyl radical ($\sim 10^5$ mol^{-1} l s^{-1}) or the aminoalkyl radical of NPG (9.7×10^5 mol^{-1} l s^{-1}).

Contrary to previous studies devoted to other aryl thiyl radicals, it was found that the addition rate constants of radicals derived from mercaptobenzoxazole (MBO) (S_1), mercaptobenzothiazole MBT (S_3), and mercaptobenzimidazole (MBI) (S_2) to four double bonds (MA, AN, VE, and VA) [63] are governed by the polar effects associated with the very high electrophilic character of S_1–S_3. In fact, the results clearly show that the enthalpy factor does not govern the reactivity: (i) the

RS$^\bullet$/monomer couples exhibiting the highest ΔH_r are characterized by the lowest k_i (AN) and (ii) ΔH_r decreases in the series AN > MA > VE > VA, whereas the k_is decrease in the opposite order VE > VA > MA > AN. For radicals, ΔH_r usually decreases in the series AN > MA > VA > VE, that is, electron-withdrawing substituents increase the monomer electronegativity and stabilize the newly formed radical. The back fragmentation reaction as shown above in Eq. (18.4) is mainly influenced by the enthalpy effects with a direct relationship between the rate constant and ΔH_r.

Other thiyls derived from tetrazole, thiadiazole, imidazole, triazole, or pyridine moieties or from disulfides (containing phenyl, propionic acid, pyridine, tetrazole, or pyridine-N-oxide moieties) can be highly effective for the addition to alkenes. This explains the excellent co-initiator ability of these structures [71].

The reactivity of the tetrazole-derived thiyl radical S_8 4 is also particularly worthwhile. Its reactivity toward monomers (VE, VA, MA, AN, allylbutylether (ABE), acrylamide (AA), N-vinylpyrolidone (NVP), dimethyl fumarate (FU), dimethyl maleate (MAL), and VC), has been studied in depth [72]. The considered double bonds possessing very different electron acceptor/donor properties and strong enthalpy/polar effects have been observed in the addition reactions. The low selectivity and the high reactivity of S_8 toward both electrophilic and nucleophilic monomers were outlined. The rate constants are higher than 10^7 mol^{-1} l s^{-1} for the complete range of monomer electronegativity going from the electron-rich π system (VE, NVP, and VC) to the electron-deficient monomers (AN, MA, and AA), except for the two bisubstituted acrylates FU and MAL (in that case, the steric hindrance destabilizes the addition reaction and the process is less exothermic than expected from their electronegativity). Comparatively, an aminoalkyl radical is much less selective, with an enhanced reactivity toward electron-deficient monomers (particularly for monomer exhibiting $\chi > 4.5$ eV) and a very low reactivity toward electron-rich structures ($k_i \ll 5 \times 10^4$ mol^{-1} l s^{-1}). The high reactivity associated with a very low selectivity of S_8 is remarkable.

In thiyl radicals formed in Barton esters such as O-acyl-N-hydroxy-pyridine-2(1H)-thione and O-acyl-N-hydroxy-thiazole-2(3H)- thione derivatives (Section 8.24), a strong difference in the spin localization is noted. The spin density of the pyridine-thiyl radical formed after the triplet state cleavage is clearly localized on the sulfur atom, whereas it is delocalized on the five-ring moiety in the thiazole thione derivatives (Figure 18.6) [74]. This expected difference in the thiyl radical reactivity can partly explain the observed different initiating abilities [74].

18.8
Sulfonyl and Sulfonyloxy Radicals

In β-sulfonyl (or β-sulfonyloxy) ketones, for example, in bifunctional ketone-sulfonyl ketone derivatives [75, 76], a β-cleavage occurs after irradiation

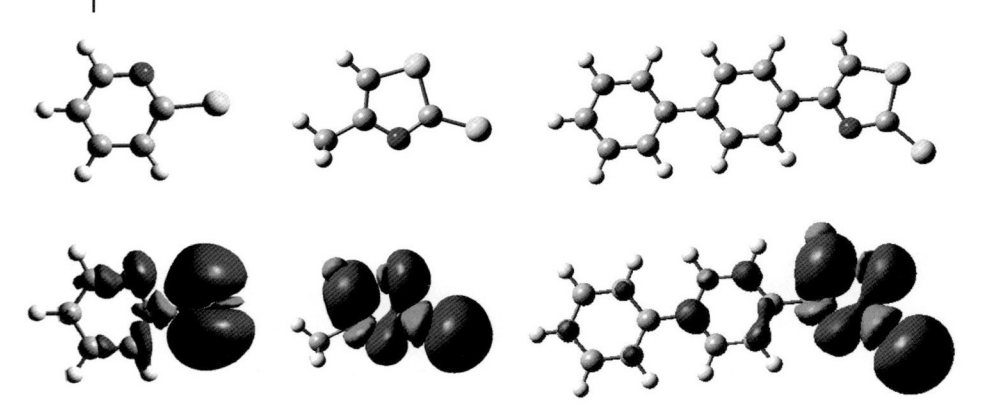

Figure 18.6 Spin densities on the thiyl radical formed in Barton esters (computed at a B3LYP/6–31 G* level) (Source: From [74].)

(Section 8.5.4). The sulfonyl RSO$_2^{\bullet}$ and sulfonyloxyl RSO$_3^{\bullet}$ radicals formed can be observed in LFP [77–79] and ESR-spin trapping experiments [54].

These radicals efficiently add to double bonds and are considered as electrophilic structures [80–82]. The addition rate constants to MMA were reported as $\sim 10^4$ and $\sim 10^6$ mol^{-1} l s^{-1} for RSO$_2^{\bullet}$ or RSO$_3^{\bullet}$ (with R = aryl), respectively [83]. The addition of RSO$_2^{\bullet}$ or RSO$_3^{\bullet}$ to electron-rich monomers is favorable. The addition of sulfonyl radicals to O$_2$ is reversible. These structures are also powerful agents in RSO$_2^{\bullet}$ (or RSO$_3^{\bullet}$)/H-donor hydrogen abstraction reactions, leading to sulfinic (or sulfonic) acids [84]. Therefore, these compounds can also be used as photoacids (Section 8.1.10).

18.9
Silyl Radicals

Despite a widespread use in organic chemistry for hydrosilylation and reduction reactions, silyl radicals have recently emerged as highly effective initiating structures [85, 86]. The photodegradation of polysilane polymers or oligomers occurring from a Si-Si cleavage has been reported in the 1980s to be efficient. However, the photoinitiation ability of the associated silyl radicals was low, as exemplified by the determined low initiation quantum yields Φ_i ($\Phi_i \sim 0.001$ compared to 0.25 for a benzoin ether) [87–92].

Silyl radicals can be observed by (i) LFP at $\lambda < 350$ nm (usually hard), (ii) steady-state ESR ($g \sim 2.0053$ for tris(trimethylsilyl)silyl radical TMS)$_3$Si$^{\bullet}$; the ESR spectra show a central set of lines due to ^{1}H *hfc* and weaker satellites because of the coupling with ^{29}Si), and (iii) ESR-spin trapping ($a_N \sim 15.1$ G; $a_H \sim 5.5$–6.0 G for (TMS)$_3$Si$^{\bullet}$ using PBN [21, 93, 94] (Figure 18.7). These *hfc* give some insight into

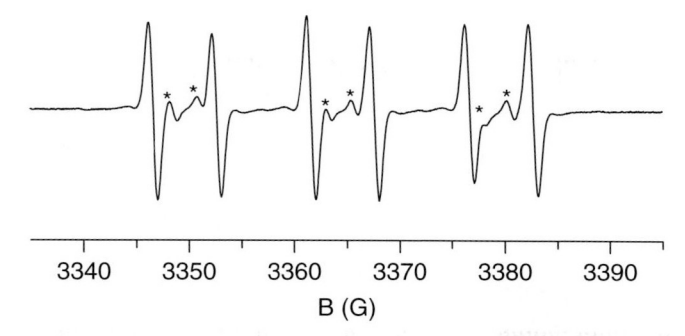

Figure 18.7 ESR-spin trapping experiments for $tBuO^\bullet$/$(TMS)_3SiH$ with PBN; $(TMS)_3Si^\bullet$ is characterized by: $a_N \sim 15.1$ G; $a_H \sim 5.5$. A minor signal ascribed to $tBuO^\bullet$ is observed (stars). (Source: From [94].)

the radical structure (shape and electronic properties) [95, 96], the high a_H being a usual feature.

The shapes of the silyl radicals are usually considered to be strongly bent out of the plane (σ type). For example, the structure of $(TMS)_3Si^\bullet$ is characterized by a pyramidal angle $\alpha = 156°$ (Figure 18.8) [97].

This property arises from the fact that, contrary to carbon-centered radicals that can only use 2s and 2p atomic orbitals to accommodate the valence electrons, d orbitals are involved in $R_3Si^\bullet$ **5**. This σ structure of silyls contributes to their high reactivity.

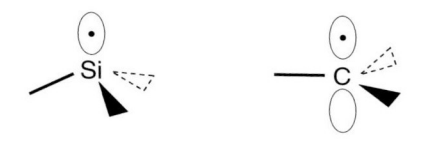

Silyls radicals **6** can be generated by the hydrogen abstraction reaction from the corresponding silane (R_3Si-H).

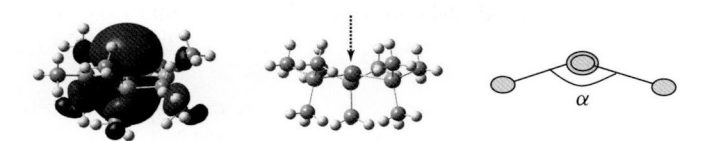

Figure 18.8 Singly occupied molecular orbital (SOMO) and structure of the tris(trimethylsilyl)silyl radical at UB3LYP/6–31 + G* level (the angle of the pyramidal structure involving the three Si atoms is represented by α). (Source: From [97].)

The hydrogen abstraction rate constants between ketone triplet states or alkoxyl radicals and silanes shown in Table 18.7 are high [98]. They range from 4.3×10^6 to 1×10^8 mol^{-1} l s^{-1} and 10^5 to 4.1×10^7 mol^{-1} l s^{-1} for ^{3}BP and ^{3}ITX, respectively, and outline the labile hydrogen character of the silanes. Much higher hydrogen abstraction rate constants (in the 10^9 mol^{-1} l s^{-1} range) were also measured for the ^{3}BP (or ^{3}ITX)/amine interaction (see above), which corresponds to an electron/proton transfer reaction. The oxidation potentials of silanes are usually >1.7 V. Using a reduction potential of -1.79 V and a triplet-state energy of 2.98 eV for ^{3}BP, an endothermic electron-transfer reaction (>+0.61 eV) is expected. This result supports the fact that the ^{3}BP (and ^{3}ITX)/silane reaction probably corresponds

Table 18.7 Interaction rate constants k_H (in mol^{-1} l s^{-1}) of silanes with BP and ITX.

Produced radical	BP[a]		ITX[a]		MA
	k_H	Φ_{Si}	k_H	Φ_{Si}	k_i
Si$_1$	4.3×10^6	0.81	7.2×10^5	0.3	2.4×10^8
Si$_2$	2.5×10^7	0.65	3.4×10^7	0.28	4.5×10^8 (3.65×10^{8b})
Si$_3$	3.1×10^7	0.75	7.7×10^6	0.21	5.1×10^8
Si$_4$	1.02×10^8	0.95	4.1×10^7	0.7	2.2×10^7
Si$_5$	1.1×10^9	0.9	1.5×10^8	0.3	3.3×10^7
Si$_6$	2.2×10^7	0.78	1.6×10^6	0.34	1.5×10^8
Si$_7$	6.1×10^8	0.24	9.1×10^7	0.1	7.9×10^7
Si$_8$	2.1×10^7	–	10^5	0.2	2.0×10^8
Si$_9$	5.0×10^6	–	nd	nd	5.1×10^7

nd, not determined.
Quantum yields in silyl radicals Φ_{Si}. Addition rate constants k_i (in mol^{-1} l s^{-1}) of the silyls to MA.
[a] In benzene.
[b] For methyl methacrylate (MMA).
From [98j], see text.

to a pure HT process and not to an electron/proton transfer sequence. This also holds true for the tBuO$^{\bullet}$/silane interaction that exhibits the same values of rate constants. The ketyl radical quantum yields in the BP/silane combinations (which are obviously equal to the silyl radical quantum yields Φ_{Si}) are usually close to unity. Lower Φ_{Si} are observed in ITX/silane. In acetonitrile, a solvent that increases the $n\pi^*$-character of ^{3}ITX, the Φ_{Si} values are about three times higher than in benzene: this result evidences the sensitivity of the silanes in the hydrogen abstraction reaction toward the spectroscopic character of the lowest lying triplet state.

Silyl radicals (i) abstract halogen atoms from alkyl halides, (ii) add to double or triple bonds, (iii) react with O_2 with a nearly diffusion-controlled rate constant with formation of silylperoxyls, and (iv) are easily oxidized to form silylium cations R_3Si^+. A detailed review of these processes can be found in [85, 86].

The addition rate constants k_i to acrylates (that range from 2.2×10^7 mol^{-1} l s^{-1} to 4.5×10^8 mol^{-1} l s^{-1} for MA) are among the highest values reported for well-known initiating radicals such as the benzoyl ($\sim 10^5$ mol^{-1} l s^{-1}), the aliphatic aminoalkyl ($\sim 2 \times 10^7$ mol^{-1} l s^{-1}), the hydroxy isopropyl ($\sim 10^7$ mol^{-1} l s^{-1}) or the phosphinoyl radicals ($\sim 2 \times 10^7$ mol^{-1} l s^{-1}). This explains the high efficiency of the silyls in radical photopolymerization (Section 9.1.2).

18.9.1
The Particular Behavior of the Tris(Trimethylsilyl)Silyl Radical

The tris(trimethylsilyl)silyl radical Si_4 is not only an excellent co-initiator in Type II systems but also a good additive to Type I PIs for polymerization reactions in aerated conditions (Section 9.1.2.7). The reactivity of Si_4 and Si_1 toward various electron-poor and electron-rich monomers (AN, MA, VA, STY, VC, and VE) was studied in detail (Table 18.8) and compared to that of the well-known triethylamine-derived aminoalkyl radical A_1 2 [99]. The high reactivity and the low selectivity of Si_4 is perfectly explained by antagonist polar and enthalpy effects. The striking feature is that the k_i values for Si_4 and Si_1 are always higher than that for A_1. The carbon-centered structure A_1 is highly specific and exhibits an enhanced reactivity toward electron-deficient monomers having $\chi > 4.5$ eV and a very low

Table 18.8 Addition reaction rate constants k_i (in mol^{-1} l s^{-1}) of silyl radicals to different monomers.

Radical	k_i				
	Styrene	Acrylonitrile	Methylacrylate	Vinylacetate	Vinylether
Si_4	5.1×10^7	5.1×10^7	2.2×10^7	1.2×10^6	2.1×10^5
Si_1	2.1×10^8	1.1×10^9	2.4×10^8	3.5×10^6	9×10^4
A_1	–	6×10^7	3×10^7	$<5 \times 10^4$	$<5 \times 10^4$

The values for the aminoalkyl A_1/monomers are reported for comparison [99].

Table 18.9 Reaction rate constants in (mol^{-1} l s^{-1}) of the radical Si$_4$ with various additives in di-*tert*-butylperoxide [100].

Radical	k$_i$			
	O$_2$	CBr$_4$	Φ_2I$^+$	Cyclopentylacetylene
Si$_4$	200 × 10^7	34 × 10^7	0.26 × 10^7	0.42 × 10^7

reactivity toward electron-rich structures ($\ll 5 \times 10^4$ mol^{-1} l s^{-1}); k_i decreases by at least a factor of 1000 when going from AN to VA. For Si$_4$ and Si$_1$, the factor is only 50 and 300, respectively. Silyl Si$_4$ is the less selective with a factor lower than 200 over the complete scale of double bonds.

For Si$_4$, the exothermicity is systematically ~40 kJ mol^{-1} lower than for Si$_1$ as resulting from the higher stabilization of Si$_4$ (the BDE(Si–H) are about 80 and 90 kcal mol^{-1} for tris(trimethylsilyl)silane and triethylsilane, respectively). The exothermicity decreases in the series AN – STY > MA > VA > VC > VE. A decrease of 1 eV of the monomer absolute electronegativity χ_{DB} is associated with a lowering of the exothermicity ~30–40 kJ mol^{-1}. Therefore, the barrier is expected to decrease with the exothermicity when going from AN to VE for a given radical. As a higher exothermicity leads to a better reactivity, the higher k_i values of Si$_1$ are ascribed to this factor. The decrease in the exothermicity leads to a strong increase in the barrier and, concomitantly, the k_is increase by four orders of magnitude from VE to AN. Thus, the enthalpy factor is the main driving factor for Si$_1$.

On the contrary, for Si$_4$, despite a strong decrease in the exothermicity, the barrier change appears as fairly low. Therefore, the enthalpy factor alone does not govern the barrier. The low change of the barrier for Si$_4$ and the lower selectivity can probably be ascribed to the polar effects. Indeed, the calculation of the amount of charge transfer in the TS outlines a high polar effect for Si$_4$, which appears to be more electrophilic with a net VE to radical charge transfer (0.08 e$^-$ vs only 0.03 e$^-$ for Si$_1$). The polar effect is much less important for Si$_1$.

The reactivity of Si$_4$ toward different additives has been investigated (Table 18.9) [100]; Si$_4$ efficiently reacts with O$_2$, alkyl halides, and iodonium salts (the silyl oxidation potentials are unknown, but the values of the rate constants clearly suggest a rather low E_{ox}). Interestingly, Si$_4$ also adds to triple bonds (here cyclopentylacetylene) with formation of vinyl radicals.

18.9.2
Reactivity and Photoinitiation under Air

In PI/R$_3$SiH, a plausible and classical initiation mechanism in laminates is displayed in Scheme 18.4 where S$_0$, S$_1$, and T$_1$ represent the singlet ground state

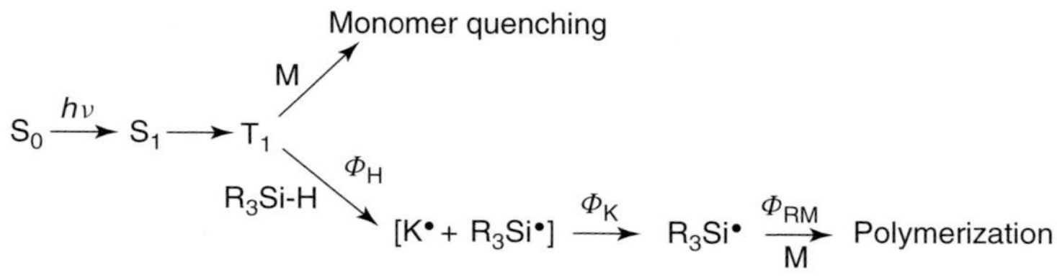

```
                    Monomer quenching
                           M
         hν             ↗
    S₀ ──→ S₁ ──→ T₁
                    ↘  Φ_H
                  R₃Si-H          Φ_K              Φ_RM
                        ↘ [K• + R₃Si•] ──→ R₃Si• ─────→ Polymerization
                                                    M
```

Scheme 18.4

and the excited singlet and triplet states of the PI, respectively; K• stands for the ketyl-type radical [98, 99] (Section 9.1.2).

The production of the initiating silyls is dramatically enhanced under air, according to the set of reactions shown in Eq. (18.5) [98, 100].

$$R_3Si^\bullet + O_2 \rightarrow R_3Si-O_2^\bullet \quad (1)$$
$$R_3Si-O_2^\bullet + R_3Si-H \rightarrow R_3Si-O_2H + R_3Si^\bullet \quad (2)$$
$$R_3Si-O_2^\bullet \rightarrow [R_2Si(-OR)-O^\bullet] \rightarrow RSi^\bullet(OR)_2 \quad (3)$$
$$RM-O_2^\bullet + R_3Si-H \rightarrow RM-O_2H + R_3Si^\bullet \quad (4)$$
$$R_3Si-O-O-R' \rightarrow R_3Si-O^\bullet + {}^\bullet O-R'\,(h\nu) \quad (5)$$
$$(18.5)$$

The peroxylation process (1) consumes oxygen with high rate constants close to $3-4 \times 10^9$ mol^{-1} l s^{-1} (Table 18.9). The reaction (2) regenerates another silyl radical. For comparison, reaction (2) for the α-aminoalkylperoxyl radical/amine interaction is rather slow (typical values < 20 mol^{-1} l s^{-1}; see Section 18.5). A particular interest of $R_3Si-O_2^\bullet$ is the possibility of a rearrangement (3) that recreates a silyl radical. The rather short lifetime (2.5 μs for the tris(trimethylsilyl)silylperoxyl) compared to the values reported for usual peroxyl (in the millisecond range for α-aminoalkylperoxyls) supports this very efficient reorganization. The process is highly exothermic in agreement with the formation of two Si-O bonds. It involves a transient intermediate $R_2Si(-OR) - O^\bullet$. The addition of $RSi^\bullet(OR)_2$ to an acrylate double bond remains very efficient as exemplified by the addition of the triethoxysilyl radical to MA (5.5×10^8 mol^{-1} l s^{-1}). Lastly, reaction (4) efficiently contributes to (i) the destruction of the peroxides formed on the growing polymer chains and (ii) the generation of new initiating radicals. The conversion of peroxyls (ROO•, R being an initiating or a propagating radical) to new initiating silyl radicals (2) and (4) is the major pathway explaining the ability of organosilanes to overcome the oxygen inhibition. A direct relationship between the ROO•/R_3SiH rate constants and the polymerization-initiating performance in aerated conditions was found in [97, 100]. Under polychromatic irradiation, the photodecomposition of the formed peroxides or hydroperoxides (5) can be achieved. Then, the generated oxyl radicals easily abstract a hydrogen atom from the silane, thereby regenerating silyl radicals.

Reaction (1) competes with the initiating silyl radical/monomer reaction. According to the rate constants of the different primary routes, many situations (either beneficial or detrimental) concerning the oxygen effect might be encountered as reactions (1–5) are affected by the silyl structures. All the reactions shown in (18.5) account well for the usually observed oxygen low sensitivity and beneficial effect in photopolymerization (Section 9.1.2.7).

18.9.3
Other Sources of Silyl Radicals

Silyl radicals can also be generated by the cleavage of disilane ($R_3Si-SiR_3$) as Type I PIs. The Si–Si photodissociation is well known in polysilanes [85–92] but leads to quite low polymerization initiation quantum yields. This behavior was ascribed to (i) the low efficiency of the cleavage process and (ii) the presence of by-side reactions of the silyls, preventing any efficient initiation. Some organic precursors that show convenient light absorptions in the 300–400 nm range were more recently proposed 7 [101–103].

For example, the efficiency of PI-Si$_1$ (for the same amount of absorbed light) is higher than that of the well-known PI DMPA. This was explained as resulting from a combination of factors: (i) a high dissociation quantum yield (often close to 1; two silyl radicals were generated for each absorbed photon), (ii) a high intrinsic reactivity of the generated radicals (i.e., the addition rate constant to MA is 8.1×10^8 mol^{-1} l s^{-1} for the silyl radical of PI-Si$_1$ compared to $\sim 10^5$ mol^{-1} l s^{-1} for the benzoyl radical generated in DMPA), and (iii) a quite intense UV absorption.

For PIs incorporating a ketone chromophore (BP in PI-Si$_6$ or TX in PI-Si$_9$), a more complex initiation mechanism was found. Silyl radicals are still generated by a Si–Si bond cleavage. However, a second pathway involving a hydrogen abstraction reaction between the BP triplet state and a SiCH$_2$-H fragment occurs. The PI ability of PI-Si$_6$ is very similar to that of DMPA.

The cleavage of Si-S and Si-C bonds was also reported in PI-Si$_3$ and PI-Si$_4$. Recently, other PIs based on silyloxyamines were mentioned [104]. As already stated, the input of the silyl radical chemistry to the PI modification should bring new promising developments.

18.10
Oxyl Radicals

Oxyl RO$^\bullet$ and peroxyl ROO$^\bullet$ radicals have considerable interest as their reactions lead to chain propagation, chain branching, or chain termination. They play an important role in polymer chemistry (e.g., in (photo)polymerization reactions under air, (photo)oxidation, and (photo)degradation of polymer materials). In the condensed phase, several kinds of processes between these radicals and additives are known to be decisive, for example, electron transfer, addition to double bonds, and hydrogen abstraction.

Peroxyl radicals appear in any reaction involving free radicals in an aerated medium. Peroxides ROORs and hydroperoxydes ROOHs are generated from the recombination of peroxyls in the polymer matrix and in HT reactions between a peroxyl and an H-donor, respectively (Section 18.11). On photolysis or thermolysis, ROOR' and ROOH (Eq. (18.6)) undergo a O-O bond cleavage. The decomposition of (hydro)peroxides is an interesting way to generate new initiating species (OH$^\bullet$, RO$^\bullet$, and D$^\bullet$ radicals formed in RO$^\bullet$ + DH → ROH + D$^\bullet$) and to overcome the oxygen inhibition in a polymerization.

$$ROOR'(ROOH) \rightarrow RO^\bullet + R'O^\bullet \,(OH^\bullet)\, (h\nu) \tag{18.6}$$

The photodecomposition of ROOR' and ROOH can also be sensitized by energy or electron transfer (Eq. (18.7)) assuming that the excited state energy and/or the redox properties of the photosensitizer are adequately selected.

$$PS^* + ROOR' \rightarrow PS +^* ROOR' \rightarrow PS + RO^\bullet + R'O^\bullet$$
$$PS^* + ROOR' \rightarrow PS^{\bullet+} + RO^- + R'O^\bullet \tag{18.7}$$

Alkoxyl radicals can hardly be investigated by steady-state ESR (these radicals are often ESR silent) or LFP (transient absorption at $\lambda_{max} < 300$ nm). These structures are, however, easily characterized by ESR-spin trapping (*hfc* being

18 Reactivity of Radicals toward Oxygen, Hydrogen Donors, Monomers, and Additives

Table 18.10 Interaction rate constants between $tBuO^\bullet$ and different H-donors: amines (2), thiols (4), and silanes (6).

Produced radical	k_H (10^7 mol^{-1} l s^{-1})	Produced radical	k_H (10^7 mol^{-1} l s^{-1})	Produced radical	k_H (10^7 mol^{-1} l s^{-1})
A_1	11	S_1	100	Si_1	1.0
A_2	20	S_2	220	Si_2	1.5
A_3	0.3	S_3	210	Si_3	6.7
A_4	0.8	–	–	Si_4	8.5
A_5	6	–	–	Si_5	2.5
A_6	6	–	–	Si_6	2.5
A_7	18	–	–	Si_7	2.2
A_8	15	–	–	Si_8	3.0
A_9	23	–	–	Si_9	1.5
A_{10}	38	–	–	–	–
A_{11}	20	–	–	–	–

See text. For ligated boranes, some typical values are given below in Table 18.15.

$a_N \sim 14$ G; $a_H \sim 1.8$ G for $tBuO^\bullet$/PBN [105–107]). A further β-fragmentation of $RO^\bullet$ is sometimes observed, leading to a ketone and an alkyl radical (Section 18.1).

Alkoxyl radicals are usually π structures with a high spin density at the oxygen atom (Figure 18.10; see a comparison with alkylperoxyls in Section 18.11). $RO^\bullet$ are electrophilic radicals and can hardly be oxidized (e.g., by iodonium salts). These structures do not abstract the halogen atom from alkylhalides but add to monomers. The $tBuO^\bullet$/MA addition rate constant remains relatively low: $k_i = 5 \times 10^2$ mol^{-1} l s^{-1} [11]. Peroxides can initiate a photopolymerization at $\lambda <$ 320 nm : $RO^\bullet$ and alkyl radicals generated from the $RO^\bullet$ fragmentation (Section 18.1) being the initiating species. $RO^\bullet$ are mainly involved in hydrogen abstraction reactions with H-donors (alkanes, amines, thiols, silanes, and boranes). Some selected rate constants are given in Table 18.10 in the case of $tBuO^\bullet$ [108].

When considering phenols used as stabilizers, high rate constants are found for the phenoxy radical formation ($>10^8$ M^{-1} s^{-1} for the $tBuO^\bullet$/phenol interaction). The generated phenoxy radicals are quite stable and do not participate in the initiation process.

18.11
Peroxyl Radicals

Peroxyls can be investigated by steady-state ESR ($g \sim 2.015$, usually single line) and many rate constants were determined by kinetic ESR [109–114] (Figure 18.9). Peroxyls are hardly observed by LFP ($\lambda_{max} <$ 320 nm). On the contrary, α-aminoalkyl peroxyls exhibit an easily detectable near-visible absorption [115]. These structures

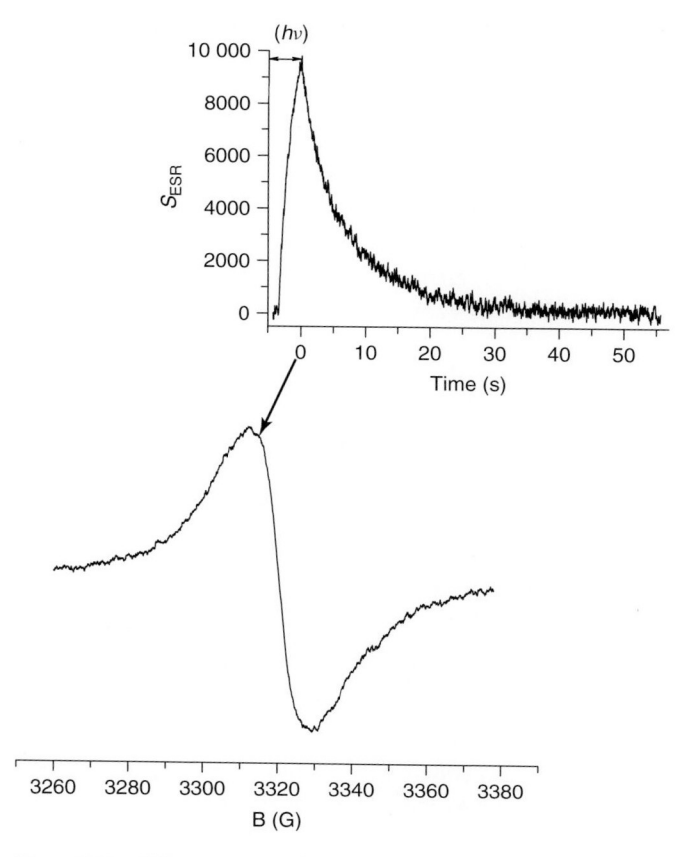

Figure 18.9 ESR spectrum and decay of tBuOO• at RT. (For more detail, see [118].)

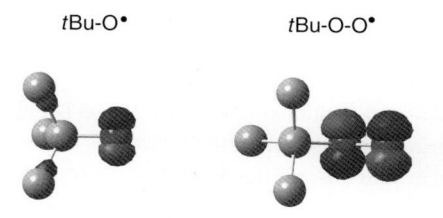

tBu-O•   tBu-O-O•

Figure 18.10 SOMOs for representative alkoxyl or peroxyl radicals (DFT level) (J. Lalevée, unpublished data).

are also characterized by ESR-spin trapping (*hfc* being $a_N \sim 13.5$ G; $a_H \sim 1.8$ G for tBuOO• using PBN [105–107, 116, 117]).

Most (carbon-centered) radicals react with molecular oxygen with rate constants close to the diffusion-controlled limit ($2-4 \times 10^9$ mol^{-1} l s^{-1}) to form ROO• (Eq. (18.8)) [15]. After hydrogen abstraction with a hydrogen donor YH, the generated ROOH can further cleave into RO• and OH• (Section 18.10), whereas the concomitantly formed carbon-centered radical Y$^\bullet$ adds to oxygen, leading to a new peroxyl radical YOO$^\bullet$.

$$R^\bullet + O_2 \rightarrow ROO^\bullet \quad (1)$$
$$ROO^\bullet + Y\text{-}H \rightarrow ROOH + Y^\bullet \quad (2)$$
$$Y^\bullet + O_2 \rightarrow YOO^\bullet \quad (3)$$
$$ROOH \rightarrow RO^\bullet + OH^\bullet \quad (4)$$
$$RO^\bullet(OH^\bullet) + Y\text{-}H \rightarrow ROH\,(H_2O) + Y^\bullet \quad (5)$$
$$ROO^\bullet + ROO^\bullet \rightarrow ROOR + O_2 \quad (6) \quad\quad (18.8)$$

Peroxyl radicals are stabilized by a π orbital structure with a significant delocalization between the two oxygen atoms (Figure 18.10). The spin density is about 60–70% for the terminal atom and 40 to 30% for the adjacent oxygen. The situation is quite different for alkoxyl radicals, which exhibit strong spin localization on the oxygen. This explains quite well the change of the alkoxyls versus alkylperoxyls reactivity (see below).

For thermodynamic reasons, peroxyl radicals do not all have the same reactivity as the O–H BDEs of hydroperoxides ROO–H are not all equal. This is readily understood on consideration of the two canonical structures ROO$^\bullet$ and RO$^{\bullet+}$O$^-$, where R has a strong influence on the spin density for the terminal oxygen (δ). In tBuOO$^\bullet$, δ decreases along with the the reactivity of the corresponding peroxyl [113, 114] as a result of the inductive electron-donating ability of tBu and the conjugative electron delocalization in RO$^{\bullet+}$O$^-$. On the contrary [119], the reactivity increases with a more important contribution of ROO$^\bullet$ (increase in the withdrawing character of R and the consequent increased localization of the unpaired electron). For example, when going from Cl$_3$COO$^\bullet$ to (CH$_3$)$_3$COO$^\bullet$, the hydrogen abstraction rate constant on cyclohexane decreases by six orders of magnitude.

18.11.1
Interaction with H-Donors

Reaction mechanisms of peroxyl radicals in solution and in films have been extensively studied. Data gained from various indirect methods as well as some absolute rate constants (mostly determined in aqueous solution by pulsed radiolysis) have been compiled [120–122]. Recently, the α-aminoalkylperoxyls radical were shown to be good representatives of alkylperoxyls: in fact, the spin density for the terminal oxygen are similar in TEA$-$OO$^\bullet$ and tBuOO$^\bullet$ ($\delta = 0.669$ and 0.682, respectively). The direct observation of TEA$-$OO$^\bullet$ and tBuO$^\bullet$ in LFP [115, 119] allows direct and easy measurement of the interaction rate constants ($k_{RO^\bullet}$, $k_{ROO^\bullet}$) with different additives (Table 18.11). This has been achieved with antioxidants (HQME and Vitamin E (VIE)), compounds bearing abstractable hydrogens (thiol MBO, laurylaldehyde (LH), ethylbenzene (EtBz)), monomers (ethyl linoleate (EL) known to polymerize by an autoxidation process, MA, VE), radical inhibitors (triphenylphosphine (TPP) and diphenyliodonium salt).

Table 18.11 Interaction rate constants ($k_{RO\bullet}$, $k_{ROO\bullet}$) in $mol^{-1}\ l\ s^{-1}$ between $tBuO^\bullet$ or $TEA-OO^\bullet$ with various substrates. See text.

	HQME	VIE	LH	MBO	EtBz	EL	TPP
$tBuO^\bullet$	9×10^8	3.1×10^9	2.8×10^7	1.0×10^9	2.1×10^6	9.5×10^6	1.1×10^9
$TEA-OO^\bullet$	6.1×10^4	1.1×10^6	190	1.8×10^5	<300	<740	1.4×10^4

From [115].

Interestingly, $tBuO^\bullet$ radicals are much more reactive than $TEA-OO^\bullet$ peroxyls (between 3000 and 15 000 times for the hydrogen abstraction reaction from HQME or VIE). A similar trend is observed when considering MBO, LH, EtBz, and EL: the rate constant ratios $k_{RO\bullet}/k_{ROO\bullet}$ are 5500, 147 000, >7000, and >12 800, respectively. This result was ascribed to the lower exothermicity of the hydrogen abstraction with peroxyls compared to alkoxyls [115], which is connected with the lower BDE of TEA-OOH (~86 kcal mol^{-1}) versus $tBuOH$ (104.8 kcal mol^{-1}). Interestingly, a rough $k_{RO\bullet}$ versus $k_{ROO\bullet}$ correlation is noted, thereby demonstrating that a similar factor (ΔH_r) affects their reactivity. The lack of a quantitative correlation can be ascribed to the already reported polar effects [120–123], the HT step being separated in an electron-transfer/proton-transfer sequence.

Recently, kinetic ESR was used to follow the $tBuOO^\bullet$/H-donor DH interaction at room temperature (Table 18.12) [118] for couples exhibiting low rate constants. This interaction still affected the BDE(D-H). In the series of investigated DHs (amines,

Table 18.12 $tBuOO^\bullet$/H-donor interaction rate constants and BDE values for the H-donor cleavable bond.

H-donors	$k\ (tBuOO^\bullet)(mol^{-1}\ l\ s^{-1})$	BDE (kcal mol^{-1})
Triethylamine	29	91.2
Methyldiethanolamine	23	87.1
Ethyldimethylaminobenzoate	6	92.8
Triisobutylamine	22	92.8
Tris(trimethylsilyl)silane	590	79.8
Diphenylsilane	20	–
Tris(trimethylsilyl)germane	1790	75.7
Triphenylgermane	260	79.9
Triphenylstannane	5200	73.8

From KESR experiments (at RT).
From [115, 118, 119].

silanes, germanes, stannanes, phosphites, phosphine oxide, and phosphines), the lower the BDE, the higher the tBuOO$^\bullet$/DH reaction rate constant is. Some polar effects, however, likely take place and explain the deviation observed when plotting $k_{tBuOO^\bullet}$ versus BDE.

18.11.2
Interaction with Monomers

In the addition reaction to double bonds (MA, VE), the TEA−OO$^\bullet$ lifetime does not change even for [MA] and [VE] = 2 M. This leads to upper values for the addition to both double bonds $\sim 1 \times 10^2$ mol^{-1} l s^{-1}. However, the MA and VE being electron-poor and electron-rich monomers, respectively, the suspected electrophilic character of the peroxyl radicals should likely lead to a reactivity difference.

18.11.3
Interaction with Triphenylphosphine

Triphenylphosphine (TPP), which is often used to reduce the oxygen inhibition in radical polymerizations (Section 7.5), rapidly forms a pentavalent oxide (O=PR$_3$) in the presence of oxygen through a free-radical-chain process, oxyl and peroxyl radicals being the intermediates as described in Eq. (18.9). The tBuO$^\bullet$ radical is 10^5 more reactive than TEA−OO$^\bullet$ ($k \sim 10^9$ and 10^4 mol^{-1} l s^{-1}, respectively). These reaction converts peroxyls into oxyls and then oxyls into phosphoranyls RO−P$^\bullet$ R$_3$ [124] that are inefficient radicals for the addition to double bonds ($k_i <$ 10^4 mol^{-1} l s^{-1}) but further consume oxygen ($k \sim 4.1 \times 10^9$ mol^{-1} l s^{-1}) [119]. Phosphoranyls can be readily observed by LFP [125]. The formation of RO−P$^\bullet$ R$_3$ explains the ability of TPP to overcome the oxygen inhibition in free radical photopolymerization (FRP). Quantum mechanical calculations show that (i) the direct formation of O=PR$_3$ is feasible in reaction (1) of Eq. (18.9): this suggests that the cleavage process in the ROO−PR$_3^\bullet$ radical adduct is probably ultrafast and (ii) the phosphoranyl radical RO−PR$_3^\bullet$ in (2) and (3) is a stable intermediate that reacts with O$_2$.

$$R'OO^\bullet + PR_3 \to [ROO\text{-}PR_3^\bullet] \to RO^\bullet + O\text{=}PR_3 \quad (1)$$
$$R'O^\bullet + PR_3 \to RO\text{-}PR_3^\bullet \quad (2)$$
$$RO\text{-}PR_3^\bullet + O_2 \to ROO^\bullet + O\text{=}PR_3 \quad (3) \quad\quad (18.9)$$

18.11.4
S$_H$2 Substitution

Homolytic substitution reactions (S$_H$2) can be a way for the decomposition of peroxyls. Indeed, it is known that the direct addition of ROO$^\bullet$ to boranes, organosilanes, and zirconium-based compounds YRR'R'' (Eq. (18.10)) is followed by a Y−R bond cleavage [126]. The driving force is often the formation of a metal–oxygen bond which is more energetic than the metal-carbon bond.

$$ROO^\bullet + YRR'R'' \rightarrow [ROO\text{-}YR'R] + R^\bullet \tag{18.10}$$

A low rate constant is measured ($<10^2$ mol^{-1} l s^{-1}) when employing butyldiisopropoxyborane. The oxygen atoms lead to an increase in the electron density at the boron atom through a back coordination from oxygen to boron to form a π bond (B–OR $\leftrightarrow$ B$^-$=O$^+$R). Although peroxyl radicals are normally considered as electrophilic species, the attack on the boron atom via its p-vacant orbital (which is involved in the π bond system) is less favorable. This leads to a slower process using butyldiisopropoxyborane contrary to tributylborane where a rate constant of 2×10^6 mol^{-1} l s^{-1} has been determined.

A very efficient process takes place between bis(cyclopentadienyl) dimethylzirconium and TEA–OO$^\bullet$ (1.5×10^4 mol^{-1} l s^{-1}) or tBuO$^\bullet$ (5.7×10^7 mol^{-1} l s^{-1}). In the zirconium propoxide Zr(OPr)$_4$/TEA–OO$^\bullet$ system, a relatively high rate constant is found (5×10^2 mol^{-1} l s^{-1}): this process is unexpectedly efficient as in that case, the cleavage of the Zr–O bond regenerates an oxygen-centered radical.

18.11.5
Other Oxyls and Peroxyls

Oxyl and peroxyl radicals generated in other atom-centered structures, for example, Si, Ge, and Sn have been studied. The peroxyl ESR spectra are affected by the metal center, that is, for silylperoxyls or germylperoxyls, $g \sim 2.02\text{–}2.025$ [127]. The interaction rate constants for germylperoxyls (e.g., TMS$_3$GeOO$^\bullet$) are relatively similar for the hydrogen abstraction with phenols to those of alkylperoxyls (e.g., $k = 1.1 \times 10^5$ and 5.5×10^6 mol^{-1} l s^{-1} for TMS$_3$GeOO$^\bullet$/MEHQ and TMS$_3$GeOO$^\bullet$/VIE) [100]. The procedure could easily be extended to a large range of reactions involving such peroxyl radicals encountered in photopolymerization (FRP and FRPCP).

18.12
Aminyl Radicals

Aminyl radicals R$_2$N$^\bullet$ with R = alkyl are usually rather difficult to observe by LFP. On the other hand, compounds with R = Ar exhibit a quite convenient visible light transient absorption [128]. R$_2$N$^\bullet$ structures were also characterized by ESR-spin trapping experiments [36].

Aminyl radicals can be generated by a N–H hydrogen abstraction reaction. This situation is often observed in hindered amine light stabilizer (HALS) derivatives (Section 7.6) [129] and is well exemplified by 2,2,6,6-tetramethylpiperidine (TMP) which does not possess a α(C–H) bond. Aminyl radicals N$_1$–N$_3$ in 8 are usually considered as electrophilic. These species can hardly be oxidized and are characterized by quite low addition rate constants to multiple bonds (see below).

A regioselectivity of the hydrogen abstraction reaction is noted when using compounds bearing both α(C–H) and N–H bonds (Section 18.5). The hydrogen abstraction in co-initiators such as primary or secondary amines could take place at the nitrogen or the α-carbon, leading to aminyl and aminoalkyl radicals, respectively. Both (i) the relative yield in aminoalkyls and aminyls and (ii) the aminyl versus aminoalkyl reactivity (C_1 vs N_1 and C_2 vs N_2 in **8**) must be considered. A recent paper [36] based on ESR and LFP clearly shows the presence of the aminoalkyl radical and the significant absence of any aminyl radical in primary and secondary amines.

From quantum mechanical calculations, it appears that aminoalkyl radicals C_1 and C_2 are 10^4 and 10^6 times more reactive toward MA than the corresponding aminyl radicals N_1 and N_2, respectively: in a general way, k_i (aminoalkyls) $\sim 10^7$ mcl^{-1} 1 s^{-1}, k_i (aminyls) $< 10^4$ mol^{-1} 1 s^{-1}. Aminyl radicals where the spin density δ values are systematically very low do not exhibit any particular nucleophilic or electrophilic character toward MA. The aminoalkyl radicals can thus be safely considered as the most efficient initiating species to start a polymerization.

Aminyl radicals can be involved in hydrogen abstraction reaction (particularly with phenols where high rate constants are usually observed). Radical N_3 is characterized by a very high reactivity (at least 200 times higher compared to that of the aminoalkyls) toward a stabilizer (4-methoxyphenol or HQME); $k = 1.1 \times 10^8$ mol^{-1} 1 s^{-1} [36]. This result is reflected by a higher value of the calculated BDE: 95.9 versus $\sim$91 kcal mol^{-1} for the BDE(N–H) of TMP and the BDE(αC–H) of cyclohexylamine or diethylamine, respectively. The BDE(O–H) of HQME being 82.8 kcal mol^{-1}, this ensures a higher exothermicity and in turn, a higher interaction rate constant.

18.13
Germyl and Stannyl Radicals

The behavior of germyl $R_3Ge^\bullet$ and stannyl $R_3Sn^\bullet$ radicals shown in **9** are strongly related to that of silyl radicals. These species can efficiently abstract halogen atoms and are also often used in the reduction process of alkyl halides in organic chemistry. Germyls are interesting polymerization-initiating species (Sections 8.17 & 9.1.2.11).

18.13 Germyl and Stannyl Radicals

[Structures: Ge₁, Ge₂, Sn₁ labeled as 9]

These radicals can be generated by hydrogen abstraction reactions on the parent hydrogenated compounds R_3XH (X = Ge or Sn), the BDE(X-H) being relatively low (e.g., BDE(Ph_3Ge-H) and BDE(Ph_3SN-H) are 79.9 and 73.8 kcal mol^{-1}, respectively) and usually lower than those of the corresponding silanes (e.g., BDE(Ph_3Si-H) ~92 kcal mol^{-1}) [6]. The formation of germyl radicals also arises in the cleavage of acylgermanes (or mono(germyl)ketones) or bis(germyl)ketones [101, 130–135] (Part 2). Germyls can also be observed and characterized by steady-state ESR (g ~2.009–2.01) [136] and ESR-spin trapping (hfc being a_N ~14.8 G; a_H ~ 6.5 G for $Ph_3Ge^{\bullet}$ using PBN [137]). They are also easily detected by LFP as their absorption spectra are red-shifted. Few kinetic data are, however, available.

18.13.1
Reactivity

Some representative data for the formation of germyl and stannyl radicals 9 are given in Table 18.13. The high hydrogen abstraction rate constants k_H (with ^{3}BP, $tBuO^{\bullet}$, or $tBuOO^{\bullet}$) and ketyl radical quantum yields clearly outline the labile hydrogen character of R_3XH. The k_H values are similar to those usually found with amines (in the 10^8 mol^{-1} l s^{-1} range) and higher for R_3GeH and R_3SnH compared to R_3SiH in agreement with the lower BDE (Section 18.11).

Germyl radicals are very reactive, as shown in Table 18.14, toward O_2 (almost a diffusion-controlled reaction) and MA (k_i ranging from 3.4×10^7 to 1.8×10^8 mol^{-1} l s^{-1}). They exhibit a rather low selectivity. Their k_i are very high when compared to classical carbon-centered radicals (Section 18.1). Owing to

Table 18.13 Rate constants and radical quantum yields in ^{3}BP (or $tBuO^{\bullet}$, $tBuOO^{\bullet}$)/germane (or stannane) couples in benzene.

Produced radical	BDE^a (kcal/mol)	k_H($tBuO^{\bullet}$) (10^7 mol^{-1} l s^{-1})	k_H($tBuOO^{\bullet}$) (mol^{-1} l s^{-1})	k_H(^{3}BP) (10^7 mol^{-1} l s^{-1})	$\Phi_R{}^b$
Ge₁	75.7	45.0	1800	105	0.81
Ge₂	79.9	34.0	260	6.5	0.69
Sn₁	73.8	153.0	5200	46.0	0.74

^{a}Bond dissociation energy calculated at the UB3LYP/LANL2DZ level.
^{b}Quantum yields in germyl or stannyl radicals (equal to the ketyl radical quantum yields).

Table 18.14 Reaction rate constants of the germyl and stannyl radicals with O_2, MA, and Ph_2I^+ PF_6^-.

Radical	O_2 k_{O_2} (10^9 mol^{-1} l s^{-1})	MA k_i (10^7 mol^{-1} l s^{-1})	Ph_2I^+ PF_6^- k_{ox} (10^7 mol^{-1} l s^{-1})
Ge_1	0.39	3.4	<2.0
Ge_2	6.0	18	5.5
Sn_1	4.5	4.8	3.1

their low ionization potential (e.g., IP = 5.6 eV for $Ph_3Ge^\bullet$), germyls are easily oxidized by Ph_2I^+ and can be used in FRPCP (see Sections 12.5). All these results demonstrate the high potential of the $R_3Ge^\bullet$ to act as radical initiating species. The toxicity concern for stannyls prevents any large use of this radical chemistry.

18.13.2
Reactivity under Air

The most striking result is the reduced oxygen inhibition when using ketone (e.g., camphorquinone)/R_3GeH couples [138]). This observation can be put together with the similar behavior already noted for silanes (Section 18.9): (i) formation of a germylperoxyl $R_3GeOO^\bullet$, (ii) hydrogen abstraction between $R_3GeOO^\bullet$ or a peroxyl $ROO^\bullet$ (with R a propagating radical) and a germane, and (iii) rearrangement of $R_3GeCO^\bullet$ recreating a germyl radical $Ge^\bullet$ (the rate constants are usually lower than for the silylperoxyls: this was ascribed to the lower rearrangement exothermicity for the Ge derivatives [100, 139]). These reactions ensure fast oxygen consumption and the concomitant generation of efficient germyl radicals. Better cured surfaces are obtained.

18.13.3
Reactivity and Structure of $(TMS)_3Ge^\bullet$ versus $(TMS)_3Si^\bullet$

A comparison of the tris(trimethylsilyl)silyl Si_4 **6** and tris(trimethylsilyl)germyl Ge_1 **9** radical reactivity was done to highlight the difference between the germyl and the silyl radical chemistries [100]. Whereas $R_3C^\bullet$ radicals only need the 2s and 2p atomic orbitals to accommodate the valence electrons, $R_3Si^\bullet$ and $R_3Ge^\bullet$ can use the s, p, and d orbitals. As characterized by DFT calculations, the Ge_1 and Si_4 radicals exhibit a pyramidal structure. The bent out-of-plane (σ-type structure) of the germyl radical is more pronounced than in a silyl. This was ascribed to the 3d orbitals that are already occupied and the possible extension to the 4d orbital. The calculated spin density on the radical center is 0.85 and 0.98 in Si_4 and Ge_1, respectively. This demonstrates that the maximum of delocalization is observed in the Si derivative.

The starting compound $(TMS)_3GeH$ is more reactive than $(TMS)_3SiH$ toward t-butoxyl, t-butylperoxyl, and phosphinoyl radicals as well as aromatic ketone triplet states (Table 18.13). This is in line with the higher labile character of R_3GeH reflected by the lower BDE ($\Delta BDE \sim 5$ kcal mol^{-1} between $(TMS)_3GeH$ and $(TMS)_3SiH$).

Ge_1 and Si_4 exhibit a comparable reactivity toward electron-poor monomers (AN and MA). Si_4 is more reactive than Ge_1 toward electron-rich monomers (VA and VE) in agreement with its higher absolute electronegativity (4.06 vs 3.70 eV). Therefore, Si_4 has a higher electrophilic character. In the halogen abstraction from CBr_4, the rate constant is higher with Si_4 (34×10^7 vs 8×10^7 mol^{-1} l s^{-1}) in line with the higher BDE(Si–Br) rendering the abstraction process more exothermic. Germyl radicals are oxidized by aryliodonium salts (see Section 12.5). A quite high rate constant (0.26×10^7 mol^{-1} l s^{-1}) is obtained for Si_4/Ph_2I^+. Ionization potentials are low (6.0 and 6.36 eV for Ge_1 and Si_4, respectively). The interaction rate constants of germyl and silyls with O_2 are very high (2×10^9 and 3.9×10^8 mol^{-1} l s^{-1} for Si_4 and Ge_1). The lifetime of $(TMS)_3SiOO^\bullet$ is short (~ 2.5 μs) compared to $(TMS)_3GeOO^\bullet$ (>5 ms): the rearrangement of the germylperoxyl is slower than that of the silylperoxyl. This was ascribed to the TMS group migration, which is more exothermic for the silylperoxyl (-110.5 kcal mol^{-1}) than for the germylperoxyl (-81.3 kcal mol^{-1}).

18.14
Boryl Radicals

Boron centered radicals (or boryls) L $\rightarrow$ $BH_2^\bullet$ have been recently introduced in polymerization photoinitiating systems. Boryls were first generated in the 80th by a hydrogen abstraction reaction from the corresponding ligated amine-($R_3N \rightarrow BH_3$) or phosphine-($R_3P \rightarrow BH_3$) boranes [140–145]. These boranes have, however, a high BDE(B-H) that prevents any efficient hydrogen abstraction. To overcome this limitation, for example, when used as co-initiators, other boranes have been recently proposed: (i) N-heterocyclic carbene-boranes (NHC-boranes) [146–149] and (ii) N-heteroaryl-boranes [150, 151]. Through adequate substitutions, the BDE(B-H) can be let down to about 70–85 kcal mol^{-1} compared to 92–105 kcal mol^{-1} for amine or phosphine-boranes.

Amine-boryls (e.g., $Et_3N \rightarrow BH_2^\bullet$) are σ radicals as evidenced by ESR [140–145]. In NHC-boryls and N–Heteroaryl-boryls, π structures are found [146–151] in agreement with a significant delocalization of the spin density in the ligand (Figure 18.11).

Boryl radicals are (i) observed by steady-state ESR at relatively low temperatures [140–145, 149], (ii) easily characterized by ESR-spin trapping (hfc being e.g., $a_N \sim 15.2$ G; $a_H \sim 3.2$ G and $a_B \sim 3.2$ G for $Et_3N-BH_2^\bullet$ with PBN [140, 152]), and (iii) detected in LFP experiments: as a function of the ligand, the transient absorptions occur in the UV (for L = amine) or the visible (L = pyridine) wavelength range [153].

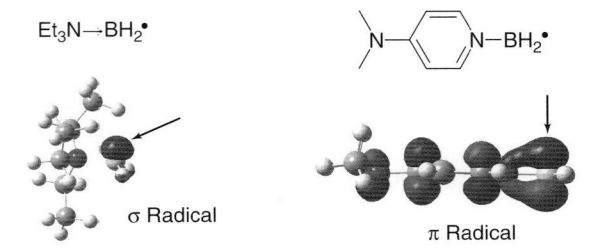

Figure 18.11 SOMOs for representative boryls at DFT level. Boron is indicated by arrow. (Source: From [150, 151].)

18.14.1
Reactivity

The interaction of the borane complexes B_1H to B_8H, where B_1 to B_8 are the corresponding boryl radicals **10** with a ketone triplet state (^3BP or ^3TX), leads to an efficient hydrogen abstraction and generates a ketyl radical and L → $BH_2^\bullet$ [150–152]. Quite high boryl radical quantum yields were measured.

The $^3BP/L$ → BH_3 interaction probably corresponds to a PHT rather than an electron/proton transfer sequence. Compared to classic aliphatic or aromatic amines, the lower values measured for compounds having a N → B bond (Table 18.15) demonstrate that the complexation depletes the electron density on the amine moiety by engaging the nitrogen lone pair. For B_5H to B_7H, the reaction occurs on the borane moiety as aminoalkyl radicals are not observed in ESR-ST. In NHC boranes (e.g., B_8), the BDE(B-H) is also quite low [146–149].

The hydrogen abstraction reactions of the boranes with oxygen-centered radicals (both oxyl and peroxyl) are quite efficient in agreement with the low BDE(B-H) for B_2H, B_3H, B_5H to B_8H: only B_1H and B_4H characterized by BDE > 90 kcal mol^{-1} exhibit interaction rate constants with $tBuO^\bullet$ < 10^8 mol^{-1} l s^{-1}. Interestingly, the

Table 18.15 Formation rate constants k_H of L → $BH_2^{\bullet}$; oxidation potentials E_{ox}; boryl radical quantum yields ($\Phi_{B\bullet}$) and B-H bond dissociation energy. See text.

Produced radical	E_{ox} (eV)	BDE (B-H) (kcal/mol)	k_H (tBuO$^{\bullet}$) (10^7 mol^{-1} l s^{-1})	k_H (^{3}BP) (10^7 mol^{-1} l s^{-1})	k_H (tBuOO$^{\bullet}$) (mol^{-1} l s^{-1})
B_1	1.5	101.1	5.9	9.1 (0.73^a)	~8
B_2	1.0	71.9	27	28 (0.99^a)	200
B_3	0.6	67.3	23	720 (0.71^a)	420
B_4	>1.8	90.7	6.6	17 (0.5^a)	180
B_5	0.55	81.3	28	160 (0.97^a)	–
B_6	0.44	82.2	31	–	–
B_7	0.57	81.9	22	–	–
B_8	–	81.9	26	96	–

^{a}Boryl or ketyl radical quantum yield ($\Phi_{B\bullet}$).

Table 18.16 Interaction rate constants of L → $BH_2^{\bullet}$ radicals with various additives at RT in acetonitrile/di-tert-butylperoxide solvent (k in mol^{-1} l s^{-1}).

Radical	k_i^a (MA) 10^7	k_i^a (EVE) 10^5	k_i' (O$_2$) 10^8	k_{Cl} (CHCl$_3$) 10^6	k_{Cl} (CH$_2$Cl$_2$) 10^6	k_{Cl} (C$_3$H$_7$I) 10^7	k_{ox} (Ph$_2$I$^+$) 10^9
B_1	13	<1	>7	–	120	–	2.4
B_2	<1	<1	~5	–	<0.1	–	1.1
B_3	<0.5	<0.5	>7	–	<0.1	–	1.9
B_4	0.75	0.75	>10	–	–	–	–
B_5	3.5	<1	>7	1.5	<0.1	3.2	2.5
B_6	14	–	>7	3.8	<0.5	5.5	2.0
B_7	2.4	–	>7	0.4	–	0.3	0.5
B_8	3.8	–	>6	6	–	18	0.8

^{a}MA and EVE stand for methylacrylate and ethylvinylether, respectively.

BDE(B-H) can be finely tuned by an appropriate selection of the ligand. The different Lewis base coordination can strongly stabilize the formed boryls leading to a strong decrease in the BDE(B-H) (ΔBDE = 33.8 kcal mol^{-1} from B_1H to B_3H).

Boryls react fast (i) with oxygen (almost diffusion-controlled reactions) to generate borylperoxyls (Table 18.16), (ii) iodonium salts (in the $10^8 - 10^9$ mol^{-1} l s^{-1} range with formation of a borenium L → BH_2^+), and (iii) electron-poor monomers (MA and AN). The addition to electron-rich monomers is quite hard. Boryls have low ionization potentials (4.7–6 eV) [150, 152, 153] like the aminoalkyl radicals [36] and accordingly are considered as nucleophilic radicals.

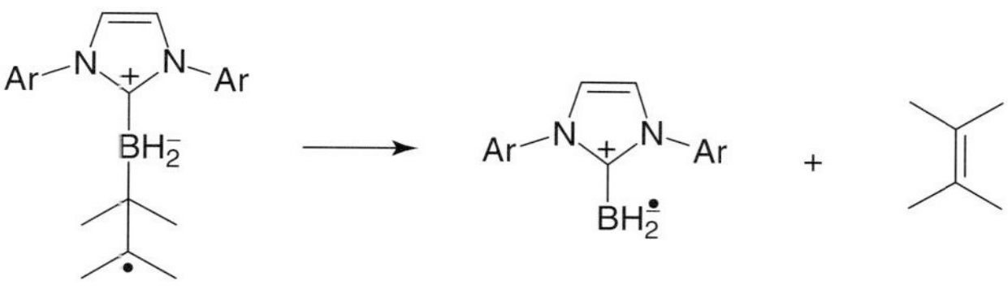

Scheme 18.5

The addition rate constant to MA ranges from 1.3×10^8 for B_1 to $<10^5$ mol^{-1} l s^{-1} for B_3. This strongly evidences that the ligand dramatically governs the boryl reactivity. This is in line with the delocalization of the boryl radical: for example, when going from the triethylamine borane to the quinoline borane complex, the boron spin density strongly decreases (1.04 vs 0.19), leading to more stabilized and less reactive structures [152].

The reversibility of the boryl radical addition to double bonds was recently questioned. The addition to electron deficient monomers (e.g., acrylates) leads to a stabilized radical adduct preventing the β-fragmentation process. In the addition to tetramethylethylene, the radical adduct is destabilized (endothermic reaction) and the β-fragmentation occurs (Scheme 18.5). The addition to an alkyne is efficient as exemplified by the Et$_3$B $\rightarrow$ BH$_2^\bullet$/cyclohexylacetylene reaction, where the rate constant is relatively high (2.2×10^6 mol^{-1} l s^{-1}). Boryls can also abstract a halogen atom from alkyl halides (RI, RBr, RCl) and can be used in organic reduction processes [154].

18.14.2
Reactivity under Air

In the presence of oxygen (Table 18.16), borylperoxyls are produced fast (diffusion-controlled reaction). The k_Hs between tBuOO$^\bullet$ and L $\rightarrow$ BH$_3$, measured by K-ESR (Table 18.15) are noticeably higher ($\sim$200–400 mol^{-1} l s^{-1} for B$_2$H to B$_4$H) than those found for tBuOO$^\bullet$/amine (Section 18.5). This demonstrates the ability of these compounds to convert the alkylperoxyl created during the course of the polymerization under aerated conditions into new initiating boryls. The same reaction should likely occur between a borylperoxyl and L $\rightarrow$ BH$_3$ and explains the excellent polymerization efficiency using these initiating systems under air (Section 9.1.2).

18.14.3
Photoinitiation under Air

Scheme 18.6 shows the plausible picture for the photoinitiation step of the acrylate M polymerization. In laminates, the BP/L $\rightarrow$ BH$_3$ works as a usual Type II photoinitiating system. In the presence of air, specific reactions occur (as with silyl and germyls; see above) and overcome the classic oxygen inhibition. In the BP triplet state, monomer and oxygen quenching may compete with the boryl

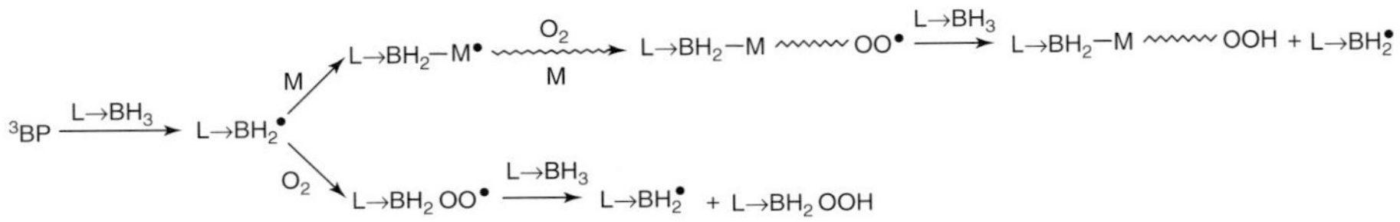

$$^3BP \xrightarrow{L \to BH_3} L \to BH_2^\bullet$$

Scheme 18.6

radical formation but in a viscous matrix, the rate constants level off. Owing to the acylate/oxygen relative concentration, the addition to acrylate is the predominant process for the boryl. In the same way, when introducing the concentrations, the ^{3}BP/monomer deactivation pathway probably remains important compared to the ^{3}BP/L → BH$_3$ interaction. Boranes can also be used as additives for Type I PIs in photopolymerization under aerated conditions.

18.15
Lophyl Radicals

The presence of a source of lophyl radicals L$^\bullet$ (Cl-HABI derivatives) in different multicomponent photoinitiating systems is well known (Sections 10 and 9.3.3). These radicals are easily observed by steady-state ESR (intense single band for $g \sim 2.003$) [155] or LFP (intense visible light absorption) [156, 157].

Lophyl radicals are rather stabilized and exhibit a low reactivity toward O$_2$ (the lifetime under air remains usually long (>1 s)). The addition to monomers is very slow. A significant reactivity (Table 18.17) [71] is noted for the hydrogen abstraction reaction from amines or thiols (with rate constants in the $10^2 - 10^6$ mol^{-1} l s^{-1} range closely related to the oxidation potentials of the hydrogen donors): this probably evidences at least a partial charge-transfer character for the reaction. The L$^\bullet$/silane (or ligated borane) reaction was recently mentioned albeit the rate constants not being determined. In Cl-HABI/silane (or borane), the silyl and boryl radicals are the initiating structures [152].

18.16
Iminyl Radicals

Iminyl radicals R$_1$R$_2$ C=N$^\bullet$ generated in PIs based on benzoyl oxime esters (Sections 8.1.11, 14.1 & 15.1) can be easily observed by steady-state ESR [158] or LFP [159]. They were also generated from (i) the hydrogen abstraction reaction from the corresponding imine (the thermolysis of O-phenyl oxime ethers through a N–O cleavage was reported as an efficient route to iminyl radicals [160]), (ii) the photolysis of acyloximes used as photobase generators, or (iii) the decomposition of iminosulfonates employed as photoacids [161, 162]. The main evolution pathway

Table 18.17 Interaction rate constants k (in $mol^{-1}\ l\ s^{-1}$) between L$^{\bullet}$ and different H-donors (thiols, amine).

Produced radical	k^a	Produced radical	k^a
A_9	$<10^5$	S_5	1.2×10^5
S_1	5×10^4	S_6	1.9×10^5
S_2	9.5×10^4	S_8	1.0×10^4
S_3	1.7×10^5	S_9	1.6×10^5
S_4	2.4×10^5	–	–

The generated thiyl radicals S_1-S_9 are shown in (6) and the aminoalkyl A_9 in (2).
^{a} In acetonitrile.
From [71].

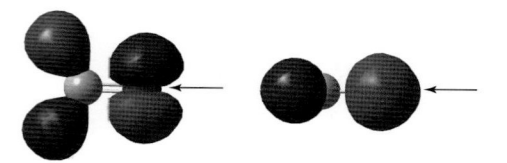

Figure 18.12 SOMOs for $H_2C=N^{\bullet}$ at DFT level (two different views). Nitrogen is indicated by an arrow. (J. Lalevée, unpublished data).

is a β-fragmentation of the iminyl (18.11) that yields a nitrile.

$$R_1R_2C=N^{\bullet} \rightarrow R_1^{\bullet} + R_2CN \tag{18.11}$$

Iminyls exhibit a rather low reactivity for the addition to double bonds (except for monomolecular processes, i.e., cyclization reactions) [163, 164]. The stability of iminyls partly arises from a powerful hyperconjugative interaction as deduced from the large value of a_H (about 87 G) in $H_2C=N^{\bullet}$ (Figure 18.12).

18.17
Metal-Centered Radicals

Some metal-centered radicals can be observed by LFP, for example, Re$^{\bullet}$(CO)$_5$ and Mn$^{\bullet}$(CO)$_5$ absorb at 550 and 800 nm, respectively [165]. These structures can hardly be observed by ESR-spin trapping (the addition to the spin trap is not favorable). The formation of metal-centered radicals was investigated in [166–172]. However, metal-centered PIs are rarely encountered in photoinitiating systems for industrial applications [173–177] except, for example, the titanocenes (Section 8.27.1) [178, 179]. This fact is mostly related to the lack of (i) kinetic data for these species and (ii) efficient pathways for a free radical formation. The thermal stability of

Table 18.18 Reaction rate constants k (in mol^{-1} l s^{-1}) of metal-centered radicals toward O_2, methylacrylate (MA), tris(trimethylsilyl)silane (TTMSS), diphenyl iodonium hexafluorophosphate (Ph_2I^+), and cumylhydroperoxide CumOOH. In tert-butylbenzene.

			k		
Radical	O_2	MA	TTMSS	Ph_2I^{+a}	CumOOH
Mn$^{\bullet}$(CO)$_5$	2.7×10^9	$<10^5$	$<2 \times 10^5$	$<1.0 \times 10^6$	–
Re$^{\bullet}$(CO)$_5$	4.1×10^9	$<10^5$	$<7 \times 10^5$	1.5×10^9	–
CpFe$^{\bullet}$(CO)$_2$	nd	$<10^5$	$<2 \times 10^5$	nd	–
CpMo$^{\bullet}$(CO)$_3$	3.2×10^9	$<10^{5a}$	$<2 \times 10^{5a}$	$<1.0 \times 10^{7a}$	2.2×10^6; 1.5×10^{6a} (3.7×10^{7b})
CpRu$^{\bullet}$(CO)$_2$	–	$<10^5$	$<2 \times 10^5$	$<5.0 \times 10^{6a}$	–
PhSe$^{\bullet}$	$<10^7$	$<10^6$	–	–	–
PhTe$^{\bullet}$	$<10^7$	$<10^6$	–	–	–

nd, not determined.
^{a}In acetonitrile.
^{b}For 3-chloroperbenzoic acid.

organometallic compounds under air must also be considered to avoid degradation reactions. The cleavage processes are usually related to the $\sigma - \sigma^*$ transition, which ensures the formation of a dissociative excited state (Eq. (18.12)). The cleavage of a Co–C bond can also occur in Co(III) complexes [180, 181].

$$Ph_3Y-YPh_3 \rightarrow 2\ Ph_3Y^{\bullet} \qquad Y = Sn$$
$$PhY'-Y'Ph \rightarrow 2\ PhY'^{\bullet} \qquad Y' = Se, Te$$
$$(CO)_5Z-Z(CO)_5 \rightarrow 2\ (CO)_5Z^{\bullet} \qquad Z = Mn, Re$$
$$Cp_2A_2(CO)_4 \rightarrow 2\ CpA^{\bullet}(CO)_2 \qquad A = Fe, Ru$$
$$Cp_2Mo_2(CO)_6 \rightarrow 2\ CpMo^{\bullet}(CO)_3$$
$$Co(III) \cdot R \rightarrow Co(II) + R^{\bullet} \qquad (18.12)$$

Some kinetic data on recently reinvestigated metal-centered radicals are given in Table 18.18. These structures are very reactive with O_2 but the addition to MA is usually slow ($k_i < 10^5$ mol^{-1} l s^{-1}). The oxidation of these radicals by iodonium salts is strongly dependent on the considered metal, that is, fast for Re and slow for Ru. The hydrogen abstraction rate constants with silanes are usually low, but the formation of silyl radicals has been well evidenced by ESR.

A key feature of these structures is their ability to decompose hydroperoxides (Eq. (18.13)) [182]. This process, which is quite fast (k in the $10^6 - 10^7$ mol^{-1} l s^{-1} range; see Table 18.18) and generates initiating alkoxyl radicals, was recently used to improve the acrylate polymerization profiles under aerated conditions as the conversion of the hydroperoxides generated during the polymerization leads to new initiating RO$^\bullet$ structures.

$$CpMo^\bullet(CO)_3 + CumOOH \rightarrow CpMoOH(CO)_3 + CumO^\bullet \qquad (18.13)$$

ArCr(CO)$_3$, which exhibits a convenient absorption in the 350–450 nm range, has been proposed as an interesting PI [183]. On laser excitation, a bleaching of ArCr(CO)$_3$ is observed and weak absorptions corresponding to ArCr(CO)$_2$ are noted at about 280 and 400 nm (Eq. (18.14)) [182]. ArCr(CO)$_2$ does not correspond to a free radical. It nevertheless reacts with alkyl halides [174] (e.g., CCl$_4$) or silanes [183] to generate free radicals (e.g., CCl$_3^\bullet$ or R$_3$Si$^\bullet$) through a quite complex set of reactions. This species is also able to decompose hydroperoxides.

$$ArCr(CO)_3 \rightarrow ArCr(CO)_2 + CO \qquad (18.14)$$

Se- or Te-centered radicals PhSe$^\bullet$ and PhTe$^\bullet$ (which can be generated on photolysis of diselenides or ditellurides), slowly interact with O$_2$. These structures cannot really initiate a polymerization ($k < 10^6$ mol^{-1} l s^{-1}; Table 18.18) and are preferably used as control agents, (PhSe)$_2$ and (PhTe)$_2$ being reported as photoiniferters (Section 18.19) [184].

18.18
Propagating Radicals

As in Section 18.4, the addition rate constants of TEA−M$^\bullet$ to another monomer (k_2) as well as the recombination of two radicals (k_1) has been investigated (Eq. (18.15)) [26–31].

$$\begin{aligned} \text{TEA-M}^\bullet + \text{TEA-M}^\bullet &\rightarrow \text{TEA-M-M-TEA} \quad (k_1) \\ \text{TEA-M}^\bullet + \text{M} &\rightarrow \text{TEA-M - M}^\bullet \quad (k_2) \end{aligned} \qquad (18.15)$$

The data obtained for representative monomers (MA, MMA, BA, AN, and HDDA) are given in Table 18.19. When available, k_2 and k_1 were found almost similar to

Table 18.19 Addition rate constants k_1 (in mol^{-1} l s^{-1}) of monomer derived radicals (MA, MMA, BA, AN, and HDDA) with a monomer unit. Recombination rate constants k_2 (in mol^{-1} l s^{-1}) of these monomer derived radicals.

	$10^{-9} k_2$	$10^{-4} k_1$	$10^{-4} k_p{}^a$
TEA−MA$^\bullet$	1.1	1.6–2.4	~3.0
TEA−MMA$^\bullet$	1.0	<0.5	0.0323; 0.0296
TEA−BA$^\bullet$	0.75	1.7–2.7	1.61; 1.37–1.54
TEA−HDDA$^\bullet$	0.45	1.9–4.3	1.0–2.0
TEA-AN$^\bullet$	1.2	<0.5	–

Propagation rate constants k_p (in mol^{-1} l s^{-1}).
^{a}Extracted from [186–191].

the usual propagation k_p and termination k_t rate constants recently determined by PLP techniques. Basically, k_1 and k_2 are representative of k_t and k_p in the early stages of the polymerization reaction in fluid media. The k_1 values are nearly diffusion controlled in solution (value close to $10^9\,\mathrm{mol^{-1}\,l\,s^{-1}}$). This approach was also extended to acryloxy-β,β-dimethyl-γ-butyrolactone LacA (known as a *highly reactive acrylate monomer*) and 2-(2-ethoxy-ethoxy) ethyl acrylate (DEEA) [185]. Other monomers (e.g., AA, styrene, vinylpyrrolidone, VC, vinyl ether, and allyl ether) might be likely studied by carefully selecting other initiating radicals. Although the method is primarily suited to self propagation, cross polymerization should also be investigated (e.g., the addition of TEA−M$^\bullet$ to VE since TEA$^\bullet$ does not react with VE) and copolymerization ratios might be accessible at low conversion.

18.19
Radicals in Controlled Photopolymerization Reactions

A detailed description of the persistent radical effect is given in [192–200] (see Section 4.2.11). Dithiocarbamyl radicals were originally proposed for the control of polymerization reactions. Nitroxides that are probably now the most widely used persistent radicals play an important role in the well-known nitroxide-mediated polymerization (NMP) [201] and the newly proposed nitroxide-mediated photopolymerization NMP2 (see a review in [202]).

18.19.1
Photoiniferters and Dithiocarbamyl Radicals

The formation of dithiocarbamyl radicals from the photolysis of benzyldithiocarbamates (Scheme 18.7) has been reported as an efficient process for the control of the photopolymerization reaction of different monomers, for example, acrylates, methacrylates, and styrene ([203, 204] and references therein).

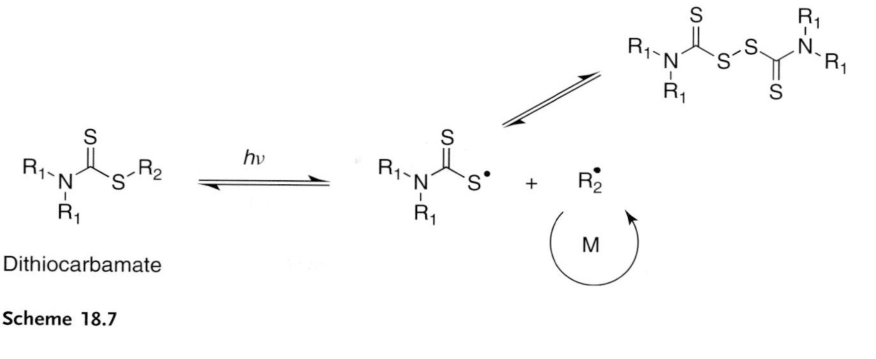

Dithiocarbamate

Scheme 18.7

Table 18.20 Reactivity of different sulfur-centered radical as control agents: reaction rate constants (k in $mol^{-1} \, l \, s^{-1}$) with O_2, MA, and TEMPO.

		k		δ_S
	O_2	MMA	TEMPO	
D_1	$<10^5$	$<10^3$	5.5×10^7	0.522
D_2	$<10^5$	$<10^3$	6.5×10^7	0.523
D_3	$<10^5$	$<10^3$	1.0×10^9	0.589
D_4	$<10^3$	$<10^3$	$<1.1 \times 10^5$	0.51

[a]The sulfur spin density δ_S is given in brackets (calculated at UB3LYP/6–31 G* level).
From [204].

The reaction of dithiocarbamyl radicals $D_1 - D_4$ with O_2 and MMA (Table 18.20) is relatively slow ($k < 10^5 \, mol^{-1} \, l \, s^{-1}$). The interaction rate constants with TEMPO spread over four orders of magnitude ($10^5 - 10^9 \, mol^{-1} \, l \, s^{-1}$). These results are related to the sulfur spin density δ_S, that is, the decrease of δ_S indicates a stabilization of the radical in agreement with a lower interaction rate constant with TEMPO. This allows a significant improvement in the control mechanism with a higher persistent character for the controlling radical. Indeed, by-side reactions of dithiocarbamyls (recombination and slow addition to double bond) [205] lead to a poorer control than that observed with other persistent radicals (e.g., nitroxides). The more stabilized sulfur-centered controlling radical can be useful to overcome this limitation. This explains quite well an increase of the control in the series $D_3 < D_2 \sim D_1 \ll D_4$.

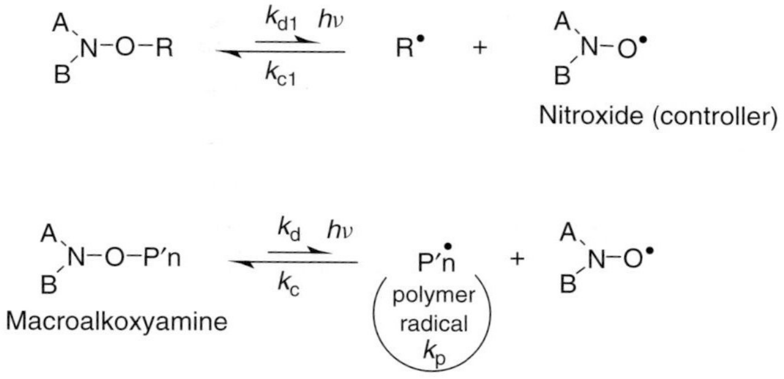

```
A\
  N—O—R   ⇌ (k_{d1}, hν / k_{c1})   R• +   A\
B/                                         B/N—O•
                                        Nitroxide (controller)


A\
  N—O—P'n   ⇌ (k_d, hν / k_c)   P'n•  +   A\
B/                              (polymer    B/N—O•
Macroalkoxyamine                 radical)
                                   k_p
```

Scheme 18.8

18.19.2
Light-Sensitive Alkoxyamines and Generation of Nitroxides

Thermolysis of alkoxyamines through a O–C bond cleavage process leads to nitroxides. The BDE(O–C) is typically in the 20–35 kcal mol^{-1} range depending on both the alkyl and nitroxide fragments, for example, in TEMPO, 2,2,5-tri-methyl-4-phenyl-3-azahexane-3-nitroxide (TIPNO) and N-(2-methylpropyl)-N-(1-diethylphosphono-2,2-dimethylpropyl)-N-oxyl (SG1) [206, 207]. Using new alkoxyamines as PI-controller radical polymerization mediators, initiation and polymerization occur on light exposure (Scheme 18.8).

The photolysis of alkoxyamines A–D (**11**) carrying a chromophore linked to the aminoxyl function [208–211] leads to different nitroxides through a homolytic C–O bond cleavage (see Chapter 2). The bond dissociation energy being lower than the triplet energy level ($\sim$280 kJ mol^{-1}), the cleavage process is exothermic (Table 18.21). The triplet-state lifetimes of A–D are relatively short compared to that of BP ($\sim$5μs) in agreement with this cleavage reaction. The cleavage is also strongly affected by the distance between the chromophore and the aminoxy function.

Table 18.21 Parameters characterizing the cleavage processes for A–D.

Alkoxyamines	BDE (C-O) (kJ mol^{-1})	τ (triplet state) τ (μs)
A	122	1.5
B	118	0.39
C	80	0.02
D	115	0.9

From [212].

The nitroxides generated from A to D are persistent (lifetime >60 s). They are not reactive with O_2 but efficiently combine with carbon-centered radicals (acrylate- or styrene-monomer-derived radicals) for the formation of (macro)alkoxyamines (Scheme 18.8): the rate constants for TEMPO or SG1 are in the 10^8 mol^{-1} l s^{-1} range (Table 18.3 and [33]).

18.20
Radicals in Hydrosilylation Reactions

In thermal hydrosilylation reactions, the addition of Si–H bonds across unsaturated bonds requires a catalyst such as the Speier's catalyst H_2PtCl_6 [213]. In photochemical hydrosilylation reactions where Pt complexes are used (Section 4.5), the processes are not associated with any free radical chemistry.

Recently, metal-free hydrosilylation procedures were presented [214–216]. This new approach still appears as a great challenge avoiding the use of highly expensive

Scheme 18.9

Initiator ⟶ In• $\xrightarrow{R_3Si-H}$ R$_3$Si• $\xrightarrow{C=C\ (2)}$ R$_3$Si-C-C• $\xrightarrow{(3)\ R_3Si-H}$ R$_3$Si-C-CH + R$_3$Si•

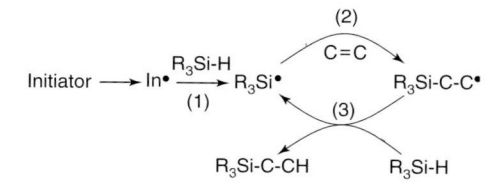

Scheme 18.9

metal catalyst (Pt or Rh). Free-radical-based hydrosilylation processes are surely promising (Scheme 18.9).

Silyl radicals are generated from the initiating radicals In• (1) associated with the decomposition of a thermal initiator (e.g., peroxides or AIBN). This decomposition can also be a photoinitiated process. The addition of the silyl radical to the monomer leads to a carbon-centered adduct (2) which can further react with the silane to regenerate a silyl and the final hydrogenated compound (3). However, this process is only efficient for silanes characterized by a low BDE(Si–H), that is, possessing a very labile hydrogen. Different rate constants R$_3$C•/R$_3$SiH are gathered in [85, 86]. For benzyl radicals, the rate constants with Et$_3$SiH and TMS$_3$SiH are $\sim 3 \times 10^3$ and 3×10^5 mol^{-1} l s^{-1}, respectively. For triethylsilane (BDE(Si–H) = 95 kcal mol^{-1}), the process is slower than for tris(trimethylsilyl)silane (BDE(Si–H) = 83 kcal mol^{-1}) [86]. Recently, hydrosilylation procedures incorporating tris(trimethylsilyl)silane were proposed for very soft synthesis conditions (low temperature, reaction under air, and presence of water). Remarkably, O$_2$ was used as an initiator [215, 216], the hydrogen abstraction reaction (1) in Scheme 18.9 corresponding to the process (Eq. (18.16)).

$$O_2 + R_3SiH \rightarrow HOO• + R_3Si• \qquad (18.16)$$

References

1. Kochi, J.K. (ed.) (1973) *Free Radicals*, vol. 2, John Wiley & Sons, Inc., New York.
2. Nonhebel, D.C. and Walton, J.C. (1974) *Free-Radical Chemistry*, Cambridge University Press, Cambridge.
3. Renaud, P. and Sibi, M.P. (eds) (2001) *Radicals in Organic Synthesis*, Wiley-VCH Verlag GmbH, Weinheim.
4. Zard, S.Z. (2003) *Radical Reactions in Organic Synthesis*, Oxford University Press, New York.
5. Alfassi, Z.B. (ed.) (1999) *General Aspects of the Chemistry of Radicals*, John Wiley & Sons, Ltd, Chichester.
6. Luo, Y.R. (2003) *Handbook of Bond Dissociation Energy in Organic Compounds*, CRC Press, Boca Raton, FL.
7. Newcomb, M. (2012) in *Encyclopedia of Radicals in Chemistry, Biology and Materials* (eds A. Studer and C. Chatgilialoglu), Wiley-VCH Verlag GmbH, pp. 92–108.
8. Jin, J. and Newcomb, M. (2008) *J. Org. Chem.*, **73**, 4740–4742.
9. Fischer, H. and Radom, L. (2001) *Angew. Chem. Int. Ed.*, **40**, 1340–1345.
10. Lalevée, J., Allonas, X., Genet, S., and Fouassier, J.P. (2003) *J. Am. Chem. Soc.*, **125**, 9377–9381.

11. Beckwith, A.L.J. and Poole, J.S. (2002) *J. Am. Chem. Soc.*, **124**, 9489–9497.
12. Walbiner, M., Wu, J.Q., and Fischer, H. (1995) *Helv. Chim. Acta*, **78**, 910–916.
13. Fischer, H., Baer, R., Hany, R., Verhoolen, I., and Walbiner, M. (1990) *J. Chem. Soc., Perkin Trans. 2*, 787–793.
14. Schnabel, W. (1991) in *Laser in Polymer Science and Technology: Applications*, vol. 2 (eds J.P. Fouassier and J.F. Rabek), CRC Press, Boca Raton, FL, pp. 95–144.
15. Maillard, B., Ingold, K.U., and Scaiano, J.C. (1983) *J. Am. Chem. Soc.*, **105**, 5095–5099.
16. Franchi, P., Lucarini, M., Pedulli, G.F., Valgimigli, L., and Lunelli, B. (1999) *J. Am. Chem. Soc.*, **121**, 507–514.
17. Lalevée, J. and Fouassier, J.P. (2012) in *Encyclopedia of Radicals in Chemistry, Biology and Materials* (eds A. Studer and C. Chatgililoglu), John Wiley & Sons, Inc., pp. 34–57.
18. Lalevée, J., Allonas, X., and Fouassier, J.P. (2004) *J. Phys. Chem. A*, **108**, 4326–4334.
19. Alfassi, Z.B., Khaikin, G.I., and Neta, P. (1995) *J. Phys. Chem.*, **99**, 265–268.
20. Alberti, A., Benaglia, M., Macciantelli, D., Rossetti, S., and Scoponi, M. (2008) *Eur. Polym. J.*, **44**, 3022–3027.
21. Fischer, H. (ed.) (2005) *Landolt Bornstein: Magnetic Properties of Free Radicals*, vol. **26d**, Springer-Verlag, Berlin.
22. Colley, C.S., Grills, D.C., Besley, N.A., Jockusch, S., Matousek, P., Parker, A.W., Town, N.J., Gill, P.M.W., and George, M.W. (2002) *J. Am. Chem. Soc.*, **124**, 14952–14958.
23. Allonas, X., Lalevée, J., Morlet-Savary, F., and Fouassier, J.P. (2006) *Polimeri*, **51**, 491–498.
24. Brown, C.E., Neville, A.G., Rayner, D.M., and Ingold, K.U. (1995) *J. Lusztyk Austr. J. Chem.*, **48**, 363–379.
25. Voll, D., Junkers, T., and Barner-Kowollik, C. (2011) *Macromolecules*, **44**, 2542–2551.
26. Knolle, W., Muller, U., and Mehnert, R. (2000) *Phys. Chem. Chem. Phys.*, **2**, 1425–1431.
27. Knolle, W. and Mehnert, R. (1995) *Nucl. Instrum. Methods Phys. Res. B*, **105**, 154–159.
28. Takacs, E. and Wojnarovits, L. (1995) *Nucl. Instrum. Methods Phys. Res. B*, **105**, 282–286.
29. Kujawa, P., Mohid, N., Zaman, K., Manshol, W., Ulanski, P., and Rosiak, J.M. (1998) *Radiat. Phys. Chem.*, **53**, 403–409.
30. Martschke, R., Farley, R.D., and Fischer, H. (1997) *Helv. Chim. Acta*, **80**, 1363–1368.
31. Lalevée, J., Allonas, X., and Fouassier, J.P. (2006) *J. Polym. Sci., A: Polym. Chem.*, **44**, 3577–3587.
32. Lalevée, J., Allonas, X., and Fouassier, J.P. (2005) *Chem. Phys. Lett.*, **415**, 287–290.
33. Lalevée, J., Gigmes, D., Bertin, D., Allonas, X., and Fouassier, J.P. (2007) *Chem. Phys. Lett.*, **449**, 231–235.
34. Matyjaszewski, K. and Davis, T.P. (eds) (2002) *Handbook of Radical Polymerization*, John Wiley & Sons, Inc., Hoboken.
35. Lalevée, J., Graff, B., Allonas, X., and Fouassier, J.P. (2007) *J. Phys. Chem. A*, **111**, 6991–6998.
36. Lalevée, J., Gigmes, D., Bertin, D., Graff, B., Allonas, X., and Fouassier, J.P. (2007) *Chem. Phys. Lett.*, **438**, 346–350.
37. Lalevée, J., Allonas, X., and Fouassier, J.P. (2002) *J. Am. Chem. Soc.*, **124**, 9613–9621.
38. Jockusch, S. and Turro, N.J. (1999) *J. Am. Chem. Soc.*, **121**, 3921–3927.
39. Wayner, D.D.M., Dannenberg, J.J., and Griller, D. (1986) *Chem. Phys. Lett.*, **131**, 189–191.
40. Lalevée, J., Allonas, X., and Fouassier, J.P. (2009) *Chem. Phys. Lett.*, **468**, 227–230.
41. Jacques, P. and Allonas, X. (1997) *Chem. Phys.*, **215**, 371–378.
42. Heuts, J.P.A., Gilbert, R.G., and Radom, L. (1995) *Macromolecules*, **28**, 8771–8777.
43. Coote, M.L., Davis, T.P., and Radom, L. (1999) *Macromolecules*, **32**, 5270–5276.
44. Scaiano, J.C. and Stewart, L.C. (1983) *J. Am. Chem. Soc.*, **105**, 3609–3613.

45. Recupero, F., Bravo, A., Bjorsvik, H.R., Fontana, F., Minisci, F., and Piredda, M. (1997) *J. Chem. Soc., Perkin Trans. 2*, 2399–2404.
46. Lalevée, J., Allonas, X., and Fouassier, J.P. (2008) *Chem. Phys. Lett.*, **454**, 415–418.
47. Decker, C. and Jenkins, A.D. (1985) *Macromolecules*, **18**, 1241–1248.
48. Davidson, R.S. (1993) in *Radiation Curing in Polymer Science and Technology*, vol. **3** (eds J.P. Fouassier and J. Rabek), Elsevier Science Publishers, Ltd, London, pp. 153–176.
49. Hoyle, C.E. and Kim, K.J. (2003) *J. Appl. Polym. Sci.*, **33**, 2985–2991.
50. Griller, D., howard, J.A., Marriott, P.R., and Scaiano, J.C. (1981) *J. Am. Chem. Soc.*, **103**, 619–624.
51. Sluggett, G.W., McGarry, P.F., Koptyug, I.V., and Turro, N.J. (1996) *J. Am. Chem. Soc.*, **118**, 7367–7372.
52. Khudyakov, I.V. and Turro, N.J. (2010) in *Carbon Centered Free Radicals and radical Cations* (ed. M.D.E. Forbes), John Wiley & Sons, Inc., New York, pp. 249–279.
53. Brunton, G., Gilbert, B.C., and Mawby, R.J. (1976) *J. Chem. Soc., Perkin Trans. 2*, 650–658.
54. Criqui, A., Lalevée, J., Allonas, X., and Fouassier, J.P. (2008) *Macromol. Chem. Phys.*, **209**, 2223–2231.
55. Sluggett, G.W., Turro, C., Georges, M.W., Koptyung, I.V., and Turro, N.J. (1995) *J. Am. Chem. Soc.*, **117**, 5148–5152.
56. Sumiyoshi, T., Schnabel, W., Henne, A., and Lechtken, P. (1985) *Polymer*, **26**, 141–147.
57. Sumiyoshi, T. and Schnabel, W. (1985) *Makromol. Chem.*, **186**, 1811–1817.
58. Schnabel, W. (1985) in *New Trends in Photochemistry of Polymers* (eds J.F. Rabek and N.S. Allen), Elsevier, London, pp. 69–82.
59. Majima, T. and Schnabel, W. (1989) *J. Photochem.*, **50**, 31–38.
60. Andrzejewska, E., Zych-Tomkowiak, D., and Andrzejewski, M. (2006) *Macromolecules*, **39**, 3777–3782.
61. Andrzejewska, E., Zych-Tomkowiak, D., Bogacki, M.B., and Andrzejewski, M. (2004) *Macromolecules*, **37**, 6346–6351.
62. Ito, O. and Matsuda, M. (1984) *Bull. Chem. Soc. Jpn.*, **57**, 1745–1749.
63. Lalevée, J., Allonas, X., Morlet-Savary, F., and Fouassier, J.P. (2006) *J. Phys. Chem. A*, **110**, 11605–11612.
64. Alfassi, Z.B. (1999) *The Chemistry of Free Radicals: S-Centered Radicals*, John Wiley & Sons, Ltd.
65. Ito, O., Nogami, K., and Matsuda, M.J. (1981) *Phys. Chem.*, **85**, 1365–1369.
66. Ito, O. and Matsuda, M.J. (1979) *Am. Chem. Soc.*, **101**, 1815–1819.
67. Ito, O. and Matsuda, M.J. (1979) *Am. Chem. Soc.*, **101**, 5732–5736.
68. Ito, O. and Matsuda, M.J. (1981) *Am. Chem. Soc.*, **103**, 5871–5874.
69. Ito, O. and Matsuda, M.J. (1982) *Am. Chem. Soc.*, **104**, 568–573.
70. Ito, O. and Matsuda, M.J. (1982) *Am. Chem. Soc.*, **104**, 1701–1705.
71. Lalevée, J., Morlet-Savary, F., El Roz, M., Allonas, X., and Fouassier, J.P. (2009) *Macromol. Chem. Phys.*, **210**, 311–319.
72. Lalevée, J., Allonas, X., and Fouassier, J.P. (2006) *J. Org. Chem.*, **71**, 9723–9727.
73. Ito, O., Omori, R., and Matsuda, M. (1982) *J. Am. Chem. Soc.*, **104**, 3934–3937.
74. Dietlin, C., Allonas, X., Morlet-Savary, F., Fouassier, J.P., Visconti, M., Norcini, G., and Romagnano, S. (2008) *J. Appl. Polym. Sci.*, **109**, 825–831.
75. Bartmann, E. and Ohngemach, J. (1991), Merck Patent, German Off., 3, 921, 459.
76. Fouassier, J.P., Lougnot, D.J., and Scaiano, J.C. (1989) *Chem. Phys. Lett.*, **160**, 335–340.
77. Ortica, F., Coenjarts, C., Scaiano, J.C., Liu, H., Pohlers, G., and Cameron, J.F. (2001) *Chem. Mater.*, **13**, 2297–2304.
78. Coenjarts, C., Ortica, F., Cameron, J., Pohlers, G., Zampini, A., Desilets, D., Liu, H., and Scaiano, J.C. (2001) *Chem. Mater.*, **13**, 2305–2312.
79. Chatgilialoglu, C., Griller, D., and Guerra, M.J. (1987) *Phys. Chem.*, **91**, 3747–3472.
80. Yoshimatsu, M., Hayashi, M., Tanabe, G., and Muraoka, O. (1996) *Tetrahedron Lett.*, **37**, 4161–4164.

81. Gozdz, A.S. and Maslak, P. (1991) *J. Org. Chem.*, **56**, 2179–2189.
82. Percec, V., Barboiu, B., and Kim, H.J. (1998) *J. Am. Chem. Soc.*, **120**, 305–310.
83. Schnabel, W. (1991) in *Laser in Polymer Science and Technology: Applications*, vol. 2 (eds J.P. Fouassier and J.F. Rabek), CRC Press, pp. 95–135.
84. Markoviæ, D., Varela-Álvarez, A., Sordo, J.A., and Vogel, P. (2006) *J. Am. Chem. Soc.*, **128**, 7782–7795.
85. Chatgilialoglu, C. (2004) *Organosilanes in Radical Chemistry*, John Wiley & Sons, Ltd, Chichester.
86. Chatgilialoglu, C. (2008) *Chem. Eur. J.*, **14**, 2310–2315.
87. Arsu, N., Hizai, G., and Yagci, Y. (1995) *J. Macromol.Chem., Macromol. Rep.*, **A32**, 1257–1263.
88. Kminek, I., Yagci, Y., and Schnabel, W. (1992) *Polym. Bull.*, **277**, 29–33.
89. Peinado, C., Alonso, A., Catalina, F., and Schnabel, W. (2001) *J. Photochem. Photobiol. A: Chem.*, **141**, 85–92.
90. West, R., Wolff, A.R., and Peterson, D.J. (1986) *J. Radiat. Curing*, **13**, 35–39.
91. Michl, J., Downing, J.W., and Karatsu, T. (1988) *Pure Appl. Chem.*, **60**, 959–966.
92. Peinado, C., Alonso, A., Catalina, F., and Schnabel, W. (2000) *Macromol. Chem. Phys.*, **201**, 1156–1163.
93. Chandra, H., Davidson, I.M.T., and Symons, M.C.R. (1983) *J. Chem. Soc., Faraday Trans. 1*, **79**, 2705–2711.
94. Lalevée, J., Blanchard, N., Tehfe, M.A., and Fouassier, J.P. (2010) *Macromolecules*, **43**, 10191–10195.
95. Chatgilialoglu, C., Guerrini, A., and Lucarini, M. (1992) *J. Org. Chem.*, **57**, 3405–3409.
96. Alberti, A. and Pedulli, G.F. (1987) *Rev. Chem. Intermed.*, **8**, 207–215.
97. Lalevée, J., El-Roz, M., Allonas, X., and Fouassier, J.P. (2009) in *Organosilanes: Properties, Performance and Applications* (eds E.B. Wyman and M.C. Skief), Nova Science Publishers, Inc., pp. 174–198.
98. Lalevée, J., Dirani, A., El-Roz, M., Allonas, X., and Fouassier, J.P. (2008) *Macromolecules*, **41**, 2003–2010.
99. Lalevée, J., Allonas, X., and Fouassier, J.P. (2007) *J. Org. Chem.*, **72**, 6434–6439.
100. Lalevée, J., Blanchard, N., Graff, B., Allonas, X., and Fouassier, J.P. (2008) *J. Organomet. Chem.*, **693**, 3643–3649.
101. Lalevée, J., El-Roz, M., Morlet-Savary, F., Graff, B., Allonas, X., and Fouassier, J.P. (2007) *Macromolecules*, **40**, 8527–8530.
102. Lalevée, J., Blanchard, N., El-Roz, M., Graff, B., Allonas, X., and Fouassier, J.P. (2008) *Macromolecules*, **41**, 4180–4186.
103. Lalevée, J., Tehfe, M.A., Allonas, X., and Fouassier, J.P. (2010) in *Polymer Initiators* (ed. W.J. Ackrine), Nova Science Publishers, Inc., pp. 201–229.
104. Versace, D.L., Tehfe, M.A., Lalevée, J., Casarotto, V., Blanchard, N., Morlet-Savary, F., and Fouassier, J.P. (2011) *J. Phys. Org. Chem.*, **24**, 342–350.
105. Janzen, E.G. (1976) *Creat. Detect. Excited State*, **4**, 83–138.
106. Haire, D.L. and Janzen, E.G. (1982) *Can. J. Chem.*, **60**, 1514–1522.
107. Niki, E., Yokoi, S., Tsuchiya, J., and Kamiya, Y. (1983) *J. Am. Chem. Soc.*, **105**, 1498–1503.
108. Howard, J.A. and Scaiano, J.C. (1984) in *Landolt Bornstein Numerical Data and Functional Relationships in Science and Technology*, Group II, vol. 13 (ed. H. Fischer), Springer, Berlin, pp. 425–457.
109. Ingold, K.U. (1969) *Acc. Chem. Res.*, **2**, 1–8.
110. Furimsky, E. and Howard, J.A. (1973) *J. Am. Chem. Soc.*, **95**, 369–373.
111. Burton, B.W., Doba, T., Gabe, E.J., Hughes, L., Lee, F.L., Prasad, L., and Ingold, K.U. (1985) *J. Am. Chem. Soc.*, **107**, 7053–7054.
112. Fukuzumi, S., Shimoosako, K., Suenobu, T., and Watanabe, Y. (2003) *J. Am. Chem. Soc.*, **125**, 9074–9080.
113. Adamic, K., Ingold, K.U., and Morton, J.R. (1970) *J. Am. Chem. Soc.*, **92**, 922–923.
114. Howard, J.A. (1972) *Can. J. Chem.*, **50**, 1981–1983.

115. Lalevée, J., Allonas, X., and Fouassier, J.P. (2007) *Chem. Phys. Lett.*, **445**, 62–67.
116. Abe, Y., Seno, S.Y., Sakakibara, K., and Hirota, M. (1991) *J. Chem. Soc., Perkin Trans. 2*, 897–903.
117. Zubarev, V.E., Belevskii, V.N., and Yarkov, S.P. (1979) *Dokl. Akad. Nauk SSSR*, **244**, 1392–1396.
118. Criqui, A., Lalevée, J., Allonas, X., and Fouassier, J.P. (2009) *J. Adv. Oxid. Technol.*, **12**, 215–219.
119. Lalevée, J., Allonas, X., Fouassier, J.P., and Ingold, K.U. (2008) *J. Org. Chem.*, **73**, 6489–6496.
120. Howard, J.A. (1997) *Landolt Bornstein Numerical Data and Functional Relationships in Science and Technology*, Group II, vol. 18, sub vol. D2, Springer, Berlin.
121. Alfassi, Z. (1997) *Peroxyl Radicals*, John Wiley & Sons, Ltd, Chichester.
122. Denisov, E.T. and Denisova, T.G. (2000) *Handbook of Antioxidants*, 2nd edn, CRC Press, New York.
123. Korcek, S., Chenier, J.H.B., Howard, J.A., and Ingold, K.U. (1972) *Can. J. Chem.*, **50**, 2285–2297.
124. Marque, S. and Tordo, P. (2005) *Top. Curr. Chem.*, **250**, 43–76.
125. Roberts, B.P. and Scaiano, J.C. (1981) *J. Chem. Soc., Perkin Trans. 2*, 905–911.
126. Ingold, K.U. and Roberts, B.P. (1971) *Free Radical Substitution Reactions*, John Wiley & Sons, Inc., New York.
127. Howard, J.A., Tait, J.C., and Tong, S.B. (1979) *Can. J. Chem.*, **57**, 2761–2768.
128. DiLabio, G.A., Litwinienko, G., Lin, S., Pratt, D.A., and Ingold, K.U. (2002) *J. Phys. Chem. A*, **106**, 11719–11725.
129. Gijsman, P. (2010) in *Photochemistry and Photophysics of Polymer Materials* (ed. N.S. Allen), John Wiley & Sons, Inc., Hoboken, NJ, pp. 627–679.
130. Ganster, B., Fischer, U.K., Moszner, N., and Liska, R. (2008) *Macromol. Rapid Commun.*, **29**, 57–62.
131. Hayashi, H. and Mochida, K. (1983) *Chem. Phys. Lett.*, **101**, 307–311.
132. Wakasa, M., Mochida, K., Sakaguchi, Y., Nakamura, J., and Hayashi, H. (1991) *J. Phys. Chem.*, **95**, 2241–2246.
133. Tehfe, M.A., Blanchard, N., Fries, C., Lalevée, J., Allonas, X., and Fouassier, J.-P. (2010) *Macromol. Rapid Commun.*, **31**, 473–478.
134. Lalevée, J., Allonas, X., and Fouassier, J.P. (2009) *Chem. Phys. Lett.*, **469**, 298–303.
135. Ganster, B., Fischer, U.K., Moszner, N., and Liska, R. (2008) *Macromolecules*, **41**, 2394–2402.
136. Mochida, K. (1984) *Bull. Chem. Soc. Jpn.*, **57**, 796–801.
137. Haire, D.L., Oehler, U.M., Krygsman, P.H., and Janssen, E.G. (1988) *J. Org. Chem.*, **53**, 4535–4542.
138. Lalevée, J., El-Roz, M., Dirani, A., Allonas, X., and Fouassier, J.P. (2008) *J. Polym. Sci. Part A: Polym. Chem.*, **46**, 3042–3047.
139. Zaborovskiy, A.B., Lutsyk, D.S., Prystansky, R.E., Kopylets, V.I., Timokhin, V.I., and Chatgilialoglu, C. (2004) *J. Organomet. Chem.*, **689**, 2912–2918.
140. Marti, V.P.J. and Roberts, B.P. (1986) *J. Chem. Soc., Perkin Trans. 2*, 1613–1621.
141. Baban, J.A., Marti, V.P.J., and Roberts, B.P. (1985) *Tetrahedron Lett.*, **26**, 1349–1352.
142. Baban, J.A., Marti, V.P.J., and Roberts, B.P. (1985) *J. Chem. Soc., Perkin Trans. 2*, 1723–1733.
143. Baban, J.A. and Roberts, B.P. (1983) *J. Chem. Soc., Chem. Commun.*, 1224–1226.
144. Johnson, K.M., Kirwan, J.N., and Roberts, B.P. (1990) *J. Chem. Soc., Perkin Trans. 2*, 1125–1132.
145. Lucarini, M., Pedulli, G.F., and Valgimigli, L. (1996) *J. Org. Chem.*, **61**, 4309–4313.
146. Ueng, S.H., Makhlouf Brahmi, M., Derat, E., Fensterbank, L., Lacote, E., Malacria, M., and Curran, D.P. (2008) *J. Am. Chem. Soc.*, **130**, 10082–10083.
147. Ueng, S.H., Solovyev, A., Yuan, X., Geib, S.J., Fensterbank, L., Lacôte, E., Malacria, M., Newcomb, M., Walton, J.C., and Curran, D.P. (2009) *J. Am. Chem. Soc.*, **131**, 11256–11262.
148. Matsumoto, T. and Gabbaí, F.P. (2009) *Organometallics*, **28**, 4252–4253.

149. Walton, J.C., Makhlouf Brahmi, M., Fensterbank, L., Lacôte, E., Malacria, M., Chu, Q., Ueng, S.-H., Solovyev, A., and Curran, D.P. (2010) *J. Am. Chem. Soc.*, **132**, 2350–2358.
150. Lalevée, J., Blanchard, N., Chany, A.-C., Tehfe, M.A., Allonas, X., and Fouassier, J.P. (2009) *J. Phys. Org. Chem.*, **22**, 986–993.
151. Lalevée, J., Tehfe, M.A., Allonas, X., and Fouassier, J.P. (2008) *Macromolecules*, **41**, 9057–9062.
152. Lalevée, J., Blanchard, N., Tehfe, M.-A., Chany, A.-C., and Fouassier, J.-P. (2010) *Chem. –A Eur. J.*, **16**, 12920–12927.
153. Sheeller, B. and Ingold, K.U. (2001) *J. Chem. Soc., Perkin Trans. 2*, 480–486.
154. Ueng, S.-H., Fensterbank, L., Lacôte, E., Malacria, M., and Curran, D.P. (2011) *Org. Biomol. Chem.*, **9**, 3415–3420.
155. Caspar, J.V., Khudyakov, I.V., Turro, N.J., and Weed, G.C. (1995) *Macromolecules*, **28**, 636–641.
156. Liu, A.D., Trifunac, A.D., and Krongrauz, V. (1991) *J. Phys. Chem.*, **95**, 5822–5827.
157. Strehmel, V., Wishart, J.F., Polyansky, D.E., and Strehmel, B. (2009) *Chem. Phys. Chem.*, **10**, 3112–3118.
158. Griller, D., Mendenhall, G.D., Van Hoof, W., and Ingold, K.U. (1974) *J. Am. Chem. Soc.*, **96**, 6068–6070.
159. Lalevée, J., Allonas, X., Fouassier, J.P., Tachi, H., Izumitani, A., Shirai, M., and Tsunooka, M. (2002) *J. Photochem. Photobiol. A: Chem.*, **151**, 27–37.
160. Blake, A., Pratt, D.A., Lin, S., Walton, J.C., Mulder, P., and Ingold, K.U. (2004) *J. Org. Chem.*, **69**, 3112–3120.
161. Arnold, P.A., Fratesi, L.E., Bejan, E., Cameron, J., Pohlers, G., Liu, H., and Scaiano, J.C. (2004) *Photochem. Photobiol. Sci.*, **3**, 864–869.
162. Lalevée, J., Allonas, X., Fouassier, J.P., Shirai, M., and Tsunooka, M. (2003) *Chem. Lett.*, **32**, 178–179.
163. Bowman, W.R., Bridge, C.F., and Brookes, P. (2000) *Tetrahedron Lett.*, **41**, 8989–8994.
164. Boivin, J., Fouquet, E., and Zard, S.Z. (1994) *Tetrahedron*, **50**, 1745–1756.
165. Tehfe, M.A., Lalevée, J., Gigmes, D., and Fouassier, J.P. (2010) *J. Polym. Sci. A: Chem.*, **48**, 1830–1837.
166. Hudson, A., Lappert, M.F., Lednor, P.W., and Nicholson, B.K. (1974) *J. Chem. Soc., Chem. Commun.*, 966–967.
167. Hudson, A., Lappert, M.F., and Nicholson, B.K. (1977) *J. Chem. Soc., Dalton Trans.*, 551–554.
168. Macyk, W., Herdegen, A., Karocki, A., Stochel, G., Stasicka, Z., Sostero, S., and Traverso, O. (1997) *J. Photochem. Photobiol. A: Chem.*, **103**, 221–226.
169. Balla, J., Bakac, A., and Espenson, J.H. (1994) *Organometallics*, **13**, 1073–1074.
170. Bitterwolf, T.E. (2004) *J. Organomet. Chem.*, **689**, 3939–3952.
171. Espenson, J.H. (1995) *J. Mol. Liq.*, **65**, 205–212.
172. Scott, S.L., Espenson, J.H., and Zhu, Z. (1993) *J. Am. Chem. Soc.*, **115**, 1789–1797.
173. Cunningham, A.F. Jr. and Desobry, V. (1993) in *Radiation Curing in Polymer Science and Technology*, vol. **2** (eds J.P. Fouassier and J.F. Rabek), Elsevier Science Publishers, Ltd, London, pp. 323–374.
174. Kunding, E.P., Xu, L.H., Kondratenko, M., Cunningham, A.F., and Kuntz, M. Jr. (2007) *Eur. J. Inorg. Chem.*, **2007**, 2934–2943.
175. Yagci, Y. and Hepuzer, Y. (1999) *Macromolecules*, **32**, 6367–6370.
176. Bamford, C.H. and Mullik, S.U. (1977) *J. Chem. Soc., Faraday Trans. 1*, **73**, 1260–1270.
177. Bamford, C.H. and Mullik, S.U. (1976) *J. Chem. Soc., Faraday Trans. 1*, **72**, 368–375.
178. Klingert, B., Roloff, A., Urwyler, B., and Wirz, J. (1988) *Helv. Chim. Acta*, **71**, 1858–1867.
179. Roloff, A., Meier, K., and Riediker, M. (1986) *Pure Appl. Chem.*, **58**, 1267–1272.
180. Murakami, Y., Hisaeda, Y., Song, X.-M., Takasaki, K., and Ohno, T. (1991) *Chem. Lett.*, **6**, 977–980.
181. Jaworska, M., Lodowski, P., Andruniów, T., and Kozlowski, P.M. (2007) *J. Phys. Chem. B*, **111**, 2419–2422.

182. Creaven, B.S., George, M.W., Ginzburg, A.G., Hughes, C., Kelly, J.M., Long, C., McGrath, I.M., and Pryce, M.T. (1993) *Organometallics*, **12**, 3127–3131.
183. Lalevée, J., Tehfe, M.-A., Gigmes, D., and Fouassier, J.-P. (2010) *Macromolecules*, **43**, 6608–6615.
184. Rathore, K., Raghunatha Reddy, K., Tomer, N.S., Desai, S.M., and Singh, R.P. (2004) *J. Appl. Polym. Sci.*, **93**, 348–355.
185. Lalevée, J., Allonas, X., and Fouassier, J.P. (2006) in *Photochemistry and UV Curing: New Trends* (ed. J.P. Fouassier), Research Signpost, Trivandrum, pp. 117–126.
186. Buback, M. and Kowollik, C. (1999) *Macromolecules*, **32**, 1445–1451.
187. Buback, M., Kurz, C.H., and Schmultz, C. (1998) *Macromol. Chem. Phys.*, **199**, 1721.
188. Asua, J.M., Bauermann, S., Buback, M., Castignolles, P., Charleux, B., Gilbert, R.G., Hutchinson, R.A., Leiza, J.R., Nikitin, A.N., Vairon, J.P., and van Herk, A.M. (2004) *Macromol. Chem. Phys.*, **205**, 2151–2159.
189. Decker, C. and Elzaouk, B. (1995) *Eur. Polym. J.*, **31**, 1155–1161.
190. Beuermann, S., Buback, M., Davis, T.P., Gilbert, R.G., Hutchinson, R.A., Olaj, O.F., Russell, G.T., Schweer, F., and Van Herk, A.M. (1997) *Macromol. Chem. Phys.*, **198**, 1545–1551.
191. Zammit, M.D., Coote, M., Davis, T.P., and Willett, G. (1998) *Macromolecules*, **31**, 995–1002.
192. Matyjaszewski, K. (2002) *Advances in Controlled/Living Radical Polymerization*, ACS Symposium Series.
193. Matyjaszewski, K. (1998) *Controlled Radical Polymerization*, ACS Symposium Series.
194. Bertin, D., Gigmes, D., Le Mercier, C., Marque, S.R.A., and Tordo, P. (2004) *J. Org. Chem.*, **69**, 4925–4930.
195. Matsuda, T. (2006) *Surface-Initiated Polymerization I*, Advances in Polymer Science, Vol. **197**, Springer, Berlin.
196. Nakayama, Y., Miyamura, M., Hirano, Y., Goto, K., and Matsuda, T. (1999) *Biomaterials*, **20**, 963–970.
197. Sellergren, B., Ruckert, B., and Hall, A.J. (2002) *Adv. Mater.*, **14**, 1204–1210.
198. Ishizu, K., Katsuhara, H., Kawauchi, S., and Furo, M. (2005) *J. Appl. Polym. Sci.*, **95**, 413–422.
199. de Boer, B., Simon, H.K., Werts, M.P.L., van der Vegte, E.W., and Hadziioannou, G. (2000) *Macromolecules*, **33**, 349–355.
200. Rahane, S.B., Metters, A.T., and Kilbey, S.M. (2006) *Macromolecules*, **39**, 8987–8962.
201. Matyjaszewski, K., Gnanou, Y., and Leibler, L. (eds) (2007) *Macromolecular Engineering: From Precise Macromolecular Synthesis to Macroscopic Materials Properties and Applications*, vol. **1**, Wiley-VCH Verlag GmbH, Weinheim, pp. 643–672.
202. Gigmes, D., Lalevée, J., and Fouassier, J.P. (to be published).
203. Otsu, T. (2000) *J. Polym. Sci. A: Polym. Chem.*, **38**, 2121–2136.
204. Lalevée, J., Blanchard, N., El-Roz, M., Allonas, X., and Fouassier, J.P. (2008) *Macromolecules*, **41**, 2347–2352.
205. Bertin, D., Boutevin, B., Gramain, P., Fabre, J.M., and Montginoul, C. (1998) *Eur. Polym. J.*, **34**, 85–90.
206. Benoit, D., Grimaldi, S., Robin, S., Finet, J.P., Tordo, P., and Gnanou, Y.J. (2000) *Am. Chem. Soc.*, **122**, 5929–5935.
207. Gigmes, D., Vinas, J., Chagneux, N., Lefay, C., Phan, T.N.T., Trimaille, T., Dufils, P.E., Guillaneuf, Y., Carrot, G., Boue, F., and Bertin, D. (2009) *ACS Symp. Ser.*, **1024**, 245.
208. Scaiano, J.C., Connolly, T.J., Mohtat, N., and Pliva, C.N. (1997) *Can. J. Chem.*, **75**, 92–98.
209. Hu, S., Malpert, J.H., Yang, X., and Neckers, D.C. (2000) *Polymer*, **41**, 445–451.
210. Goto, A., Scaiano, J.C., and Maretti, L. (2007) *Photochem. Photobiol. Sci.*, **6**, 833–839.
211. Guillaneuf, Y., Bertin, D., Gigmes, D., Versace, D.L., Lalevee, J., and Fouassier, J.P. (2010) *Macromolecules*, **43**, 2204–2210.
212. Guillaneuf, Y., Versace, D.L., Bertin, D., Lalevee, J., Gigmes, D., and

Fouassier, J.P. (2010) *Macromol. Rapid Commun.*, **31**, 1909–1912.

213. Elschenbroich, C. (2006) *Organometallics*, Wiley-VCH Verlag GmbH, Weinheim.

214. Amrein, S., Timmermann, A., and Studer, A. (2001) *Org. Lett.*, **3**, 2357–2360.

215. Postigo, A., Kopsov, S., Zlotsky, S.S., Ferreri, C., and Chatgilialoglu, C. (2009) *Organometallics*, **28**, 3282–3287.

216. Postigo, A. and Nudelman, N.S. (2010) *J. Phys. Org. Chem.*, **23**, 910–914.

19
Reactivity of Radicals: Towards the Oxidation Process

19.1
Reactivity of Radicals toward Metal Salts

In different approaches, free radicals were used in combination with metal salts (Eq. (19.1)) for the formation of metal nanoparticles (NPs) (1) and/or cations (2) to initiate a cationic photopolymerization.

$$PI \longrightarrow R^\bullet \ (h\nu)$$
$$R^\bullet + M^+ \longrightarrow \longrightarrow R^+ + M(0) \ NP \qquad (1)$$
$$R^+ + monomer \longrightarrow cationic\ polymerization \qquad (2) \qquad (19.1)$$

In the *in situ* manufacture of silver and gold NPs (Eq. (19.2)), the benzophenone ketyl (BPH$^\bullet$) and the 2-hydroxy isopropyl radicals (generated from ^{3}BP/H-donor and classical Type I photoinitiators (PIs), respectively) can behave as oxidizable radicals [1–3]. The oxidation rate constants of BPH$^\bullet$ by Au(I) and Au(III) determined by LFP are $k_{ox} = 2.5 \times 10^8$ and 9.9×10^8 mol^{-1} l s^{-1}, respectively [4]. The BPH$^\bullet$/Ag$^+$ electron-transfer rate constant is not diffusion limited [5]. The ketyl radicals derived from benzoin derivatives can also be used. Silyl radicals (from the R$_3$Si–H in Type II PIs or polysilanes in Type I PIs) are easily oxidized by a silver salt. The rate constant is unknown, but Ag(0)NP was observed [6]. Aminoalkyl radicals (e.g., produced in the eosin/amine system) can also lead to a metal salt reduction, but this procedure is usually avoided as the presence of an amine and a metal salt (Ag or Au) is often associated with a slow degradation reaction [1–3].

$$M^{n+} + BPH^\bullet \longrightarrow BP + H^+ + M^{(n-1)+}$$
$$(CH_3)_2 C^\bullet OH + Au(III) \longrightarrow (CH_3)_2 C=O + H^+ + Au(II)$$
$$(CH_3)_2 C^\bullet OH + Au(II) \longrightarrow (CH_3)_2 C=O + H^+ + Au(I)$$
$$(CH_3)_2 C^\bullet OH + Au(I) \longrightarrow (CH_3)_2 C=O + H^+ + Au(NP) \qquad (19.2)$$

Such a radical/metal salt interaction can be used (Sections 9.3.16 and 12.5) in (i) FRP due to the presence of initiating radicals or (ii) free-radical-promoted cationic photopolymerization (FRPCP, R$^+$ being the initiating species) [7, 8]. The basic interest is the concomitant formation of the polymer network and the *in situ* generation of metal (Ag, Au, and Pd) NPs. For example, the *in situ* synthesis of

Photoinitiators for Polymer Synthesis: Scope, Reactivity and Efficiency, First Edition.
Jean-Pierre Fouassier and Jacques Lalevée.
© 2012 Wiley-VCH Verlag GmbH & Co. KGaA. Published 2012 by Wiley-VCH Verlag GmbH & Co. KGaA.

silver–epoxy nanocomposites by FRPCP has been reported in [9]. Where possible, the SbF_6^- counter anion that leads to higher rate constants for the propagation of the cationic polymer chain growth is preferred (e.g., in $AgSbF_6$).

19.2
Radical/Onium Salt Reactivity in Free-Radical-Promoted Cationic Photopolymerization

Among the different strategies developed for long wavelength excitation in cationic photopolymerization (Part 3) [10–14], FRPCP appears as an interesting alternative (see Section 12.5 and references therein). In this mode, a compound is required to photochemically generate a radical $R^\bullet$ [15]. This latter has to be oxidized by an onium salt (On^+), the resulting cation (R^+) being the initiating structure. The initiation quantum yield (Φ_i) is defined as $\Phi_i = \Phi_{Rad}\Phi_{R+}\Phi_{M+}$, where Φ_{Rad}, Φ_{R+}, and Φ_{M+} refer to the yield in free radicals and the yields in cations and monomer cations, respectively (Scheme 12.19).

The Φ_{R+} yields are expected to be high, as suitable free radicals can be easily oxidized. Different $R^\bullet/Ph_2I^+$ oxidation rate constants are given in Table 19.1; the ionization potentials (IPs) are also given when available. As expected, free radicals exhibiting quite low IPs (<7 eV) are characterized by high oxidation rate constants k_{ox}. For benzoyl, lophyl, thiyl, alkoxyl, peroxyl, and phenyl radicals, the oxidation is usually quite hard ($k_{ox} < 10^5$ mol^{-1} l s^{-1}) (see above). These radicals cannot be considered as reactive structures in FRPCP.

The Φ_{M+} yields represent the ability of the R^+ cation to react with the monomer M and form the propagating monomer cation $R - M^+$. For cyclic compounds (e.g., epoxides), the reaction corresponds to a ring-opening process as confirmed by the presence of, for example, terminal phosphinoyl groups in the polyether chains (through an analysis of the polymer chain ends by NMR [17]). For vinyl monomers, $R-M^+$ results from the addition of R^+ to the double bond.

Recently, the interaction energies E_{int} (expressed as the exothermicity of the R^++M reaction) calculated by molecular orbital calculations (e.g., UB3LYP/6-31 + G* level in [16]) for various cation/cyclohexene oxide (CHO) couples are reported in Table 19.1. The E_{int}s are strongly affected by the generated cationic species: E_{int} is lower for a carbon-centered structure than for a silyl. This behavior is related to the very high Si–O bond dissociation energy compared to that of the C–O bond [16], which ensures a high reactivity of R_3Si^+ toward the epoxy function. A similar behavior is found for germyl radicals. The E_{int}s in phosphinoyl radical/CHO and silyl radical/CHO couples are quite close. The oxidation of the phosphinoyls is, however, much difficult as evidenced by both higher IP and lower oxidation rate constant by Ph_2I^+. The THF radical can also be oxidized by Ph_2I^+, but the rate constant is not available.

When using a different R^+, the observed further formation of protons (Eq. (19.3)) is related to the fragmentation of the cation (e.g., for the ketyl cation BPH^+) or the reaction of H_2O traces with the cations (e.g., for the silyls; however, this remains a minor pathway [16]). Then, the proton participates to some extent in the

Table 19.1 Ionization potentials (IPs) and oxidation rate constants (mol^{-1} l s^{-1}) of different free radicals with diphenyl iodonium salt.

Radicals	IP (eV)	$k_{ox}{}^a$	E_{int} (cation/CHO) (kJ mol^{-1})
(TMS)$_3$Si• [tris(trimethylsilyl)silyl radical]	6.2	2.6×10^6	152
H–Si(Ph)$_2$–Si(Ph)$_2$•	5.6	2.2×10^8	151
Ph$_3$Ge•	–	5.5×10^7	–
(4-dimethylaminopyridine)→BH$_2$•	–	2.5×10^9	–
Ph$_2$P(=O)•	6.6	$< 1.0 \times 10^6$	145
(2,4,6-trimethylbenzoyl)P(=O)Ph•	–	$< 1.6 \times 10^7$	–
Ph$_2$C•(OH)	5.9	3×10^7	105
9-hydroxyphenanthren-10(9H)-on-9-yl•	–	5×10^6	–

(continued overleaf)

Table 19.1 (continued)

Radicals	IP (eV)	k_{ox}	E_{int} (cation/CHO) (kJ mol^{-1})
Et–O–C(=O)–C$_6$H$_4$–N(CH$_3$)(CH$_2^\bullet$)	5.8	1.6×10^9	72
(4-Me$_2$N-C$_6$H$_4$)C(=O)(C$_6$H$_4$-N(CH$_3$)(CH$_2^\bullet$))	–	1.6×10^7	–
Ph–C$^\bullet$(OCH$_3$)$_2$	5.8	–	85
(HO)C$^\bullet$	6.2	–	109

Interaction energy of the different corresponding ions with cyclohexene oxide (CHO) used as a model epoxide.
^{a}From [16, 17].

initiation process.

$$BPH^+ \longrightarrow BP + H^+$$
$$R_3Si^+ + H_2O \longrightarrow R_3Si\text{–}OH + H^+ \quad (19.3)$$

Sulfonium salts are dramatically less efficient than iodonium salts in FRPCP, the reduction potentials of Ph$_2$I$^+$ and Ph$_3$S$^+$ being about −0.2 and −1.1 V, respectively [18]. This behavior is well exemplified by the low initiating ability of camphorquinone/R$_3$SiH/Ph$_3$S$^+$ compared to CQ/R$_3$SiH/Ph$_2$I$^+$ in the cationic photopolymerization of epoxides under visible lights [19]. A weak tris(trimethylsilyl)silyl/Ph$_3$S$^+$ interaction is noted, that is, this system leading to a quite efficient polymerization [19]. Vinyl-cation-based initiating systems of epoxy ring-opening photopolymerization have also been proposed. Vinyl radicals can be oxidized by sulfonium salts. In these systems, photochemical free radical sources can be also used in conjunction with diphenyl acetylene (DPA) and triphenyl sulfonium hexafluorophosphate, the addition of radicals to the alkyne undergoing the formation of vinyl radicals [20].

In contrast to sulfonium salts, ethyl-(tetrahydrothiophenium methyl)acrylate hexafluoroantimonate ETM$^+$SbF$_6^-$ and N-ethoxy-2-methylpyridinium (EMP$^+$) that are characterized by lower reduction potentials than Ph$_3$S$^+$ (−0.58 V) can be more easily used in FRPCP [21].

References

1. Scaiano, J.C., Billone, P., Gonzalez, C.M., Maretti, L., Marin, M.L., McGilvray, K.L., and Yuan, N. (2009) *Pure Appl. Chem.*, **81**, 635–647.
2. McGilvray, K.L., Decan, M.R., Wang, D., and Scaiano, J.C. (2006) *J. Am. Chem. Soc.*, **128**, 15980–15981.
3. Marin, M.L., McGilvray, K.L., and Scaiano, J.C. (2008) *J. Am. Chem. Soc.*, **130**, 16572–16584.
4. Lalevée, J., Gigmes, D., Bertin, D., Graff, B., Allonas, X., and Fouassier, J.P. (2007) *Chem. Phys. Lett.*, **438**, 346–350.
5. Stamplecoskie, K.G. and Scaiano, J.C. (2011) *J. Am. Chem. Soc.*, **133**, 3913–3920.
6. Souane, R., Tehfe, M.A., Lalevée, J., Gigmes, D., and Fouassier, J.P. (2010) *Macromol. Chem. Phys.*, **211**, 1441–1445.
7. Sangermano, M., Yagci, Y., and Rizza, G. (2007) *Macromolecules*, **40**, 8827–8829.
8. Yagci, Y., Sangermano, M., and Rizza, G. (2008) *Polymer*, **49**, 5195–5198.
9. Yagci, Y., Sangermano, M., and Rizza, G. (2008) *Chem. Commun.*, 2771–2777.
10. Allen, N.S. (1996) *J. Photochem. Photobiol. A Chem.*, **100**, 101–109.
11. Saeva, F.D., Breslin, D.T., and Martic, P.A. (1989) *J. Am. Chem. Soc.*, **111**, 1328–1334.
12. Hacker, N.P. and Larson, C.E. (1988) US Patent 4, 760, 013.
13. Crivello, J.V. and Lee, J.L. (1981) *Macromolecules*, **14**, 1141–1147.
14. Yagci, Y., Lukac, I., and Schnabel, W. (1993) *Polymer*, **34**, 1130–1133.
15. Ledwith, A. (1978) *Polymer*, **19**, 1217–2122.
16. Lalevée, J., Tehfe, M.A., Morlet-Savary, F., Graff, B., and Fouassier, J.P. (2011) *Prog. Org. Coat.*, **70**, 23–31.
17. Dursun, C., Degirmenci, M., Yagci, Y., Jockusch, S., and Turro, N.J. (2003) *Polymer*, **44**, 7389–7396.
18. Yagci, Y. and Reetz, I. (1998) *Prog. Polym. Sci.*, **23**, 1485–1538.
19. Lalevée, J., El-Roz, M.A., Allonas, X., and Fouassier, J.P. (2008) *J. Polym. Sci. A: Polym. Chem.*, **46**, 2008–2014.
20. Okan, A., Serath, I.E., and Yagci, Y. (1996) *Polym. Bull.*, **37**, 723–728.
21. Denizligil, S., Yagci, Y., and Mc Ardle, C. (1995) *Polymer*, **36**, 3093–3098.

Conclusion

In this book, we have presented the state-of-the-art in photoinitiators and photosensitizers of free radical, cationic, anionic polymerization as well as photoacid- and photobase-induced photo-cross-linking reactions carried out under very different experimental conditions for many various applications in a lot of industrial sectors. It outlines the huge and continuously emerging development of tailor-made systems due to the wide opportunities in organic synthesis, the better knowledge of the excited state processes, and the fruitful approach of the reactivity/efficiency relationships. Photopolymerization reactions are not restricted now to the radiation curing area. Many other new applications are found in (laser) imaging technologies, microelectronics, medicine, life sciences, optics, optoelectronics, holography, and telecommunication. Trends in research and development will reflect both the needs of industry for new products and the identification of new topics for attention and will shape the future works. They are obviously inscribed in the current topics detailed throughout this book. More specifically, nonexhaustive examples of promising trends can be gathered in three main directions[1]:

1) The development of smart compounds:
   a. presenting a high (photo)chemical reactivity for fast curing speeds, well adapted to polychromatic UV or/and visible lights (Hg or Xe lamps and doped Hg lamps), or (quasi)monochromatic lights (high-power LED arrangements, lasers, and diode laser arrays);
   b. keeping a high photosensitivity and a high reactivity under soft irradiation conditions (low intensity, under air) and/or on specific visible wavelengths for the polymerization of low viscosity monomers;
   c. leading to new photosensitizers, cleavable-bond-containing photoinitiators, co-initiators based on new chemistries (e.g., Si, Ge, B, and metal-centered radicals), organometallic derivatives having specific properties, long wavelength-absorbing compounds, green and red laser-light-photosensitive formulations, and tailor-made multicomponent photoinitiating systems;

---

1) For the future of the specific RadTech market, see the papers of D. Harbourne and Shi Weng Fang in Proc. RadTech Asia, 2011, S. Peeters in Proc. RadTech Europe, 2011.

Photoinitiators for Polymer Synthesis: Scope, Reactivity and Efficiency, First Edition.
Jean-Pierre Fouassier and Jacques Lalevée.
© 2012 Wiley-VCH Verlag GmbH & Co. KGaA. Published 2012 by Wiley-VCH Verlag GmbH & Co. KGaA.

d. developing new functionalities: tunable photoinitiators, dual radical/cationic photoinitiators, photoinitiators behaving as photocatalysts, one-pot photoinitiators, photolatent catalysts for two-pack resin coatings, acid or base releasing systems (PAG and PBG);
e. proposing new initiating cations in free-radical-promoted cationic photopolymerization (FRPCP);
f. decreasing the oxygen inhibition: this is particularly of interest for (i) free radical photopolymerization (FRP) under soft conditions (LED bulbs etc.) and (ii) the really interesting FRPCP route in cationic photopolymerization;
g. exhibiting new handling or end-use properties (using, e.g., macro-, hyperbranched or immobilized photoinitiators), improved solubility, low toxicity concern, low cost, and low carbon footprint;
h. introducing novel final properties through (i) the design of new radical and cationic monomers/oligomers, for example, low ceiling temperature compounds, sol–gel hybrid monomers, silicon-containing polymers, and nanoparticle-incorporating formulations, (ii) the improvement of existing strategies: thiol-ene (or thiol-yne), silyl-ene, dual cure, hybrid cure, water-based resin curing, and FRPCP, and (iii) the proposal of self-healing materials under light irradiation;
i. allowing a biocompatibility in (photoadaptable) hydrogels, tissue-engineering applications, and bone replacement materials;
j. providing a control of the photopolymerization (using photoiniferters, nitroxide-mediated photopolymerization NMP2, photo-RAFT and photo-ATRP) for achieving the manufacture of complex macromolecular architectures or a controlled grafting (e.g., for formation of covalently bound multilayer coatings);
k. strengthening new interesting fields: (i) for applications, for example, in automotive, flexible electronics, optoelectronics, photovoltaics, displays, coil coatings, UV flexo, screen printing, secure printing, UV drop-on-demand inkjet, food packaging, and fiberglass composites, (ii) UV-curable textile finishing; easy cleanable protective coatings; coatings with improved hardness, scratch resistance, antimicrobial and antireflective properties, and direct-to-metal UV coatings, (iii) UV LED curing (e.g., of inks), (iv) *in situ* incorporation of nanoparticles (for magneto-optics, optoelectronics and biosensors); manufacture of nanocomposites (improvement of the coating properties, etc.); (v) development of photopolymerized materials possessing specific properties (shape memory, optical, electrical, and magnetic), and (vi) photopolymerization in various experimental conditions; photopolymerization of (or in) ionic liquids; and molecular self-assembly bottom-up strategies.

2) The development of a green chemistry:
a. reducing the volatile organic compounds (VOC) emissions, improving air quality, developing solvent-less technologies, and manufacturing durable products;

b. involving renewable (vegetable-oil-based resins, etc), biocompatible (biodegradable compounds, and biophotoresists), environmental friendly (polycaprolactone, polylactide, etc.), safe (low volatility, negligible toxicity, etc.), and reworkable (de-cross-linkable polymers) materials;
  c. using sun or new light sources (household fluorescent bulbs, white or colored LED bulbs, and diode lasers).
3) The development of nanoscale photopolymerization reactions:
  a. ensuring an *in situ* incorporation of metal nanoparticles using well-designed photoinitiating systems;
  b. allowing the manufacture of micro-/nanostructured materials (nanorods, nanowires, nanospheres, nanogels, and sol–gel hybrid nanomaterials), lab-on-a-chip devices, anisotropic surfaces, and carbon nanotubes/photopolymerized lipidic assemblies (CNTs/PLAs) constructs;
  c. using two-photon absorption excitation processes (3D writing, nanolithography, and stereolithography), near-field optical techniques (integration of optical devices), and plasmon-mediated photopolymerization;
  d. introducing new properties using, for example, antimicrobial Ag(0)-nanoparticle-containing films, polymer–clay or silica nanocomposites, metal-polymer conductive coatings, and photoluminescent materials for OLED applications.

The researcher ingenuity has no limit. There is no doubt that ever more efficient and well-adapted new photoinitiating systems as well as new innovative applications will be proposed in the coming years. We sincerely hope that in the future authors of books dealing with this subject will face a mountain of references!

November 2011   *Prof. Jean Pierre Fouassier and Prof. Jacques Lalevée*

Index

a

absorption of light by a molecule   74
acetals   204
acrylamide photopolymerization   376
acrylate and methacrylate radicals   403–404
acrylates   45–47
acylsulfonium salts   301
addition/fragmentation reaction   331
addition of a monomer radical to a monomer unit   442
addition rate constants to monomer   392
addition reaction, reversibility   393
Ag(NP)   455
alcohols and tetrahydrofuran (THF)   214
aldehydes   204, 229, 237–238
aliphatic ketones   229
alkoxyamines   183, 212–213
alkyl and carbon-centered radicals   399–401
alkylphenylglyoxylates   225, 228
aluminate complexes   178
amines   201–203
aminoalkyl radicals
 – amine class role   408
 – aminoalkyl radicals and the halogen abstraction reaction   410–411
 – chain length effect   409
 – hydrogen abstraction reaction regioselectivity   409–410
 – N-phenyl glycine derivatives   408–409
 – reactivity   404–408
 – – under air   411
amino ketones   137
aminobenzophenones   218
aminyl radicals   431–432
ammonium tetraorganyl borate salts   351–352
anionic photoinitiators   343

 – cyano derivative/amine system   344
 – inorganic complexes   343
 – organometallic complexes   343–344
 – photosensitive systems for living anionic polymerization   344
anionic photopolymerization   55, 59
anthraquinones   228
Arrhenius's law   98
aryl radicals   401–402
ATRP   57
Au(NP)   455
azides and aromatic bis-azides   168
azo derivatives   168

b

BAPO   162
Barton's ester derivatives   174
benzodioxinone/iodonium salt   320
benzoin derivatives   129–132
 – ether derivatives   132–136
 – ether series   149
benzophenone- and thioxanthone-moiety-based cleavable systems   158
 – benzophenone phenyl sulfides   158
 – benzophenone-sulfonyl ketones   159
 – benzophenone thiobenzoates   159
 – cleavable benzophenone, xanthone, and thioxanthone derivatives   160–161
 – halogenated derivatives   160
 – ketosulfoxides   159
benzophenone phenyl sulfides   158
benzophenones   218–219, 237
 – absorption   215
 – aminobenzophenones   218
 – excited states   215–217
 – photolysis   218
 – photopolymerization activity   219

benzophenone-sulfonyl ketones 159
benzophenone thiobenzoates 159
benzoxazines 204
benzoyl-chromophore-based photoinitiators 127–129
– benzoin derivatives 129–132
– benzoin ether derivatives 132–136
– benzoin ether series 149
– dialkoxyacetophenones and diphenylacetophenones 136
– halogenated ketones 136
– hydroxy alkyl acetophenones 137, 139, 151–154
– ketone sulfonic esters 140
– macrophotoinitiators 156–157
– modified sulfonyl ketones 154
– morpholino and amino ketones 137, 149–151
– oxime esters 148
– oxysulfonyl ketones 147–148
– substituent effect limit 154–156
– sulfonyl ketones 144–147
– supported cleavable photoinitiators 157
– thiobenzoate derivatives 140, 144
benzoyl phosphine oxide derivatives
– absorption properties and photolysis 164
– bis-acyl phosphine oxide and phenolic compound interaction 165
– compounds 161–162
– excited state processes 162–163
benzoyl radicals 402–403
N-benzylated-structure-based photobases 352–353
biacetyl 223
biradical-generating ketones 166
bis-acyl phosphine oxide and phenolic compound interaction 165
bisarylimidazole derivative and additive 244
bond dissociation energy 388
borane complexes 211–212
boryl radicals 435–436
– photoinitiator under air 438–439
– reactivity 436–438
– – under air 438

c

cage effects 85–86
camphorquinone 224–225
carbamates 351
carbon-centered radicals 399
carbon–germanium cleavable-bond-based derivatives 170–172
carbon–silicon and germanium–silicon cleavable–bond-based derivatives 172

cationic photinitiating systems 289
– diazonium salts 289
– free-radical-promoted cationic photopolymerization
– – addition and fragmentation reaction 331–332
– – applications 331
– – one-component radical source-onium salt system 330
– – onium salt role 329–330
– – photoinitiation step 330–331
– – radical/metal-salt-based systems 331
– – radical/onium salt interaction 318
– – radical source/iodonium salt three-component 320–329
– – radical source/iodonium salt two-component 318–320
– onium salts 289, 303–304
– – absorption 304–305
– – amphifunctionality 308
– – benzene release 307
– – iodonium and sulfonium salts 289–303
– – odor 307
– – and photosensitizer systems 311–318
– – solubility 305
– – stability 305, 307
– – toxicity 307–308
– organometallic derivatives 308
– – inorganic transition metal complexes 310
– – non-transition-metal complexes 310
– – organometallic complexes transition 308–310
– photosensitive systems
– – for hybrid cure 333
– – for living cationic polymerization 332–333
cationic photopolymerization 58, 105–106
chain length effect 409
charge-transfer photopolymerization 53–54
chemical amplification 349
chemical reactivity 89
chromium complexes 177
CIDEP 92
CIDNP 92
cleavable benzophenone, xanthone, and thioxanthone derivatives 160–161
cleavable ketones as type II photoinitiators 229
cleavable photoinitiators
– conventional, and silicon chemistry 172–173
– in living polymerization

– – cleavable C–S- and S–S-bond-based photoiniferters   179–181
– – TEMPO-based alkoxyamines   182–183
– nanoparticle-formation-mediated   185
– supported   157
– for two-photon absorption   184
cleavage process   94
coating properties   24
composites, manufacture of   25
computer-to-plate technology   27
controlled polymerization   43
controlled radical polymerization, photosensitive systems, for   182–183, 212, 443–444
coumarin/amine/ferrocenium salt   277–278
coumarins   225
CW lasers   17–18
cyano derivative/amine system   344

d

decatunstate   326
dialkoxyacetophenones and diphenylacetophenones   136
diazonium salts   289
4,4′-diethylamino benzophenone (EAB)   218
diffusion, role of   84–85
difunctional compounds   152–154
digermane and distannane derivatives   170
diketones   167
– aromatic   223–224
– camphorquinone   224–225
diode lasers   359, 362, 363
diselenide and diphenylditelluride derivatives   170
disilane derivatives   169–170
disulphide derivatives   168–169
dithiocarbamyl radicals   443
donor and acceptor systems   243
doped lamps   14, 15
dual cure photopolymerization   54–55
dye/amine/metal salt   27–273
dye/amine/onium salt and bromo compound   270
dye/amine/triazine derivative   271
dye-based systems
– amine systems   238–239, 240–241
– amine water-soluble systems   241
– coinitiator systems   239–240
– kinetic data   241
dye/borate/additive   273
dye/ferrocenium salt/amine/hydroperoxide   278
dye/ferrocenium salt/hydroperoxide   277
dye/ketone/amine   272

e

efficiency of photopolymerization reaction   103
– environment role   117–18
– kinetic laws
– – cationic photopolymerization   105–106
– – laser-induced photopolymerization   107–108
– – photopolymerization kinetics in bulk   109
– – photopolymerization rate dependence   106–107
– – radical photopolymerization   103–105
– light absorption by pigment   112–113
– light stabilizers absorption   115–117
– monitoring of photopolymerization reaction   109, 110–111
– – FTIR analysis   109–110
– – optical pyrometry   110
– – photocalorimetry   110
– oxygen inhibition   113–115
– versus reactivity   111–112
efficiency versus reactivity   111
electromagnetic radiation   11–12
electron and proton hydrogen transfer (EPHT)   95
electron transfer   95–96, 97
– photosensitization through
– – photoinitiaton step   317–318
– – photosensitizers   312–317
emission spectra   12, 13, 14, 15, 16, 17, 19
energy transfer   96–97, 312
ESR   92
excimer lamps   14–15
excited state processes   78, 89
– kinetics   78–80
experimental devices and application examples   21
– computer-to-plate technology   27
– conventional printing plates   25
– holography   27
– laser direct imaging   26–27
– medical applications   28–29
– microelectronics applications   26
– nano-object fabrication   29
– objects and composite manufacture   25
– optics   28
– photopolymerization
– – and nanotechnology   32–33
– – using near-field optical techniques   29
– search for green chemistry   33–34
– search for new properties and new end uses   30–32
– stereolithography   25–26

experimental devices and application examples (contd.)
– UV curing area
– – coating properties   24
– – end uses   21–24
– – equipment   21
exposure reciprocity law (ERL)   76–77

f

femtosecond lasers   18
ferrocenium salts   248–250
film cationic photopolymerization   48–49
filters   12
fluorenones   229
free-radical-promoted cationic photopolymerization
– addition and fragmentation reaction   331–332
– applications   331
– one-component radical source-onium salt system   330
– onium salt role   329–330
– photoinitiation step   330–331
– radical/metal-salt-based systems   331
– radical/onium salt interaction   318
– radical source/iodonium salt
– – three-component   320–329
– – two-component systems   318–320
frontal photopolymerization   43
FTIR analysis   109

g

germane/iodonium salt   320
germanes and stannanes   210–211
germyl and stannyl radicals   432–433
– reactivity   433–434
– – under air   434
– (TMS)$_3$Ge$^•$ versus (TMS)$_3$Si$^•$ reactivity and structure   434–435
gradient photopolymerization   43
green chemistry   33–34

h

HABI derivatives   244, 276
halogen abstraction reaction   410
halogenated derivatives   160
halogenated ketones   136
heterogeneous (micro) media   42
Hg lamps   359, 362
highly reactive acrylate monomers (HRAMs)   46
hindered amine light stabilizer (HALS)   202
holography   27
household lamps   18–19

hybrid cure, photosensitive systems for   333
hybrid cure photopolymerization   55
hybrid sol–gel photopolymerization   59
hydrocarbon and amine   257
hydrogels   30
hydrogen donors   201
– acetals   204
– alcohols and tetrahydrofuran (THF)   214
– aldehydes   204
– alkoxyamines   212–213
– amines   201–203
– benzoxazines   204
– borane complexes   211–212
– germanes and stannanes   210–211
– hydroperoxides   205
– metal-(IV) and amine-containing structures   208–210
– monomers   213–214
– phosphorus-containing compounds   211
– photoinitiator   214
– polymer substrate   214
– silanes   205–207
– silicon-hydride-terminated surface   214–215
– silylamines   207–208
– silyloxyamines   210
– thio derivatives   203
hydrogen transfer processes   95–96
hydroperoxides   205
hydroxamic and thiohydroxamic acids and esters   174–176
hydroxy alkyl acetophenones   137, 139
– difunctional compounds   152–154
– oil- and water-soluble compounds   151–152
hydroxy alkyl conjugated ketones   158
hydroxy alkyl heterocyclic ketones   157–158

i

iminosulfonates and oximesulfonates   347–348
iminyl radicals   439–440
initiating radicals   404, 405, 421, 423, 443, 447
inner filter effect   81
inorganic complexes   343
inorganic transition metal complexes   310
interpenetrate polymer network   333
iodonium and sulfonium salts   289
– absorption properties   293–294
– acylsulfonium salts   301
– anion role   292–293
– compounds   289–290
– iodonium salts decompostion process

– – iodonium salt/additive combination 298–299
– – iodonium salts behavior in photopolymerization 298
– – mechanism 294–295
– – phenyl iodinium cation radical reaction 296–297
– – phenyl radical role 297
– – photoinitiation step 298
– – photolysis 296
– photopolymerization reaction 290–292
– substituted salt derivatives 301–303
– sulfonium salts decomposition process
– – mechanism 299–300
– – photoinitiation step 301
– – photolysis 300–301
iridium complexes 279, 327
isopropylthioxanthone 220

j
Jablonski's diagram 78

k
ketocoumarins 225
ketone
– amine 271
– amine/imide derivatives 270–271
– amine/onium salt and bromo compound 269–270
– iodonium salt 318–319
– and ketone-based systems 246–247
– sulfonic esters 140
ketone-and hydrogen-donor-based systems
– aldehydes 229
– aliphatic ketones 229
– alkylphenylglyoxylates 225, 228
– anthraquinones 228
– basic mechanisms 199–200
– benzophenone derivatives 215–219
– cleavable ketones as type II photoinitiators 229
– coumarins 225
– diketones 223–225
– fluorenones 229
– hydrogen donors 201–215
– ketocoumarins 225
– low- and high-molecular-weight compounds and macrophotoinitiators 229–232
– low-molecular-weight one-component systems 236–238
– oxygen self-consuming thioxanthone derivatives 236
– photomasked photoinitiator 235–236

– reactivity 232–233
– thioxanthone derivatives 219–223
– two-photon absorption photoinitiators 235
– water-soluble compounds 233–234
ketone/germane/iodonium, salt 323
ketone/silane/iodonium salt 322
ketosulfoxides 159
ketyl radicals 399, 456

l
laser direct imaging 26–27
laser-induced photoacoustic calorimetry (LIPAC) 91
laser-induced photopolymerization 107–108
laser sources 17–18
LED 359, 363
light absorbing amine and monomer 257
light absorbing transients (LATs) 218
light-emitting diodes (LEDS) 16–17
light intensity effect 362
light-sensitive alkoxyamines and generation of nitroxides 445–446
light sources 5, 11
– characteristics of 12
– conventional and unconventional 13
– – doped lamps 14, 15
– – excimer lamps 14–15
– – household lamps 18–19
– – laser sources 17–18
– – light-emitting diodes (LEDS) 16–17
– – mercury arc lamp 14, 15
– – microwave lamps 14, 15
– – pulsed light sources 17
– – sun 18
– – UV plasma source 19
– – Xenon lamp 13
– electromagnetic radiation 11–12
light stabilizers 115–117
light stabilizers, absorption of 115
living anionic polymerization, photosensitive systems for 344
living cationic polymerization, photosensitive systems for 332
lophyl radicals 439
low- and high-molecular-weight compounds and macrophotoinitiators 229–232
low-molecular-weight one-component systems 236–238

m
macrophotoinitiators 156–157
magnetic field 44
maleimide and amine and photoinitiator 241, 243

Marcus theory 95
medical applications 28
mercaptans 203
mercury arc lamp 14, *15*
metal-(IV) and amine-containing structures 208–210
metal carbonyl
– and silane 250
– silane/hydroperoxide 275–276
metal carbonyl/silane/iodonium salt 324
metal-centered radicals 440–442
metallopolymers 30
metal-releasing compound 178–179
metal salts 455
– and metallic salt complexes 178
metathesis photopolymerization 56
methacrylate radicals 403
microgels 30
microelectronics, applications in 26
microwave lamps 14, *15*
Mischler's ketone (MK) 218
molecular orbital calculations 98
monomers 213–214
– oligomer film photopolymerization 373, 376
– photopolymerization in organic solvents and bulk media 367, 370–371
– quenching 378–380
morpholino and amino ketones 137, 149–151
multicomponent photoinitiating systems 269
– coumarin/amine/ferrocenium salt 277–278
– dye/amine/metal salt 272–273
– dye/amine/onium salt and bromo compound 270
– dye/amine/triazine derivative 271
– dye/borate/additive 273
– dye/ferrocenium salt/amine/hydroperoxide 278
– dye/ferrocenium salt/hydroperoxide 277
– dye/ketone/amine 272
– ketone/amine/imide derivatives 270–271
– ketone, amine/onium salt and bromo compound 269–270
– ketone/ketone/amine 271
– metal carbonyl compound/silane/hydroperoxide 275–276
– organometallic compound/silane/iodonium salt 279
– photosensitizer/amine/HABI derivatives/onium salt 276–277
– photosensitizer/Cl-HABI/additive 273–275
– triazine-derivative-containing three-component systems 272
– type II photoinitiator/silane 279–281
multilayers 44
multiphotonic absorption 77

n
nanogels 30
nano-object fabrication through two-photon absorption polymerization 29
nano-objects, fabrication of 29
nanoparticle-formation-mediated cleavable photoinitiators 185
nanoparticle-formation-mediated type II photoinitiators 256–257
nanoparticles (NPs) 455
nanosecond laser flash photolysis 89
nanotechnology 32
naphthalimides 348–349
near-field optical techniques 29
nitro compound and amine 257
nitroxides 445
nonvertical energy transfer (NVET) 96–97
NMP2 57
non-transition-metal complexes 310

o
oil- and water-soluble compounds 151–152
oil-soluble monomers photopolymerization in direct micelles 376
one-component photoinitiating systems
– azides and aromatic bis-azides 168
– azo derivatives 168
– Barton's ester derivatives 174
– benzophenone- and thioxanthone-moiety-based cleavable systems 158
– – benzophenone phenyl sulfides 158
– – benzophenone-sulfonyl ketones 159
– – benzophenone thiobenzoates 159
– – cleavable benzophenone, xanthone, and thioxanthone derivatives 160–161
– – halogenated derivatives 160
– – ketosulfoxides 159
– benzoyl-chromophore-based photoinitiators 127–129
– – benzoin derivatives 129–132
– – benzoin ether derivatives 132–136
– – dialkoxyacetophenones and diphenylacetophenones 136
– – halogenated ketones 136
– – hydroxy alkyl acetophenones 137, 139

– – ketone sulfonic esters   140
– – morpholino and amino ketones   137
– – oxime esters   148
– – oxysulfonyl ketones   147–148
– – sulfonyl ketones   144–147
– – thiobenzoate derivatives   140, 144
– benzoyl phosphine oxide derivatives
– – absorption properties and photolysis   164
– – bis-acyl phosphine oxide and phenolic compound interaction   165
– – compounds   161–162
– – excited state processes   162–163
– biradical-generating ketones   166
– carbon–germanium cleavable-bond-based derivatives   170–172
– carbon–silicon and germanium–silicon cleavable–bond-based derivatives   172
– cleavable photoinitiators for two-photon absorption   184
– cleavable photoinitiators in living polymerization
– – cleavable C–S- and S–S-bond-based photoiniferters   179–181
– – TEMPO-based alkoxyamines   182–183
– digermane and distannane derivatives   170
– diketones   167
– diselenide and diphenylditelluride derivatives   170
– disilane derivatives   169–170
– disulphide derivatives   168–169
– hydroxamic and thiohydroxamic acids and esters   174–176
– hydroxy alkyl conjugated ketones   158
– hydroxy alkyl heterocyclic ketones   157–158
– metal-releasing compound   178–179
– metal salts and metallic salt complexes   178
– miscellaneous systems   185
– nanoparticle-formation-mediated cleavable photoinitiators   185
– organoborates   176
– organometallic compounds   176
– – aluminate complexes   178
– – chromium complexes   177
– – titanocenes   177
– oxyamines
– – alkoxyamines   183
– – silyloxyamines   184
– peresters   173
– peroxides   166–167
– phosphine oxide derivatives   165

– silicon chemistry and conventional cleavable photoinitiators   172–173
– substituted benzoyl-chromophore-based photoinitiators   148–149
– – benzoin ether series   149
– – hydroxy alkyl acetophenone series   151–154
– – macrophotoinitiators   156–157
– – modified sulfonyl ketones   154
– – morpholino ketone and amino ketone series   149–151
– – substituent effect limit   154–156
– – supported cleavable photoinitiators   157
– sulfur–carbon cleavable-bond-based derivatives   173
– sulfur–silicon cleavable-bond-based derivatives   173
– trichloromethyl triazines   165
– UV-light-cleavable bonds   185–186
onium salt   303–304, 329–330, 456–458. *See also under individual entries*
– absorption   304–305
– amphifunctionality   308
– benzene release   307
– iodonium and sulfonium salts   289–303
– odor   307
– and photosensitizer systems   311–312
– – photosensitization through electron transfer   312–318
– – photosensitization through energy transfer   312
– solubility   305
– stability   305, 307
– toxicity   307–308
optical pyrometry   110
optical transitions
optics   28
organoborates   176
organometallic complexes   343–344
organometallic compounds   176
– aluminate complexes   178
– chromium complexes   177
– and ketone-based systems   247
– silane/iodonium salt   279
– titanocenes   177
organometallic derivatives   248, 308
– inorganic transition metal complexes   310
– non-transition-metal complexes   310
– organometallic complexes transition   308–310
oxime esters   148, 351
oximesulfonates   347

oxyamines
– alkoxyamines   183
– silyloxyamines   184
oxygen inhibition   114
oxygen effect   362
oxygen self-consuming thioxanthone derivatives   236
oxyl radicals   425–426
oxysulfonyl ketones   147–148

p

penultimate unit effect   409
peresters   173
peroxides   166–167
peroxyl radicals   426–428
– interaction with H-donors   428–430
– interaction with monomers   430
– interaction with triphenylphosphine   430
– S_H2 substitution   430–431
phenacylsulfonium salts   301
phenolic compounds interaction with   165, 399
phenolic compounds (POHs)   203
phenols   117
N-phenyl glycine derivatives   408–409
phosphine oxide derivatives   165
phosphorus-centered radicals   412–413
phosphorus-containing compounds   211
photoacid generators (PAG) systems   347
– chemical amplification   349
– iminosulfonates and oximesulfonates   347–348
– naphthalimides   348–349
photoactivated hydrosilylation reactions   61
photoassisted oxypolymerization   43
photobase generators (PBG) systems   351, 353–354
– ammonium tetraorganyl borate salts   351–352
– N-benzylated-structure-based photobases   352–353
– carbamates   351
– oxime esters   351
– photobases and base proliferation processes   354
photobases and base proliferation processes   354
photocalorimetry   110
photocatalysts   279
photochemical reactivity   89
– cleavage process   94–95
– energy transfer   96–97
– excited-state process analysis

– – CIDNP, CIDEP, and ESR spectroscopy   92–93
– – direct detection of radicals   92
– – nanosecond laser flash photolysis   89
– – photothermal techniques   90–91
– – picosecond-pump-probe-spectroscopy   90
– – time-resolved FTIR spectroscopy   91–92
– hydrogen transfer processes   95–96
– quantum mechanical calculations   93
– radical reactivity   98–99
photo-cross-linking   3–6
– reactions, in photobases and photoacids presence   59–60
photocuring   5
photoinduced polymerization initiation step
– cage effects   85–86
– competitive reactions in excited states   83
– diffusion role   84–85
– initiating species production   82–83
photoiniferters and dithiocarbamyl radicals   443–444
photoinitiating systems
– photoinitiator-free systems and self-initiating monomers   283
– self-assembled photoinitiator monolayers   284–285
– semiconductor nanoparticles   284
photoinitiation step   82
photoinitiation under air   422–424
photoinitiator   214
– alcohol/iodonium salt   328–329
– alkylhalide/iodonium salt   321
– amine/iodonium salt   321
– and disulfide and group 4B dimetal derivatives   252–253
– and onium salts   254
– and peroxide- and hydroperoxide-based systems   251–252
– and phosphorus-containing compounds   253–254
– and photosensitizer   80–81
– silane/iodonium salt   321–327
photoinitiator-free systems and self-initiating monomers   283
photolatent systems   59
photomasked photoinitiator   235–236
photomaterials   6
photopolymerization reactions   3–6. See also individual entries
– anionic photopolymerization   55, 59
– cationic photopolymerization reactions   58
– charge-transfer photopolymerization   53–54
– dual cure photopolymerization   54–55

- encountered reactions, media, and experimental conditions 41–44
- film cationic photopolymerization 48–49
- film radical of acrylates 45–7
- hybrid cure photopolymerization 55
- hybrid sol–gel photopolymerization 59
- metathesis photopolymerization 56
- nanotechnology and 32–33
- using near-field optical techniques 29
- photoactivated hydrosilylation reactions 61
- photo-cross-linking reactions in photobases and photoacids presence 59–60
- photopolymerization
- – – of powder formulations 52–53
- – – of water-borne light curable systems 52
- radical photopolymerization reactions 56–58
- remote curing 60–61
- thiol-ene-photopolymerization 49–51
- two-photon absorption-induced polymerization 60

photosensitive systems 73
- absorption 81–82
- excited state process kinetics 78–80
- exposure reciprocity law (ERL) 76–77
- general properties 73
- for hybrid cure 333
- Jablonski's diagram 78
- light absorption and optical transitions 74–76
- for living anionic polymerization 344
- for living cationic polymerization 332–333
- molecular orbitals and energy levels 74
- multiphotonic absorption 77
- photoinduced polymerization initiation step
- – – cage effects 85–86
- – – competitive reactions in excited states 83
- – – diffusion role 84–85
- – – initiating species production 82–83
- – photoinitiator and photosensitizer 80–81
- reactivity 86–87

photosensitizer 80
- amine/HABI derivatives/onium salt 276–277
- Cl-HABI/additive 273–275
- -linked photoinitiator and coinitiator-based systems 250–251
- and titanocenes 254–255
- and triazine derivative 257–258

photostabilizers, see light stabilizers
photothermal techniques 90–91
picosecond-pump-probe-spectroscopy 90
pigment, absorption of light 112
polymerization rate 106

polymer substrate 214
polysilane/iodonium salt 319–320
posteffect 104
powder formulations 52–53
propagating radicals 442–443
pulsed laser-induced polymerization (PLP) 42
pulsed lasers 18
pulsed light sources 17
pure hydrogen transfer (PHT) 95
pyridinium salts 303
pyrylium and thiopyrylium salts and additive 245–246

q
quantum mechanical calculations 93

r
radical/metal-salt-based systems 331
radical photoinitiators
- excited-state reactivity
- – – monomer quenching 378–380
- – – radical generation 377–378
- medium role on photoinitiator reactivity 381
- – – reactivity in bulk 386–388
- – – reactivity in microheterogeneous solution 383–286
- – – reactivity in solution 381–383
- performance 359–360
- – – light intensity role 361–364
- – – surrounding atmosphere role 361
- – – viscosity role 360–361
- reactivity and efficiency of 367
- relative efficiency 367
- – – acrylamide photopolymerization in water 376
- – – MMA photopolymerization in organic solvents 367
- – – monomer/oligomer film photopolymerization 373, 376
- – – monomers photopolymerization in organic solvents and bulk media 367, 370–371
- – – oil-soluble monomers photopolymerization in direct micelles 376
- – – TMPTA photopolymerization in film 371–373, 374–376
- – – water-soluble monomers photopolymerization in reverse micelles 377
- structure/property relationships in photoinitiating systems 388

radical photoinitiators (contd.)
– – addition rate constants to monomer 392
– – bond dissociation energy in cleavable systems 388–389
– – bond dissociation energy in noncleavable systems 389–391
– – first monomer radical generation 391–392
– – radical/monomer reactivity description 392–394
radical photopolymerization 56–58, 103–105
radical reactivity 399
– acrylate and methacrylate radicals 403–404
– alkyl and carbon-centered radicals 399–401
– aminoalkyl radicals
– – amine class role 408
– – aminoalkyl radicals and the halogen abstraction reaction 410–411
– – chain length effect 409
– – hydrogen abstraction reaction regioselectivity 409–410
– – N-phenyl glycine derivatives 408–409
– – reactivity 404–408
– – reactivity under air 411
– aminyl radicals 431–432
– aryl radicals 401–402
– benzoyl radicals 402–403
– boryl radicals 435–436
– – photoinitiator under air 438–439
– – reactivity 436–438
– in free-radical-promoted cationic photopolymerization 456–458
– germyl and stannyl radicals 432–433
– – reactivity 433–434
– – $(TMS)_3Ge^\bullet$ versus $(TMS)_3Si^\bullet$ reactivity and structure 434–435
– iminyl radicals 439–440
– kinetic data 399, 415, 433, 441
– lophyl radicals 439
– metal-centered radicals 440–442
– metal salts and 455–456
– oxyl radicals 425–426
– peroxyl radicals 426–428
– – interaction with H-donors 428–430
– – interaction with monomers 430
– – interaction with triphenylphosphine 430
– – S_H2 substitution 430–431
– phosphorus-centered radicals 412–413
– propagating radicals 442–443
– radicals in controlled photopolymerization reactions 443

– – light-sensitive alkoxyamines and generation of nitroxides 445–446
– – photoiniferters and dithiocarbamyl radicals 443–444
– radicals in hydrosilylation reactions 446–447
– silyl radicals 418–421, 424–425
– – reactivity and photoinitiation under air 422–424
– – tris(trimethylsilyl)silyl radical 421–422
– sulfonyl and sulfonyloxy radicals 417–418
– thiyl radicals 413, 415–417
radicals
– in controlled photopolymerization reactions 443
– – light-sensitive alkoxyamines and generation of nitroxides 445–446
– – photoiniferters and dithiocarbamyl radicals 443–444
– in hydrosilylation reactions 446–447
radical source/onium salt based systems 318
Raft 57
reactivity
– under air 411,438
– in bulk 84, 386
– in microheterogeneous solution 383
– in solution 381
– of a photoinitiator 86
– of radicals 98, 367–456
– of radicals towards monomer 392
– of radicals towards onium salts 456
Reciprocity law 76
recombination of monomer radicals 442
regioselectivity of the hydrogen abstraction reaction 409
Rehm–Weller approach 96
remote curing 60–61
renewable monomers 46
reworkable materials 47
Rose Bengal 278
ruthenium complexes 279, 326

S

self-assembled photoinitiator monolayers 284–285
self-initiating monomers 283
semiconductor-based lasers and diode lasers 18
semiconductor nanoparticles 284
sesamin 204
S_H2 substitution 430
silanes 205
– as coinitiators 205–206
– future 207

– leading to *in situ* hydrophobic coating   207
– and oxygen inhibition   206
– and silane-ene and silane-acrylate chemistry   207
silicon chemistry and conventional cleavable photoinitiators   172–173
silicon-hydride-terminated surface   214–215
silylamines   207–208
silyloxyamines   184, 210
silyl radicals   418–421
– other sources of   424–425
– reactivity and photoinitiation under air   422–424
– tris(trimethylsilyl)silyl radical   421–422
solvatochromic comparison (SCM) method   76
spatially controlled photopolymerization   44
stannyl radicals   432
stereolithography   25–26
substituent effect limit   154–156
sulfonyl and sulfonyloxy radicals   417–418
sulfonyl ketones   144–147
– modified   154
sulfonyloxy radicals   417
sulfur–carbon cleavable-bond-based derivatives   173
sulfur–silicon cleavable-bond-based derivatives   173
sunlight   18
– exposure   363
supported cleavable photoinitiators   157
surface graft photopolymerization   43

t

template photopolymerization   43
TEMPO-based alkoxyamines   182–183
thermal lens spectroscopy (TLS)   91
thioanthrenium salts   289
thiobenzoate derivatives   140, 144
thio derivatives   203
thiol   203
thiol-ene-photopolymerization   49–51
thiopyrilium salts   245–246, 323
thioxanthone-moiety-based cleavable systems   160
thioxanthones   219–220, 236–237
– substituted
– – compounds   220–221
– – interaction rate constants   221–223
– – photolysis   221
thiyl radicals   413, 415–417
time-resolved FTIR spectroscopy   91–92
titanocenes   177
TPO   161

transition state theory (TST)   95
triazine-derivative
– containing three-component systems   272
– photosensitizer   257
trichloromethyl triazines   165
trimethylolpropane triacrylate (TMPTA) photopolymerization, in film   371–373, 374–376
triphenylphosphine   430
tris(trimethylsilyl)germyl radical   421
tris(trimethylsilyl)silyl radical   205, 421–422
trithianes   203
two-component photoinitiating systems   199
– bisarylimidazole derivative and additive   244
– donor and acceptor systems   243
– dye-based systems
– – amine systems   238–239, 240–241
– – amine water-soluble systems   241
– – coinitiator systems   239–240
– – kinetic data   241
– ferrocenium salts   248–250
– hydrocarbon and amine   257
– ketone-and hydrogen-donor-based systems
– – aldehydes   229
– – aliphatic ketones   229
– – alkylphenylglyoxylates   225, 228
– – anthraquinones   228
– – basic mechanisms   199–200
– – benzophenone derivatives   215–219
– – cleavable ketones as type II photoinitiators   229
– – coumarins   225
– – diketones   223–225
– – fluorenones   229
– – hydrogen donors   201–215
– – ketocoumarins   225
– – low- and high-molecular-weight compounds and macrophotoinitiators   229–232
– – low-molecular-weight one-component systems   236–238
– – oxygen self-consuming thioxanthone derivatives   236
– – photomasked photoinitiator   235–236
– – reactivity   232–233
– – thioxanthone derivatives   219–223
– – two-photon absorption photoinitiators   235
– – water-soluble compounds   233–234
– ketone and ketone-based systems   246–247
– light absorbing amine and monomer   257
– maleimide and amine and photoinitiator   241, 243

two-component photoinitiating systems (contd.)
- metal carbonyl and silane 250
- nanoparticle-formation-mediated type II photoinitiators 256–257
- nitro compound and amine 257
- organometallic compound and ketone-based systems 247
- organometallic derivatives 248
- photoinitiator
- – – disulfide and group 4B dimetal derivatives 252–253
- – – onium salts 254
- – – peroxide- and hydroperoxide-based systems 251–252
- – – phosphorus-containing compounds 253–254
- photosensitizer
- – – -linked photoinitiator and coinitiator-based systems 250–251
- – – titanocenes 254–255
- – – triazine derivative 257–258
- pyrylium and thiopyrylium salts and additive 245–246
- type I photoinitiator and additive 255–256
two-photon absorption-induced polymerization 60

two-photon absorption photoinitiators 235
type II photoinitiator/silane 279–281

u
UV curing 5
- area
- – coating properties 24
- – end uses 21–24
- – equipment 21
UV-light-cleavable bonds 185–186
UV plasma source 19

v
vertical energy transfer (VET) 96
vinyl halide/iodonium salt 320
violanthrone 281, 326
viscosity effect 360

w
water-borne light curable systems 52
water-soluble compounds 233–234
water-soluble monomers photopolymerization in reverse micelles 377

x
xanthone-moiety-based cleavable systems 160
Xenon lamp 13